HANDBOOK OF METAHEURISTICS

INTERNATIONAL SERIES IN
OPERATIONS RESEARCH & MANAGEMENT SCIENCE

Frederick S. Hillier, Series Editor
Stanford University

Weyant, J. / *ENERGY AND ENVIRONMENTAL POLICY MODELING*
Shanthikumar, J.G. & Sumita, U. / *APPLIED PROBABILITY AND STOCHASTIC PROCESSES*
Liu, B. & Esogbue, A.O. / *DECISION CRITERIA AND OPTIMAL INVENTORY PROCESSES*
Gal, T., Stewart, T.J., Hanne, T. / *MULTICRITERIA DECISION MAKING: Advances in MCDM Models, Algorithms, Theory, and Applications*
Fox, B.L. / *STRATEGIES FOR QUASI-MONTE CARLO*
Hall, R.W. / *HANDBOOK OF TRANSPORTATION SCIENCE*
Grassman, W.K. / *COMPUTATIONAL PROBABILITY*
Pomerol, J.-C. & Barba-Romero, S. / *MULTICRITERION DECISION IN MANAGEMENT*
Axsäter, S. / *INVENTORY CONTROL*
Wolkowicz, H., Saigal, R., & Vandenberghe, L. / *HANDBOOK OF SEMI-DEFINITE PROGRAMMING: Theory, Algorithms, and Applications*
Hobbs, B.F. & Meier, P. / *ENERGY DECISIONS AND THE ENVIRONMENT: A Guide to the Use of Multicriteria Methods*
Dar-El, E. / *HUMAN LEARNING: From Learning Curves to Learning Organizations*
Armstrong, J.S. / *PRINCIPLES OF FORECASTING: A Handbook for Researchers and Practitioners*
Balsamo, S., Personé, V., & Onvural, R. / *ANALYSIS OF QUEUEING NETWORKS WITH BLOCKING*
Bouyssou, D. et al. / *EVALUATION AND DECISION MODELS: A Critical Perspective*
Hanne, T. / *INTELLIGENT STRATEGIES FOR META MULTIPLE CRITERIA DECISION MAKING*
Saaty, T. & Vargas, L. / *MODELS, METHODS, CONCEPTS and APPLICATIONS OF THE ANALYTIC HIERARCHY PROCESS*
Chatterjee, K. & Samuelson, W. / *GAME THEORY AND BUSINESS APPLICATIONS*
Hobbs, B. et al. / *THE NEXT GENERATION OF ELECTRIC POWER UNIT COMMITMENT MODELS*
Vanderbei, R.J. / *LINEAR PROGRAMMING: Foundations and Extensions, 2nd Ed.*
Kimms, A. / *MATHEMATICAL PROGRAMMING AND FINANCIAL OBJECTIVES FOR SCHEDULING PROJECTS*
Baptiste, P., Le Pape, C. & Nuijten, W. / *CONSTRAINT-BASED SCHEDULING*
Feinberg, E. & Shwartz, A. / *HANDBOOK OF MARKOV DECISION PROCESSES: Methods and Applications*
Ramík, J. & Vlach, M. / *GENERALIZED CONCAVITY IN FUZZY OPTIMIZATION AND DECISION ANALYSIS*
Song, J. & Yao, D. / *SUPPLY CHAIN STRUCTURES: Coordination, Information and Optimization*
Kozan, E. & Ohuchi, A. / *OPERATIONS RESEARCH/MANAGEMENT SCIENCE AT WORK*
Bouyssou et al. / *AIDING DECISIONS WITH MULTIPLE CRITERIA: Essays in Honor of Bernard Roy*
Cox, Louis Anthony, Jr. / *RISK ANALYSIS: Foundations, Models and Methods*
Dror, M., L'Ecuyer, P. & Szidarovszky, F. / *MODELING UNCERTAINTY: An Examination of Stochastic Theory, Methods, and Applications*
Dokuchaev, N. / *DYNAMIC PORTFOLIO STRATEGIES: Quantitative Methods and Empirical Rules for Incomplete Information*
Sarker, R., Mohammadian, M. & Yao, X. / *EVOLUTIONARY OPTIMIZATION*
Demeulemeester, R. & Herroelen, W. / *PROJECT SCHEDULING: A Research Handbook*
Gazis, D.C. / *TRAFFIC THEORY*
Zhu, J. / *QUANTITATIVE MODELS FOR PERFORMANCE EVALUATION AND BENCHMARKING*
Ehrgott, M. & Gandibleux, X. / *MULTIPLE CRITERIA OPTIMIZATION: State of the Art Annotated Bibliographical Surveys*
Bienstock, D. / *Potential Function Methods for Approx. Solving Linear Programming Problems*
Matsatsinis, N.F. & Siskos, Y. / *INTELLIGENT SUPPORT SYSTEMS FOR MARKETING DECISIONS*
Alpern, S. & Gal, S. / *THE THEORY OF SEARCH GAMES AND RENDEZVOUS*
Hall, R.W. / *HANDBOOK OF TRANSPORTATION SCIENCE—2nd Ed.*

HANDBOOK OF METAHEURISTICS

edited by

Fred Glover
Leeds School of Business
University of Colorado at Boulder

Gary A. Kochenberger
College of Business
University of Colorado at Denver

SPRINGER SCIENCE+BUSINESS MEDIA, LLC

Library of Congress Cataloging-in-Publication Data

Handbook of metaheuristics / edited by Fred Glover, Gary A. Kochenberger.
p. cm. – (International series in operations research & management science; 57)
Includes bibliographical references and index.

DOI 10.1007/978-0-306-48056-0
1. Mathematical optimization–Handbooks, manuals, etc. 2. Operations Research–Handbooks, manuals, etc. I. Glover, Fred. II. Kochenberger, Gary A. III. Series.

T57.H34 2002
658.4'034–dc21 2002034082

Originally published by Kluwer Academic Publishers in 2003
MyCopy version of the original edition 2003

Printed on acid-free paper.
www.springer.com/mycopy

To our wives, Diane and Ann, whose meta-patience and meta-support have sustained us through this effort!

CONTENTS

LIST OF CONTRIBUTORS

Rama Akkiraju
IBM T.J. Watson Labs
E-mail: akkiraju@us.ibm.com

Edmund Burke
University of Nottingham
E-mail: ekb@cs.nott.ac.uk

Carlos Cotta
Universidad de Malaga
E-mail: ccottap@lcc.uma.es

Teodor Gabriel Crainic
University of Montreal
E-mail: theo@crt.umontreal.ca

Marco Dorigo
Universite Libre de Bruxelles
E-mail: mdorigo@ulb.ac.be

Andreas Fink
Universitat Braunschweig
E-mail: a.fink@tu-bs.de

Filippo Focacci
ILOG S. A.
E-mail: ffocacci@ilog.fr

Eugene C. Freuder
University of New Hampshire
E-mail: ecf@cs.unh.edu

Michel Gendreau
University of Montreal
E-mail: michelg@crt.umontreal.ca

Fred Glover
University of Colorado
E-mail: fred.glover@colorado.edu

Pierre Hansen
University of Montreal
E-mail: pierreh@crt.umontreal.ca

Emma Hart
Napier University
E-mail: emmah@dcs.napier.ac.uk

Darrall Henderson
US Military Academy
E-mail: darrall@stanfordalumni.org

Sheldon H. Jacobson
University of Illinois
E-mail: shj@uiuc.edu

Alan W. Johnson
US Military Academy
E-mail: aa2895@usma.edu

Graham Kendall
University of Nottingham
E-mail: gxk@cs.nott.ac.uk

John R. Koza
Stanford University
E-mail: koza@stanford.edu

François Laburthe
Bouygues e-Lab
E-mail: flaburthe@bouygues.com

Manuel Laguna
University of Colorado
E-mail: manuel.laguna@Colorado.edu

Andrea Lodi
University of Bologna
E-mail: alodi@deis.unibo.it

Helena R. Lourenço
Universitat Pompeu Fabra
E-mail: helena.ramalhinho@econ.upf.es

Rafael Martí
Universitat de Valencia
E-mail: rafael.marti@uv.es

Olivier C. Martin
Universite Paris-Sud
E-mail: martino@ipno.in2p3.fr

Nenad Mladenović
Serbian Academy of Science
E-mail: nenad@mi.sanu.ac.yu

Pablo Moscato
Universidade Estadual de Campinas
E-mail: moscato@densis.fee.unicamp.br

Sesh Murthy
IBM T.J. Watson Labs
E-mail: murthy@watson.ibm.com

Jim Newall
University of Nottingham
E-mail: jpn@cs.nott.ac.uk

Jean-Yves Potvin
University of Montreal
E-mail: potvin@iro.unmontreal.ca

Colin Reeves
Coventry University
E-mail: c.reeves@coventry.ac.uk

Mauricio G.C. Resende
AT&T Labs Research
E-mail: mgcr@research.att.com

Celso C. Ribeiro
Catholic University of Rio de Janeiro
E-mail: celso@inf.puc-rio.br

Peter Ross
Napier University
E-mail: peter@dcs.napier.ac.uk

Sonia Schulenburg
Napier University
E-mail: s.schulenberg@napier.ac.uk

Kate A. Smith
Monash University
E-mail: kate.smith@infotech.monash.edu.au

Thomas Stützle
Darmstadt University of Technology
E-mail: stuetzle@informatik.tu-darmstadt.de

Sarosh Talukdar
Carnegie Mellon University
E-mail: talukdar@ece.cmu.edu

Michel Toulouse
University of Manitoba
E-mail: toulouse@cs.umanitoba.ca

Edward P.K. Tsang
University of Essex
E-mail: edward@essex.ac.uk

Stefan Voß
Universitat Braunschweig
E-mail: stefan.voss@tu-bs.de

Cristos Voudouris
BTexact Technologies
E-mail: chris.voudouris@bt.com

Mark Wallace
Imperial College
E-mail: mgw@icparc.ic.ac.uk

David L. Woodruff
University of California at Davis
E-mail: dlwoodruff@ucdavis.edu

PREFACE

Metaheuristics, in their original definition, are solution methods that orchestrate an interaction between local improvement procedures and higher level strategies to create a process capable of escaping from local optima and performing a robust search of a solution space. Over time, these methods have also come to include any procedures that employ strategies for overcoming the trap of local optimality in complex solution spaces, especially those procedures that utilize one or more neighborhood structures as a means of defining admissible moves to transition from one solution to another, or to build or destroy solutions in constructive and destructive processes.

The degree to which neighborhoods are exploited varies according to the type of procedure. In the case of certain population-based procedures, such as genetic algorithms, neighborhoods are implicitly (and somewhat restrictively) defined by reference to replacing components of one solution with those of another, by variously chosen rules of exchange popularly given the name of "crossover." In other population-based methods, based on the notion of path relinking, neighborhood structures are used in their full generality, including constructive and destructive neighborhoods as well as those for transitioning between (complete) solutions. Certain hybrids of classical evolutionary approaches, which link them with local search, also use neighborhood structures more fully, though apart from the combination process itself. Meanwhile, "single thread" solution approaches, which do not undertake to manipulate multiple solutions simultaneously, run a wide gamut that not only manipulate diverse neighborhoods but incorporate numerous forms of strategies ranging from thoroughly randomized to thoroughly deterministic, depending on the elements such as the phase of search or (in the case of memory-based methods) the history of the solution process.[1]

A number of the tools and mechanisms that have emerged from the creation of metaheuristic methods have proved to be remarkably effective, so much so that metaheuristics have moved into the spotlight in recent years as the preferred line of attack for solving many types of complex problems, particularly those of a combinatorial nature. While metaheuristics are not able to certify the optimality of the solutions they find, exact procedures (which theoretically can provide such a certification, if allowed to run long enough)[2] have often proved incapable of finding solutions whose

[1] Methods based on incorporating collections of memory-based strategies, invoking forms of memory more flexible and varied than those used in approaches such as tree search and branch and bound, are sometimes grouped under the name Adaptive Memory Programming. This term, which originated in the tabu search literature where such adaptive memory strategies were first introduced and continue to be the primary focus, is also sometimes used to encompass other methods that have more recently adopted memory-based elements.

[2] Some types of problems seem quite amenable to exact methods, particularly to some of the methods embodied in the leading commercial software packages for mixed integer programming. Yet even by these approaches the "length of time" required to solve many problems exactly appears to exceed all reasonable measure, including in some cases measures of astronomical scale. It has been conjectured that metaheuristics

quality is close to that obtained by the leading metaheuristics—particularly for real world problems, which often attain notably high levels of complexity. In addition, some of the more successful applications of exact methods have come about by incorporating metaheuristic strategies within them. These outcomes have motivated additional research and application of new and improved metaheuristic methodologies.

This handbook is designed to provide the reader with a broad coverage of the concepts, themes and instrumentalities of this important and evolving area of optimization. In doing so, we hope to encourage an even wider adoption of metaheuristic methods for assisting in problem solving, and to stimulate research that may lead to additional innovations in metaheuristic procedures.

The handbook consists of 19 chapters. Topics covered include Scatter Search, Tabu Search, Genetic Algorithms, Genetic Programming, Memetic Algorithms, Variable Neighborhood Search, Guided Local Search, GRASP, Ant Colony Optimization, Simulated Annealing, Iterated Local Search, Multi-Start Methods, Constraint Programming, Constraint Satisfaction, Neural Network Methods for Optimization, Hyper-Heuristics, Parallel Strategies for Metaheuristics, Metaheuristic Class Libraries, and A-Teams. This family of metaheuristic chapters, while not exhaustive of the many approaches that have sprung into existence in recent years, encompasses the critical strategic elements and their underlying ideas that represent the state-of-the-art of modern metaheuristics.

This book is intended to provide the communities of both researchers and practitioners with a broadly applicable, up to date coverage of metaheuristic methodologies that have proven to be successful in a wide variety of problem settings, and that hold particular promise for success in the future. The various chapters serve as stand alone presentations giving both the necessary underpinnings as well as practical guides for implementation. The nature of metaheuristics invites an analyst to modify basic methods in response to problem characteristics, past experiences, and personal preferences and the chapters in this handbook are designed to facilitate this process as well.

The authors who have contributed to this volume represent leading figures from the metaheuristic community and are responsible for pioneering contributions to the fields they write about. Their collective work has significantly enriched the field of optimization in general and combinatorial optimization in particular. We are especially grateful to them for agreeing to provide the first-rate chapters that appear in this handbook. We would also like to thank our graduate students, Gyung Yung and Rahul Patil, for their assistance. Finally, we would like to thank Gary Folven and Carolyn Ford of Kluwer Academic Publishers for their unwavering support and patience throughout this project.

---

succeed where exact methods fail because of their ability to use strategies of greater flexibility than permitted to assure that convergence will inevitably be obtained.

Chapter 1

SCATTER SEARCH AND PATH RELINKING: ADVANCES AND APPLICATIONS

Fred Glover and Manuel Laguna
Leeds School of Business, Campus Box 419,
University of Colorado, Boulder,
CO 80309-0419, USA
E-mail: Fred.Glover@Colorado.edu, Manuel.Laguna@Colorado.edu

Rafael Marti
Dpto. de Estadística e Investigación Operativa, Facultad de Matemáticas, Universitat de Valencia, Dr. Moliner, 50, 46100 Burjassot, Valencia, Spain
E-mail: Rafael.Marti@uv.es

Abstract Scatter search (SS) is a population-based method that has recently been shown to yield promising outcomes for solving combinatorial and nonlinear optimization problems. Based on formulations originally proposed in the 1960s for combining decision rules and problem constraints, SS uses strategies for combining solution vectors that have proved effective in a variety of problem settings. Path relinking (PR) has been suggested as an approach to integrate intensification and diversification strategies in a search scheme. The approach may be viewed as an extreme (highly focused) instance of a strategy that seeks to incorporate attributes of high quality solutions, by creating inducements to favor these attributes in the moves selected. The goal of this paper is to examine SS and PR strategies that provide useful alternatives to more established search methods. We describe the features of SS and PR that set them apart from other evolutionary approaches, and that offer opportunities for creating increasingly more versatile and effective methods in the future. Specific applications are summarized to provide a clearer understanding of settings where the methods are being used.

1 INTRODUCTION

Scatter search (SS), from the standpoint of metaheuristic classification, may be viewed as an evolutionary (population-based) algorithm that constructs solutions by combining others. It derives its foundations from strategies originally proposed for combining decision rules and constraints in the context of integer programming. The goal of this methodology is to enable the implementation of solution procedures that can derive new solutions from combined elements in order to yield better solutions than those procedures that base their combinations only on a set of original elements. For example, see the overview by Glover (1998).

The antecedent strategies for combining decision rules were first introduced in the area of scheduling, as a means to obtain improved local decisions. Numerically weighted combinations of existing rules, suitably restructured so that their evaluations embodied a common metric, generated new rules (Glover, 1963). The approach was motivated by the conjecture that information about the relative desirability of alternative choices is captured in different forms by different rules, and that this information can be exploited more effectively when integrated than when treated in isolation (i.e., by choosing selection rules one at a time). Empirical outcomes disclosed that the decision rules created from such combination strategies produced better outcomes than standard applications of local decision rules. The strategy of creating combined rules also proved superior to a "probabilistic learning approach" that used stochastic selection of rules at different junctures, but without the integration effect provided by the combined rules (Crowston et al., 1963).

The associated procedures for combining constraints likewise employed a mechanism of generating weighted combinations. In this case, nonnegative weights were introduced to create new constraint inequalities, called *surrogate constraints*, in the context of integer and nonlinear programming (Glover, 1965, 1968). The approach isolated subsets of (original) constraints that were gauged to be most critical, relative to trial solutions that were obtained based on the surrogate constraints. This critical subset was used to produce new weights that reflected the degree to which the component constraints were satisfied or violated. In addition, the resulting surrogate constraints served as source constraints for deriving new inequalities (cutting planes) which in turn provide material for creating further surrogate constraints.

Path relinking has been suggested as an approach to integrate intensification and diversification strategies (Glover and Laguna, 1997) in the context of tabu search. This approach generates new solutions by exploring trajectories that connect high-quality solutions, by starting from one of these solutions and generating a path in the neighborhood space that leads toward the other solutions.

Recent applications of both methods (and of selected component strategies within them) that have proved highly successful are:

- **The Linear Ordering Problem** (Campos, Laguna and Martí)
- **The Bipartite Drawing Problem** (Laguna and Martí)
- **The Graph Coloring Problem** (Hamiez and Hao)
- **Capacitated Multicommodity Network Design** (Ghamlouche, Crainic and Gendreau)
- **The Maximum Clique Problem** (Cavique, Rego and Themido)
- **Assigning Proctor to Exams** (Ramalhinho, Laguna and Martí)
- **Periodic Vehicle Loading** (Delgado, Laguna and Pacheco)
- **Job Shop Scheduling** (Nowicki and Smutnicki)
- **The Arc Routing Problem** (Greistorfer)
- **Resource Constrained Project Scheduling** (Valls, Quintanilla and Ballestín)
- **Multiple Criteria Scatter Search** (Beausoleil)
- **Meta-Heuristic Use of Scatter Search via OptQuest** (Hill)

- **Pivot Based Search Integrated with Branch and Bound for Binary MIPs** (Løkketangen and Woodruff)
- **Scatter Search to Generate Diverse MIP Solutions** (Glover, Løkketangen and Woodruff)
- **Path Relinking to Improve Iterated Start Procedures** (Ribeiro and Resende)

A number of these applications are described in Section 4 where a collection of vignettes is presented. They provide a diverse range of settings where SS and PR have made useful contributions, and suggest the form of additional applications where similar successes may be anticipated.

2 SCATTER SEARCH

Scatter search is designed to operate on a set of points, called *reference points*, which constitute good solutions obtained from previous solution efforts. Notably, the basis for defining "good" includes special criteria such as diversity that purposefully go beyond the objective function value. The approach systematically generates combinations of the reference points to create new points, each of which is mapped into an associated feasible point. The combinations are generalized forms of linear combinations, accompanied by processes to adaptively enforce feasibility conditions, including those of discreteness (Glover, 1977).

The SS process is organized to (1) capture information not contained separately in the original points, (2) take advantage of auxiliary heuristic solution methods (to evaluate the combinations produced and to actively generate new points), and (3) make dedicated use of strategy instead of randomization to carry out component steps. SS basically consist of five methods:

1. A *Diversification Generation Method* to generate a collection of diverse trial solutions, using one or more arbitrary trial solutions (or seed solutions) as an input.
2. An *Improvement Method* to transform a trial solution into one or more enhanced trial solutions. (Neither the input nor the output solutions are required to be feasible, though the output solutions are typically feasible. If the input trial solution is not improved as a result of the application of this method, the "enhanced" solution is considered to be the same as the input solution.)
3. A *Reference Set Update Method* to build and maintain a *reference set* consisting of the b "best" solutions found (where the value of b is typically small, e.g., no more than 20), organized to provide efficient accessing by other parts of the solution procedure. Several alternative criteria may be used to add solutions to the reference set and delete solutions from the reference set.
4. A *Subset Generation Method* to operate on the reference set, to produce a subset of its solutions as a basis for creating combined solutions. The most common subset generation method is to generate all pairs of reference solutions (i.e., all subsets of size 2).
5. A *Solution Combination Method* to transform a given subset of solutions produced by the Subset Generation Method into one or more combined solutions.

The combination method is analogous to the crossover operator in genetic algorithms although it should be capable of combining more than two solutions. (The combination processes proposed in the original SS paper included forms of "crossover" not envisioned in the GA literature until a decade later, and combination processes proposed since then, as in Glover (1994,1995) utilize principles and constructions that remain beyond the scope embraced by GA approaches.)

The basic procedure in Figure 1.1 starts with the creation of an initial reference set of solutions (*RefSet*). The Diversification Generation Method is used to build a large set of diverse solutions P. The size of P (*PSize*) is typically 10 times the size of *RefSet*. Initially, the reference set *RefSet* consists of b distinct and maximally diverse solutions from P. The solutions in *RefSet* are ordered according to quality, where the best solution is the first one in the list. The search is then initiated by assigning the value of TRUE to the Boolean variable *NewSolutions*. In step 3, *NewSubsets* is constructed and *NewSolutions* is switched to FALSE. For illustrative purposes we focus attention on subsets of size 2. Hence the cardinality of *NewSubsets* corresponding to the initial reference set is given by $(b^2 - b)/2$, which accounts for all pairs of solutions in *RefSet*. (Special conditions are imposed on subsets of larger sizes to ensure a suitable composition is achieved while generating no more than a restricted number of these subsets.) The pairs in *NewSubsets* are selected one at a time in lexicographical order and the Solution Combination Method is applied to generate one or more solutions in step 5. If a newly created solution improves upon the worst solution currently in

1. Start with $P = \emptyset$. Use the Diversification Generation Method to construct a solution x. If $x \notin P$ then add x to P (i.e., $P = P \cup x$), otherwise, discard x. Repeat this step until $|P| = PSize$. Build $RefSet = \{x^1, \ldots, x^b\}$ with b diverse solutions in P.
2. Evaluate the solutions in *RefSet* and order them according to their objective function value such that x^1 is the best solution and x^b the worst. Make *NewSolutions* = TRUE.

while (*NewSolutions*) **do**

   3. Generate *NewSubsets*, which consists of all pairs of solutions in *RefSet* that include at least one new solution. Make *NewSolutions* = FALSE.

   while (*NewSubset* $\neq \emptyset$) **do**

      4. Select the next subset s in *NewSubSets*.
      5. Apply the Solution Combination Method to s to obtain one or more new solutions x.

      if (x is not in *RefSet* and $f(x) < f(x^b)$) **then**

         6. Make $x^b = x$ and reorder *RefSet*.
         7. Make *NewSolutions* = TRUE.

      end if

      8. Delete s from *NewSubsets*.

   end while

end while

Figure 1.1. Outline of basic scatter search for a minimization objective.

RefSet, the new solution replaces the worst and *RefSet* is reordered in step 6. The *NewSolutions* flag is switched to TRUE and the subset *s* that was just combined is deleted from *NewSubsets* in steps 7 and 8, respectively.

This basic design can be expanded and improved in different ways. The SS methodology is very flexible, since each of its elements can be implemented in a variety of ways and degrees of sophistication. Different improvements and designs from this basic SS algorithm are given in Glover (1998), Glover et al. (1999, 2000), Laguna (2000) and Laguna and Armentano (2001).

3 PATH RELINKING

One of the main goals in any search method is to create a balance between search intensification and search diversification. Path relinking has been suggested as an approach to integrate intensification and diversification strategies (Glover and Laguna, 1997). Features that have been added to Scatter Search, by extension of its basic philosophy, are also captured in the Path Relinking framework. This approach generates new solutions by exploring trajectories that connect high-quality solutions—by starting from one of these solutions, called an *initiating solution*, and generating a path in the neighborhood space that leads toward the other solutions, called *guiding solutions*. This is accomplished by selecting moves that introduce attributes contained in the guiding solutions.

The approach may be viewed as an extreme (highly focused) instance of a strategy that seeks to incorporate attributes of high quality solutions, by creating inducements to favor these attributes in the moves selected. However, instead of using an inducement that merely encourages the inclusion of such attributes, the path relinking approach subordinates other considerations to the goal of choosing moves that introduce the attributes of the guiding solutions, in order to create a "good attribute composition" in the current solution. The composition at each step is determined by choosing the best move, using customary choice criteria, from a restricted set—the set of those moves currently available that incorporate a maximum number (or a maximum weighted value) of the attributes of the guiding solutions.

The approach is called path relinking either by virtue of generating a new path between solutions previously linked by a series of moves executed during a search, or by generating a path between solutions previously linked to other solutions but not to each other. Figure 1.2 shows two hypothetical paths (i.e., a sequence of moves) that link solution A to solution B, to illustrate relinking of the first type. The solid line indicates an original path produced by the "normal" operation of a procedure that produced a series of moves leading from A to B, while the dashed line depicts the relinking path. The paths are different because the move selection during the normal operation does not "know" where solution B lies until it is finally reached, but simply follows a trajectory whose intermediate steps are determined by some form of evaluation function. For example, a commonly used approach is to select a move that minimizes (or maximizes) the objective function value in the local sense. During path relinking, however, the main goal is to incorporate attributes of the guiding solution (or solutions) while at the same time recording the objective function values.

The effort to represent the process in a simple diagram such as the one preceding creates some misleading impressions, however. First, the original (solid line) path,

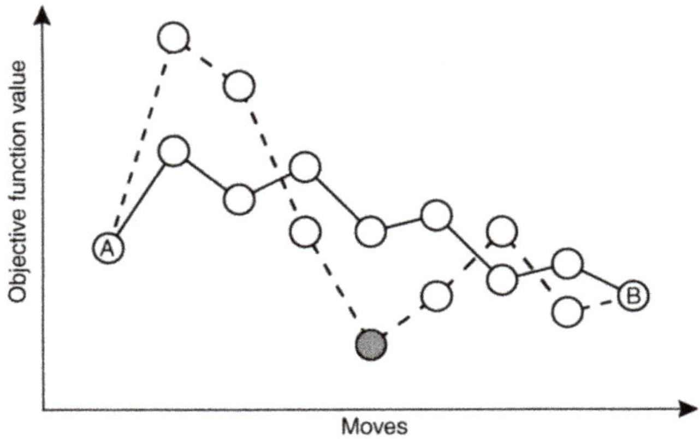

Figure 1.2. Path relinking illustration.

whch is shown to be "greedy" relative to the objective function, is likely to be significantly more circuitous along dimensions we are not able to show, and by the same token to involve significantly more steps (intervening solutions)—an aspect not portrayed in Figure 1.2. Second, because the relinked path is not governed so strongly by local attraction, but instead is influenced by the criterion of incorporating attributes of the guiding solution, it opens the possibility of reaching improved solutions that would not be found by a "locally myopic" search. Figure 1.2 shows one such solution (the darkened node) reached by the dotted path. Beyond this, however, the relinked path may encounter solutions that may not be better than the initiating or guiding solution, but that provide fertile "points of access" for reaching other, somewhat better, solutions. For this reason it is valuable to examine neighboring solutions along a relinked path, and keep track of those of high quality which may provide a starting point for continued search.

The incorporation of attributes from elite parents in partially or fully constructed solutions was foreshadowed by another aspect of scatter search, embodied in an accompanying proposal to assign preferred values to subsets of *consistent* and *strongly determined* variables. The theme is to isolate assignments that frequently or influentially occur in high quality solutions, and then to introduce compatible subsets of these assignments into other solutions that are generated or amended by heuristic procedures. (Such a process implicitly relies on a form of frequency-based memory to identify and exploit variables that qualify as consistent.)

Multiparent path generation possibilities emerge in path relinking by considering the combined attributes provided by a set of guiding solutions, where these attributes are weighted to determine which moves are given higher priority. The generation of such paths in neighborhood space characteristically "relinks" previous points in ways not achieved in the previous search history, hence giving the approach its name. This multiparent Path relinking approach generates new elements by a process that emulates the strategies of the original Scatter Search approach at a higher level of generalization. The reference to neighborhood spaces makes it possible to preserve desirable solution properties (such as complex feasibility conditions in scheduling and routing), without requiring artificial mechanisms to recover these properties in situations where they may otherwise become lost.

The PR approach benefits from a **tunneling strategy** that often encourages a different neighborhood structure to be used than in the standard search phase. For example, moves for Path relinking may be periodically allowed that normally would be excluded due to creating infeasibility. Such a practice is protected against the possibility of becoming "lost" in an infeasible region, since feasibility evidently must be recovered by the time the guiding solution is reached.

A natural variation of path relinking occurs by using **constructive neighborhoods** for creating offspring from a collection of parent solutions. In this case the guiding solutions consist of subsets of elite solutions, as before, but the initiating solution begins as a partial (incomplete) solution or even as a null solution, where some of the components of the solutions, such as subsets of free variables, are not yet assigned. The use of a constructive neighborhood permits such an initiating solution to "move toward" the guiding solutions, by a neighborhood path that progressively introduces elements contained in the guiding solutions, or that are evaluated as attractive based on the composition of the guiding solutions.

4 SS/PR VIGNETTES

This section provides a collection of "vignettes" that briefly summarize applications of SS and PR in a variety of settings. These vignettes are edited versions of reports by researchers and practitioners who are responsible for the applications. A debt of gratitude is owed to the individuals whose contributions have made this summary possible.

4.1 The Linear Ordering Problem

Given a matrix of weights $E = \{e_{ij}\}_{m \times m}$, the linear ordering problem (LOP) consists of finding a permutation p of the columns (and rows) in order to maximize the sum of the weights in the upper triangle. In mathematical terms, we seek to maximize:

$$C_E(p) = \sum_{i=1}^{m-1} \sum_{j=i+1}^{m} e_{p(i)p(j)},$$

where $p(i)$ is the index of the column (and row) in position i in the permutation. In the LOP, the permutation p provides the ordering of both the columns and the rows. The equivalent problem in graphs consists of finding, in a complete weighted graph, an acyclic tournament with a maximal sum of arc weights. In economics, the LOP is equivalent to the so-called *triangulation problem for input-output tables* (Reinelt, 1985).

Campos et al. (1999) propose a solution method for the linear ordering problem based on the scatter search template in Glover (1998). The procedure combines the following elements:

(a) Diversification Generator

(b) Improvement Method

(c) Reference Set Update Method
(d) Subset Generation Method
(e) Solution Combination Method

where (a), (b) and (e) are context dependent and (c) and (d) are "generic" elements.

The authors developed and tested 10 **Diversification Generation Methods**. These methods included a completely random method, several versions of GRASP, a deterministic method that disregards the objective function, and a method using frequency-based memory as proposed in tabu search. The diversification approach using TS frequency-based memory was found to clearly outperform the competing methods.

The **Improvement Method** is based on the neighborhood search developed for the Tabu Search algorithm for the LOP in Laguna, Martí and Campos (1999).

The **Solution Combination Method** uses a min-max construction based on votes. The method scans each reference permutation to be combined, and uses the rule that each reference permutation votes for its first element that is still not included in the combined permutation (referred to as the "incipient element").

In a set of **Computational Testing Experiments**, the authors compare the performance of two variants of the scatter search implementation with three methods: Chanas and Kobylanski (CK, 1996), Tabu Search (TS, Laguna, Martí and Campos, 1998) and a greedy procedure especially designed for the LOP. The scatter search procedures were tested on four sets of instances.

The tabu search method and the two scatter search instances dominate the other approaches in terms of solution quality. The TS method is the fastest of the high quality methods, running 3–5 times faster than the first scatter search variant, but the scatter search variants give the best overall solution quality, indicating their value where quality is the dominant consideration.

4.2 The Bipartite Drawing Problem

The problem of minimizing straight-line crossings in layered bipartite graphs consists of aligning the two shores V_1 and V_2 of a bipartite graph $G = (V_1, V_2, E)$ on two parallel straight lines (layers) such that the number of crossing between the edges in E is minimized when the edges are drawn as straight lines connecting the end-nodes. The problem is also known as the *bipartite drawing problem* (BDP).

The main application of this problem is found in automated drawing systems, where drawing speed is a critical factor. Simple heuristics are very fast but result in inferior solutions, while high-quality solutions have been found with meta-heuristics that demand an impractical amount of computer time. Laguna and Martí (1999) propose a method that combines GRASP and Path Relinking to develop a procedure that can compete in speed with the simple heuristics and in quality with the complex meta-heuristics.

The hybrid procedure proposed for the BDP utilizes GRASP as the multistart method to be augmented, and stores a small set of high quality (elite) solutions to be used for guiding purposes.

In a set of **Computational Testing Experiments**, the authors compare the performance of the GRASP and Path Relinking implementations with two methods: the iterated barycenter (BC, Eades and Kelly, 1986) and a version of the Tabu Search algorithm (TS, Martí, 1996). The former is the best of the simple heuristics for the BDP (Martí and Laguna, 1997), while the later has been proven to consistently provide the best solutions in terms of quality. For these experiments 3,200 instances have been generated with the `random_bigraph` code of the Stanford GraphBase by Knuth (1993).

The first experiment shows that the best solution quality is obtained by the tabu search method, which is able to match the 900 known optimal solutions, while GRASP matches 750 and PR matches 866. However, in contrast to some other applications (such as the Linear Ordering problem previously described), TS employs more computational time (15 s) than the other methods reported. GRASP performs quite well, considering its average percent deviation from optima of 0.44% achieved on an average of 0.06 s. Notably, path relinking achieves a significantly improved average percent deviation of 0.09% with a modest increase in computer time (0.28 s). Finally, iterated BC from 10 random starts turns in a substantially less attractive performance, with an average percent deviation of 3.43% achieved in 0.08 s.

The second experiment is devoted to sparse graphs. It is shown that the path relinking algorithm achieves the best average deviation of less than 1%. The computational effort associated with the PR variant is very reasonable (with a worst case average of 1.61 s).

A third experiment was performed to assess the efficiency of the proposed procedures in denser graphs (relative to the second experiment). The results show that the tabu search procedure outperforms both the BC and the GRASP variants. The average deviation from the best known values is 1.41% for the TS procedure, while PR obtains an average deviation of 8.96%, using similar computational time (i.e., 26 s for TS versus 22 s for PR).

The enhancements produced by path relinking suggests the potential merit of joining the PR guidance strategies with other multistart methods.

4.3 The Graph Coloring Problem

Graph k-coloring can be stated as follows: given an undirected graph G with a set V of vertices and a set E of edges connecting vertices, k-coloring G means finding a partition of V into k classes $V_1, \ldots, V_k$, called *color classes*, such that no couple of vertices $(u, v) \in E$ belongs to the same color class. Formally, $\{V_1, \ldots, V_k\}$ is a valid k-coloring of the graph $G = (V, E)$ if $\forall i \in [1, \ldots, k]$ and $\forall (u, v) \in V_i, (u, v) \notin E$. The *graph coloring problem* (GCP) is the optimization problem associated with k-coloring. It aims at searching for the minimal k such that a proper k-coloring exists. This minimum is the chromatic number $X(G)$ of graph G.

Graph coloring has many real applications, e.g., timetable construction, frequency assignment, register allocation or printed circuit testing. There are many resolution methods for this problem: greedy constructive approaches (DSATUR, RLF), hybrid strategies (HCA for instance), local search metaheuristics (simulated annealing, tabu), neural network attempts, ... Despite the fact that the literature on graph coloring is always growing, there exists, to our knowledge, no approach relying on scatter search for the graph coloring problem. We summarize here such an experimental investigation following the scatter search template of Glover (1998).

Our **Diversification Generation Method** uses *independent sets* to build initial configurations. Color classes are built one by one by selecting vertices in a random order to insure diversity.

The **Improvement Method** is based on the tabu search algorithm of Dorne and Hao (1998). This algorithm iteratively changes the current color of a conflicting vertex to another one, until achieving a proper coloring. A tabu move leading to a configuration better than the best configuration found so far, within the same execution of the improvement method or within the overall scatter search procedure, is always accepted (*aspiration criterion*).

Although the **Reference Set Update Method** is usually a "generic" element of scatter search, we provide here the way configurations are compared in terms of *diversity*. This point is crucial since, in the context of graph coloring, the Hamming distance is not well suited to compare two configurations c_1 and c_2. The *distance* between c_1 and c_2 is the minimum number of moves necessary to transform c_1 into c_2. The *fitness* of any configuration is naturally its number of conflicting edges.

The **Solution Combination Method** uses a generalization of the powerful *greedy partition crossover* (GPX), proposed by Galinier and Hao (1999) within an evolutionary algorithm. GPX has been especially developed for the graph coloring problem with results reaching, and sometimes improving, those of the best known algorithms for the GCP. Given a subset p generated by the subset generation method, the generalized combination operator builds the k color classes of the new configuration one by one. First, choose a configuration $c \in p$. Remove from c a minimal set of conflicting vertices such as c becomes a partial proper k-coloring. Next, fill in a free color class of the new configuration with all conflict-free vertices of the color class having maximum cardinality in c. Repeat these steps until the k color classes of the new configuration contain at least one vertex. Finally, to complete the new configuration if necessary, assign to each free vertex a color such that it minimizes the conflicts over the graph.

Computational Testing Experiments has been carried out on some of the well-known DIMACS benchmark graphs (Johnson and Trick, 1996). The scatter search procedure (SSGC) was compared with the generic tabu search (GTS) algorithm of Dorne and Hao (1998) together with the best-known methods available for the graph coloring problem: two local search algorithms based on particular neighborhoods and a distributed population-based algorithm (Morgenstern, 1996), and an hybrid method including a descent algorithm and a tabu procedure with various heuristics mixed with a greedy construction stage and the search for a maximum clique (Funabiki and Higashino, 2000).

The scatter search approach SSGC managed to reach the results of the best-known algorithms in quality (minimal number of colors used), except on the r1000.5 graph for which a 237-coloring has been published recently (Funabiki and Higashino, 2000). (The sophisticated algorithm used to reach this coloring includes, among other components, the search for a maximum clique.) Nevertheless, SSGC obtained here a better coloring (240) than GTS (242) and outperformed the previous best result (241) for this graph (Morgenstern, 1996). Our scatter search approach also improves in quality on the results obtained with tabu search (GTS) on a few graphs. This means that tabu search, the *improvement method* we used within scatter search, surely benefits from the other general components of scatter search.

4.4 Capacitated Multicommodity Network Design

The fixed-charge capacitated multicommodity network design formulation (CMND) represents a generic model for a wide range of applications in planning the construction, development, improvement, and operations of transportation, logistics, telecommunication, and production systems, as well as in many other major areas. The problem is usually modeled as a combinatorial optimization problem and is NP-hard in the strong sense. Thus, not only the generation of optimal solutions to large problem instances constitutes a significant challenge, but even identifying efficiently good feasible solutions has proved a formidable task not entirely mastered.

The goal of a CMND formulation is to find the optimal configuration—the links to include in the final design—of a network of limited capacity to satisfy the demand of transportation of different commodities sharing the network. The objective is to minimize the total system cost, computed as the sum of the link fixed and routing costs.

The paper Ghamlouche, Crainic and Gendreau (2001) proposed a new class of cycle-based neighborhood structures for the CMND and evaluated the approach within a very simple tabu-based local search procedure that currently appears as the best approximate solution method for the CMND in terms of robust performance, solution quality, and computing efficiency. Still more recently, Ghamlouche, Crainic and Gendreau (2002) explore the adaptation of path relinking to the CMND. This work evaluates the benefits of combining the cycle-based neighborhood structures and the path relinking framework into a better meta-heuristic for this difficult problem.

The method proceeds with a sequence of cycle-based tabu search phases that investigate each visited solution and add elite ones to the reference set ***R***. When a predefined number of consecutive moves without improvement is observed, the method switches to a path relinking phase.

What solutions are included in the reference set, how good and how diversified they are, has a major impact on the quality of the new solutions generated by the path relinking method. We study six strategies corresponding to different ways to build ***R***.

In strategy ***S*1**, ***R*** is built using each solution that, at some stage of the tabu search phase, improves the best overall solution and become the best one.

—Strategy ***S*2** retains the "*best*" *local minima* found during the tabu search phase. This strategy is motivated by the idea that local minimum solutions share characteristics with optimum solutions.

—Strategy ***S*3** selects ***R***-*improving* local minima, that is local minimum solutions that offer a better evaluation of the objective function than those already in ***R***.

—Strategy ***S*4** allows solutions to be retained in ***R*** not only according to an attractive solution value but also according to a *diversity*, or *dissimilarity* criterion.

—Strategy ***S*5** aims to ensure both the *quality* and the *diversity* of solutions in ***R***. Starting with a large set of "good" solutions, ***R*** is partially filled with the best solutions found, to satisfy the purpose of quality. It is then extended with solutions that change significantly the structure of the solutions already in ***R*** to ensure diversity.

—Strategy ***S*6** proceeds similarly to ***S*5** with the difference that ***R*** is extended with solutions *close* to those already in ***R***.

During the path relinking phase, moves from the initial to a neighboring one direct the search towards the guiding solution. Due to the nature of the neighborhoods used, there is no guarantee that the guiding solution will be reached. One cannot, therefore, stop the process only if the current and the guiding solutions are the same. We then define Δ_{IG} as the number of arcs with different status between the initial and the guiding solutions and we allow the search to explore a number of solutions not larger than Δ_{IG}.

Initial and guiding solutions are chosen from the reference set. This choice is also critical to the quality of the new solutions and, thus, the performance of the procedure. We investigate the effect of the following criteria:

—***C*1**: Guiding and initial solutions are defined as the *best* and *worst* solutions, respectively.
—***C*2**: Guiding solution is defined as the *best* solution in the reference set, while the initial solution is the *second best* one.
—***C*3**: Guiding solution is defined as the *best* solution in the reference set, while the initial solution is defined as the solution with *maximum Hamming distance* from the guiding solution.
—***C*4**: Guiding and initial solutions are chosen *randomly* from the reference set.
—***C*5**: Guiding and initial solutions are chosen as the *most distant* solutions in the reference set.
—***C*6**: Guiding and initial solutions are defined respectively as the *worst* and the *best* solutions in the reference set.

The path relinking phase stops when the reference set becomes empty (cardinality ≤ 1). Then, either stopping conditions are verified, or the procedure is repeated to build a new reference set.

Extensive computational experiments, conducted on one of the 400 MHz processors of a Sun Enterprise 10000, indicate that the path relinking procedure offers excellent results. It systematically outperforms the cycle-based tabu search method in both solution quality and computational effort. On average, for 159 problems path relinking obtains a gap of 2.91% from the best solutions found by branch-and-bound versus a gap of 3.69% for the cycle-based tabu search. (However, the branch and bound code, CPLEX 6.5, was allowed to run for 10 CPU hours.) Thus, path relinking offers the best current meta-heuristic for the CMND.

4.5 A Scatter Search for the Maximum Clique Problem

The maximum clique problem (MCP) can be defined as follows. Given an undirected graph $G = (V, A)$ and $A(v_i)$ denoting the set of vertices $\mathrm{v_j}$ such that $(v_i, v_j) \in A$, then a graph $G1 = (V1, A1)$ is called a subgraph of G if $V1 \subseteq V$, and for every $v_j \in V1$, $A1(v_i) = A(v_i) \cap V1$. A graph $G1$ is said to be complete if there is an arc for each pair of vertices. A complete subgraph is also called a clique. A clique is maximal, if it is not contained in any other clique. In the MCP the objective is to find a complete subgraph of largest cardinality in a graph. The clique number is equal to the cardinality of the largest clique of G.

The MCP is an important problem in combinatorial optimization with many applications which include: market analysis, project selection, and signal transmission. The interest for this problem led to the algorithm thread challenge on experimental analysis and algorithm performance promoted by Second DIMACS Implementation Challenge (Johnson and Trick, 1996).

Cavique et al. (2001) developed an experimental study for solving the Maximum Clique Problem (MCP) using a Scatter Search framework. The proposed algorithm considers structured solution combinations weighted by a "filtering vector" playing the role of linear combinations. For the heuristic improvement a simple tabu search procedure based on appropriate neighborhood structures is used. Some special features and techniques have been introduced for the implementation to this problem. The algorithm implementation is structured into five basic methods.

4.5.1 Diversification Generation Method

The aim of the diversification generation method is to create a set of solutions as scattered as possible within the solution space while also using as many variables (or solution attributes) as possible. In the MCP, all vertices in the graph G should be present in RS.

When the algorithm starts, RS is initialized with a set of diverse solutions obtained by a constructive procedure, which starting from a single vertex, each step adds a new vertex to the current clique until a maximal clique is found. Starting from a different vertex not yet included in RS, the procedure is repeated as many times as the cardinality of the reference set. The clique value is used to order the solutions in RS.

4.5.2 Improvement Method

The improvement method has two phases: given a solution that is typically infeasible, the method first undertakes to recover to a feasible one; and afterward it attempts to increase the objective function value. Neighborhood structures based on add, drop, and node swap moves are used in the local search. The method allows for the solutions being infeasible by temporarily dealing with non complete subgraphs, which implements a strategic oscillation allowing trajectories to cross infeasible regions of solutions.

4.5.3 Reference Set Update Method

The reference set update method must be carefully set up with diverse and high quality solutions to avoid the phenomenon of *premature convergence* of RS, which occurs when all the solutions are similar. To prevent this "pitfall", the reference set RS is divided into two groups: the set of best solutions and the set of diverse solutions.

Regarding the replacement policy, a combination of the best replacement policy and the worst replacement policy called "journal replacement" policy is used, which replaces the worst solution with the new best solution found, reporting all the "hits" of the search.

4.5.4 Subset Generation Method

This method generates the following types of solution subsets which are combined in the next method. The method generates subsets with two, three or more elements in a relatively reduced computational effort. To eliminate repetition of the elements in the subsets, the reference set with diverse solutions is used for the two by two combinations, instead of the complete reference set. The method also includes a new feature by adding a distant (or diverse) solution maximizing the distance from the region defined as the union of the vertices in the solution's subset. In this way, a new point "far from" the solution cluster is obtained at each iteration to maintain an appropriate diversity of solutions in the reference set.

4.5.5 Solution Combination Method

This method uses each subset generated in the subset generation method and combines the subset solutions, returning one or more trial solutions. Solution combinations are created using a filter vector applied to the union of solutions, called λ-filter. The λ-filters are used in Scatter Search as a form of structured combinations of solutions. Instead of drifting within the solution space defined by the reference set, the SS procedure searches each region extensively by applying different λ-filters. Each λ-filter generates a trial solution to be improved by the Improvement Method. A sequence of previously planned λ-filters generates a set of solutions within and beyond regions defined by two or more solutions in which new trial solutions will be chosen for updating the reference set in an evolutionary fashion.

Applying λ-filters in subsets with diverse solutions, a bypass strategy is created. Instead of finding solutions between two others, it is possible to bypass the path using a intermediate reference solution.

Computational results obtained on a set of the most challenging clique DIMACS benchmark instances shown the scatter search algorithm can be advantageously compared with some of the most competitive algorithms for the MCP.

4.6 Assigning Proctor to Exams with Scatter Search

Several real assignment problems can be viewed as a generalization of the well-known Generalized Assignment Problem. One of these problems is the proctor assignment problem (PAP), which consists in the assignment of proctors to final exams at a school or university, with respect to some objective function as for example the maximization of the total preferences of proctors to exams' dates.

Martí, Lourenço and Laguna (2000) presented a Scatter Search to solve particular instances of the PAP, based on the real data from a Spanish University. The problem was formulated as a multiobjective integer program with a total preference and workload-fairness objective functions, and can be stated as follows: consider a set of proctors at a large university. Each proctor has a maximum number of hours that he/she can devote to proctor final exams. This limit depends on his/her contract and teaching load. Each final exam requires a given number of proctors for proctoring. Since the most of the proctors are graduate students and Teaching Assistants (TAs), they also have final exams and therefore they cannot proctor exams during periods that conflict with their own exams. The constraints can be summarized as follows:

- Each exam must be proctored by a specified number of TAs.
- A TA cannot exceed his/her maximum number of proctor hours.
- A TA cannot proctor more than one exam at the same time.
- A TA cannot proctor a final exam that conflicts with one of his/her own.
- A TA should proctor the exams of the courses he/she taught.

The last constraint can be handled before formulating the model by simply assigning proctors to the exams of the courses they taught and adjusting the associated input data accordingly (e.g., reducing the total number of proctor hours and the exam requirements). Teaching assistants have preferences for some exams, which reflect their desire for proctoring on a given day or avoiding certain days. For example, some TAs would like to avoid proctoring an exam the day before one of their own exams. As a result of these preferences, one objective of the problem is to make assignments that maximize a function of the total preferences.

Another important criterion is the assignment of proctor to exams such that the workload is evenly distributed among TA's. Unfair workloads are likely to generate conflicts among TA's and between TA's and the administration. Several objective functions can be formulated to measure the workload-fairness of a given assignment. One possibility is to maximize the minimum workload associated with each TA. Since the number of available hours for each TA varies, the workload can be expressed as the ratio of assigned hours to available hours.

Martí et al. (2000) considered a weighted function to deal with the multiobjective model and proposed a scatter search method based on the work by Glover (1998), Laguna (1999) and Campos et al. (1998). The diversification generation method generates the population solutions using the preferences values modified by a frequency function. This frequency function is used to bias the potential assignment of TAs to exams during subsequent constructions of solutions, and therefore to induce diversity in the new solutions with respect to the solutions already in the population. TAs are assigned to exams in order to maximize the modified preference values. The reference set is constructed by using the best solutions and a distance function between solutions to diversify the solutions in this set. The solution combination method is applied to each subset generated as in Glover (1998). It is based on a voting system, where each solution votes for specific assignment of TAs to exams. The resulting solution may be infeasible with respect to some constraints, and in this case, a repair mechanism is applied. The method outputs the best solution with respect to the weighting function.

The data used for these experiments correspond to real instances of the proctor assignment problem at the Universitat Pompeu Fabra in Barcelona (Spain). The results were compared with manually generated assignments also with assignments found by solving the mixed-integer programming formulation with Cplex 6.5 (some of which are optimal). For the set of test problems that utilize the utility function, the scatter search solutions are often slightly sub-optimal. However, this is offset by the advantage that the scatter search reference set contains a number of high-quality solutions, allowing the decision-maker to choose the one to implement, based on non-quantitative elements. The maximum standard deviation of the utility function value for solutions in the final reference set was 0.000407 for all problem instances. This indicates that practically all of the solutions in the final reference set have the same quality with respect to the objective function value. Since the utility function is a mathematical representation of some subjective measure of performance associated with a given assignment, the ability to choose among solutions that have similar objective function values is an important feature of a decision support system designed for this managerial situation.

Since scatter-search is a population-based search, the method is a useful solution technique to solve multiobjective problems by finding an approximation of the set of Pareto-optimal solutions. A multiobjective scatter search for the solving the PAP is investigated in Lourenço et al. (2001). The main features of this approach are the construction and updating of the reference set using the set of non-dominated solutions. Also, the cardinality of the reference set varies with respect to the size of the set of non-dominated solutions. An improvement method is applied to improve the solutions obtained by the greedy heuristic, the diversification method and the solution combination method. This improvement method consists of a simple local search method, where the neighborhood is obtained by exchanging one TA for another one from the list of proctors. Finally, the method is restarted with the set of non-dominated solutions in the reference set. Preliminary results for this new approach indicate that multiobjective scatter search with restarting gives the best results with respect to the weighting function, across different versions of the method. Also, multiobjective scatter search enables the user to analyze a collection of very good solutions and make the final decision.

4.7 Periodic Vehicle Loading

Delgado et al. (2002) address a logistical problem of a manufacturer of auto parts in the north of Spain. The manufacturer stores auto parts in its warehouse until customers retrieve them. The customers and the manufacturer agree upon an order pickup frequency. The problem is to find the best pickup schedule, which consists of the days and times during the day that each customer is expected to retrieve his/her order. For a given planning horizon, the optimization problem is to minimize the labor requirements to load the vehicles that the customers use to pick up their orders.

Heuristically, the authors approach this situation as a decision problem in two levels. In the first level, customers are assigned to a calendar, consisting of a set of days with the required frequency during the planning horizon. Then, for each day, the decision at the second level is to assign each customer to a time slot. The busiest time slot determines the labor requirement for a given day. Therefore, once customers have been assigned to particular days in the planning horizon, the second-level decision problem is equivalent to a multiprocessor scheduling problem (MSP), where each time slot is the equivalent of a processor, and wherethe objective is to minimize the makespan.

A scatter search procedure is developed for the problem of minimizing labor requirements in this periodic vehicle-loading problem and artificial as well as real data are used to assess its performance. The scatter search constructs and combines calendar assignments and uses a heuristic to solve the MSP's for each day in the planning horizon and thus obtain a complete solution.

The diversification method is based on GRASP constructions. The greedy function calculates the increase in labor requirements from assigning a previously unassigned order to a calendar. The procedure starts with all the orders in the "unassigned" set. The orders are considered one by one, from the largest to the smallest (i.e., from the one that requires the most amount of labor to the one that requires the least amount of labor).

The improvement method is based on a procedure that changes the assignment of an order from its current calendar to another. Preliminary experiments showed that the performance of the improving method with simple moves (i.e., the change of calendars for one order only) was not as good as the performance of a local search employing composite moves. A composite move is a chain of simple moves. Therefore, while a simple move prescribes the change of one order from one calendar to another, a composite move prescribes the change of several orders from their current calendars to others. It may seem that a local search based on simple moves should be capable of finding sequences of moves that are equivalent to composite moves. However, this is not necessarily the case because the local search based on simple moves is greedy and searches for the best exchange and performs the exchange only if it results in an improving move. A local search with composite moves, on the other hand, may perform some non-improving simple moves that lead to a large improving move.

The combination method generates new solutions by combining the calendar assignments of two reference solutions. The objective function values of the reference solutions being combined are used to probabilistically assign orders to calendars in the new trial solution. That is, on the average, most of the assignments come from the reference solution with the better objective function value. The procedure uses a static update of the reference set.

Using both randomly generated data adapted from the literature and real data from a manufacturer, the authors were able to show the merit of the scatter search design. In particular, extensive experiments show that significant savings may be realized when replacing the manufacturer's current rules of thumb with the proposed procedure for planning purposes.

4.8 Tabu and Scatter Search in Job-Shop Scheduling

The job-shop scheduling problem is known as a particularly hard combinatorial optimization case. It arises from OR practice, has a relatively simple formulation, excellent industrial applications, a finite but potentially astronomical number of solutions and unfortunately is strongly NP-hard. It is also considered an indicator of practical efficiency of advanced scheduling algorithms. In the early nineties, after a series of works dealing with optimization algorithms of the B&B type, it became clear that pure optimization methods for this problem had a ceiling on their performance. In spite of important advances over the past two decades, the best B&B methods cannot solve instances with more than 200 operations in a reasonable time (hours, days, weeks).

A new era started when job-shop algorithms based on the TS approach appeared. The simple and almost ascetic Algorithm TSAB (Nowicki and Smutnicki, 1996), designed originally in 1993, found the optimal solution of the notorious job-shop instance FT10 (100 operations) in a few seconds on a PC. This instance had waited 26 years, since 1963, to be solved by an optimization algorithm. But going far beyond the solution of FT10, the TSAB approach made it possible to solve, in a very short time on a PC, instances of size up to 2,000 operations with unprecedented accuracy—producing a deviation from an optimality bound of less than 4% on average. This is considerably better than the deviation of approximately 20% for special insertions technique, 35% for standard priority rules and over 130% for random solutions. Another highly effective tabu search method for the job shop problem has recently been introduced by Grabowski and Wodecki (2001).

Further exploration of the ideas underlying TSAB focuses on two independent subjects: (1) acceleration of the speed of the algorithm or some its components, and (2) a more sophisticated diversification mechanism, the key for advanced search scattering. Recent papers by Nowicki and Smutnicki (2001a,b), provide some original proposals located precisely in these research streams. They refer to a new look at the landscape and valleys in the solution space, set against the background of theoretical properties of various distance measures. There are proposed *accelerators* based on theoretical properties, which, by means of skillful decomposition and aggregation of calculations, speed up significantly search process, namely: (a) INSA accelerator (advanced implementation of insertion algorithm used for starting solution in TSAB), (b) tabu status accelerator, (c) NSP accelerator (fast single neighborhood search). Next, in order to diversify the search, TSAB has been embedded in the Scatter Search and Path Relinking framework. The resulting algorithm ***i***-TSAB described in Nowicki and Smutnicki (2001a), the powerful successor of TSAB, works with elite centers of local search areas forming a MILESTONE structure, modified by space explorations conducted from VIEWPOINTS located on GOPS (a class of goal oriented paths).

As the immediate practical result of this new approach, 24 better upper bounds (new best solutions) have been found for 24 of the 35 instances from the common benchmark set of Taillard, attacked by all job-shop algorithms designed till now. The proposed algorithm still runs on a standard PC in a time of minutes.

4.9 A Tabu Scatter Search Metaheuristic for the Arc Routing Problem

The problem treated in Greistorfer (2001a) is the so-called *capacitated Chinese postman problem* (CCPP). The goal of the (undirected) CCPP is to determine a least-cost schedule of routes in an undirected network under the restriction of a given fleet of vehicles with identical capacity, which operates from a single depot node. In the standard version of the CCPP the number of vehicles is unlimited, i.e. it is a decision variable. The CCPP is a special instance of the general class of arc routing problems, a group of routing problems where the demand is located on arcs or edges (one-way or two-way roads) connecting a pair of nodes (junctions). Relevant practical examples of the CCPP are postal mail delivery, school bus routing, road cleaning, winter gritting or household refuse collection. But applications are not limited to the routing of creatures or goods. There are also cases in industrial manufacturing, e.g. the routing of automatic machines that put conducting layers or components on to a printed circuit board.

The algorithmic backbone of the *tabu scatter search* (TSS) metaheuristic introduced is a tabu search (TS) which operates with a set of neighborhood operators (edge exchange and insert moves) on a long-term diversification strategy guided by frequency counts. The short-term tabu memory works with edges and simply prohibits reversal moves within a dynamically varied tenure period. Additionally, the procedure has a pool component which accompanies the TS by maintaining a set of elite solutions found in the course of the optimization. If the classic genetic algorithm can be understood as a pure parallel pool method because it always works on a set of high quality solutions, then the TSS follows a sequential pool design, where periods of isolated and single-solution improvements of the TS alternate with multi-solution combinations. With respect to the type of encoding, the solution combination method (SCM) is purely phenotypical, which turns this pool method into a type of scatter search (SS) algorithm. The TSS architecture as proposed here does not exactly follow the template ideas of Glover (1998), although there are many common features as will be outlined below.

The SCM component combines elite solutions which have been collected by the TS. As suggested in the template paper and further literature, the combination of solutions is supported by generalized rounding procedures which may follow heuristic rules or approaches based on linear programming (LP). The underlying principle of the SCM proposed follows an adapted transportation problem (TPP) formulation, which is the generalization of the assignment operator of Cung et al. (1997).

The TSS and its SCM benefit from a data structure which, generally, can be used for problems described by pure permutations (customers) or by permutations where sub-strings (routes) have to be considered as well. The SCM works as follows. Given a set of elite solutions $S_1, \ldots, S_c$, the TPP coefficients a_{ij} denote the number of times a customer j is assigned to a route i in this pool subset. The TPP coefficient matrix can be interpreted as an *assignment frequency matrix* (AFM), being the linear combination of the c individual assignment matrices. Unit TPP demands are a consequence of the need that every edge has to be serviced by a single vehicle. The supply of a route is approximated by the average number of customers that can be serviced respecting the vehicle capacity whereas a dummy column picks up the oversupply. Maximizing this TPP results in customer-route assignments which maximize the total number of desirable assignments while simultaneously minimizing the total Euclidean distance to the *AFM*, which represents the (infeasible) combination of the initial trial points. Although the outcome of this SCM can be directly used, it is clearly improvable since the optimal clusters (sets) provided do not imply any guidance of how the vehicle routes (sequences) should be formed. Therefore, a *greedy sequencing heuristic* (GSH) is used for post-optimization to put the customers of all routes into a cost-convenient order.

The overall TSS starts from a random pool whose elements are exchanged for solutions which have been improved in the TS phase. The SCM is occasionally called and forms a combined solution which is then returned to the TS for further inspections. This alternating process between TS and the SCM stops after a pre-defined period of iterations.

The TSS is tested on several classes of CCPP instances: there are a number of planar Euclidean grid-graph instances, Euclidean random instances and the well-known DeArmon data set as used for the CARPET arc routing heuristic of Hertz et al. (2000). In a direct comparison with an old TS method (see Greistorfer (1995)) the TSS significantly improves the results for the Euclidean classes (in 54% of all cases) and is clearly able to keep up with CARPET regarding the instances from literature. Here the number of (known) optimal solutions is identical, while the TSS finds one more best-known solution. Its worst average deviation (due to a single instance) is only 1.29% higher than the one of CARPET. The total running times, which are scaled with respect to the CARPET-PC, are longer. However, it is shown that on average the TSS obtains its best results faster than the CARPET heuristic. Thus, adding a pool component to a TS and using an advanced SCM has obvious merits.

4.9.1 Testing Population Designs

The theme of Greistorfer (2001a) is continued in Greistorfer (2001b), where the focus is more on the methodology. The task is to work out relevant pool strategies and to evaluate them by means of thorough computational comparisons. Test results again refer to a sample of CCPP arc routing instances but, as mentioned above, the encoding offers a certain ability to generalize the algorithmic findings for a number of different problems. From the manifold design options for heuristic pool methods, the discussion concentrates on three basic components: the *input* and *output* functions, which are responsible for pool maintenance and which determine the transfer of elite solutions, and a solution *combination* method which must effectively combine a set of elite solutions provided by the output function.

The heuristic design variants of the TSS comprise four input strategies, $I_{0,\ldots,3}$, four output strategies, $O_{0,\ldots,3}$, and three SCMs, namely $M_{0,\ldots,2}$ (including the settings of Greistorfer (2001a), indexed with a 0).

Input strategy I_0 reduces the quality aspect to the cost dimension of a solution and does not deal with structural properties of a solution or their relations to each other. I_1 overcomes this disadvantage by including full duplication checks between potential elite solutions and pool members. I_2, known as the *reference set update method* of Glover (1998), is additionally linked with preceding hash comparisons. In I_3 an attempt is made to find a compromise by skipping the full duplication part in I_2, i.e. I_3 only relies on hashing. Input functions $I_{1,2,3}$ build on ordered structures which are provided by a *sorting algorithm*. A corresponding procedure is suggested by the author as well as a relevant *hash function*.

By analogy to I_0, O_0 does not utilize structural information and simply refers to a random selection. O_1 uses frequency counts and selects those solutions for combinations which have not been used before or have been used rarely. Such a tracking of the number of involvements of a solution in a combination process introduces a certain memory effect. By contrast, the *subset generation method* of Glover (1998) explicitly makes use of an algorithmic structure which completely avoids a duplicate subset selection. Output strategies O_2 and O_3 select solutions which have the smallest and largest distance to each other, respectively. For that purpose a *distance function* is proposed which aggregates the customer positions and their route membership.

The combination strategy M_0 is the LP-based transportation method which is described in detail in Greistorfer (2001a). The last two SCM variants are based on deriving average solutions. M_1 constructs average customer labels whereas M_2 determines average customer positions (see also Campos et al., 1999). Both approaches relax the capacity restriction whose validity has to be secured afterwards by splitting the permutation sequence into a set of route clusters. The final offspring is obtained after applying the post-optimizing GSH.

The computational investigation of the results for the different TSS designs was performed by checking all possible $4 \cdot 4 \cdot 3 = 48$ TSS configurations against each other. Each configuration was run over the whole set of test instances and evaluated by its corresponding average objective function value derived from the best solutions found. The best configuration turned out to be (I_3, O_3, M_0). In order to evaluate the specific effects of an input, output or combination variant, all results were checked according to a *variance analysis* by means of SPSS.

Generally, effects of variations tended to be smaller at the input side of the pool since all tests did not indicate any significant difference between the input strategies described. One model of explanation is that duplications are effectively prevented by each of the input procedures $I_{1,2,3}$, while the (empirical) probability of collision (hashing error) is smaller than 0.195% for $I_{2,3}$. Another reason might be that strongly diversifying input strategies are not adequately utilized by the other variable components. The results generated by the straightforward algorithmic setting of I_0 cannot keep up with the results of the other methods.

The picture completely changes when output strategies or different SCMs are looked upon. It was found that the min-distance approach in O_2 is definitely an inferior option. The argument that good solutions are mostly found in the vicinity of the best solution cannot be upheld, which is clearly in agreement with the SS philosophy of selecting diverse solutions to be combined. This fact is also underlined by the superiority of the max-distance function O_3 over O_2. The expected memory effect in O_1 turned out to be too small to guide the selection process. Random sampling in O_0 can be justified in an isolated view which ignores input and combination method effects.

As an SCM, the LP approach of M_0 significantly contributes to finding better solutions than M_1 and M_2. While there are no significant relations between the latter ones, the individual best choice for M_0 proves its very useful role in the collective optimal design (I_3, O_3, M_0).

4.10 A Population Based Approach to the Resource Constrained Project Scheduling

Valls et al. (2001) propose a population-based approach to the resource-constrained project-scheduling problem (RCPSP), where n activities have to de processed, taking into account the precedence relations and the resource restrictions. The procedure incorporates different strategies for generating and improving a population of schedules. The method has two phases. Phase 1 (figure) can be interpreted in terms of the scatter search methodology.

Figure. Phase_1 outline

1. POP = INITIAL_SET_1(20)

2. POP' = HIA(POP)

3. **For** $[i = 1, 2]$

  3.1. POP5 = {the 5 best schedules in POP'}

  3.2. POP = CSA(4,POP5)

  3.3. POP40 = {the 40 best schedules in POP}

  3.4. POP' = HIA(POP40)

4. **Return** the best schedule obtained

INITIAL_SET_1(*size*) is the Diversification Generation Method, which generates a collection of diverse trial solutions. To achieve quality and diversity, different well-known priority rules and random procedures are used. The best *size* solutions are stored in the set POP.

The Improvement Method is called HIA, the Homogenous Interval Algorithm. It is applied to the solutions in POP and is an iterative procedure for improving the local use of resources. It incorporates an oscillatory mechanism that alternatively searches two different regions of the schedule space (strategic oscillation).

The Subset Generation Method and the Solution Combination Method are carried out by the Convex Search Algorithm (CSA). CSA(k, SET) generates all pairs of reference solutions of SET and combine each of them with a procedure that integrates path relinking characteristics.

First of all, CSA codifies each schedule by a topological order (TO) representation. A TO representation of a schedule S is a special priority value vector γ, the one that fulfills the conditions (1) $\{\gamma(i), i = 1, \ldots, n\} = \{1, \ldots, n\}$, (2) $s_i < s_j \rightarrow \gamma(i) < \gamma(j)$, being s_i the beginning of activity i in S, and (3) $s_i = s_j$ and $i < j \rightarrow \gamma(i) < \gamma(j)$, i.e., the label is used to order activities with the same beginning. The Serial schedule generation scheme can be used to decodify a TO representation γ and obtain an active schedule $S(\gamma)$, by selecting at each stage the eligible activity j with the lowest priority $\gamma(j)$.

Afterwards, for each pair of reference schedules A and B, CSA calculates the priority value vectors γ^p, defined by $\gamma^p(j) = (1 - p/2^k)\gamma_A(j) + (p/2^k)\gamma_B(j), p = 1, \ldots, 2^k - 1$, where k is an integer and γ_A and γ_B are the TO representations of A and B, respectively. Although γ^p is not a TO representation, is a vector of priorities compatible with the precedence relations that can be easily transformed into a TO representation that gives the same schedule as γ^p when decodified. Moreover, the priority vectors $\gamma^p, p = 1, \ldots, 2^k - 1$ are uniformly distributed in the geometric segment joining γ_A and γ_B and it can be statistically proved that something similar happens to the schedules $S(\gamma^p), p = 1, \ldots, 2^k - 1$, in the path between $S(\gamma_A)$ and $S(\gamma_B)$. This means that each $S(\gamma^p)$ incorporates more attributes of $S(\gamma_B)$ and less of $S(\gamma_A)$ as p increases, so this procedure builds a trajectory between $S(\gamma_A)$ and $S(\gamma_B)$.

The best 40 schedules calculated by CSA are introduced in POP40, and the improvement procedure HIA is applied to each of them. The Reference Set Update Method is quite simple: the first reference set contains the best 5 schedules calculated in step 2 and the following reference sets are formed by the best 5 solutions obtained after having applied CSA and HIA. So, the reference set is totally replaced in each iteration, looking for a fast convergence, since step 3 is applied only twice.

At the end of phase 1, the best solution obtained so far is generally of high quality. Experience seems to indicate that good candidate schedules are usually to be found "fairly close" to other good schedules. The objective of the second phase is therefore to closely explore regions near high quality sequences. Exploring such a region means, first, generating a population by taking a random sample from a region near a good sequence, and second, applying to the population the improving procedure used in the first phase—but with a variation. The on-going search is interrupted when a better sequence is obtained and a fresh search starts from the point of the newer sequence. Phase two begins from the best solution obtained in phase one.

Computational experiments have been carried out on the standard j120 set generated using ProGen. They show that the indicated SS algorithm produces higher quality solutions than state-of-the-art heuristics for the RCPSP in an average time of less than 5 s in a PC at 400 MHz.

4.11 Multiple Criteria Scatter Search

Beausoleil (2001) has developed a scatter search procedure that uses the concept of Pareto optimality to obtain a good approximate Pareto frontier. Tabu Search is used to obtain an initial set of reference points. Different frequency memories are used to diversify the search. In order to designate a subset of strategies to generate a reference solutions a choice function called Kramer Selection is used. A Kemen-Snell measure is applied in order to find a diverse set to complement the subset of high quality current Pareto solutions. Path Relinking and Extrapolated Path Relinking are used as a Combination Method.

Structured weighted combination is used in a special case to obtain weighted solutions inside the convex region spanned by selected reference points. To implement the process, in the Tabu Search phase, memory is maintained of selected attributes of recent moves and their associated solutions. A thresholding aspiration guides the selection of an initial set of solutions. Solution quality is measured by introducing an Additive

Value Function in this phase. A study involving multiple cases demonstrates the ability of the algorithm to find a diverse Pareto frontier.

The results of the experiments show that the first TS phase generates an initial good Pareto frontier. The combined method using path relinking and extrapolated path relinking as an intensification-diversification method proves an effective mechanism to generate new Pareto points, yielding a good approximation to the Pareto frontier in a relatively small number of iterations.

4.12 Meta-Heuristic Use of Scatter Search via OptQuest

Optimization and simulation models play an important role in Defense Analyses. A crucial component of a model-based analysis effort is agreement upon the planning scenario upon which the analysis is conducted. For example in military force structuring, the model might suggest a prescribed force structure necessary to best meet the demands of a planning scenario. Conversely, given some proposed force structure, a model might provide insight into a "best use" of that force within the specified scenario. A particularly perplexing challenge for military analysts occurs when they must suggest a single overall force structure after considering multiple competing scenarios, each suggestive of potentially differing optimal force structures.

Hill and McIntyre (2000) addressed this particular vexing military force structure problem. They define a **robust force structure (solution)** as that force structure (solution) "providing the best overall outcome as evaluated with respect to some set of scenarios each of which has an associated likelihood of occurrence." Their approach considered the multi-scenario optimization problem within which each particular scenario solution becomes a component of an aggregate multi-scenario solution. They treat the multi-scenario space as a composite of the component scenario spaces where each component space contributes relative to its likelihood of occurring or relative importance weight. Using a meta-heuristic to guide a search using the combat model as an evaluation function provides a means to find a single solution, potentially optimal in the multi-scenario space, and by definition, robust across each of the individual scenario spaces.

The Hill and McIntyre approach is presented graphically in the Figure above. Central to the approach is a CONTROLLER interface between the meta-heuristic module

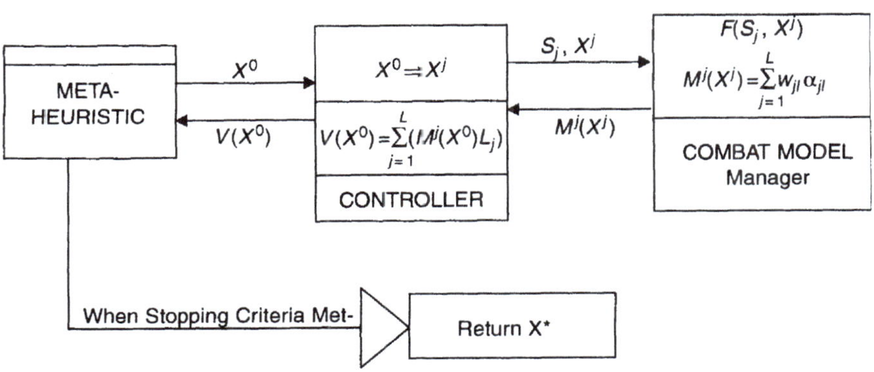

and the combat models conducting the evaluations. The META-HEURISTIC guides the search process providing the CONTROLLER potential solutions (input force structure) and receiving from the CONTROLLER evaluations of those solutions. The COMBAT MODEL receives its input (and scenario) from the CONTROLLER, evaluates the input, and returns the requisite quality measure from the combat model assessment associated with the input. The CONTROLLER accepts the potential solutions, provides those to each of the scenario evaluators in the COMBAT MODEL, and combines each measure into the final value or fitness of the potential solution. This process continues until predefined stopping conditions are satisfied at which time the best, or set of best, solutions are returned.

Bulut (2001) applied scatter search, implemented within the OptQuest callable library (Laguna and Marti, 2002) to solve a multi-scenario optimization problem based on the United States Air Force's Combat Forces Assessment Model (CFAM), a large-scale linear programming model for weapons allocation analyses. Three notional planning scenarios were used and a robust solution sought to the multi-scenario problem. He compared OptQuest results with previous results obtained using a genetic algorithm (for the same scenarios). His results indicated that better overall solutions, a greater diversity of solutions, and quicker convergence results were obtained using the OptQuest scatter search approach.

The methodology proposed by Hill and McIntyre (2000) and implemented by Bulut (2001) using OptQuest is directly applicable to any analytical situation involving competing "scenarios" within which one needs a single solution that is "robust" relative to any differences among the scenarios.

4.13 Pivot Based Search Integrated with Branch and Bound for Binary MIPs

Linear programming models with a mixture of real-valued and binary variables are often appropriate in strategic planning, production planning with significant setup times, personnel scheduling and a host of other applications. The abstract formulation for linear problems with binary integers takes as data a row vector **c**, of length n, a $m \times n$ matrix A and a column vector $\boldsymbol{b}$ of length m. Let D be the index set $1, \ldots, n$. The problem is to select a column vector, $\boldsymbol{x}$ of length n so as to

$$\min \sum_{i \in I} c_i x_i$$

subject to

$$Ax \geq b$$
$$x_i \in \{0, 1\} \quad i \in I$$
$$x_i \geq 0 \quad i \in D \backslash I,$$

where the index set I gives the variables that must take on zero-one values.

Issues related to the behavior of a pivot based tabu search integrated with branch and bound algorithm, using path relinking and chunking are discussed by Løkketangen and Woodruff (2000).

The integration takes place primarily in the form of local searches launched from the nodes of the branch and bound tree. These searches are terminated when an integer feasible solution is found or after some number of pivots, *NI*. Any time a new best is found, the search is continued for an additional *NI* pivots. Chunking (see Woodruff 1996, 1998) is used to detect solutions that should be used for special path relinking searches that begin at the LP relaxation and to determine when the use of pivot searches should be discontinued. (See also Glover et al., 2000, for another application of chunking to the same kind of problems.)

As the search is launched from nodes in a B&B tree, there are some special considerations that come into play that sets this use of the pivot based search somewhat apart from other implementations. First, the chunking mechanism and the path relinking based target searches, respectively, fulfill the functions of diversification or intensification. Second, the purpose, or focus, of the search is somewhat different from the stand-alone search, in that for some of the searches, the emphasis is shifted more towards obtaining integer feasibility quickly. This focus is controlled by a separate parameter *skew*, that is used to adjust the relative importance of obtaining feasibility versus maintaining a good objective function value.

Chunking addresses the questions of when the launching of pivot based searches should be terminated, and when the path relinking searches should be launched. More specifically, path relinking searches are used to exploit "unique" or "outlying" solutions. The meaning of "unique" and "outlying" can be supplied by chunking.

Two types of local searches can be launched at a node. The first are the normal TS pivot-based searches launched from nodes in the B&B tree (see Løkketangen and Glover, 1995, 1996, 1998, 1999).

The other type are the Path Relinking searches. After a best-so-far solution $\boldsymbol{x}^*$ has been found, the chunking mechanisms try to identify distant solutions, $\mathbf{x}'$ (w.r.t. the current sample). When such a distant solution has been identified, a *2-target* search is launched. This is a variant of path relinking, with the purpose of launching a new search into unknown territory, while at the same time keeping good parts of the solutions. This integrates the considerations of intensification and diversification.

The starting point of this search is the relaxed root node solution LP^* (being an upper bound on the objective function value), and the target for the path relinking is the hyperplane defined by the common integer solution values of $\boldsymbol{x}^*$ and $\mathbf{x}'$. All integer variables are freed. To allow the search to focus on this hyperplane, the integer infeasibility part of the move evaluation function is temporarily modified. (The objective function value component is unaltered, as is the aspiration criterion—see Løkketangen and Woodruff, 2000.) Instead of using the normal integer infeasibility measure of summing up over all the integer variables the distance to the nearest integer, the authors use the following scheme:

- Sum up over all the integer variables.
- If the two targets have the same integer solution value for the variable, use the distance to this value.
- If the two targets differ, use the normal integer infeasibility measure (i.e., the closest integer value).

When the search reaches the hyperplane connecting $\boldsymbol{x}^*$ and $\mathbf{x}'$, the normal move evaluation function is reinstated, and the search continues in normal fashion for *NI* iterations.

Computational testing was done on problems from Miplib and Dash Associates, consisting of a mix of MIP's and IP's. The testing showed that the local searches had a beneficial effect on the overall search time for a number of problem instances, particularly those that were harder to solve.

4.14 Scatter Search to Generate Diverse MIP Solutions

Often, scatter search and star path algorithms (Glover, 1995), generate diverse sets of solutions as a means to an end. In a recent paper by Glover, Løkketangen and Woodruff (2000) diversity is the ultimate goal for which scatter search and star paths are employed. This paper presents methods of systematically uncovering a diverse set of solutions for 0–1 mixed integer programming problems. These methods can be applied to instances without any special foreknowledge concerning the characteristics of the instances, but the absence of such knowledge gives rise to a need for general methods to assess diversity.

When the objective function is only an approximation of the actual goals of the organization and its stakeholders, the one solution that optimizes it may be no more interesting than other solutions that provide good values. However, information overload can be a problem here as well. It is not desirable to swamp the decision maker with solutions. Highly preferable is to identify a set of solutions that are decently good and, especially, diverse. One can reasonably rely on the objective function to quantify the notion of "decently good". The diversification methods given by Glover, Løkketangen and Woodruff are based on the idea of generating extreme points in a polyhedral region of interest and then using these points and the paths between them in a variety of ways. The methods examine points on the polyhedron, within and "near" it. Their algorithm proceeds in two phases: first it generates a set of *centers* and then connects them using *star paths*.

The description of the generation of centers can also be broken into two phases. First a diversification generator is used to create points. In the second phase, these points are provided as data to an optimization problem that results in extreme points that are averaged to create the centers.

Although a diverse set of good solutions is clearly desirable, it is not clear in advance how to measure the property of diversity. In spite of the fact that the objective function is not exact, it presumably gives a reasonable way to assess the relative "goodness" of a set of solutions. No such simple mapping is known from solution vectors to a one-dimensional measure of diversity. Diversity measures are required both for the design of practical software and for research purposes. For practical software, it is important to know if the user should be "bothered" with a particular solution vector—that is, to know if a vector adds enough diversity to warrant adding it to the set of solutions that are displayed. For research purposes, one might want to compare the set of vectors generated by one method with a set of vectors generated by another.

There are a number of advantages to the quadratic metric known in this context as *Mahalanobis* distances. This metric is scale invariant and can take correlations into account if based on a covariance matrix. Furthermore, this type of distance connects naturally with a scalar measure of the diversity of a set of vectors, which is the determinant of the covariance matrix of the set. Under the assumption of multivariate normality, the covariance matrix defines ellipsoids of constant Mahalanobis distances that constitute probability contours. Large covariance determinants correspond to large volumes in the ellipsoids. The assumption of multivariate normality is not needed to use the covariance determinant to put an order on sets of vectors and furthermore it is not needed to see that adding points with large Mahalanobis distances will increase the covariance determinant.

However, there is a major difficulty. In order to calculate a covariance matrix for a set of vectors of length $p = n$ one must have a set of vectors that does not lie entirely in a subspace. This means that at a minimum the set must contain $n + 1$ vectors and for MIP solutions, more vectors will often be required to span the full n dimensions. For even modest sized MIPs this is not good. In order to have a working definition of diversity, one must have thousands of solution vectors. A remedy for this difficulty that also increases the plausibility of multivariate normality has been referred to as *chunking* by Woodruff (1998). A generalization based on principal components has also been proposed by Woodruff (2001).

As general purpose optimization methods are embedded in decision support systems, there will unquestionably be an increased need not only for optimal solutions, but also for a diverse set of good solutions. Scatter search and star paths can be an effective means to this end.

Results of computational experiments demonstrate the efficacy of the "scatter-star-path" method for generating good, diverse vectors for MIP problems. Furthermore, the results show that the method offers particular advantages when used in conjunction with brand and bound. The creation of these results illustrates the use of methods for measuring the diversity for a set of solutions.

4.15 Path Relinking to Improve Iterated Re-start Procedures

Research has been performed to investigate the ability of path relinking to improve the performance of iterated re-start procedures, with attention focused in particular on the GRASP method (Ribeiro and Resende, 2002). One possible shortcoming of the standard GRASP algorithm is the independence of its iterations, i.e., the fact that it does not learn from the history of solutions found in previous iterations. This is so because it discards information about any solution encountered that does not improve the incumbent. Information gathered from good solutions can be used to implement extensions based on path-relinking.

Path relinking was originally proposed in the context of tabu search as an intensification strategy which explores trajectories connecting high-quality solutions. The use of path relinking within a GRASP procedure was first proposed by Laguna and Martí (1999), being followed by several extensions, improvements, and successful applications (e.g., Canuto et al., 2001; Aiex et al., 2002; Ribeiro et al., 2002). Path relinking and a very short term memory used within the local search were instrumental to make a recently proposed GRASP heuristic for the capacitated minimum spanning tree problem competitive with other approaches in the literature (Souza et al., 2002). Two basic strategies are used to apply path relinking in the context of a GRASP heuristic:

- apply path relinking as a post-optimization step to all pairs of elite solutions; and
- apply path relinking as an intensification strategy to each local optimum obtained after the local search phase

Both strategies maintain and handle a pool with a limited number Max_Elite of elite solutions found along the search (we used Max_Elite ranging from 10 to 20 in most implementations). The pool is originally empty. Each locally optimal solution obtained by local search is considered as a candidate to be inserted into the pool if it is sufficiently different from every other solution currently in the pool. If the pool already has Max_Elite solutions and the candidate is better than the worst of them, then the former replaces the latter. If the pool is not full, the candidate is simply inserted.

Applying path relinking as an intensification strategy to each local optimum seems to be more effective than simply using it as a post-optimization step. In this context, path relinking is applied to pairs X–Y of solutions, where X is the locally optimal solution obtained after local search and Y is one of a few elite solutions randomly chosen from the pool (usually only one elite solution is selected). The algorithm starts by computing the symmetric difference between X and Y, resulting in a set Δ of moves which should be applied to one of them (the initial solution) to reach the other (the guiding solution). Starting from the initial solution, the best move still in Δ is applied, until the guiding solution is attained. The best solution found along this trajectory is also considered as a candidate for insertion in the pool and the incumbent is updated. Several alternatives have been considered and combined in recent implementations to explore trajectories connecting X and Y:

- do not apply path relinking at every GRASP iteration, but instead only periodically;
- explore two different trajectories, using first X, then Y as the initial solution;
- explore only one trajectory, starting from either X or Y; and
- do not follow the full trajectory, but instead only part of it.

All these alternatives involve trade-offs between computation time and solution quality. Ribeiro et al. (2002) observed that exploring two different trajectories for each pair X–Y takes approximately twice the time needed to explore only one of them, with very marginal improvements in solution quality. They also observed that if only one trajectory is to be investigated, better solutions are found when path relinking starts from the best among X and Y. Since the neighborhood of the initial solution is much more carefully explored than that of the guiding one, starting from the best of them gives to the algorithm a better chance to investigate with more details the neighborhood of the most promising solution. For the same reason, the best solutions are usually found closer to the initial solution than to the guiding one, allowing pruning the relinking trajectory before the latter is reached. The same findings were also observed on a recent implementation of a GRASP heuristic for a multicommodity flow problem arising from PVC rerouting in frame relay services. Detailed computational results and implementation strategies are described by Resende and Ribeiro (2002).

Path relinking is a quite effective strategy to introduce memory in GRASP, leading to very robust implementations. This is illustrated by the results obtained with the hybrid GRASP with path relinking for the Steiner problem in graphs described in Ribeiro et al. (2002) which in particular improved the best known solutions for 33 out of the 41 still open problems in series i640 of the SteinLib repository (Voss et al., 2001) on April 6, 2001.

Even though parallelism is not yet systematically used to speed up or to improve the effectiveness of metaheuristics, parallel implementations are very robust and abound in the literature (see e.g. Cung et al., 2001 for a recent survey). Most parallel implementations of GRASP follow the *independent-thread multiple-walk* strategy, based on the distribution of the iterations over the processors.

The efficiency of multiple-walk independent-thread parallel implementations of metaheuristics, running multiple copies of the same sequential algorithm, has been addressed by some authors. A given target value τ for the objective function is broadcasted to all processors which independently run the sequential algorithm. All processors halt immediately after one of them finds a solution with value at least as good as τ. The speedup is given by the ratio between the times needed to find a solution with value at least as good as τ, using respectively the sequential algorithm and the parallel implementation with ρ processors. These speedups are linear for a number of metaheuristics, including simulated annealing, iterated local search, and tabu search. This observation can be explained if the random variable *time to find a solution within some target value* is exponentially distributed (Verhoeven and Aarts, 1995). In this case, the probability of finding a solution within a given target value in time pt with a sequential algorithm is equal to that of finding a solution at least as good as the former in time t using p independent parallel processors, leading to linear speedups. An analogous proposition can be stated for a two parameter (shifted) exponential distribution.

Aiex et al. (2002) have shown experimentally that the solution times for GRASP also have this property, showing that they fit a two-parameter exponential distribution. This result was based on computational experiments involving GRASP procedures applied to 2400 instances of five different problems: maximum independent set, quadratic assignment, graph planarization, maximum weighted satisfiability, and maximum covering. The same result still holds when GRASP is implemented in conjunction with a post-optimization path relinking procedure.

In the case of *multiple-walk cooperative-thread* parallel strategies, the threads running in parallel exchange and share information collected along the trajectories they investigate. One expects not only to speed up the convergence to the best solution but, also, to find better solutions than independent-thread strategies. Cooperative-thread strategies may be implemented using path-relinking, combining elite solutions stored in a central pool with the local optima found by each processor at the end of each GRASP iteration. Canuto et al. (2000, 2001) used path relinking to implement a parallel GRASP for the prize-collecting Steiner tree problem. A similar approach was recently adopted by Aiex et al. (2000) for the 3-index assignment problem. Each processor, upon completing its iterations, applies path relinking to pairs of elite solutions stored in a pool, and each processor keeps its own local pool of elite solutions. The strategy used in Canuto (2000) is truly cooperative, since pairs of elite solutions from a centralized unique central pool are distributed to the processors which perform path relinking in parallel. Computational results obtained with implementations using MPI and running on a cluster of 32 Pentium II-400 processors and on a SGI Challenge computer with 28 196-MHz MIPS R10000 processor (Aiex et al., 2000) show linear speedups and further illustrate the effectiveness of path relinking procedures used in conjunction with GRASP to improve the quality of the solutions found by the latter.

ACKNOWLEDGEMENTS

Research by F. Glover and M. Laguna was partially supported by the U.S. Office of Naval Research grant N00014-02-0151, and research by R. Martí partially supported by the Ministerio de Ciencia y Tecnología of Spain: TIC2000-1750-C06-01.

REFERENCES

Aiex, R.M., Resende, M.G.C., Pardalos, P.M. and Toraldo, G. (2000) GRASP with path relinking for the three-index assignment problem (submitted for publication).

Aiex, R.M., Resende, M.G.C. and Ribeiro, C.C. (2002) Probability distribution of solution time in GRASP: An experimental investigation. *Journal of Heuristics*, **8**, 343–373.

Alvarez, A., González, J.L. and De Alba, K. (2001a) Un algoritmo de búsqueda para un problema de red capacitada multiproducto. In: C. Zozaya, M. Mejía, P. Noriega, A. Sánchez (eds.), *Proceedings of 3rd International Meeting of Computational Sciences*, Aguascalientes México, pp. 105–114.

Alvarez, A., González, J.L. and De Alba, K. (2001b) Scatter search for the multicommodity capacitated network design problem. Proceedings of the 6th Annual International Conference on Industrial Engineering—Theory, Applications and Practice. San Francisco, CA, USA.

Beausoleil, R.P. (2001) *Multiple criteria scatter search.* 4th Metaheuritics International Congress, Porto, Portugal, pp. 539–543.

Bulut, G. (2001) Robust multi-scenario optimization of an air expeditionary force force structure applying scatter search to the combat forces assessment model. Masters Thesis, Department of Operational Sciences, Air Force Institute of Technology, AFIT/GOR/ENS/01M-05.

Campos, V., Laguna, M. and Martí, R. (1999) Scatter search for the linear ordering problem. In: David Corne, Marco Dorigo and Fred Glover (eds.), *New Ideas in Optimization*, McGraw-Hill, pp. 331–340.

Canuto, S.A. (2000) *Local Search for the Prize-collecting Steiner Tree Problem* (in Portuguese), M.Sc. Dissertation, Department of Computer Science, Catholic University of Rio de Janeiro.

Canuto, S.A., Resende, M.G.C. and Ribeiro, C.C. (2001) Local search with perturbations for the prize-collecting Steiner tree problem in graphs. *Networks*, **38**, 50–58.

Cavique, L., Rego, C. and Themido, I. (2001) A Scatter Search Algorithm for the Maximum Clique Problem. *Essays and Surveys in Metaheuristics*, Kluwer Academic Publishers.

Chanas, S. and Kobylanski, P. (1996) A new heuristic algorithm solving the linear ordering problem. *Computational Optimization and Applications*, **6**, 191–205.

Crowston, W.B., Thompson, G.L. and Trawick, J.D. (1963) Probabilistic learning combinations of local job shop scheduling rules. Chapters II and III, ONR Research Memorandum No. 117, GSIA, Carnegie Mellon University, Pittsburgh, PA.

Cung, V.-D., Mautor, T., Michelon, P. and Tavares, A. (1997) A scatter search based approach for the quadratic assignment problem. In: T. Bäck, Z. Michalewicz and X. Yao (eds.), *Proceedings of IEEE-ICEC-EPS'97, IEEE International Conference on Evolutionary Computation and Evolutionary Programming Conference*. Indianapolis, pp. 165–170.

Cung, V.D., Martins, S.L., Ribeiro, C.C. and Roucairol, C. (2001) Strategies for the parallel implementation of metaheuristics. In: C.C. Ribeiro and P. Hansen (eds.), *Essays and Surveys in Metaheuristics*. Kluwer, pp. 263–308.

Delgado, C., Laguna, M. and Pacheco, J. (2002) Minimizing Labor Requirements in a Periodic Vehicle Loading Problem. University of Burgos, Spain.

Dorne, R. and Hao, J.-K. (1998) Tabu search for graph coloring, T-colorings and set T-colorings. In: S. Voss, S. Martello, I.H. Osman and C. Roucairol (eds.), *Meta-heuristics: Advances and Trends in Local Search Paradigms for Optimization*. Kluwer, pp. 77–92.

Eades, P. and Kelly, D. (1986) Heuristics for drawing 2-layered networks. *Ars Combinatoria*, **21**, 89–98.

Festa, P. and Resendee, M.G.C. (2001) GRASP: an annotated bibliography. In: P. Hansen and C. Riberio (eds.), Essays and Surveys on Meta-Heuristics, Kluwer Academic Publishers, Boston, USA.

Fleurent, C. and Glover, F. (1999) Improved constructive multi-start strategies for the quadratic assignment problem using adaptive memory. *INFORMS Journal on Computing*, **11**, 198–204.

Funabiki, N. and Higashino, T. (2000) A minimal-state processing search algorithm for graph colorings problems. *IEICE Transactions on Fundamentals*, **E83-A(7)**, 1420–1430.

Galinier, P. and Hao, J.-K. (1999) Hybrid evolutionary algorithms for graph coloring. *Journal of Combinatorial Optimization*, **3**(4), 379–397.

García, F. , Melián, B., Moreno, J.A. and Moreno, J.M. (2001) Hybrid metaheuristics based on the scatter search. Proceeding of EUNITE 2001. pp. 479–485. ISBN: 3-89653-916-7. (European Symposium on Intelligent Technologies, Hybriis Systems and their implementation on Smart Adaptive Systems. December 13–14. Puerto de la Cruz, Tenerife, Spain.)

Garey, M. and Johnson, D. (1979) *Computers and Intractability. A Guide to the Theory of NP-Completeness*. W.H. Freeman and Company, New York.

Ghamlouche, I., Crainic, T.G. and Gendreau, M. (2002) Path relinking, cycle-based neighbourhoods and capacitated multicommodity network design. Publication CRT-2002-01, Centre de recherche sur les transports, Université de Montréal.

Ghamlouche, I., Crainic, T.G. and Gendreau, M. (2001) Cycle-based neighbourhoods for fixed-charge capacitated multicommodity network design. Publication CRT-2001-01, Centre de recherche sur les transports, Université de Montréal.

Glover, F. (1963) Parametric combinations of local job shop scheduling rules. Chapter IV, ONR Research Memorandum No. 117, GSIA, Carnegie Mellon University, Pittsburgh, PA.

Glover, F. (1965) A multiphase dual algorithm for the zero-one integer programming problem. *Operations Research*, **13**(6), 879–919.

Glover, F. (1968) Surrogate constraints. *Operations Research*, **16**(4), 741–749.

Glover, F. (1977) Heuristics for integer programming using surrogate constraints. *Decision Sciences*, **8**(1), 156–166.

Glover, F. (1994) Tabu search for nonlinear and parametric optimization (with links to genetic algorithms). *Discrete Applied Mathematics*, **49**, 231–255.

Glover, F. (1995) Scatter search and star paths: beyond the genetic metaphor. *OR Spektrum*, **17**, 125–137.

Glover, F. (1997) Tabu search and adaptive memory programming—advances, applications and challenges. In. R. Barr, Helgason and Kennington (Co-eds.), *Advances in Meta-heuristics, Optimization and Stochastic Modeling Techniques*. Kluwer Academic Publishers, Boston, USA, pp. 1–175.

Glover, F. (1998) A template for scatter search and path relinking. In: J.-K. Hao, E. Lutton, E. Ronald, M. Schoenauer and D. Snyers (eds.), *Artificial Evolution, Lecture Notes in Computer Science* 1363. Springer, pp. 3–51.

Glover, F. and Laguna, M. (1997) *Tabu Search*. Kluwer Academic Publishers, Boston.

Glover, F., Løkketangen, A. and Woodruff, D.L. (2000) Scatter search to generate diverse MIP solutions. In: M. Laguna and J.L. González-Velarde (eds.), *OR Computing Tools for Modeling, Optimization and Simulation: Interfaces in Computer Science and Operations Research*, pp. 299–317.

Grabowski, J. and Wodecki, M. (2001) A new very fast tabu search algorithm for the job shop problem. Preprint 21/2001, Instytut Cybernetyki Techncznej Politechniki Wroclawskiej, Wroclaw.

Greistorfer, P. (1995) Computational experiments with heuristics for a capacitated arc routing problem. In: U. Derigs, A. Bachem and A. Drexl (eds.), *Operations Research Proceedings 1994*. Springer-Verlag, Berlin, pp. 185–190.

Greistorfer, P. (2001a) A tabu scatter search metaheuristic for the arc routing problem. *Computers & Industrial Engineering* (forthcoming).

Greistorfer, P. (2001b) Testing population designs. *4th Metaheuristics International Conference (MIC'2001)*. Porto, pp. 713–717 (17 pages submitted as "Experimental pool design").

Hamiez, J.P. and Hao, J.K. (2001) Scatter search for graph coloring. To appear in the *LNCS* series (Springer).

Herrmann J.W., Ioannou, G., Minis, I. and Proth, J.M. (1996) A dual ascent approach to the fixed-charge capacitated network design problem. *European Journal of Operational Research*, **95**, 476–490.

Hill, R.R. and McIntyre, G. (2000) A methodology for robust, multi-scenario optimization. *Phalanx*, **33**(3).

Johnson, D.S. and Trick, M.A. (eds.) (1996) *Cliques, Coloring, and Satisfiability: 2nd DIMACS Implementation Challenge, 1993*, DIMACS Series in Discrete Mathematics and Theoretical Computer Science, Vol. 26, American Mathematical Society.

Knuth, D.E. (1993) *The Stanford GraphBase: A Platform for Combinatorial Computing*. Addison Wesley, New York.

Laguna, M. (1999) Scatter search. In: P.M. Pardalos and M.G.C. Resende (eds.), *Handbook of Applied Optimization*. Oxford Academic Press (to appear).

Laguna, M. and Martí, R. (1999) GRASP and path relinking for 2-layer straight line crossing minimization. *INFORMS Journal on Computing*, **11**(1), 44–52.

Laguna, M. and Martí, R. (2002) The OptQuest callable library. In: S. Voss and D.L. Woodruff (eds.), *Optimization Software Class Libraries*, Kluwer, Boston.

Laguna, M., Martí, R. and Campos, V. (1999) Intensification and diversification with Elite Tabu search solutions for the linear ordering problem. *Computers and Operations Research*, **26**, 1217–1230

Løkketangen, A. and Woodruff (2000) Integrating pivot based search with branch and bound for binary MIP's. *Control and Cybernetics, Special Issue on Tabu Search*, **29**(3), 741–760.

Løkketangen, A. and Glover, F. (1998) Solving zero-one mixed integer programming problems using tabu search. *European Journal of Operational Research*, **106**, 624–658.

Løkketangen, A. and Glover, F. (1995) Tabu search for zero/one mixed integer programming with advanced level strategies and learning. *International Journal of Operations and Quantitative Management*, **1**(2), 89–109.

Løkketangen, A. and Glover, F. (1996) Probabilistic move selection in Tabu search for 0/1 mixed integer programming problems. In: *Metaheuristics: Theory and Applications*, by Kluwer, March 96. An earlier version is in the conference proceedings from MIC '95.

Løkketangen, A. and Glover, F. (1999) Candidate list and exploration strategies for solving 0/1 MIP problems using a Pivot neighborhood. In: S. Voß, S. Martello, I.H. Osman and C. Roucairol (eds.), *Meta-Heuristics: Advances and Trends in Local Search Paradigms for Optimization*. Kluwer Academic Publishers, pp. 141–155.

Magnanti, T. and Wong, R. (1984) Network design and transportation planning: models and algorithms. *Transportation Science*, **18**, 1–55.

Martí, R. (1998) A Tabu search algorithm for the Bipartite drawing problem. *European Journal of Operational Research*, **106**, 558–569.

Martí, R. and Laguna, M. (1997) Heuristics and metaheuristics for 2-layer straight line crossing minimization. *Discrete and Applied Mathematics* (to appear).

Martí, R., Lourenço, L. and Laguna, M. (2000) Assigning proctors to exams with scatter search. In: M. Laguna and J.L. González Velarde (eds.), *Computing Tools for Modeling, Optimization and Simulation*, Kluwer Academic Publishers, pp. 215–228.

Martí, R., Lourenço, L. and Laguna, M. (2001) *Assigning proctors to exams with scatter search (second part)*, Economic Working Papers Series, Department of Economics and Business, Universitat Pompeu Fabra, no. 534.

Morgenstern, C.A. (1996) Distributed coloration neighborhood search. In Johnson and Trick (1996), pp. 335–357.

Nowicki, E. and Smutnicki, C. (1996) A fast tabu search algorithm for the job-shop problem. *Management Science*, **42**(6), 797–813.

Nowicki, E., and Smutnicki, C. (2001a) New ideas in TS for job-shop scheduling. Technical Report 50/2001. In: C. Rego and B. Alidaee (eds.), *Adaptive Memory and Evolution: Tabu Search and Scatter Search*. Kluwer Academic Publishers.

Nowicki, E. and Smutnicki, C. (2001b) New tools to solve the job-shop problem. Technical Report 51.

Osman, I.H. (2000) Meta-heuristics: A general framework. In: *Proceedings of the Workshop on "algorithm engineering as a new paradigm: a challenge to hard computation problems"* October 30–November 2, Research Institute for Mathematical Science, Kyoto University, Japan, pp. 117–118.

Osman, I.H. and Samad Ahmadi (2001) Guided Construction search Meta-Heuristics for the Capacitated Clustering Problem. Working Paper, School of Business, American University of Beirut.

Osman, I.H. (1994) Capacitated clustering problems by hybrid simulated annealing and tabu search. *International Transactions in Operational Research*, **1**, 317–336.

Rego, Cesar and Pedro Leão (2000) A Scatter Search Tutorial for Graph-Based Permutation Problems. Hearin Center for Enterprise Science, U of Mississippi, Report Number: HCES-10-00.

Reinelt, G. (1985) *The Linear Ordering Problem: Algorithms and Applications*, Research and Exposition in Mathematics, Vol. 8, H.H. Hofmann and R. Wille (eds.), Heldermann Verlag Berlin.

Resende, M.G.C. and Ribeiro, C.C. (2001) A GRASP with path relinking for permanent virtual circuit routing. *Research Report* (submitted for publication).

Resende, M.G.C. and C.C. Ribeiro (2002) "GRASP", In: F. Glover and G. Kochenberger (eds.), *State-of-the-Art Handbook of Metaheuristics*. Kluwer (to appear).

Ribeiro, C.C., ~Uchoa, E. and Werneck, R.F. (2002) A hybrid GRASP with perturbations for the Steiner problem in graphs. *INFORMS Journal on Computing* (to appear)

Rusell, R.A. and Igo, W. (1979) An assignment routing problem. *Networks*, **9**, 1–17.

Souza, M.C., Duhamel, C. and Ribeiro, C.C. (2002) A GRASP heuristic using a path-based local search for the capacitated minimum spanning tree problem. *Research Report* (submitted for publication).

Valls, V., Quintanilla, S. and Ballestín, F. (2001) A Population Based Approach to the Resource Constrained Project Scheduling. TR06-2001, Departamento de Estadística e Investigación Operativa, Facultad de Matemáticas, Universitat de València (Spain).

Verhoeven, M.G.A. and Aarts, E.H.L. (1995) Parallel local search. *Journal of Heuristics*, **1**, 43–65.

Voss, S., Martin, A. and Koch, T. (2001) SteinLib testdata library. online document at http://elib.zib.de/steinlib/steinlib.html.

Woodruff, D.L. (1998) Proposals for chunking and Tabu search. *European Journal of Operation Research*, **106**, 585–598.

Woodruff, D.L. (2001) General purpose metrics for solution variety. Technical Report, Graduate School of Management, UC Davis, Davis CA 95616.

Woodruff, D.L. (1996) Chunking applied to reactive Tabu search. In: I.H. Osman and J.P. Kelly (eds.), *Metaheuristics: Theory and Applications*. pp. 555–570.

Chapter 2

AN INTRODUCTION TO TABU SEARCH

Michel Gendreau
Centre de recherche sur les transports
Département d'informatique et de recherche opérationnelle
Université de Montréal
Case postale 6128, Succursale "Centre-ville"
Montréal, Canada H3C 3J7
E-mail: michelg@crt.umontreal.ca

Abstract This chapter presents the fundamental concepts of Tabu Search (TS) in a tutorial fashion. Special emphasis is put on showing the relationships with classical Local Search methods and on the basic elements of any TS heuristic, namely, the definition of the search space, the neighborhood structure, and the search memory. Other sections cover other important concepts such as search intensification and diversification and provide references to significant work on TS. Recent advances in TS are also briefly discussed.

1 INTRODUCTION

Over the last fifteen years, well over a hundred papers presenting applications of Tabu Search (TS), a heuristic method originally proposed by Glover in 1986, to various combinatorial problems have appeared in the operations research literature. In several cases, the methods described provide solutions very close to optimality and are among the most effective, if not the best, to tackle the difficult problems at hand. These successes have made TS extremely popular among those interested in finding good solutions to the large combinatorial problems encountered in many practical settings. Several papers, book chapters, special issues and books have surveyed the rich TS literature (a list of some of the most important references is provided in a later section). In spite of this abundant literature, there still seem to be many researchers who, while they are eager to apply TS to new problem settings, find it difficult to properly grasp the fundamental concepts of the method, its strengths and its limitations, and to come up with effective implementations. The purpose of this chapter is to address this situation by providing an introduction in the form of a tutorial focusing on the fundamental concepts of TS. Throughout the chapter, two relatively straightforward, yet challenging and relevant, problems will be used to illustrate these concepts: the Classical Vehicle Routing Problem (CVRP) and the Capacitated Plant Location Problem (CPLP). These will be introduced in the following section. The remainder of the chapter is organized as follows. The basic concepts of TS (search space, neighborhoods, and short-term tabu lists) are described and illustrated in Section 3. Intermediate, yet critical, concepts, such as intensification and diversification, are described in Section 4. This is followed in Section 5 by a brief discussion of advanced topics and recent trends in TS, and in Section 6 by a short list of key references on TS and its applications. Section 7 provides practical tips for newcomers struggling with unforeseen problems as they first try to apply TS to their favorite problem. Section 8 concludes the chapter with some general advice on the application of TS to combinatorial problems.

2 ILLUSTRATIVE PROBLEMS

2.1 The Classical Vehicle Routing Problem

Vehicle Routing Problems have very important applications in the area of distribution management. As a consequence, they have become some of the most studied problems in the combinatorial optimization literature and large number of papers and books deal with the numerous procedures that have been proposed to solve them. These include several TS implementations that currently rank among the most effective. The *Classical Vehicle Routing Problem* (CVRP) is the basic variant in that class of problems. It can formally be defined as follows. Let $G = (V, A)$ be a graph where V is the vertex set and A is the arc set. One of the vertices represents the *depot* at which a fleet of m identical vehicles of capacity Q is based, and the other vertices customers that need to be serviced. With each customer vertex v_i are associated a demand q_i and a service time t_i. With each arc (v_i, v_j) of A are associated a cost c_{ij} and a travel time t_{ij}. The CVRP consists in finding a set of routes such that:

1. Each route begins and ends at the depot;
2. Each customer is visited exactly once by exactly one route;
3. The total demand of the customers assigned to each route does not exceed Q;
4. The total duration of each route (including travel and service times) does not exceed a specified value L;
5. The total cost of the routes is minimized.

A feasible solution for the problem thus consists in a partition of the customers into m groups, each of total demand no larger than Q, that are sequenced to yield routes (starting and ending at the depot) of duration no larger than L.

2.2 The Capacitated Plant Location Problem

The *Capacitated Plant Location Problem* (CPLP) is one of the basic problems in Location Theory. It is encountered in many application settings that involve locating facilities with limited capacity to provide services. The CPLP can be formally described as follows. A set of customers I with demands $d_i, i \in I$, for some product are to be served from plants located in a subset of sites from a given set J of "potential sites". For each site $j \in J$, the fixed cost of "opening" the plant at j is f_j and its capacity is K_j. The cost of transporting one unit of the product from site j to customer i is c_{ij}. The objective is to minimize the total cost, i.e., the sum of the fixed costs for open plants and the transportation costs.

Letting x_{ij} $(i \in I, j \in J)$ denote the quantity shipped from site j to customer i (the x_{ij}'s are the so-called *flow variables*) and y_j $(j \in J)$ be a 0–1 variable indicating whether or not the plant at site j is open (the y_j's are the *location variables*), the problem can be formulated as the following mathematical program:

$$\text{(CPLP)} \quad \textit{Minimize } z = \sum_{j\in J} f_j y_j + \sum_{i\in I}\sum_{j\in J} c_{ij}x_{ij}$$

$$\textit{subject to} \sum_{j\in J} x_{ij} = di, \quad i \in I$$

$$\sum_{i\in I} x_{ij} \leq K_j y_j, \quad j \in J$$

$$x_{ij} \geq 0, \quad i \in I, \; j \in J$$

$$y_j \in \{0, 1\}, \quad j \in J$$

Remark 2.1. *For any vector* $\tilde{\mathbf{y}}$ *of location variables, optimal (w.r.t. to this plant configuration) values for the flow variables* $\mathbf{x}(\tilde{\mathbf{y}})$ *can be retrieved by solving the associated transportation problem:*

$$\text{(TP)} \quad \textit{Minimize } z(\tilde{\mathbf{y}}) = \sum_{i\in I}\sum_{j\in J} c_{ij}x_{ij}$$

$$\textit{subject to} \sum_{j\in J} x_{ij} = d_i, \quad i \in I$$

$$\sum_{i\in I} x_{ij} \leq K_j \tilde{y}_j, \quad j \in J$$

$$x_{ij} \geq 0, \quad i \in I, \quad j \in J$$

If $\tilde{\mathbf{y}} = \mathbf{y}^*$*, the optimal location variable vector, the optimal solution to the original CPLP problem is simply given by* $(\mathbf{y}^*, \mathbf{x}(\mathbf{y}^*))$.

Remark 2.2. *An optimal solution of the original CPLP problem can always be found at an extreme point of the polyhedron of feasible flow vectors defined by the constraints:*

$$\sum_{j\in J} x_{ij} = d_i, \quad i \in I$$

$$\sum_{i\in I} x_{ij} \leq K_j, \quad j \in J$$

$$x_{ij} \geq 0, \quad i \in I, \; j \in J$$

This property follows from the fact that the CPLP can be interpreted as a fixed-charge problem defined in the space of the flow variables. This fixed-charge problem has a concave objective function that always admits an extreme point minimum. The optimal values for the location variables can easily be obtained from the optimal flow vector by setting y_j *equal to* 1 *if* $\sum_{i\in I} x_{ij} > 0$*, and to* 0 *otherwise.*

To our knowledge, no TS heuristic has ever been proposed for the CPLP, but we will see in the following that this problem can be used to illustrate many important concepts related to the approach.

3 BASIC CONCEPTS

3.1 Historical Background

Before introducing the basic concepts of TS, we believe it is useful to go back in time to try to better understand the genesis of the method and how it relates to previous work.

Heuristics, i.e., approximate solution techniques, have been used since the beginnings of operations research to tackle difficult combinatorial problems. With the development of complexity theory in the early 70's, it became clear that, since most of these problems were indeed *NP-hard*, there was little hope of ever finding efficient exact solution procedures for them. This realization emphasized the role of heuristics for solving the combinatorial problems that were encountered in real-life applications and that needed to be tackled, whether or not they were *NP-hard*. While many different approaches were proposed and experimented with, the most popular one was based on Local Search (LS) improvement techniques. LS can be roughly summarized as an iterative search procedure that, starting from an initial feasible solution, progressively improves it by applying a series of local modifications (or *moves*). At each iteration, the search moves to an improving feasible solution that differs only slightly from the current one (in fact, the difference between the previous and the new solutions amounts to one of the local modifications mentioned above). The search terminates when it encounters a local optimum with respect to the transformations that it considers, an important limitation of the method: unless one is extremely lucky, this local optimum is often a fairly mediocre solution. In LS, the quality of the solution obtained and computing times are usually highly dependent upon the "richness" of the set of transformations (moves) considered at each iteration of the heuristic.

In 1983, the world of combinatorial optimization was shattered by the appearance of a paper in which the authors (Kirkpatrick, Gelatt and Vecchi) were describing a new heuristic approach called *Simulated Annealing* (SA) that could be shown to converge to an optimal solution of a combinatorial problem, albeit in infinite computing time. Based on analogy with statistical mechanics, SA can be interpreted as a form of controlled random walk in the space of feasible solutions. The emergence of SA indicated that one could look for other ways to tackle combinatorial optimization problems and spurred the interest of the research community. In the following years, many other new approaches, mostly based on analogies with natural phenomena, were proposed (TS, *Ant Systems, Threshold Methods*) and, together with some older ones, such as *Genetic Algorithms* (Holland, 1975), they gained an increasing popularity. Now collectively known under the name of *Meta-Heuristics* (a term originally coined by Glover in 1986), these methods have become over the last fifteen years the leading edge of heuristic approaches for solving combinatorial optimization problems.

3.2 Tabu Search

Building upon some of his previous work, Fred Glover proposed in 1986 a new approach, which he called Tabu Search, to allow LS methods to overcome local optima. (In fact, many elements of this first TS proposal, and some elements of later TS elaborations, were introduced in Glover, 1977, including short term memory to prevent the reversal of recent moves, and longer term frequency memory to reinforce attractive components.) The basic principle of TS is to pursue LS whenever it encounters a local optimum by allowing non-improving moves; *cycling* back to previously visited solutions is prevented by the use of *memories*, called *tabu lists*, that record the recent history of the search, a key idea that can be linked to Artificial Intelligence concepts. It is interesting to note that, the same year Hansen proposed a similar approach, which he named *steepest ascent/mildest descent*. It is also important to remark that Glover did not see TS as a proper heuristic, but rather as a Meta-Heuristic, i.e., a general strategy for guiding and controlling "inner" heuristics specifically tailored to the problems at hand.

3.3 Search Space and Neighborhood Structure

As we just mentioned, TS is an extension of classical LS methods. In fact, basic TS can be seen as simply the combination of LS with short-term memories. It follows that the two first basic elements of any TS heuristic are the definition of its *search space* and its *neighborhood structure*.

The search space of an LS or TS heuristic is simply the space of all possible solutions that can be considered (visited) during the search. For instance, in the CVRP example described in Section 2, the search space could simply be the set of feasible solutions to the problem, where each point in the search space corresponds to a set of vehicles routes satisfying all the specified constraints. While in that case the definition of the search space seems quite natural, it is not always so. Consider now the CPLP example of Section 2: the feasible space involves both integer location and continuous flow variables that are linked by strict conditions; moreover, as has been indicated before, for any feasible set of values for the location variables, one can fairly easily retrieve optimal values for the flow variables by solving the associated transportation problem. In this context, one could obviously use as a search space the full feasible space; this would involve manipulating both location and flow variables, which is not an easy task. A more attractive search space is the set of feasible vectors of location variables, i.e., feasible vectors in $\{0, 1\}^{|J|}$, any solution in that space being "completed" to yield a feasible solution to the original problem by computing the associated optimal flow variables. It is interesting to note that these two possible definitions are not the only ones. Indeed, on the basis of Remark 2.2, one could also decide to search instead the set of extreme points of the set of feasible flow vectors, retrieving the associated location variables by simply noting that a plant must be open whenever some flow is allocated to it. In fact, this type of approach was used successfully by Crainic, Gendreau and Farvolden (2000) to solve the Fixed Charge Multi-commodity Network Design Problem, a more general problem that includes the CPLP as a special case. It is also important to note that it is not always a good idea to restrict the search space to feasible solutions; in many cases, allowing the search to move to infeasible solutions is desirable, and sometimes necessary (see Section 4.4 for further details).

Closely linked to the definition of the search space is that of the neighborhood structure. At each iteration of LS or TS, the local transformations that can be applied to the current solution, denoted S, define a set of neighboring solutions in the search space, denoted $N(S)$ (the neighborhood of S). Formally, $N(S)$ is a subset of the search space defined by:

$$N(S) = \{\text{solutions obtained by applying a single local transformation to } S\}.$$

In general, for any specific problem at hand, there are many more possible (and even, attractive) neighborhood structures than search space definitions. This follows from the fact that there may be several plausible neighborhood structures for a given definition of the search space. This is easily illustrated on our CVRP example that has been the object of several TS implementations. In order to simplify the discussion, we suppose in the following that the search space is the feasible space.

Simple neighborhood structures for the CVRP involve moving at each iteration a single customer from its current route; the selected customer is inserted in the same route or in another route with sufficient residual capacity. An important feature of these neighborhood structures is the way in which insertions are performed: one could use random insertion or insertion at the best position in the target route; alternately, one could use more complex insertion schemes that involve a partial re-optimization of the target route, such as GENI insertions (see Gendreau et al., 1994). Before proceeding any further it is important to stress that while we say that these neighborhood structures involve moving a *single* customer, the neighborhoods they define contain all the feasible route configurations that can be obtained from the current solution by moving *any* customer and inserting it in the stated fashion. Examining the neighborhood can thus be fairly demanding.

More complex neighborhood structures for the CVRP, such as the λ-interchange of Osman (1993), are obtained by allowing simultaneously the movement of customers to different routes and the swapping of customers between routes. In Rego and Roucairol (1996), moves are defined by *ejection chains* that are sequences of coordinated movements of customers from one route to another; for instance, an ejection chain of length 3 would involve moving a customer v_1 from route R_1 to route R_2, a customer v_2 from R_2 to route R_3 and a customer v_3 from R_3 to route R_4. Other neighborhood structures involve the swapping of sequences of several customers between routes, as in the Cross-exchange of Taillard et al. (1997). These types of neighborhoods have seldom be used for the CVRP, but are common in TS heuristics for its time-windows extension, where customers must be visited within a pre-specified time interval. We refer the interested reader to Gendreau, Laporte and Potvin (2002) and Bräysy and Gendreau (2001) for a more detailed discussion of TS implementations for the CVRP and the Vehicle Routing Problem with Time-Windows.

When different definitions of the search space are considered for any given problem, neighborhood structures will inevitably differ to a considerable degree. This can be illustrated on our CPLP example. If the search space is defined with respect to the location variables, neighborhood structures will usually involve the so-called "*Add/Drop*" and "*Swap*" moves that respectively change the status of one site (i.e., either opening a closed facility or closing an open one) and move an open facility from one site to another (this move amounts to performing simultaneously an Add move and a Drop move). If, however, the search space is the set of extreme points of the set of feasible flow vectors, these moves become meaningless. One should instead consider moves defined by the application of pivots to the linear programming formulation of the transportation problem, since each pivot operation moves the current solution to an adjacent extreme point.

The preceding discussion should have clarified a major point: choosing a search space and a neighborhood structure is by far the most critical step in the design of any TS heuristic. It is at this step that one must make the best use of the understanding and knowledge he/she has of the problem at hand.

3.4 Tabus

Tabus are one of the distinctive elements of TS when compared to LS. As we already mentioned, tabus are used to prevent cycling when moving away from local optima through non-improving moves. The key realization here is that when this situation occurs, something needs to be done to prevent the search from tracing back its steps to where it came from. This is achieved by declaring *tabu* (disallowing) moves that reverse the effect of recent moves. For instance, in the CVRP example, if customer v_1 has just been moved from route R_1 to route R_2, one could declare tabu moving back v_1 from R_2 to R_1 for some number of iterations (this number is called the *tabu tenure* of the move). Tabus are also useful to help the search move away from previously visited portions of the search space and thus perform more extensive exploration.

Tabus are stored in a *short-term memory* of the search (the *tabu list*) and usually only a fixed and fairly limited quantity of information is recorded. In any given context, there are several possibilities regarding the specific information that is recorded. One could record complete solutions, but this requires a lot of storage and makes it expensive to check whether a potential move is tabu or not; it is therefore seldom used. The most commonly used tabus involve recording the last few transformations performed on the current solution and prohibiting reverse transformations (as in the example above); others are based on key characteristics of the solutions themselves or of the moves.

To better understand how tabus work, let us go back to our reference problems. In the CVRP, one could define tabus in several ways. To continue our example where customer v_1 has just been moved from route R_1 to route R_2, one could declare tabu specifically moving back v_1 from R_2 to R_1 and record this in the short-term memory as the triplet (v_1, R_2, R_1). Note that this type of tabu will not constrain the search much, but that cycling may occur if v_1 is then moved to another route R_3 and then from R_3 to R_1. A stronger tabu would involve prohibiting moving back v_1 to R_1 (without consideration for its current route) and be recorded as (v_1, R_1). An even stronger tabu would be to disallow moving v_1 to any other route and would simply be noted as (v_1).

In the CPLP, when searching the space of location variables, tabus on Add/Drop moves should prohibit changing the status of the affected location variable and can be recorded by noting its index; tabus for Swap moves are more complex: they could be declared with respect to the site where the facility was closed, to the site where the facility was opened, to both locations (i.e., changing the status of both location variables is tabu), or to the specific swapping operation. When searching the space of flow variables, one can take advantage of the fact that a pivot operation is associated with a unique pair of entering and leaving variables to define tabus; while here again several combinations are possible, experience has shown that when dealing with pivot neighborhood structures, tabus imposed on leaving variables (to prevent them from coming back in the basis) are usually much more effective.

Multiple tabu lists can be used simultaneously and are sometimes advisable. For instance, in the CPLP, if one uses a neighborhood structure that contains both Add/Drop and Swap moves, it might be a good idea to keep a separate tabu list for each type of moves.

Standard tabu lists are usually implemented as circular lists of fixed length. It has been shown, however, that fixed-length tabus cannot always prevent cycling, and some authors have proposed varying the tabu list length during the search (Glover, 1989, 1990; Skorin-Kapov, 1990; Taillard, 1990, 1991). Another solution is to randomly generate the tabu tenure of each move within some specified interval; using this approach requires a somewhat different scheme for recording tabus that are then usually stored as *tags* in an array (the entries in this array will usually record the iteration number until which a move is tabu; see Gendreau, Hertz and Laporte, 1994, for more details).

3.5 Aspiration Criteria

While central to TS, tabus are sometimes too powerful: they may prohibit attractive moves, even when there is no danger of cycling, or they may lead to an overall stagnation of the searching process. It is thus necessary to use algorithmic devices that will allow one to *revoke* (cancel) tabus. These are called *aspiration criteria*. The simplest and most commonly used aspiration criterion (found in almost all TS implementations) consists in allowing a move, even if it is tabu, if it results in a solution with an objective value better than that of the current best-known solution (since the new solution has obviously not been previously visited). Much more complicated aspiration criteria have been proposed and successfully implemented (see, for instance, de Werra and Hertz, 1989, or Hertz and de Werra, 1991), but they are rarely used. The key rule in this respect is that if cycling cannot occur, tabus can be disregarded.

3.6 A Template for Simple Tabu Search

We are now in the position to give a general template for TS, integrating the elements we have seen so far. We suppose that we are trying to minimize a function $f(S)$ over some domain and we apply the so-called "best improvement" version of TS, i.e., the version in which one chooses at each iteration the best available move (this is the most commonly used version of TS).

Notation

- $\boldsymbol{S}$, the current solution,
- $\boldsymbol{S^*}$, the best-known solution,
- $\boldsymbol{f^*}$, value of S^*,
- $\boldsymbol{N(S)}$, the neighborhood of S,
- $\boldsymbol{\tilde{N}(S)}$, the "admissible" subset of $N(S)$ (i.e., non-tabu or allowed by aspiration).

Initialization

Choose (construct) an initial solution S_0.
Set $S := S_0, f^* := f(S_0), S^* := S_0, T := \emptyset$.

Search

While *termination criterion not satisfied* do

- Select S in *argmin* $[f(S')]$; $S' \varepsilon \tilde{N}(S)$
- if $f(S) < f^*$, then set $f^* := f(S), S^* := S$;
- record tabu for the current move in T (delete oldest entry if necessary);

endwhile.

3.7 Termination Criteria

One may have noticed that we have not specified in our template above a termination criterion. In theory, the search could go on forever, unless the optimal value of the problem at hand is known beforehand. In practice, obviously, the search has to be stopped at some point. The most commonly used stopping criteria in TS are:

- after a fixed number of iterations (or a fixed amount of CPU time);
- after some number of iterations without an improvement in the objective function value (the criterion used in most implementations);
- when the objective reaches a pre-specified threshold value.

In complex tabu schemes, the search is usually stopped after completing a sequence of *phases*, the duration of each phase being determined by one of the above criteria.

3.8 Probabilistic TS and Candidate Lists

In "regular" TS, one must evaluate the objective for every element of the neighborhood $N(S)$ of the current solution. This can prove extremely expensive from the computational standpoint. An alternative is to instead consider only a random sample $N'(S)$ of $N(S)$, thus reducing significantly the computational burden. Another attractive feature of this alternative is that the added randomness can act as an anti-cycling mechanism; this allows one to use shorter tabu lists than would be necessary if a full exploration of the neighborhood was performed. One the negative side, it must be noted that, in that case, one may miss excellent solutions (more on this topic in Section 7.3). Probabilities may also be applied to activating tabu criteria.

Another way to control the number of moves examined is by means of *candidate list* strategies, which provide more strategic ways of generating a useful subset $N'(S)$ of $N(S)$. (The probabilistic approach can be considered to be one instance of a candidate list strategy, and may also be used to modify such a strategy.) Failure to adequately address the issues involved in creating effective candidate lists is one of the more conspicuous shortcomings that differentiates a naive TS implementation from one that is more solidly grounded. Relevant designs for candidate list strategies are discussed in Glover and Laguna (1997). We also discuss a useful type of candidate generation approach in Section 4.4.

4 INTERMEDIATE CONCEPTS

Simple TS as described above can sometimes successfully solve difficult problems, but in most cases, additional elements have to be included in the search strategy to make it fully effective. We now briefly review the most important of these.

4.1 Intensification

The idea behind the concept of search intensification is that, as an intelligent human being would probably do, one should explore more thoroughly the portions of the search space that seem "promising" in order to make sure that the best solutions in these areas are indeed found. From time to time, one would thus stop the normal searching process to perform an intensification phase. In general, intensification is based on some *intermediate-term memory*, such as a *recency memory*, in which one records the number of consecutive iterations that various "solution components" have been present in the current solution without interruption. For instance, in the CPLP application, one could record how long each site has had an open facility. A typical approach to intensification is to restart the search from the best currently known solution and to "freeze" (fix) in it the components that seem more attractive. To continue the CPLP example, one could thus freeze a number of facilities in the sites that have had them for the largest number of iterations and perform a restricted search on the other sites. Another technique that is often used consists in changing the neighborhood structure to one allowing more powerful or more diverse moves. In the CVRP example, one could therefore allow more complex insertion moves or switch to an ejection chain neighborhood structure. In the CPLP example, if Add/Drop moves were used, Swap moves could be added to the neighborhood structure. In probabilistic TS, one could increase the sample size or switch to searching without sampling.

Intensification is used in many TS implementations, but it is not always necessary. This is because there are many situations where the search performed by the normal searching process is thorough enough. There is thus no need to spend time exploring more carefully the portions of the search space that have already been visited, and this time can be used more effectively as we shall see right now.

4.2 Diversification

One of the main problems of all methods based on Local Search approaches, and this includes TS in spite of the beneficial impact of tabus, is that they tend to be too "local" (as their name implies), i.e., they tend to spend most, if not all, of their time in a restricted portion of the search space. The negative consequence of this fact is that, although good solutions may be obtained, one may fail to explore the most interesting parts of the search space and thus end up with solutions that are still pretty far from the optimal ones. *Diversification* is an algorithmic mechanism that tries to alleviate this problem by forcing the search into previously unexplored areas of the search space. It is usually based on some form of *long-term memory* of the search, such as a *frequency memory*, in which one records the total number of iterations (since the beginning of the search) that various "solution components" have been present in the current solution or have been involved in the selected moves. For instance, in the CPLP application, one could record during the number of iterations during which each site has had an open facility. In the CVRP application, one could note how many times each customer has been moved from its current route. In cases where it is possible to identify useful "regions" of the search space, the frequency memory can be refined to track the number of iterations spent in these different regions.

There are two major diversification techniques. The first, called *restart diversification*, involves forcing a few rarely used components in the current solution (or the best known solution) and restarting the search from this point. In CPLP procedures, one could thus open one or a few facilities at locations that have seldom had them up to that point and resume searching from that plant configuration (one could also close facilities at locations that have been used the most frequently). In a CVRP heuristic, customers that have not yet been moved frequently could be forced into new routes. The second diversification method, *continuous diversification*, integrates diversification considerations directly into the regular searching process. This is achieved by *biasing* the evaluation of possible moves by adding to the objective a small term related to component frequencies (see Soriano and Gendreau, 1996, for an extensive discussion on these two techniques). A third way of achieving diversification is *strategic oscillation* as we will see in the next subsection.

Before closing this subsection, we would like to stress that ensuring proper search diversification is possibly **the most critical issue** in the design of TS heuristics. It should be addressed with extreme care fairly early in the design phase and revisited if the results obtained are not up to expectations.

4.3 Allowing Infeasible Solutions

Accounting for all problem constraints in the definition of the search space often restricts the searching process too much and can lead to mediocre solutions. This occurs, for example, in CVRP instances where the route capacity or duration constraints are too tight to allow moving customers effectively between routes. In such cases, *constraint relaxation* is an attractive strategy, since it creates a larger search space that can be explored with "simpler" neighborhood structures. Constraint relaxation is easily implemented by dropping selected constraints from the search space definition and adding to the objective weighted penalties for constraint violations. This, however, raises the issue of finding correct weights for constraint violations. An interesting way of circumventing this problem is to use *self-adjusting penalties*, i.e., weights are adjusted dynamically on the basis of the recent history of the search: weights are increased if only infeasible solutions were encountered in the last few iterations, and decreased if all recent solutions were feasible (see, for instance, Gendreau, Hertz and Laporte, 1994, for further details). Penalty weights can also be modified systematically to drive the search to cross the feasibility boundary of the search space and thus induce diversification. This technique, known as *strategic oscillation*, was introduced as early as 1977 by Glover and used since in several successful TS procedures. (An important early variant oscillates among alternative types of moves, hence neighborhood structures, while another oscillates around a selected value for a critical function.)

4.4 Surrogate and Auxiliary Objectives

There are many problems for which the true objective function is quite costly to evaluate, a typical example being the CPLP when one searches the space of location variables. (Remember that, in this case, computing the objective value for any potential solution entails solving the associated transportation problem.) When this occurs, the evaluation of moves may become prohibitive, even if sampling is used. An effective approach to handle this issue is to evaluate neighbors using a *surrogate objective*, i.e., a function that is correlated to the true objective, but is less computationally demanding, in order to identify a (small) set of promising candidates (potential solutions achieving the best values for the surrogate). The true objective is then computed for this small set of candidate moves and the best one selected to become the new current solution. (See Crainic et al., 1993, for an example of this approach.)

Another frequently encountered difficulty is that the objective function may not provide enough information to effectively drive the search to more interesting areas of the search space. A typical illustration of this situation is the variant of the CVRP in which the fleet size is not fixed, but is rather the primary objective (i.e., one is looking for the minimal fleet size allowing a feasible solution). In this problem, except for solutions where a route has only one or a few customers assigned to it, most neighborhood structures will lead to the situation where all elements in the neighborhood score equally with respect to the primary objective (i.e., all allowable moves produce solutions with the same number of vehicles). In such a case, it is absolutely necessary to define an *auxiliary objective function* to orient the search. Such a function must measure in some way the desirable attributes of solutions. In our example, one could, for instance, use a function that would favor solutions with routes having just a few customers, thus increasing the likelihood that a route can be totally emptied in a subsequent iteration. It should be noted that coming up with an effective auxiliary objective is not always easy and may require a lengthy trial and error process. In some other cases, fortunately, the auxiliary objective is obvious for anyone familiar with the problem at hand. (See Gendreau, Soriano and Salvail, 1993, for an illustration.)

5 ADVANCED TOPICS AND RECENT TRENDS IN TABU SEARCH

The concepts and techniques described in the previous sections are sufficient to design effective TS heuristics for many combinatorial problems. Most early TS implementations, several of which were extremely successful, relied indeed almost exclusively on these algorithmic components. Nowadays, however, most leading edge research in TS makes use of more advanced concepts and techniques. While it is clearly beyond the scope of an introductory tutorial, such as this one, to review this type of advanced material, we would like to give readers some insight into it by briefly describing some current trends in TS research. Readers who wish to learn more about this topic should read our survey paper (Gendreau, 2002) and some of the references provided in the next section.

A large part of the recent research in TS deals with various techniques for making the search more effective. These include methods for exploiting better the information that becomes available during search and creating better starting points, as well as more powerful neighborhood operators and parallel search strategies. (On this last topic, see the taxonomy of Crainic, Toulouse and Gendreau, 1997, and the survey of Cung et al., 2002.) The numerous techniques for making better use of the information are of particular significance since they can lead to dramatic performance improvements. Many of these rely on *elite solutions* (the best solutions previously encountered) or on parts of these to create new solutions, the rationale being that "fragments" (elements) of excellent solutions are often identified quite early in the searching process, but that the challenge is to complete these fragments or to recombine them (Glover, 1992; Glover and Laguna, 1993, 1997; Rochat and Taillard, 1995). Other methods, such as the *Reactive Tabu Search* of Battiti and Tecchiolli (1994), attempt to find ways of making the search move away from local optima that have already been visited.

Another important trend in TS (this is, in fact, a pervasive trend in the whole meta-heuristics field) is *hybridization*, i.e., using TS in conjunction with other solution approaches such as Genetic Algorithms (Fleurent and Ferland, 1996; Crainic and Gendreau, 1999), Lagrangean relaxation (Grünert, 2002), Constraint Programming (Pesant and Gendreau, 1999), column generation (Crainic et al., 2000) and integer programming techniques (there is a whole chapter on this topic in Glover and Laguna, 1997).

TS research has also started moving away from its traditional application areas (graph theory problems, scheduling, vehicle routing) to new ones: continuous optimization (Rolland, 1996), multi-criteria optimization, stochastic programming, mixed integer programming (Lokketangen and Glover, 1996; Crainic et al., 2000), real-time decision problems (Gendreau et al., 1999), etc. These new areas confront researchers with new challenges that, in turn, call for novel and original extensions of the method.

6 KEY REFERENCES

Readers who wish to read other introductory papers on TS can choose among several ones. Let us mention the ones by de Werra and Hertz (1989), Hertz and de Werra (1991), Glover and Laguna (1993), Glover, Taillard and de Werra (1993) and, in French, by Soriano and Gendreau (1997). The book by Glover and Laguna (1997) is the ultimate reference on TS: apart from the fundamental concepts of the method, it presents a considerable amount of advanced material, as well as a variety of applications. It is interesting to note that this book contains several ideas applicable to TS that yet remain to be fully exploited. The issues of *Annals of Operations Research*, respectively devoted to "Tabu Search" (Glover et al., 1993) and "Metaheuristics in Combinatorial Optimization" (Laporte and Osman, 1996), and the books made up from selected papers presented at the Meta-Heuristics International Conferences (MIC) are also extremely valuable. At this time, the books for the 1995 Breckenridge conference (Osman and Kelly, 1996), the 1997 Sophia-Antipolis one (Voss et al., 1999) and the 1999 Angra dos Reis one (Ribeiro and Hansen, 2002) have been published. The proceedings of MIC'2001, held in Porto, are currently available electronically on the website located at URL:http://tew.ruca.ua.ac.be/eume/MIC2001.

7 TRICKS OF THE TRADE

7.1 Getting Started

Newcomers to TS trying to apply the method to a problem that they wish to solve are often confused about what they need to do to come up with a successful implementation. Basically, they do not know where to start. We believe that the following step-by-step procedure can help a lot and provides a useful framework for getting started.

A step-by-step procedure

1. **Read** one or two good introductory papers to gain some knowledge of the concepts and of the vocabulary.
2. **Read** several papers describing in detail applications in various areas to see how the concepts have been actually implemented by other researchers.
3. **Think** a lot about the problem at hand, focusing on the definition of the **search space** and the **neighborhood structure**.
4. **Implement** a **simple** version based on this search space definition and this neighborhood structure.

5. **Collect statistics** on the performance of this simple heuristic. It is usually useful at this point to introduce a variety of **memories**, such as frequency and recency memories, to really track down what the heuristic does.
6. **Analyze results** and **adjust** the procedure accordingly. It is at this point that one should eventually introduce mechanisms for search intensification and diversification or other intermediate features. Special attention should be paid to **diversification**, since this is often where simple TS procedures fail.

7.2 More Tips

It is not unusual that, in spite of following carefully the preceding procedure, one ends up with a heuristic that nonetheless produces mediocre results. If this occurs, the following tips may prove useful:

1. If there are **constraints**, consider **penalizing** them. Letting the search move to infeasible solutions is often necessary in highly constrained problems to allow for a meaningful exploration of the search space (see Section 4).
2. Reconsider the **neighborhood structure** and change it if necessary. Many TS implementations fail because the neighborhood structure is too simple. In particular, one should make sure that the chosen neighborhood structure allows for a purposeful evaluation of possible moves (i.e., the moves that seem intuitively to move the search in the "right" direction should be the ones that are likely to be selected); it might also be a good idea to introduce a **surrogate objective** (see Section 4) to achieve this.
3. **Collect more statistics**.
4. **Follow the execution of the algorithm step-by-step** on some reasonably sized instances.
5. Reconsider **diversification**. As mentioned earlier, this is a critical feature in most TS implementations.
6. **Experiment** with **parameter** settings. Many TS procedures are extremely sensitive to parameter settings; it is not unusual to see the performance of a procedure dramatically improve after changing the value of one or two key parameters (unfortunately, it is not always obvious to determine which parameters are the key ones in a given procedure).

7.3 Additional Tips for Probabilistic TS

While it is an effective way of tackling many problems, probabilistic TS creates problems of its own that need to be carefully addressed. The most important of these is the fact that, more often than not, the best solutions returned by probabilistic TS will not be local optima with respect to the neighborhood structure being used. This is particularly annoying since, in that case, better solutions can be easily obtained, sometimes even manually. An easy way to come around this is to simply perform a local improvement phase (using the same neighborhood operator) from the best found solution at the end of the TS itself. One could alternately switch to TS without sampling (again from the best found solution) for a short duration before completing the algorithm. A possibly more effective technique is to add throughout the search an intensification step without sampling; in this fashion, the best solutions available in the various regions of the search space explored by the method will be found and recorded. (Glover and Laguna, 1993, similarly proposed special aspiration criteria for allowing the search to reach local optima at useful junctures.)

7.4 Parameter Calibration and Computational Testing

Parameter calibration and computational experiments are key steps in the development of any algorithm. This is particularly true in the case of TS, since the number of parameters required by most implementations is fairly large and since the performance of a given procedure can vary quite significantly when parameter values are modified. The first step in any serious computational experimentation is to select a good set of benchmark instances (either by obtaining them from other researchers or by constructing them), preferably with some reasonable measure of their difficulty and with a wide range of size and difficulty. This set should be split into two subsets, the first one being used at the algorithmic design and parameter calibration steps, and the second reserved for performing the final computational tests that will be reported in the paper(s) describing the heuristic under development. The reason for doing so is quite simple: when calibrating parameters, one always run the risk of *overfitting*, i.e., finding parameter values that are excellent for the instances at hand, but poor in general, because these values provide too good a "fit" (from the algorithmic standpoint) to these instances. Methods with several parameters should thus be calibrated on much larger sets of instances than ones with few parameters to ensure a reasonable degree of robustness. The calibration process itself should proceed in several stages:

1. Perform exploratory testing to find good ranges of parameters. This can be done by running the heuristic with a variety of parameter settings.
2. Fix the value of parameters that appear to be "robust", i.e., which do not seem to have a significant impact on the performance of the procedure.
3. Perform systematic testing for the other parameters. It is usually more efficient to test values for only a single parameter at a time, the others being fixed at what appear to be reasonable values. One must be careful, however, for *cross effects* between parameters. Where such effects exist, it can be important to jointly test pairs or triplets of parameters, which can be an extremely time-consuming task.

The paper by Crainic et al. (1993) provides a detailed description of the calibration process for a fairly complex TS procedure and can used as a guideline for this purpose.

8 CONCLUSION

Tabu Search is a powerful algorithmic approach that has been applied with great success to many difficult combinatorial problems. A particularly nice feature of TS is that, like all approaches based on Local Search, it can quite easily handle the "dirty" complicating constraints that are typically found in real-life applications. It is thus a really practical approach. It is not, however, a panacea: every reviewer or editor of a scientific journal has seen more than his/her share of failed TS heuristics. These failures stem from two major causes: an insufficient understanding of fundamental concepts of the method (and we hope that this tutorial will help in alleviating this shortcoming), but also, more often than not, a crippling lack of understanding of the problem at hand. One cannot develop a good TS heuristic for a problem that he/she does not know well! This is because significant problem knowledge is absolutely required to perform the most basic steps of the development of any TS procedure, namely the choice of a search space and of an effective neighborhood structure. If the search space and/or the neighborhood structure are inadequate, no amount of TS expertise will be sufficient to save the day. A last word of caution: to be successful, all meta-heuristics need to achieve both *depth* and *breadth* in their searching process; depth is usually not a problem for TS, which is quite aggressive in this respect (TS heuristics generally find pretty good solutions very early in the search), but breadth can be a critical issue. To handle this, it is extremely important to develop an effective diversification scheme.

ACKNOWLEDGEMENTS

The author is grateful to the Canadian Natural Sciences and Engineering Council and the Fonds FCAR of the Province of Quebec for their financial support. The author also wishes to thank Fred Glover for his insightful comments on an earlier version of this chapter.

REFERENCES

Battiti, R. and Tecchiolli, G. (1994) The reactive tabu search. *ORSA Journal on Computing*, **6**, 126–140.

Bräysy, O. and Gendreau, M. (2001) Tabu search heuristics for the vehicle routing problem with time windows. Report STF42 A01022, SINTEF Applied Mathematics, Oslo, Norway. Forthcoming in *TOP*.

Crainic, T.G. and Gendreau, M. (1999). Towards an evolutionary method—cooperative multi-thread parallel tabu search heuristic Hybrid. In S. Voss, S. Martello, I.H. Osman and C. Roucairol (Eds.), *Meta-Heuristics: Advances and Trends in Local Search Paradigms for Optimization*, Kluwer Academic Publishers, pp. 331–344.

Crainic, T.G., Gendreau, M., and Farvolden, J.M. (2000) Simplex-based tabu search for the multicommodity capacitated fixed charge network design problem. *INFORMS Journal on Computing*, **12**, 223–236.

Crainic, T.G., Gendreau, M., Soriano, P., and Toulouse, M. (1993) A tabu search procedure for multicommodity location/allocation with balancing requirements. *Annals of Operations Research*, **41**, 359–383.

Crainic, T.G., Toulouse, M., and Gendreau, M. (1997) Toward a taxonomy of parallel tabu search heuristics. *INFORMS Journal on Computing*, **9**, 61–72.

Cung, V.-D., Martins, S.L., Ribeiro, C.C., and Roucairol, C. (2002) Strategies for the parallel implementation of metaheuristics. In C.C. Ribeiro and P. Hansen (Eds.), *Essays and Surveys in Metaheuristics*, Kluwer Academic Publishers, pp. 263–308.

Fleurent, C. and Ferland, J.A. (1996) Genetic and hybrid algorithms for graph colouring. *Annals of Operations Research*, **63**, 437–461.

Gendreau, M. (2002) Recent advances in tabu search. In C.C. Ribeiro and P. Hansen (eds.), *Essays and Surveys in Metaheuristics*, Kluwer Academic Publishers, pp. 369–377.

Gendreau, M., Guertin, F., Potvin, J.-Y. and Taillard, É.D. (1999) Parallel tabu search for real-time vehicle routing and dispatching. *Transportation Science*, **33**, 381–390.

Gendreau, M., Hertz, A., and Laporte, G. (1994) A tabu search heuristic for the vehicle routing problem. *Management Science*, **40**, 1276–1290.

Gendreau, M., Laporte, G. and Potvin, J.-Y. (2002) Metaheuristics for the capacitated VRP. In P. Toth and D. Vigo (Eds.), *The Vehicle Routing Problem*, SIAM Monographs on Discrete Mathematics and Applications, pp. 129–154.

Gendreau, M., Soriano, P. and Salvail, L. (1993) Solving the maximum clique problem using a tabu search approach. *Annals of Operations Research*, **41**, 385–403.

Glover, F. (1977) Heuristics for integer programming using surrogate constraints. *Decision Sciences*, **8**, 156–166.

Glover, F. (1986) Future paths for integer programming and links to artificial intelligence. *Computers and Operations Research*, **13**, 533–549.

Glover, F. (1989) Tabu Search—Part I. *ORSA Journal on Computing*, **1**, 190–206.

Glover, F. (1990) Tabu Search—Part II. *ORSA Journal on Computing*, **2**, 4–32.

Glover, F. (1992) Ejection chains, reference structures and alternating path methods for traveling salesman problems. University of Colorado. Shortened version published in *Discrete Applied Mathematics*, **65**, 223–253, 1996.

Glover, F. and Laguna, M. (1993) Tabu search. In C.R. Reeves (Ed.), *Modern Heuristic Techniques for Combinatorial Problems*, Blackwell, pp. 70–150.

Glover, F. and Laguna, M. (1997) *Tabu Search*, Kluwer Academic Publishers, Norwell, MA.

Glover, F., Laguna, M., Taillard, É., and de Werra, D. (Eds.) (1993) Tabu search. *Annals of Operations Research*, **41**, J.C. Baltzer Science Publishers, Basel, Switzerland.

Glover, F., Taillard, É. and de Werra, D. (1993) A user's guide to tabu search. *Annals of Operations Research*, **41**, 3–28.

Grünert, T. (2002) Lagrangean tabu search. In C.C. Ribeiro and P. Hansen (Eds.) *Essays and Surveys in Metaheuristics*, Kluwer Academic Publishers, pp. 379–397.

Hertz, A. and de Werra, D. (1991) The tabu search metaheuristic: how we used it. *Annals of Mathematics and Artificial Intelligence*, **1**, 111–121.

Holland, J.H. (1975) *Adaptation in Natural and Artificial Systems*, University of Michigan Press, Ann Arbor, MI.

Kirkpatrick, S., Gelatt Jr., C.D., and Vecchi, M.P. (1983) Optimization by simulated annealing. *Science*, **220**, 671–680.

Laporte, G. and Osman, I.H. (Eds.) (1996) Metaheuristics in combinatorial optimization. *Annals of Operations Research*, **63**, J.C. Baltzer Science Publishers, Basel, Switzerland.

Lokketangen, A. and Glover, F. (1996) Probabilistic move selection in tabu search for 0/1 mixed integer programming problems. In I.H. Osman and J.P. Kelly (Eds.), *Meta-Heuristics: Theory and Applications*, Kluwer Academic Publishers, pp. 467–488.

Osman, I.H. (1993) Metastrategy simulated annealing and tabu search algorithms for the vehicle routing problem. *Annals of Operations Research*, **41**, 421–451.

Osman, I.H. and Kelly, J.P. (Eds.) (1996) *Meta-Heuristics: Theory and Applications*, Kluwer Academic Publishers, Norwell, MA.

Pesant, G. and Gendreau, M. (1999) A constraint programming framework for local search methods. *Journal of Heuristics*, **5**, 255–280.

Rego, C. and Roucairol, C. (1996) A parallel tabu search algorithm using ejection chains for the vehicle routing problem. In I.H. Osman and J.P. Kelly (Eds.), *Meta-Heuristics: Theory and Applications*, Kluwer Academic Publishers, pp. 661–675.

Ribeiro, C.C. and Hansen, P. (Eds.) (2002) *Essays and Surveys in Metaheuristics*, Kluwer Academic Publishers, Norwell, MA.

Rochat, Y. and Taillard, É.D. (1995) Probabilistic diversification and intensification in local search for vehicle routing. *Journal of Heuristics*, **1**, 147–167.

Rolland, E. (1996) A tabu search method for constrained real-number search: applications to portfolio selection. Working Paper, The Gary Anderson Graduate School of Management, University of California, Riverside.

Skorin-Kapov, J. (1990) Tabu search applied to the quadratic assignment problem. *ORSA Journal on Computing*, **2**, 33–45.

Soriano, P. and Gendreau, M. (1996) Diversification strategies in tabu search algorithms for the maximum clique problems. *Annals of Operations Research*, **63**, 189–207.

Soriano, P. and Gendreau, M. (1997) Fondements et applications des méthodes de recherche avec tabous. *RAIRO (Recherche opérationnelle)*, **31**, 133–159 (in French).

Taillard, É. (1990) Some efficient heuristic methods for the flow shop sequencing problem. *European Journal of Operational Research*, **47**, 65–74.

Taillard, É. (1991) Robust taboo search for the quadratic assignment problem. *Parallel Computing*, **17**, 443–455.

Taillard, É.D., Badeau, P., Gendreau, M. Guertin, F. and Potvin, J.-Y. (1997) A tabu search heuristic for the vehicle routing problem with soft time windows. *Transportation Science*, **31**, 170–186.

Voss, S., Martello, S., Osman, I.H. and Roucairol, C. (Eds.) (1999) *Meta-Heuristics: Advances and Trends in Local Search Paradigms for Optimization*, Kluwer Academic Publishers, Norwell, MA.

de Werra, D. and Hertz, A. (1989) Tabu search techniques: a tutorial and an application to neural networks. *OR Spektrum*, **11**, 131–141.

Chapter 3

GENETIC ALGORITHMS

Colin Reeves
School of Mathematical and Information Sciences
Coventry University
Priory St
Coventry CV1 5FB
E-mail: C.Reeves@conventry.ac.uk
http://www.mis.coventry.ac.uk/~colinr/

Part A: Background

1 INTRODUCTION

The term *genetic algorithm*, almost universally abbreviated nowadays to GA, was first used by John Holland [1], whose book *Adaptation in Natural and Aritificial Systems* of 1975 was instrumental in creating what is now a flourishing field of research and application that goes much wider than the original GA. Many people now use the term *evolutionary computing* or *evolutionary algorithms* (EAs), in order to cover the developments of the last 10 years. However, in the context of metaheuristics, it is probably fair to say that GAs in their original form encapsulate most of what one needs to know.

Holland's influence in the development of the topic has been very important, but several other scientists with different backgrounds were also involved in developing similar ideas. In 1960s Germany, Ingo Rechenberg [2] and Hans-Paul Schwefel [3] developed the idea of the *Evolutionsstrategie* (in English, *evolution strategy*), while—also in the 1960s—Bremermann, Fogel and others in the USA implemented their idea for what they called *evolutionary programming*. The common thread in these ideas was the use of mutation and selection—the concepts at the core of the neo-Darwinian theory of evolution. Although some promising results were obtained, evolutionary computing did not really take off until the 1980s. Not the least important reason for this was that the techniques needed a great deal of computational power. Nevertheless, the work of these early pioneers is fascinating to read in the light of our current knowledge; David Fogel (son of one of the early pioneers) has documented some of this work in [4].

1975 was a pivotal year in the development of genetic algorithms. It was in that year that Holland's book was published, but perhaps more relevantly for those interested in metaheuristics, that year also saw the completion of a doctoral thesis by one of Holland's graduate students, Ken DeJong [5]. Other students of Holland's had completed theses in this area before, but this was the first to provide a thorough treatment of the GA's capabilities in optimization.

A series of further studies followed, the first conference on the nascent subject was convened in 1985, and another graduate student of Holland's, David Goldberg, produced first an award-winning doctoral thesis on his application to gas pipeline optimization, and then, in 1989, an influential book [6]—*Genetic Algorithms in Search, Optimization, and Machine Learning*. This was the final catalyst in setting off a sustained development of GA theory and applications that is still growing rapidly.

Optimization has a fairly small place in Holland's work on adaptive systems, yet the majority of research on GAs tends to assume this is their purpose. DeJong, who initiated this interest in optimization, has cautioned that this emphasis may be misplaced in a paper [7] in which he contends that GAs are not really function optimizers, and that this is in some ways incidental to the main theme of adaptation. Nevertheless, using GAs for optimization is very popular, and frequently successful in real applications, and to those interested in metaheuristics, it will undoubtedly be the viewpoint that is most useful.

Unlike the earlier evolutionary algorithms, which focused on mutation and could be considered as straightforward developments of hill-climbing methods, Holland's GA had an extra ingredient—the idea of recombination. It is interesting in this regard to compare some of the ideas being put forward in the 1960s in the field of *operational research* (OR).

OR workers had by that time begun to develop techniques that seemed able to provide 'good' solutions, even if the quality was not provably optimal (or even near-optimal). Such methods became known as *heuristics*. A popular technique, which remains at the heart of many of the metaheuristics described in this handbook, was that of *neighbourhood search*, which has been used to attack a vast range of combinatorial optimization problems. The basic idea is to explore 'neighbours' of an existing solution—these being defined as solutions obtainable by a specified operation on the base solution.

One of the most influential papers in this context was that published by Lin [8], who found excellent solutions to the traveling salesman problem by investigating neighbourhoods formed by breaking any 3 links of a tour and re-connecting them. Empirically, Lin found that these '3-optimal' solutions were of excellent quality—in the case of the (rather small) problems he investigated, often close to the global optimum. However, he also made another interesting observation, and suggested a way of exploiting it. While starting with different initial permutations gave different 3-optimal solutions, these 3-optimal solutions were observed to have a lot of features (links) in common. Lin therefore suggested that search should be concentrated on those links about which there was not a consensus, leaving the common characteristics of the solutions alone. This was not a GA as Holland was developing it, but there are clear resonances. Much later, after GAs had become more widely known, Lin's ideas were re-discovered as 'multi-parent recombination' and 'consensus operators'.

Other OR research of the same era took up these ideas. Roberts and Flores [9] (apparently independently) used a similar approach to Lin's for the TSP, while Nugent *et al.* [10] applied this basic idea for the quadratic assignment problem. However, the general principle was not adopted into OR methodology, and relatively little was done to exploit the idea until GAs came on the OR scene in the 1990s.

2 BASIC CONCEPTS

Assume we have a discrete search space χ, and a function

$$f : \chi \mapsto \mathbb{R}.$$

The general problem is to find

$$\min_{\mathbf{x} \in \chi} f.$$

Here $\mathbf{x}$ is a vector of *decision variables*, and f is the *objective function*. We assume here that the problem is one of minimization, but the modifications necessary for a maximization problem are nearly always obvious. Such a problem is commonly called a discrete or combinatorial optimization problems (COP).

One of the distinctive features of the GA approach is to allow the separation of the *representation* of the problem from the actual variables in which it was originally formulated. In line with biological usage of the terms, it has become customary to distinguish the 'genotype'—the encoded representation of the variables, from the 'phenotype'—the set of variables themselves. That is, the vector $\mathbf{x}$ is represented by a string s, of length l, made up of symbols drawn from an alphabet $\mathcal{A}$, using a mapping

$$c : \mathcal{A}^l \mapsto \chi$$

In practice, we may need to use a search space

$$\mathcal{S} \subseteq \mathcal{A}^l$$

to reflect the fact that some strings in the image of $\mathcal{A}^l$ under c may represent invalid solutions to the original problem. (This is a potential source of difficulty for GAs in combinatorial optimization—a topic that is covered in [11].) The string length l depends on the dimensions of both χ and $\mathcal{A}$, and the elements of the string correspond to 'genes', and the values those genes can take to 'alleles'. This is often designated as the *genotype–phenotype mapping*. Thus the optimization problem becomes one of finding

$$\min_{\mathbf{s} \in \mathcal{S}} g(\mathbf{s})$$

where the function

$$g(\mathbf{s}) = f(c(\mathbf{s})).$$

It is usually desirable that c should be a bijection. (The important property of a bijection is that it has an inverse, i.e., there is a unique vector $\mathbf{x}$ for every string $\mathbf{s}$, and a unique string $\mathbf{s}$ for every vector $\mathbf{x}$.) In some cases the nature of this mapping itself creates difficulties for a GA in solving optimization problems, as discussed in [11].

In using this device, Holland's ideas are clearly distinct from the similar methodology developed by Rechenberg [2] and Schwefel [3], who preferred to work with the original decision variables directly. Both Holland's and Goldberg's books claim that representing the variables by binary strings (i.e., $\mathcal{A} = \{0, 1\}$) is in some sense 'optimal', and although this idea has been challenged, it is still often convenient from a mathematical standpoint to consider the binary case. Certainly, much of the theoretical work in GAs tends to make this assumption. In applications, many representations are possible—some of the alternatives that can be used in particular COPs are discussed in [11].

The original motivation for the GA approach was a biological analogy. In the selective breeding of plants or animals, for example, offspring are sought that have certain desirable characteristics—characteristics that are determined at the genetic level by the way the parents' chromosomes combine. In the case of GAs, a *population* of strings is used, and these strings are often referred to in the GA literature as *chromosomes*. The recombination of strings is carried out using simple analogies of genetic *crossover* and *mutation*, and the search is guided by the results of evaluating the objective function f for each string in the population. Based on this evaluation, strings that have higher *fitness* (i.e., represent better solutions) can be identified, and these are given more opportunity to breed. It is also relevant to point out here that fitness is not necessarily to be identified simply with the composition $f(c(\mathbf{s}))$; more generally, fitness is $h(f(c(\mathbf{s})))$ where $h : \mathbb{R} \mapsto \mathbb{R}$ is a monotonic function.

Perhaps the most fundamental characteristic of genetic algorithms is that their use of populations of many strings. Here again, the German school of ES initially did not use populations, and focussed almost exclusively on 'mutation' operators which are generally closer in concept to the types of operator used in neighbourhood search and its extensions. Holland also used mutation, but in his scheme it is generally treated as subordinate to crossover. Thus, in Holland's GA, instead of the search moving from point to point as in NS approaches, the whole set of strings undergoes 'reproduction' in order to generate a new population.

DeJong's work established that population-based GAs using crossover and mutation operators could successfully deal with optimization problems of several different types, and in the years since this work was published, the application of GAs to COPs has grown almost exponentially.

These operators and some developments of them are described more fully in part B. At this point, however, it might be helpful to provide a very basic introduction. Crossover is a matter of replacing some of the genes in one parent by corresponding genes of the other. An example of one-point crossover would be the following. Given the parents `P1` and `P2`, with crossover point 3 (indicated by `X`), the offspring will be the pair `O1` and `O2`:

```
P1  1  0  1  0  0  1  0     O1  1  0  1  1  0  0  1
             X
P2  0  1  1  1  0  0  1     O2  0  1  1  0  0  1  0
```

The other common operator is mutation, in which a subset of genes is chosen randomly and the allele value of the chosen genes is changed. In the case of binary strings, this simply means complementing the chosen bits. For example, the string `O1` above, with genes 3 and 5 mutated, would become `1 0 0 1 1 0 1`. A simple template for the operation of a genetic algorithm is shown in Figure 3.1. The individual parts of this very general formulation will be discussed in detail in Part B.

```
Choose an initial population of chromosomes;
while termination condition not satisfied do
   repeat
      if crossover condition satisfied then
      {select parent chromosomes;
      choose crossover parameters;
      perform crossover};
      if mutation condition satisfied then
      {choose mutation points;
      perform mutation};
      evaluate fitness of offspring
   until sufficient offspring created;
select new population;
endwhile
```

Figure 3.1. A genetic algorithm template. This is a fairly general formulation, accommodating many different forms of selection, crossover and mutation. It assumes user-specified conditions under which crossover and mutation are performed, a new population is created, and whereby the whole process is terminated.

3 WHY DOES IT WORK?

Exactly how and why GAs work is still hotly debated. There are various schools of thought, and none can be said to provide a definitive answer. A comprehensive survey will be available shortly in [12]. Meanwhile, the following is a brief guide to the main concepts that have been used.

3.1 The 'traditional' view

Holland's explanation of why it is advantageous to search the space $\mathcal{A}^l$ rather than χ hinges on three main ideas. Central to this understanding is the concept of a *schema*. A schema is a subset of the space $\mathcal{A}^l$ in which all the strings share a particular set of defined values. This can be represented by using the alphabet $\mathcal{A} \cup *$; in the binary case, `1 * * 1`, for example, represents the subset of the 4-dimensional hypercube $\{0, 1\}^4$ in which both the first and last genes take the value 1, i.e., the strings {`1 0 0 1`, `1 0 1 1`, `1 1 0 1`, `1 1 1 1`}.

The first of Holland's ideas is that of *intrinsic* (or *implicit*) parallelism—the notion that information on many schemata can be processed in parallel. Under certain conditions that depend on population size and schema characteristics, Holland estimated that a population of size M contains information on $\mathcal{O}(M^3)$ schemata. However, these schemata cannot *actually* be processed in parallel, because independent estimates of their fitness cannot be obtained in general [13].

The second concept is expressed by the so-called *Schema Theorem*, in which Holland showed that if there are $N(S, t)$ instances of a given schema S in the population at time t, then at the next time step (following reproduction), the expected number of instances in the new population can be bounded by

$$E[N(S,t+1)] \geq \frac{F(S,t)}{\bar{F}(t)} N(S,t)\{1 - \epsilon(S,t)\}$$

where $F(S,t)$ is the fitness of schema S, $\bar{F}(t)$ is the average fitness of the population, and $\epsilon(S,t)$ is a term which reflects the potential for genetic operators to destroy instances of schema S.

By failing to appreciate the stochastic and dynamic nature of this relationship, somewhat extravagant conclusions have been drawn from this theorem, expressed in the frequently made statement that good schemata will receive exponentially increasing numbers of trials in subsequent generations. However, it is clear that the Schema Theorem is a result in expectation only, and then only for one generation. Thus, any attempt to extrapolate this result for more than one generation is doomed to failure because the terms are then no longer independent of what is happening in the rest of the population. Also, given a finite population, it is clear that any exponential increase cannot last very long.

Holland also attempted to model schema processing (or hyperplane competitions) by means of an analogy to stochastic two-armed bandit problems. This is a well-known statistical problem: we are given two 'levers' which if pulled give 'payoff' values according to different probability distributions. The problem is to use the results of previous pulls in order to maximize the overall future expected payoff. In [1] it is argued that a GA approximates an 'optimal' strategy which allocates an (exponentially) increasing number of trials to the observed better lever; this is then used to contend for the supposed efficiency of a GA in distinguishing between competing schemata or hyperplanes.

Early accounts of GAs suggested quite strongly that in a GA we had thus discovered an algorithm that used the best available search strategy to solve not merely one, but many hyperplane competitions at once: the 'only case where combinatorial explosion works in our favour'. Unfortunately, Wolpert and Macready's 'No-Free-Lunch' Theorem (NFLT) [14] has rather destroyed such dreams.[1]

In fact, intrinsic parallelism turns out to be of strictly limited application; it merely describes the number of schemata that are likely to be present in some numbers given certain assumptions about string length, population size and (most importantly) the way in which the population has been generated—and the last assumption is unlikely to be true except at a very early stage of the search. Even then, only in very unusual circumstances—that of orthogonal populations [13]—could the hyperplane competitions actually be processed in parallel; normally, the competitions are not independent. The two-armed bandit analogy also fails in at least two ways: first, Macready and Wolpert [15] have recently argued that there is no reason to believe that the strategy described by Holland as approximated by a GA is a optimal one. They also believe there is in any case a flaw in Holland's mathematics.

This is not to say that the Schema Theorem in particular, or the idea of a schema in general, is useless, but that what it says is of limited and mainly short-term value—principally, that certain schemata are likely to increase their presence in the next

[1]The NFLT, put simply, says that on the average, nothing—ant colonies, GAs, simulated annealing, tabu search, etc.—is better than random search. Success comes from adapting the technique to the problem at hand.

population, and that those schemata will be on the average fitter, and less resistant to destruction by crossover and mutation, than those that do not.

This brings us to the third assumption implicit in the implementation of a GA—that the re-combination of small pieces of the genotype (good schemata)into bigger pieces is indeed a sensible method of finding optimal solutions. Goldberg [6] calls this the building-block hypothesis (BBH).There is certainly some negative evidence, in that problems constructed to contain misleading building-blocks may indeed be hard for a GA to solve. The failure of the BBH is often invoked as a explanation when a GA fails to solve particular COPs.

However, the properties of these problems are not usually such that they are uniquely difficult for GAs. Holland himself, with two other co-workers, looked for positive evidence in favour of the building-block hypothesis [16] and found the results rather problematical: functions constructed precisely to provide a 'royal road' made up of building blocks of increasing size and fitness turned out to be much more efficiently solved by 'non-genetic' methods.

3.2 Other Approaches

By writing his theorem in the form of a lower bound, Holland was able to make a statement about schema S that is independent of what happens to other schemata. However, in practice what happens to schema S will influence the survival (or otherwise) of other schemata, and what happens to other schemata will affect what happens to S as is made plain by the exact models of Vose [17] and Whitley [18].

Markov chain theory [17,18] has been applied to GAs [19,20] to gain a better understanding of the GA as a whole. However, while the results are fascinating in illuminating some nuances of GA behaviour, the computational requirements are formidable for all but the smallest of problems, as shown by DeJong et al. [21], or Rees and Koehler [22], for example.

Shapiro et al. [23] first examined GAs from a statistical mechanics perspective, and there is a growing literature on this topic. Peck and Dhawan [24] have linked GAs to global randomized search methods. But one of the difficulties in analysing GAs is that there is not a single generic GA, the behaviour of which will characterize the class of algorithms that it represents. In practice, there is a vast number of ways of implementing a GA, as will be seen in the discussion in Part B, and what works in one case may not work in another. Some workers have therefore tried to look for ways of predicting algorithm performance for particular problem classes.

Reeves and Wright [13] summarize a perspective based on relating GAs to statistical methods of experimental design, which draws upon the biological concept of *epistasis*. This expresses the idea that the expression of a chromosome is not merely a sum of the effects of its individual alleles, but that the alleles located in some genes influence the expression of the alleles in others. From a mathematical viewpoint, epistasis is equivalent to the existence of interactions in the fitness function. If we knew the extent of these non-linearities, we might be able to choose an appropriate algorithm. Unfortunately, as is explained in [25], it is unlikely that this approach will be successful, although the literature surrounding the question of epistasis has produced some useful insights into GAs.

Several authors [26–28] have pointed out connections between GAs and neighbourhood search methods, and this has led to a considerable literature on the analysis of problem landscapes. The concept of a landscape has been used informally for many years, but recent work [29] has put the idea on a rigorous mathematical foundation which is still being explored. Some of its uses in the context of GAs is described in [30]. It appears that this way of thinking about algorithms has great potential for unifying different metaheuristics and increasing our understanding of them.

4 APPLICATIONS AND SOURCES

There are numerous examples of the successful application of GAs to combinatorial optimization problems. Books such as those by Davis [31] and Chambers [32,33] are useful in displaying the range of problems to which GAs have been applied. In a chapter such as this, it is impossible to give an exhaustive survey of relevant applications of GAs, but [11] lists some of the more useful and accessible references that should be of interest to people who are experimenting with metaheuristics. However, because of the enormous growth in reported applications of GAs, this list is inevitably incomplete, as well as somewhat dated already. For a time, Alander attempted to maintain a comprehensive bibliography: an early version of this is included in [33], while an updated one was provided in [34]. However, this is one area where the phenonemon of exponential growth is indubitable, and the sheer number of papers published in the last 5 years seem to have overwhelmed this enterprise.

For more information on applications, and on GAs in general, the reader has several useful books to choose from: the early ones by Holland, Goldberg and Michalewicz [1,6,35] tend to be over-committed to the schema-processing point of view, but they are all still excellent sources of information. Reeves [36] also reflects the state of the theory at the time the book was written, although it covers other heuristic methods too. More recently, Mitchell [37] and Falkenauer [38] demonstrate a more careful approach to schemata, and Bäck [39] covers the wider field of evolutionary algorithms. All are worth consulting. For a rigorous theoretical study, there is the book by Vose [40], which deals mainly with the Markov chain and dynamical systems approach, while the forthcoming text [12] will survey in some detail several other perspectives on GAs.

There are now also many conferences on GAs and related topics—too many to list in detail. The proceedings of the original biennial series of *International Conferences on Genetic Algorithms* [41–47], which is still of considerable historical interest,[2] has become an annual event (GECCO), while the IEEE has established an alternative series under the title of the *Congress on Evolutionary Computation*. In Europe, there are two biennial series of somewhat wider scope: the *Parallel Problem-Solving from Nature* series [48–53], and the *International Conference on Artificial Neural Networks and Genetic Algorithms* [54–57]. For the theoretically minded, there is a biennial workshop to consider—the *Foundations of Genetic Algorithms* [58–63].

There are also many journals now publishing GA-related research. The major GA journals are *Evolutionary Computation* (MIT Press) and *IEEE Transactions on Evolutionary Computation* (IEEE); other theoretical articles appear in journals related to AI or to complex systems. Most OR journals—*INFORMS Journal on Computing, Computers and OR, Journal of the OR Society, European Journal*

[2] Apart from the intrinsic worth of these papers, it is well worth checking to see if someone has tried your bright new idea already.

of OR etc.—have frequent papers on GAs, mainly applications. There are newsgroups on the internet (`comp.ai.genetic`) and the moderated news digest at `GA-List-Request@aic.nrl.navy.mil`.

Part B: Guidelines

The basic principles of a GA were shown in Figure 3.1, but as usual, the details are all important. The various stages involved in implementing a GA will now be described.

5 INITIAL POPULATION

The major questions to consider are firstly the size of the population, and secondly the method by which the individuals are chosen. The size of the population has been approached from several theoretical points of view, although the underlying idea is always of a trade-off between efficiency and effectiveness. Intuitively, it would seem that there should be some 'optimal' value for a given string length, on the grounds that too small a population would not allow sufficient room for exploring the search space effectively, while too large a population would so impair the efficiency of the method that no solution could be expected in a reasonable amount of time. Goldberg [64,65] was probably the first to attempt to answer this question, using the idea of schemata, as outlined in the next chapter. Unfortunately, from this viewpoint, it appeared that the population size should increase as an exponential function of the string length. Experimental evidence [66,67] suggested that populations of the size proposed by Goldberg's theory are not necessary.

A slightly different question that we could ask is regarding a *minimum* population size for a meaningful search to take place. In Reeves [68], the initial principle was adopted that, at the very least, every point in the search space should be reachable from the initial population by crossover only. This requirement can only be satisfied if there is at least one instance of every allele at each locus in the whole population of strings. On the assumption that the initial population is generated by a random sample with replacement (which is a conservative assumption in this context), the probability that at least one allele is present at each locus can be found. For binary strings this is easily seen to be

$$P_2^* = (1 - (1/2)^{M-1})^l$$

from which we can calculate that, for example, a population of size 17 is enough to ensure that the required probability exceeds 99.9% for strings of length 50. For q-ary alphabets, the calculation is somewhat less straightforward, but expressions are given in [68] that can be converted numerically into graphs for specified confidence levels. The results of this work suggested that a population size of $\mathcal{O}(\log l)$ would be sufficient to cover the search space.

Finally, as to how the population is chosen, it is nearly always assumed that initialization should be random. Rees and Koehler [22], using a model-based approach that draws on the theoretical work of Vose [17], have demonstrated that sampling without replacement is preferable in the context of very small populations. More generally, it is obvious that randomly chosen points do not necessarily cover the search space

Individual	Gene					
1	0	1	3	0	2	4
2	1	4	4	2	3	0
3	0	0	1	2	4	3
4	2	4	0	3	1	4
5	3	3	0	4	4	2
6	4	1	2	4	3	0
7	2	0	1	3	0	1
8	1	3	3	1	2	2
9	4	2	2	1	1	3
10	3	2	4	0	0	1

Figure 3.2. An example of Latin hypercube sampling for $l = 6$ and $|\mathcal{A}| = 5$. Notice that each allele occurs exactly twice for each gene.

uniformly, and there may be advantages in terms of coverage if we use more sophisticated statistical methods, especially for non-binary alphabets. One such simple idea is a generalization of the Latin hypercube which can be illustrated as follows.

Suppose each gene has 5 alleles, labelled $0, \ldots, 4$. We choose the population size to be a multiple of 5, say m, and the alleles in each 'column' are generated as an independent random permutation of $0, \ldots, (m-1)$, which is then taken modulo 5. Figure 3.2 shows an example for a population of size 10. To obtain search space coverage at this level with simple random initialization would need a much larger population.

Another point to mention here is the possibility of 'seeding' the initial population with known good solutions. Some reports (e.g. in [69,70]) have found that including a high-quality solution, obtained from another heuristic technique, can help a GA find better solutions rather more quickly than it can from a random start. However, there is also the possibility of inducing premature convergence [71,72].

6 TERMINATION

Unlike simple neighbourhood search methods that terminate when a local optimum is reached, GAs are stochastic search methods that could in principle run for ever. In practice, a termination criterion is needed; common approaches are to set a limit on the umber of fitness evaluations or the computer clock time, or to track the population's diversity and stop when this falls below a preset threshold. The meaning of diversity in the latter case is not always obvious, and it could relate either to the genotype or the phenotype, or even, conceivably, to the fitnesses, but the most common way to measure it is by genotype statistics. For example, we could decide to terminate a run if at every locus the proportion of one particular allele rose above 90%.

7 CROSSOVER CONDITION

Given the stress on recombination in Holland's original work, it might be thought that crossover should always be used, but in fact there is no reason to suppose that it has to be so. Further, while we could follow a strategy of crossover-AND-mutation to generate new offspring, it is also possible to use crossover-OR-mutation. There are many examples of both in the literature. The first strategy initially tries to carry out crossover, then attempts mutation on the off spring (either one or both). It is conceivable that in some cases nothing actually happens at all with this strategy—the offspring are simply clones of the parents. Others always do something, either crossover of mutation, but not both. (Even then, cloning is still possible with crossover if the parents are too alike.)

The mechanism for implementing such choices is customarily a randomized rule, whereby the operation is carried out if a pseudo-random uniform deviate exceeds a threshold value. In the case of crossover, this is often called the crossover rate, often denoted by the symbol χ. For mutation, we have a choice between describing the number of mutations per string, or per bit; bit-wise mutation, at a rate denoted by μ, is more common.

In the -OR- case, there is the further possibility of modifying the relative proportions of crossover and mutation as the search progresses. Davis [31] has argued that different rates are appropriate at different times: high crossover at the start, high mutation as the population converges. He has further suggested that the operator proportions could be adapted online, in accordance with their track record in finding new high-quality chromosomes.

8 SELECTION

The basic idea of selection is that it should be related to fitness, and the original scheme for its implementation is commonly known as the *roulette-wheel* method. It uses a probability distribution for selection in which the selection probability of a given string is proportional to its fitness. Figure 3.3 provides a simple example of roulette-wheel selection (RWS). Pseudo-random numbers are used one at a time to choose strings for parenthood. For example, in Figure 3.3, the number 0.13 would select string 1, the number 0.68 would select string 4.

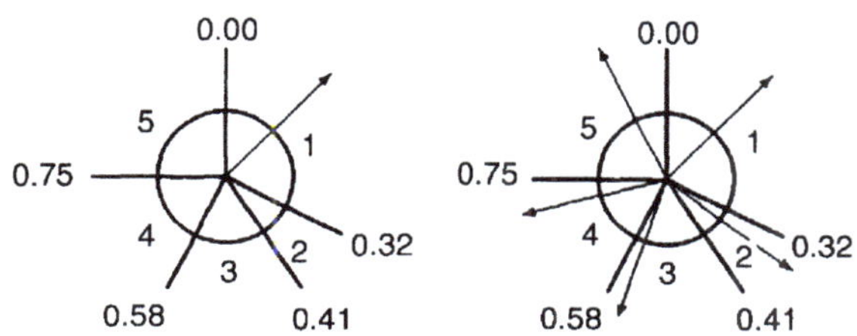

Figure 3.3. Suppose there are 5 strings in a population with fitnesses {32, 9, 17, 17, 25} respectively. The probability of selection of each individual is proportional to the area of a sector of a roulette wheel (or equivalently, to the angle subtended at the centre). The numbers on the spokes of the wheel are the cumulative probabilities for use by a pseudo-random number generator. On the left we have standard roulette-wheel selection, with a single pointer that has to be spun 5 times. On the right we have SUS, using 5 connected equally-spaced pointers; one spin provides 5 selections.

Finding the appropriate number for a given pseudo-random number r requires searching an array for values that bracket r—this can be done in $\mathcal{O}(\log M)$ time for a population of size M. However, RWS has a high stochastic variability, and the actual number of times N_C that chromosome C is selected in any generation may be very different from its expected value $E[N_C]$. For this reason, sampling *without* replacement may be used, to ensure that at least the integral part of $E[N_C]$ is achieved, with fractions being allocated using random sampling.

In practice, Baker's [73] *stochastic universal selection* (SUS) is a particularly effective way of realizing this outcome. Instead of a single choice at each stage, we imagine that the roulette wheel has an equally spaced multi-armed spinner. Spinning the wheel produces simultaneously the values N_C for all the chromosomes in the population. From the viewpoint of statistical sampling theory, this corresponds to systematic sampling [74]. Experimental work by Hancock [75] clearly demonstrates the superiority of this approach, although much published work on applications of GAs still appears to rely on the basic roulette-wheel method.[3]

An associated problem is that of finding a suitable measure of fitness for the members of the population. Simply using the objective function values $f(\mathbf{x})$ is rarely sufficient, because the scale on which $f(\mathbf{x})$ is measured is important. (For example, values of 10 and 20 are much more clearly distinguished than 1010 and 1020.) Further, if the objective is minimization rather than maximization, a transformation is clearly required.

Some sort of scaling is thus usually applied, and Goldberg [6] gives a simple algorithm to deal with both minimization and maximization. The method is cumbersome, however, and it needs continual re-scaling as the search progresses. Two alternatives provide more elegant solutions.

8.1 Ranking

Ranking the chromosomes in fitness order loses some information, but there is no need for re-scaling, and selection algorithm is simpler and more efficient. Suppose the probability of selecting the string that is ranked kth in the population is denoted by $P[k]$. In the case of linear ranking, we assume that

$$P[k] = \alpha + \beta k$$

where α and β are constants. The requirement that $P[k]$ be a probability distribution gives us one condition:

$$\sum_{k=1}^{M} (\alpha + \beta k) = 1$$

which leaves us free to choose the other parameter in a way that tunes the *selection pressure*. This term is loosely used in many papers and articles on GAs. Here, we mean the following.

[3] Note that SUS does not necessarily reduce the total of random numbers needed. Having generated a multiset of size M as our 'mating pool', we still have to use random numbers to decide which pairs mate together, whereas in RWS we can simply pair them in the order generated.

Definition 3.1. *Selection pressure*

$$\phi = \frac{Prob.[selecting\ fittest\ string]}{Prob.[selecting\ average\ string]}.$$

In the case of linear ranking, we interpret the average as meaning the median *string, so that*

$$\phi = \frac{\alpha + \beta M}{\alpha + \beta(M+1)/2}.$$

(This assumes the population size is odd—however, the analysis holds mutatis mutandis for the case of an even number.) Some simple algebra soon establishes that

$$\beta = \frac{2(\phi - 1)}{M(M-1)} \quad \text{and} \quad \alpha = \frac{2M - \phi(M+1)}{M(M-1)}$$

which implies that $1 \leq \phi \leq 2$. *With this framework, it is easy to see that the cumulative probability distribution can be stated in terms of the sum of an arithmetic progression, so that finding the appropriate* k *for a given pseudo-random number* r *is simply a matter of solving a quadratic equation for* k, *which can be done simply in* $O(1)$ *time. The formula is*

$$k = \frac{-(2\alpha + \beta) \pm \sqrt{(2\alpha + \beta)^2 + 4\beta r}}{2\beta}.$$

Other functions can be used besides linear ranking [75,76] but the above scheme is sufficiently flexible for most applications.

8.2 Tournament Selection

The other alternative to strict fitness-proportional selection is *tournament selection* in which a set of τ chromosomes is chosen and compared, the best one being selected for parenthood. This approach has similar properties to linear ranking for $\tau = 2$. It is easy to see that the best string will be selected every time it is compared, while the median string will be chosen with probability $2^{-(\tau-1)}$. Thus the selection pressure is given by $\phi = 2^{\tau-1}$, which for $\tau = 2$ is similar to linear ranking when $\alpha \to 0$.

One potential advantage of tournament selection over all other forms is that it only needs a preference ordering between pairs or groups of strings, and it can thus cope with situations where there is no formal objective function at all—in other words, it can deal with a purely *subjective* objective function!

However, we should point out again that tournament selection is also subject to arbitrary stochastic effects in the same way as roulette-wheel selection—there is no guarantee that every string will appear in a given cycle. Indeed, using sampling with replacement there is a probability of approximately $0.368(\approx e^{-1})$ that a given string will not appear at all. One way of coping with this, at the expense of a little extra computation, is to use a variance reduction technique from simulation theory. Saliby [77] distinguishes between the *set* effect and the *sequence* effect in drawing items from a finite population. In applying his ideas here, we know that we need τ items to be drawn M times, so we simply construct τ random permutations[4] of the numbers $1, \ldots, M$—the indices of the individuals in the population. These are concatenated into one long

[4] There is a simple algorithm for doing this efficiently—see Nijenhuis and Wilf [78], e.g., or look at the Stony Brook Algorithm Repository [79]

sequence which is then chopped up into M pieces, each containing the τ indices of the individuals to be used in the consecutive tournaments. If M is not an exact multiple of τ, there is the small chance of some distortion where the permutations join, but this is a relatively minor problem.

9 CROSSOVER

Crossover is simply a matter of replacing some of the genes in one parent by the corresponding genes of the other. Suppose we have 2 strings **a** and **b**, each consisting of 6 variables, i.e.

$$(a_1, a_2, a_3, a_4, a_5, a_6) \quad \text{and} \quad (b_1, b_2, b_3, b_4, b_5, b_6),$$

which represent two possible solutions to a problem. (Note that we have chosen here to leave the alphabet unspecified, to emphasize that binary representation is not a critical aspect of GAs.) One-point crossover (1X) has been described earlier; *two-point* crossover (denoted by 2X), is very similar. Two crosspoints are chosen at random from the numbers $1, \ldots, 5$, and a new solution produced by combining the pieces of the the original 'parents'. For instance, if the crosspoints were 2 and 4, the 'offspring' solutions would be

$$(a_1, a_2, b_3, b_4, a_5, a_6) \quad \text{and} \quad (b_1, b_2, a_3, a_4, b_5, b_6).$$

A similar prescription can be given for m-point crossover where $m > 1$.

An early and thorough investigation of multi-point crossovers is that by Eshelman et al. [80], who examined the biasing effect of traditional one-point crossover, and considered a range of alternatives. Their central argument is that two sources of bias exist to be exploited in a genetic algorithm: *positional* bias, *and distributional* bias. Simple crossover has considerable positional bias, in that it relies on the building-block hypothesis, and if this is invalid, the bias may prevent the production of good solutions.

On the other hand, 1X has no distributional bias, in that the crossover point is chosen randomly using the uniform distribution. But this lack of bias is not necessarily a good thing, as it limits the exchange of information between the parents. In [80], the possibilities of changing these biases, in particular by using multi-point crossover, were investigated and empirical evidence strongly supported the suspicion that one-point crossover is not the best option. In fact, the evidence was somewhat ambiguous, but seemed to point to an 8-point crossover operator as the best overall, in terms of the number of function evaluations needed to reach the global optimum.

Another obvious alternative, which removes any bias, is to make the crossover process completely random—the so-called uniform crossover. This can be seen most easily by observing that the crossover operator itself can be written as a binary string or *mask*—in fact, when implementing crossover in a computer algorithm, this is the obvious way to do it. For example, the mask

```
1 1 0 0 1 1
```

represents the 2-point crossover used above, where a 1 means that the alleles are taken from the first parent, while a 0 means they come from the second.

By generating the pattern of 0s and 1s stochastically (using a Bernoulli distribution) we thus get uniform crossover (UX), which might generate a mask such as

```
1 0 1 0 0 1
```

which implies that the 1st, 3rd and 6th alleles are taken from the first parent, the others from the second. This idea was first used by Syswerda [81], who implicitly assumed that the Bernoulli parameter $p = 0.5$. Of course, this is not necessary: we can bias UX towards one or other of the parents by choosing p appropriately.

DeJong and Spears [82] produced a theoretical analysis that was able to characterize the amount of disruption introduced by a given crossover operator exactly. In particular, the amount of disruption in UX can be tuned by choosing different values of p.

Of course, there are also many practical considerations that influence the implementation of crossover. How often do we apply it? Some always do, others use a stochastic approach, applying crossover with a probability $\chi < 1$. Do we generate one offspring or two? In many cases there are natural 'twin' offspring resulting, but in more sophisticated problems it may be that only one offspring arises. When we choose only one from two, how do we do it? In accordance with the stochastic nature of the GA, we may well decide to choose either of the offspring at random. Alternatively, we could bias the decision by making use of some other property such as the fitness of the new individuals, or the loss (or gain) in diversity that results in choosing one rather than the other.

Booker [83] reported significant gains from using an adaptive crossover rate: the rate was varied according to a characteristic called *percent involvement*. This is simply the percentage of the current population that is producing offspring—too small a value is associated with loss of diversity and premature convergence.

9.1 Non-linear Crossover

In cases of non-linear encodings, crossover has to be reinterpreted. One of the most frequently occurring problems is where the solution space is the space of permutations (Π_l) of the numbers $1, \ldots, l$—well-known examples of this include many scheduling problems, and the famous travelling salesman problem (TSP).

For instance, the simple-minded application of 1X in the following case produces an infeasible solution:

```
P1  1  6  3  4  5  2      O1  1  6  1  2  6  5
             X
P2  4  3  1  2  6  5      O2  4  3  3  4  5  2
```

If this represents a TSP, the first offspring visits cities 1 and 6 twice, and never gets to cities 3 or 4. A moment's thought is enough to realize that this type of behaviour will be the rule, not an exception. Clearly we need to think of something rather smarter if we are to be able to solve such problems.

One of the first ideas for such problems was the PMX (partially mapped crossover) operator [84]. Two crossover points are chosen uniformly at random between 1 and l. The section between these points defines an interchange mapping. Thus, in the example above, PMX might proceed as follows:

```
P1  1  6  3  4  5  2      O1  3  5  1  2  6  4
       X        Y
P2  4  3  1  2  6  5      O2  2  1  3  4  5  6
```

Here the crossover points X and Y define an interchange mapping

$$3 \leftrightarrow 1; \quad 4 \leftrightarrow 2; \quad 5 \leftrightarrow 6$$

on their respective strings, which means that the cut blocks have been swapped and now appear in different contexts from before. Another possibility is to apply a binary mask, as in linear crossover, but with a different meaning. Such a mask, generated as with UX say, might be the following

```
1  0  1  0  0  1
```

which is applied to the parents in turn. First the components corresponding to 1s are copied from one parent, and then those that correspond to 0s are taken in the order they appear from the second parent in order to fill the gaps. Thus the above example generates the following pairs of strings:

```
P1  1  6  3  4  5  2  ⇒  1  _  3  _  _  2  ⇒  O1  1  4  3  6  5  2
P2  4  3  1  2  6  5  ⇒  4  _  1  _  _  5  ⇒  O2  4  6  1  3  2  5
```

10 MUTATION

Firstly, we note that in the case when crossover-OR-mutation is used, we must first decide whether any mutation is carried out at all. Assuming that it is, the concept of mutation is even simpler than crossover, and again, this can easily be represented as a bit-string. We generate a mask such as

```
0  1  0  0  0  1
```

using a Bernoulli distribution at each locus—with a small value of p in this case. (The above example would then imply that the 2nd and 6th genes are assigned new allele values.) However, there are different ways of implementing this simple idea that can make a substantial difference to the performance of a GA. The naive idea would be to draw a random number for *every* gene in the string and compare it to μ, but this is potentially expensive in terms of computation if the strings are long and the population is large. An efficient alternative is to draw a random variate from a Poisson distribution with parameter λ, where λ is the average number of mutations per chromosome. A common value for λ is 1—in other words, if l is the string length, the (bit-wise) mutation rate is $\mu = 1/l$, which as early as 1964 [85] was shown to be in some sense an 'optimal' mutation rate. Having decided that there are (say) m mutations, we draw m random numbers (without replacement) uniformly distributed between 1 and l in order to specify the loci where mutation is to take place.

In the case of binary strings, mutation simply means complementing the chosen bit(s). More generally, when there are several possible allele values for each gene, if we decide to change a particular allele, we must provide some means of deciding what its new value should be. This could be a random choice, but if (as in some cases) there is some ordinal relation between allele values, it may be more sensible to restrict the choice to alleles that are close to the current value, or at least to bias the probability distribution in their favour.

It is often suggested that mutation has a somewhat secondary function, that of helping to preserve a reasonable level of population diversity—an insurance policy which enables the process to escape from sub-optimal regions of the solution space, but not all authors agree. Proponents of evolutionary programming, for example, consider crossover to be an irrelevance, and mutation plays the major role [86]. Perhaps it is best to say that the balance between crossover and mutation is often a problem-specific one, and definite guidelines are hard to give.

However, several authors have suggested some type of adaptive mutation: for example, Fogarty [87] experimented with different mutation rates at different loci. Reeves [69] varied the mutation probability according to the diversity in the population (measured in terms of the coefficient of variation of fitnesses). More sophisticated procedures are possible, and anecdotal evidence suggests that many authors use some sort of diversity maintenance policy.

11 NEW POPULATION

Holland's original GA assumed a *generational* approach: selection, recombination and mutation were applied to a population of M chromosomes until a new set of M individuals had been generated. This set then became the new population. From an optimization viewpoint this seems an odd thing to do—we may have spent considerable effort obtaining a good solution, only to run the risk of throwing it away and thus preventing it from taking part in further reproduction. For this reason, De Jong [5] introduced the concepts of *élitism* and *population overlaps*. His ideas are simple—an élitist strategy ensures the survival of the best individual so far by preserving it and replacing only the remaining $(M - 1)$ members of the population with new strings. Overlapping populations take this a stage further by replacing only a fraction G (the *generation gap*) of the population at each generation. Finally, taking this to its logical conclusion produces the so-called steady-state or incremental strategies, in which only one new chromosome (or sometimes a pair) is generated at each stage. Davis [31] gives a good general introduction to this type of GA.

Slightly different strategies are commonly used in the ES community, which traditionally designates them either (λ, μ) or $(\lambda + \mu)$. In the first case, $\mu(> \lambda)$ offspring are generated from λ parents, and the best λ of these offspring are chosen to start the next generation. For the + strategy, μ offspring are generated and the best λ individuals are chosen from the combined set of parents and offspring.

In the case of incremental reproduction it is also necessary to select members of the population for deletion. Some GAs have assumed that parents are replaced by their children. Many implementations, such as Whitley's GENITOR [76], delete the worst member of the population. Goldberg and Deb [88] have pointed out that this exerts a very strong selective pressure on the search, which may need fairly large populations and high mutation rates to prevent a rapid loss of diversity. A milder prescription is to delete one of the worst $P\%$ of the population (for example, Reeves [69] used $P = 50$, i.e., selection from those worse than the median). This is easily implemented when rank-based selection is used. Yet another approach is to base deletion on the *age* of the strings.

11.1 Diversity Maintenance

As hinted above, one of the keys to good performance (in nature as well as in GAs) is to maintain the diversity of the population as long as possible. The effect of selection is to reduce diversity, and some methods can reduce diversity very quickly. This can be mitigated by having larger populations, or by having greater mutation rates, but there are also other techniques that are often employed.

A popular approach, commonly linked with steady-state or incremental GAs, is to use a 'no-duplicates' policy [31]. This means that the offspring are not allowed into the population if they are merely clones of existing individuals. The downside, of course, is the need to compare each current individual with the new candidate, which adds to the computational effort needed—an important consideration with large populations. (In principle, some sort of 'hashing' approach could be used to speed this process up, but whether this was used in [31] is not clear.)

We can of course take steps to reduce the chance of cloning before offspring are generated. For instance, with `1X`, the two strings

```
1 1 0 1 0 0 1
1 1 0 0 0 1 0
```

will generate only clones if the crossover point is any of the first three positions. Booker [83] suggested that before applying crossover, we should examine the selected parents to find suitable crossover points. This entails computing an 'exclusive-OR' (XOR) between the parents, so that only positions between the outermost 1s of the XOR string (the 'reduced surrogate') should be considered as crossover points. Thus in the example above, the XOR string is

```
0 0 0 1 0 1 1
```

so that, as previously stated, only the last 3 crossover points will give rise to a different string.

12 REPRESENTATION

As remarked in part A, the focus in this handbook is on using GAs as optimizers in a search space, given a suitable encoding and fitness function. We now consider how the search space S might be constructed in some generic cases.

12.1 Binary Problems

In some problems a binary encoding might arise naturally. Consider the operational research problem known as the *knapsack* problem, stated as follows.

Example 3.1 (The 0–1 knapsack problem). *A set of n items is available to be packed into a knapsack with capacity C units. Item i has value v_i and uses up c_i units of capacity.*

Determine the subset I of items which should be packed in order to maximize

$$\sum_{i \in I} v_i$$

such that

$$\sum_{i \in I} c_i \leq C.$$

If we define

$$x_i = \begin{cases} 1 & \text{if item } i \text{ is packed} \\ 0 & \text{otherwise} \end{cases}$$

the knapsack problem can be re-formulated as an *integer program*:

$$\textit{maximize} \sum_{i=1}^{n} x_i v_i$$

$$\textit{such that} \sum_{i=1}^{n} x_i c_i \leq C$$

from which it is clear that we can define a solution as a binary string of length n. In this case there is thus no distinction between genotype and phenotype.

However, such problems are not necessarily easy to solve with a GA. In this case, the presence of constraints is likely to cause difficulties—two feasible parents may not produce feasible offspring, unless special crossover operators are constructed. In fact, such problems as these are really subset selection problems, which are best tackled by other means [89], despite the seductiveness of the binary encoding. It is now widely recognised that 'natural' binary encodings nearly always bring substantial problems for simple GAs.

12.2 Discrete (but not Binary) Problems

There are cases in which a discrete alphabet of higher cardinality than 2 might be appropriate. The rotor stacking problem, as originally described by McKee and Reed [90], is a good example.

Example 3.2. *A set of n rotors is available, each of which has k holes drilled in it. The rotors have to be assembled into a unit by stacking them and bolting them together, as in Figure 1.4. Because the rotors are not perfectly flat, stacking them in different orientations will lead to assemblies with different characteristics in terms of deviations from true symmetry, with the consequent effect (in operation) that the assembled unit will wobble as it spins. The objective is to find which of all the possible combinations of orientations produces the least deviation.*

In this case a k-ary coding is natural. A solution is represented by a string of length n, each gene corresponding to a rotor and the alleles, drawn from $\{1, \ldots, k\}$, representing the orientation (relative to a fixed datum) of the holes. Thus, the string (1322) represents a solution to a 4-rotor problem where hole 1 of the first rotor is aligned with hole 3 of the second and hole 2 of the third and fourth. Of course, it would be possible to encode the alleles as binary strings, but there seems little point in so doing—particularly if k

is not a power of 2, as there will then be some binary strings that do not correspond to any actual orientation.

This seems very straightforward, but there is a subtle point here that could be overlooked. The assignment of labels to the holes is arbitrary, and this creates the problem of 'competing conventions' as it has been called.[5] For example, given a natural order for labelling each rotor, the string (3211) represents the same solution as (1322). This can be alleviated in this case by fixing the labeling for one rotor, so that a solution can be encoded by a string of length $(n-1)$.

As far as the operators are concerned, standard crossovers can be used here, but mutation needs careful consideration, as outlined above in Section 10.

12.3 Permutation Problems

There are also some problems where the 'obvious' choice of representation is defined, not over a set, but over a permutation. The TSP is one of many problems for which this is true. As another example, consider the permutation flowshop sequencing problem (PFSP).

Example 3.3. *Suppose we have n jobs to be processed on m machines, where the processing time for job i on machine j is given by $p(i, j)$. For a job permutation $\{\pi_1, \pi_2, \dots, \pi_n\}$, we calculate the completion times $C(\pi_i, j)$ as follows:*

$$
\begin{aligned}
C(\pi_1, 1) &= p(\pi_1, 1) \\
C(\pi_i, 1) &= C(\pi_{i-1}, 1) + p(\pi_i, 1) \quad \text{for } i = 2, \dots, n \\
C(\pi_1, j) &= C(\pi_1, j-1) + p(\pi_1, j) \quad \text{for } j = 2, \dots, m \\
C(\pi_i, j) &= \max\{C(\pi_{i-1}, j), C(\pi_i, j-1)\} + p(\pi_i, j) \\
&\quad \text{for } i = 2, \dots, n;\ j = 2, \dots, m.
\end{aligned}
$$

The PFSP is then to find a permutation π^* in the set of all permutations Π such that

$$f(\pi^*) \leq f(\pi) \ \forall \pi \in \Pi.$$

(Several performance measures $f(\cdot)$ are possible; common ones are the maximum or mean completion time.)

Here the natural encoding (although not the only one) is simply the permutation of the jobs as used to calculate the completion times. So the solution (1462537), for example, simply means that job 1 is first on each machine, then job 4, job 6, etc.

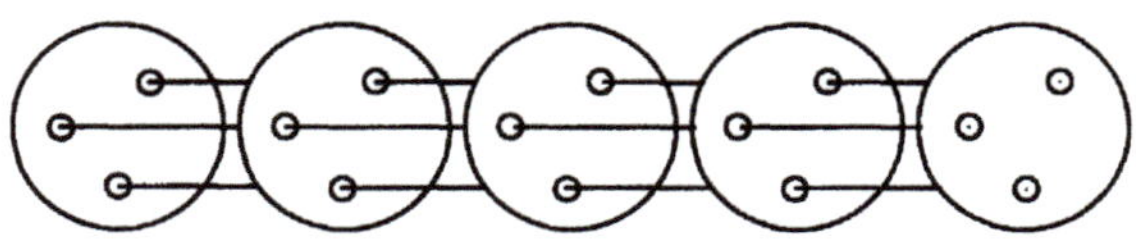

Figure 3.4. Rotor stacking problem.

[5] This phenomenon is a common one whenever the coding function $c(\cdot)$ is not injective. It has been observed in problems ranging from optimizing neural nets to the TSP. Radcliffe, who calls it 'degeneracy' [91] has presented the most thorough analysis of this problem and how to treat it.

Unfortunately, the standard crossover operators patently fail to preserve the permutation except in very fortunate circumstances, as discussed in Section 9.1. Some solutions to this problem were outlined there; more comprehensive discussion of possible methods of attack are contained in [92,93], while [69,94] describe some approaches of particular relevance to the PFSP.

12.4 Non-binary Problems

In many cases the natural variables for the problem are not binary, but integer or real-valued. In such cases a transformation to a binary string is required first. (Note that this is a different situation from the rotor-stacking example, where the integers were merely labels: here the values are assumed to be meaningful as numbers.) While the main thrust of metaheuristics research and application is directed to discrete optimization, it is perhaps appropriate to mention these other problems here.

Example 3.4. *It is required to maximize*

$$f(x) = x^3 - 60x^2 + 900x + 100$$

over the search space $\chi = \{x : x \in \mathbb{Z}; x \in \{0, 31\}\}$, *i.e., the solution* x^* *is required to be an integer in the range* $[0, 31]$.

To use the conventional form of genetic algorithm here, we would use a string of 5 binary digits with the standard binary to integer mapping, i.e., $(0,0,0,0,0) = 0, \ldots, (1,1,1,1,1) = 31$. Of course, in practice we could solve such a problem easily without recourse to encoding the decision variable in this way, but it illustrates neatly the sort of optimization problem to which GAs are often applied. Such problems assume firstly that we know the domain of each of our decision variables, and secondly that we have some idea of the precision with which we need to specify our eventual solution. Given these two ingredients, we can determine the number of bits needed for each decision variable, and concatenate them to form the chromosome. However, problems involving continuous variables can more easily be handled within the ES framework.

13 RANDOM NUMBERS

As GAs are stochastic in nature, it is clear that a reliable random number source is very important. Most computer systems have built-in `rand()` functions, and that is the usual method of generating random numbers. Not all random number generators are reliable, however, as Ross [95] has pointed out, and it is a good idea to use one that has been thoroughly tested, such as those described in the *Numerical Recipes* series [96].

14 CONCLUSIONS

While this exposition has covered the basic principles of GAs, the number of variations that have been suggested is enormous. Probably everybody's GA is unique! Many variations in population size, in initialization methods, in fitness definition, in selection and replacement strategies, in crossover and mutation are obviously possible. Some have added information such as age, or artificial tags, to chromosomes; others have allowed varying population sizes, or induced the formation of multiple populations in 'niches'. It is in the nature of GAs that parallel processing can often be used to advantage, and here again, there are many possibilities, ranging from simple parallelization of function evaluations in a generational GA, to very sophisticated implementations that add a spatial aspect to the algorithm.

The GA community has yet to reach a consensus on any of these things, and in the light of the NFLT, this is perhaps not surprising. However, some ideas do emerge as a reasonable set of recommendations. From a practitioner's viewpoint, Levine [72] made the following observations:

1. A steady-state (or incremental) approach is generally more effective and efficient than a generational method.
2. Don't use simple roulette wheel selection. Tournament selection or SUS is better.
3. Don't use one-point crossover. UX or 2X should be preferred.
4. Make use of an adaptive mutation rate—one that is fixed throughout the search (even at $1/l$) is too inflexible.
5. Hybridize wherever possible; don't use a GA as a black box, but make use of any problem-specific information that you have.

Not everyone will agree with this particular list, and there is a conflict inherent in the first two points, since SUS functions best in a generational setting. Broadly speaking, however, it is one with which many researchers could be comfortable. Two other points could be added:

6. Make diversity maintenance a priority.
7. Don't be afraid to run the GA several times.

Why this last point? Statements are frequently made that GAs can find global optima. Well, they can—but in practice there are many other 'attractors' to which they may converge, and these may be some distance from global optimality, so it makes sense to explore several alternatives.

In conclusion, this chapter has presented a brief survey of the history, theory and issues of implementation relating to one of the most popular metaheuristics. It is hoped that sufficient guidance has been given for newcomers to the field, while also perhaps suggesting new avenues to explore for those who have been acquainted with GAs for some time.

BIBLIOGRAPHY

[1] J.H. Holland (1975) *Adaptation in Natural and Artificial Systems*, University of Michigan Press, Ann Arbor, Michigan; re-issued by MIT Press (1992).

[2] I. Rechenberg (1973) *Evolutionsstrategie: Optimierung technischer Systeme nach Prinzipen der biologischen Evolution*, Frommmann-Holzboog Verlag, Stuttgart (2nd edition 1993).

[3] H.-P. Schwefel (1977) *Numerische Optimierung von Computer-modellen mittels der Evolutionsstrategie*, Birkhäuser Verlag, Basel. (English edition: *Numerical Optimization of Computer Models*, John Wiley & Sons, Chichester, 1981.)

[4] D.B. Fogel (1998) *Evolutionary Computation: The Fossil Record*, IEEE Press, Piscataway, NJ.

[5] K.A. De Jong (1975) *An analysis of the behavior of a class of genetic adaptive systems*, Doctoral dissertation, University of Michigan, Ann Arbor, Michigan.

[6] D.E. Goldberg (1989) *Genetic Algorithms in Search, Optimization, and Machine Learning*, Addison-Wesley, Reading, Massachusetts.

[7] K.A. De Jong (1993) Genetic algorithms are NOT function optimizers. In D. Whitley (ed.), *Foundations of Genetic Algorithms 2*. Morgan Kaufmann, San Mateo, CA, pp. 5–18.

[8] S. Lin (1965) Computer solutions of the traveling salesman problem. *Bell Systems Tech. J.*, **44**, 2245–2269.

[9] S.M. Roberts and B. Flores (1966) An engineering approach to the travelling salesman problem. *Man. Sci.*, **13**, 269–288.

[10] C.E. Nugent, T.E. Vollman and J.E. Ruml (1968) An experimental comparison of techniques for the assignment of facilities to locations. *Operations Research*, **16**, 150–173.

[11] C.R. Reeves (1997) Genetic algorithms for the Operations Researcher. *INFORMS Journal on Computing*, **9**, 231–250.

[12] C.R. Reeves and J.E. Rowe (2001) *Genetic Algorithms: Principles and Perspectives*, Kluwer, Norwell, MA.

[13] C.R. Reeves and C.C. Wright (1999) Genetic algorithms and the design of experiments. In L.D. Davis, K. DeJong, M.D. Vose and L.D. Whitley (eds.), *Evolutionary Algorithms: IMA Volumes in Mathematics and its Applications*, Vol. 111. Springer-Verlag, New York, pp. 207–226.

[14] D.H. Wolpert and W.G. Macready (1997) No free lunch theorems for optimization. *IEEE Trans. Ev. Comp*, **1**, 67–82.

[15] W.G. Macready and D.H. Wolpert (1996) *On 2-armed Gaussian Bandits and Optimization*. Technical Report SFI-TR-96-03-009, Santa Fe Institute, Santa Fe, New Mexico.

[16] M. Mitchell, J.H. Holland and S. Forrest (1994) When will a genetic algorithm outperform hill climbing? In J.D. Cowan, G. Tesauro and J. Alspector (eds.), *Advances in Neural Information Processing Systems 6*. Morgan Kaufmann, San Mateo, CA.

[17] M.D. Vose (1993) Modeling simple genetic algorithms. In L.D. Whitley (ed.), *Foundations of Genetic Algorithms 2*. Morgan Kaufmann, San Mateo, CA, 63–73.

[18] D. Whitley (1993) An executable model of a simple genetic algorithm. In L.D. Whitley (ed.), *Foundations of Genetic Algorithms 2*. Morgan Kaufmann, San Mateo, CA, 45–62.

[19] M.D. Vose (1994) A closer look at mutation in genetic algorithms. *Annals of Maths and AI*, **10**, 423–434.

[20] M.D. Vose and A.H. Wright (1995) Stability of vertex fixed points and applications. In D. Whitley and M. Vose (eds.), *Foundations of Genetic Algorithms 3*. Morgan Kaufmann, San Mateo, CA, 103–113.

[21] K.A. De Jong, W.M. Spears and D.F. Gordon (1995) Using Markov chains to analyze GAFOs. In D. Whitley and M. Vose (eds.), *Foundations of Genetic Algorithms 3*, Morgan Kaufmann, San Mateo, CA, 115–137.

[22] J. Rees and G.J. Koehler (1999) An investigation of GA performance results for different cardinality alphabets. In L.D. Davis, K. DeJong, M.D. Vose and L.D. Whitley (eds.), *Evolutionary Algorithms*: IMA Volumes in Mathematics and its Applications, Vol. 111. Springer-Verlag, New York, 191–206.

[23] J.L. Shapiro, A. Prügel-Bennett and M. Rattray (1994) A statistical mechanics formulation of the dynamics of genetic algorithms. *Lecture Notes in Computer Science*, Vol. 865. Springer-Verlag, Berlin, pp. 17–27.

[24] C.C. Peck and A.P. Dhawan (1995) Genetic algorithms as global random search methods: An alternative perspective. *Evolutionary Computation*, **3**, 39–80.

[25] C.R. Reeves (1999) Predictive measures for problem difficulty. In: *Proceedings of 1999 Congress on Evolutionary Computation*, IEEE Press, pp. 736–743.

[26] C.R. Reeves (1994) Genetic algorithms and neighbourhood search. In T.C. Fogarty (ed.), *Evolutionary Computing: AISB Workshop, Leeds, UK, April 1994; Selected Papers*. Springer-Verlag, Berlin.

[27] T.C. Jones (1995) *Evolutionary Algorithms, Fitness Landscapes and Search*, Doctoral dissertation, University of New Mexico, Albuquerque, NM.

[28] J.C. Culberson (1995) Mutation-crossover isomorphisms and the construction of discriminating functions. *Evolutionary Computation*, **2**, 279–311.

[29] P.F. Stadler and G.P. Wagner (1998) Algebraic theory of recombination spaces. *Evolutionary Computation*, **5**, 241–275.

[30] C.R. Reeves (2000) Fitness landscapes and evolutionary algorithms. In C. Fonlupt, J.-K. Hao, E. Lutton, E. Ronald and M. Schoenauer (eds.), *Artificial Evolution: 4th European Conference; Selected Papers*. Springer-Verlag, Berlin, pp. 3–20.

[31] L. Davis (ed.) (1991) *Handbook of Genetic Algorithms*, Van Nostrand Reinhold, New York.

[32] L. Chambers (ed.) (1995) *Practical Handbook of Genetic Algorithms: Applications, Volume I*, CRC Press, Boca Raton, Florida.

[33] L. Chambers (ed.) (1995) *Practical Handbook of Genetic Algorithms: New Frontiers, Volume II*, CRC Press, Boca Raton, Florida.

[34] J.T. Alander (1996) *An Indexed Bibliography of Genetic Algorithms*. In J.T. Alander (ed.), *Proceedings of the 2nd Nordic Workshop on Genetic Algorithms and their Applications*. University of Vaasa Press, Vaasa, Finland, pp. 293–350.

[35] Z. Michalewicz (1996) *Genetic Algorithms + Data Structures = Evolution Programs* (3rd edition), Springer-Verlag, Berlin.

[36] C.R. Reeves (ed.) (1993) *Modern Heuristic Techniques for Combinatorial Problems*, Blackwell Scientific Publications, Oxford, UK; re-issued by McGraw-Hill, London, UK (1995).

[37] M. Mitchell (1996) *An Introduction to Genetic Algorithms*, MIT Press, Cambridge, MA.

[38] E. Falkenauer (1998) *Genetic Algorithms and Grouping Problems*, John Wiley & Sons, Chichester, UK.

[39] Th. Bäck (1996) *Evolutionary Algorithms in Theory and Practice: Evolution Strategies, Evolutionary Programming, Genetic Algorithms*, Oxford University Press, Oxford.

[40] M.D. Vose (1999) *The Simple Genetic Algorithm: Foundations and Theory*, MIT Press, Cambridge, MA.

[41] J.J. Grefenstette (ed.) (1985) *Proc. of an International Conference on Genetic Algorithms and their applications*. Lawrence Erlbaum Associates, Hillsdale, NJ.

[42] J.J. Grefenstette (ed.) (1987) *Proceedings of the 2nd International Conference on Genetic Algorithms*. Lawrence Erlbaum Associates, Hillsdale, NJ.

[43] J.D. Schaffer (ed.) (1989) *Proceedings of 3rd International Conference on Genetic Algorithms*, Morgan Kaufmann, San Mateo, CA.

[44] R.K. Belew and L.B. Booker (eds.) (1991) *Proceedings of 4th International Conference on Genetic Algorithms*, Morgan Kaufmann, San Mateo, CA.

[45] S. Forrest (ed.) (1993) *Proceedings of 5th International Conference on Genetic Algorithms*, Morgan Kaufmann, San Mateo, CA.

[46] L.J. Eshelman (ed.) (1995) *Proceedings of 6th International Conference on Genetic Algorithms*, Morgan Kaufmann, San Mateo, CA.

[47] Th. Bäck (ed.) (1997) *Proceedings of 7th International Conference on Genetic Algorithms*, Morgan Kaufmann, San Francisco, CA.

[48] H.-P. Schwefel and R. Männer (eds.) (1991) *Parallel Problem-Solving from Nature*, Springer-Verlag, Berlin.

[49] R. Männer and B. Manderick (eds.) (1992) *Parallel Problem-Solving from Nature, 2*, Elsevier Science Publishers, Amsterdam.

[50] Y. Davidor, H.-P. Schwefel and R. Männer (eds.) (1994) *Parallel Problem-Solving from Nature, 3*, Springer-Verlag, Berlin.

[51] H.-M. Voigt, W. Ebeling, I. Rechenberg and H.-P. Schwefel (eds.) (1996) *Parallel Problem-Solving from Nature, 4*, Springer-Verlag, Berlin.

[52] A.E. Eiben, T. Bäck, M. Schoenauer, H.-P. Schwefel (eds.) *Parallel Problem-Solving from Nature, 5*, Springer-Verlag, Berlin.

[53] M. Schoenauer, K. Deb, G. Rudolph, X. Yao, E. Lutton, J.J. Merelo and H.-P. Schwefel (eds.) (2000) *Parallel Problem-Solving from Nature, 6*, Springer-Verlag, Berlin.

[54] R.F. Albrecht, C.R. Reeves and N.C. Steele (eds.) (1993) *Proceedings of the International Conference on Artificial Neural Networks and Genetic Algorithms*, Springer-Verlag, Vienna.

[55] D.W. Pearson, N.C. Steele and R.F. Albrecht (eds.) (1995) *Proceedings of the 2nd International Conference on Artificial Neural Networks and Genetic Algorithms*, Springer-Verlag, Vienna.

[56] G.D. Smith, N.C. Steele and R.F. Albrecht (eds.) (1997) *Proceedings of the 3rd International Conference on Artificial Neural Networks and Genetic Algorithms*, Springer-Verlag, Vienna.

[57] A. Dobnikar, N.C. Steele, D.W. Pearson and R.F. Albrecht (eds.) (1999) *Proceedings of the 4th International Conference on Artificial Neural Networks and Genetic Algorithms*, Springer-Verlag, Vienna.

[58] G.J.E. Rawlins (ed.) (1991) *Foundations of Genetic Algorithms*, Morgan Kaufmann, San Mateo, CA.

[59] L.D. Whitley (ed.) (1993) *Foundations of Genetic Algorithms 2*, Morgan Kaufmann, San Mateo, CA.

[60] D. Whitley and M. Vose (eds.) (1995) *Foundations of Genetic Algorithms 3*, Morgan Kaufmann, San Mateo, CA.

[61] R.K. Belew and M.D. Vose (eds.) (1997) *Foundations of Genetic Algorithms 4*, Morgan Kaufmann, San Francisco, CA.

[62] W. Banzhaf and C.R. Reeves (eds.) (1999) *Foundations of Genetic Algorithms 5*, Morgan Kaufmann, San Francisco, CA.

[63] W. Martin and W. Spears (eds.) (2001) *Foundations of Genetic Algorithms 6*, Morgan Kaufmann, San Francisco, CA.

[64] D.E. Goldberg (1985) *Optimal initial population size for binary-coded genetic algorithms*. TCGA Report 85001, University of Alabama, Tuscaloosa.

[65] D.E. Goldberg (1989) Sizing populations for serial and parallel genetic algorithms. In [43], 70–79.

[66] J.J. Grefenstette (1986) Optimization of control parameters for genetic algorithms. *IEEE-SMC*, **SMC-16**, 122–128.

[67] J.D. Schaffer, R.A. Caruana, L.J. Eshelman and R. Das (1989) A study of control parameters affecting online performance of genetic algorithms for function optimization. In J.D. Schaffer (ed.), *Proceedings of 3rd International Conference on Genetic Algorithms*, Morgan Kaufmann, Los Altos, CA, pp. 51–60.

[68] C.R. Reeves (1993) Using genetic algorithms with small populations. In S. Forrest (ed.) *Proceedings of 5th International Conference on Genetic Algorithms*, Morgan Kaufmann, San Mateo, CA, pp. 92–99.

[69] C.R. Reeves (1995) A genetic algorithm for flowshop sequencing. *Computers & Operations Research*, **22**, 5–13.

[70] R.K. Ahuja and J.B. Orlin (1997) Developing fitter GAs. *INFORMS Journal on Computing*, **9**, 251–253.

[71] A. Kapsalis, G.D. Smith and V.J. Rayward-Smith (1993) Solving the graphical steiner tree problem using genetic algorithms. *Journal of Operational Research Society*, **44**, 397–406.

[72] D. Levine (1997) GAs: A practitioner's view. *INFORMS Journal on Computing*, **9**, 256–257.

[73] J.E. Baker (1987) Reducing bias and inefficiency in the selection algorithm. In J.J. Grefenstette (ed.), *Proceedings of the 2nd International Conference on Genetic Algorithms*, Lawrence Erlbaum Associates, Hillsdale, New Jersey, 14–21.

[74] S.L. Lohr (1999) *Sampling: Design and Analysis*, Duxbury Press, Pacific Grove, CA.

[75] P.J.B. Hancock (1994) An empirical comparison of selection methods in evolutionary algorithms. In T.C. Fogarty (ed.), *Evolutionary Computing: AISB Workshop, Leeds, UK, April 1994; Selected Papers*. Springer-Verlag, Berlin, pp. 80–94.

[76] D. Whitley (1989) The GENITOR algorithm and selection pressure: Why rank-based allocation of reproductive trials is best. In J.D. Schaffer (ed.), *Proceedings of 3rd International Conference on Genetic Algorithms*. Morgan Kaufmann, Los Altos, CA, pp. 116–121.

[77] E. Saliby (1990) Descriptive sampling: A better approach to Monte Carlo simulation. *Journal of Operational Research Society*, **41**, 1133–1142.

[78] A. Nijenhuis and H.S. Wilf (1978) *Combinatorial Algorithms for Computers and Calculators*. Academic Press, New York.

[79] S.S. Skiena (2000) *The Stony Brook Algorithm Repository*, http://www.cs.sunysb.edu/algorith/index.html

[80] L.J. Eshelman, R.A. Caruana and J.D. Schaffer (1989) Biases in the crossover landscape. In [43], 10–19.

[81] G. Syswerda (1989) Uniform crossover in genetic algorithms. In [43], 2–9.

[82] K.A. De Jong and W.M. Spears (1992) A formal analysis of the role of multi-point crossover in genetic algorithms. *Annals of Maths. and AI*, **5**, 1–26.

[83] L.B. Booker (1987) Improving search in genetic algorithms. In L. Davis (ed.), *Genetic Algorithms and Simulated Annealing*. Morgan Kauffmann, Los Altos, CA, pp. 61–73.

[84] D.E. Goldberg and R. Lingle (1985) Alleles, loci and the traveling salesman problem. In J.J. Grefenstette (ed.), *Proceedings of an International Conference on Genetic Algorithms and Their Applications*. Lawrence Erlbaum Associates, Hillsdale, New Jersey, pp. 154–159.

[85] H.J. Bremermann, J. Rogson and S. Salaff (1964) Global properties of evolution processes. In H.H. Pattee (ed.), *Natural Automata and Useful Simulations*, pp. 3–42.

[86] D.B. Fogel (1999) An overview of evolutionary programming. In L.D. Davis, K. DeJong, M.D. Vose and L.D. Whitley (eds.), *Evolutionary Algorithms: IMA Volumes in Mathematics and its Applications*, Vol. 111. Springer-Verlag, New York, pp. 89–109.

[87] T.C. Fogarty (1989) Varying the probability of mutation in the genetic algorithm. In J.D. Schaffer (ed.), *Proceedings of 3rd International Conference on Genetic Algorithms*. Morgan Kaufmann, Los Altos, CA, 104–109.

[88] D.E. Goldberg and K. Deb (1991) A comparative analysis of selection schemes used in genetic algorithms. In G.J.E. Rawlins (ed.), *Foundations of Genetic Algorithms*. Morgan Kaufmann, San Mateo, CA.

[89] N.J. Radcliffe and F.A.W. George (1993) A study in set recombination. In S. Forrest (ed.), *Proceedings of 5th International Conference on Genetic Algorithms*. Morgan Kaufmann, San Mateo, CA, 23–30.

[90] S. McKee and M.B. Reed (1987) An algorithm for the alignment of gas turbine components in aircraft. *IMA J Mathematics in Management*, **1**, 133–144.

[91] N.J. Radcliffe and P. Surry (1995) Formae and the variance of fitness. In D. Whitley and M. Vose (eds.), *Foundations of Genetic Algorithms 3*. Morgan Kaufmann, San Mateo, CA, pp. 51–72.

[92] B.R. Fox and M.B. McMahon (1991) Genetic operators for sequencing problems. In G.J.E. Rawlins (ed.), *Foundations of Genetic Algorithms*. Morgan Kaufmann, San Mateo, CA, pp. 284–300.

[93] P.W. Poon and J.N. Carter (1995) Genetic algorithm crossover operators for ordering applications. *Computers & Operations Research*, **22**, 135–147.

[94] C.R. Reeves and T. Yamada (1998) Genetic algorithms, path relinking and the flowshop sequencing problem. *Evolutionary Computation*, **6**, 45–60.

[95] P. Ross (1997) `srandom()` anomaly. *Genetic Algorithms Digest*, http://www.aic.nrl.navy.mil/galist/ **11:23**.

[96] W.H. Press, S.A. Teukolsky, W.T. Vetterling and B.P. Flannery (1992) *Numerical Recipes in C: The Art of Scientific Computing*, Cambridge University Press, Cambridge, UK.

Chapter 4

GENETIC PROGRAMMING: AUTOMATIC SYNTHESIS OF TOPOLOGIES AND NUMERICAL PARAMETERS

John R. Koza

1 INTRODUCTION

Many mathematical algorithms are capable of solving problems by producing optimal (or near-optimal) numerical values for a prespecified set of parameters. However, for many practical problems, one cannot begin a search for the set of numerical values until one first ascertains the number of numerical values that one is seeking.

In fact, many practical problems of design and optimization entail first discovering an entire graphical structure (i.e., a topology). After the topology is identified, optimal (or near-optimal) numerical values can be sought for the elements of the structure.

For example, if one is seeking an analog electrical circuit whose behavior satisfies certain prespecified high-level design goals, one must first ascertain the circuit's topology and then discover the numerical value of each electrical component in the circuit.

Specifically, the *topology* of an electrical circuit comprises

- the total number of electrical components, in the circuit,
- the type of each component (e.g., resistor, capacitor, transistor) at each location in the circuit, and
- a list of all the connections between the leads of the components.

The *sizing* of a circuit consists of the component value(s) for each component of the circuit that requires a component value.

The automatic synthesis of the topology and sizing of analog electrical circuits is a vexatious problem. As Aaserud and Nielsen (1995) noted

> [M]ost . . . analog circuits are still handcrafted by the experts or so-called 'zahs' of analog design. The design process is characterized by a combination of experience and intuition and requires a thorough knowledge of the process characteristics and the detailed specifications of the actual product.
>
> Analog circuit design is known to be a knowledge-intensive, multiphase, iterative task, which usually stretches over a significant period of

time and is performed by designers with a large portfolio of skills. It is therefore considered by many to be a form of art rather than a science.

The purpose of a controller is to force, in a meritorious way, the actual response of a system (conventionally called the *plant*) to match a desired response (called the *reference signal*) (Dorf and Bishop, 1998). Controllers are typically composed of signal processing blocks, such as integrators, differentiators, leads, lags, delays, gains, adders, inverters, subtractors, and multipliers.

A similarly vexatious situation arises if one is seeking the design of a controller whose behavior satisfies certain prespecified high-level design goals.

The topology of a controller comprises

- the total number of signal processing blocks in the controller,
- the type of each block (e.g., integrator, differentiator, lead, lag, delay, gain, adder, inverter, subtractor, and multiplier),
- the connections between the inputs and output of each block in the controller and the external input and external output points of the controller.

The *tuning* (sizing) of a controller consists of the parameter values associated with each signal processing block.

A parallel situation arises in connection with networks of chemical reactions (metabolic pathways).

The concentrations of substrates, products, and intermediate substances participating in a network of chemical reactions are modeled by non-linear continuous-time differential equations, including various first-order rate laws, second-order rate laws, power laws, and the Michaelis–Menten equations (Voit, 2000). The concentrations of catalysts (e.g., enzymes) control the rates of many chemical reactions in living things.

The *topology* of a network of chemical reactions comprises

- the total number of reactions,
- the number of substrates consumed by each reaction,
- the number of products produced by each reaction,
- the pathways supplying the substrates (either from external sources or other reactions in the network) to each reaction,
- the pathways dispersing each reaction's products (either to other reactions or external outputs), and
- an identification of whether a particular enzyme acts as a catalyst.

The *sizing* for a network of chemical reactions consists of all the numerical values associated with the network (e.g., the rates of each reaction).

A similarly vexatious situation arises if one is seeking the design of an antenna whose behavior satisfies certain prespecified high-level design goals.

While it might seem difficult or impossible to automatically create both the topology and numerical parameters for a complex structure merely from a high-level statement of the structure's design goals, recent work has demonstrated that genetic programming can automatically create complex structures that exhibit prespecified behavior in fields where the structure's behavior is modeled by differential equations (both linear and non-linear) or by other equations (e.g., Maxwell's equations).

In this chapter, we will demonstrate that a biologically motivated algorithm (genetic programming) can automatically synthesize both a graphical structure (the topology) and a set of optimal or near-optimal numerical values for each element of

- analog electrical circuits (Section 3),
- controllers (Section 4),
- antennas (Section 5), and
- networks of chemical reactions (metabolic pathways) (Section 6).

2 GENETIC PROGRAMMING

Genetic programming progressively breeds a population of computer programs over a series of generations by starting with a primordial ooze of thousands of randomly created computer programs and using the Darwinian principle of natural selection, recombination (crossover), mutation, gene duplication, gene deletion, and certain mechanisms of developmental biology.

Genetic programming breeds computer programs to solve problems by executing the following three steps:

(1) Generate an initial population of compositions (typically random) of the functions and terminals of the problem.

(2) Iteratively perform the following substeps (referred to herein as a generation) on the population of programs until the termination criterion has been satisfied:
   (A) Execute each program in the population and assign it a fitness value using the fitness measure.
   (B) Create a new population of programs by applying the following operations. The operations are applied to program(s) selected from the population with a probability based on fitness (with reselection allowed).
      (i) Reproduction: Copy the selected program to the new population.
      (ii) Crossover: Create a new offspring program for the new population by recombining randomly chosen parts of two selected programs.
      (iii) Mutation: Create one new offspring program for the new population by randomly mutating a randomly chosen part of the selected program.
      (iv) Architecture-altering operations: Select an architecture-altering operation from the available repertoire of such operations and create one new offspring program for the new population by applying the selected architecture-altering operation to the selected program.

(3) Designate the individual program that is identified by result designation (e.g., the best-so-far individual) as the result of the run of genetic programming. This result may be a solution (or an approximate solution) to the problem.

Genetic programming is described in the book *Genetic Programming: On the Programming of Computers by Means of Natural Selection* (Koza, 1992; Koza and Rice, 1992), the book *Genetic Programming II: Automatic Discovery of Reusable Programs* (Koza, 1994a, 1994b), and the book *Genetic Programming III: Darwinian Invention and Problem Solving* (Koza, Bennett, Andre, and Keane, 1999; Koza, Bennett, Andre, Keane, and Brave, 1999).

Genetic programming is an extension of the genetic algorithm (Holland 1975) in which the population being bred consists of computer programs.

Genetic programming starts with an initial population of randomly generated computer programs composed of the given primitive functions and terminals. The programs in the population are, in general, of different sizes and shapes. The creation of the initial random population is a blind random search of the space of computer programs composed of the problem's available functions and terminals.

On each generation of a run of genetic programming, each individual in the population of programs is evaluated as to its fitness in solving the problem at hand. The programs in generation 0 of a run almost always have exceedingly poor fitness for non-trivial problems of interest. Nonetheless, some individuals in a population will turn out to be somewhat more fit than others. These differences in performance are then exploited so as to direct the search into promising areas of the search space. The Darwinian principle of reproduction and survival of the fittest is used to probabilistically select, on the basis of fitness, individuals from the population to participate in various operations. A small percentage (e.g., 9%) of the selected individuals are reproduced (copied) from one generation to the next. A very small percentage (e.g., 1%) of the selected individuals are mutated in a random way. Mutation can be viewed as an undirected local search mechanism. The vast majority of the selected individuals (e.g., 90%) participate in the genetic operation of crossover (sexual recombination) in which two offspring programs are created by recombining genetic material from two parents.

The creation of the initial random population and the creation of offspring by the genetic operations are all performed so as to create syntactically valid, executable programs. After the genetic operations are performed on the current generation of the population, the population of offspring (i.e., the new generation) replaces the old generation. The tasks of measuring fitness, Darwinian selection, and genetic operations are then iteratively repeated over many generations. The computer program resulting from this simulated evolutionary process can be the solution to a given problem or a sequence of instructions for constructing the solution.

Probabilistic steps are pervasive in genetic programming. Probability is involved in the creation the individuals in the initial population, the selection of individuals to participate in the genetic operations (e.g., reproduction, crossover, and mutation), and the selection of crossover and mutation points within parental programs.

The dynamic variability of the size and shape of the computer programs that are created during the run is an important feature of genetic programming. It is often difficult and unnatural to try to specify or restrict the size and shape of the eventual solution in advance.

The individual programs that are evolved by genetic programming are typically multi-branch programs consisting of one or more result-producing branches and zero, one, or more automatically defined functions (subroutines).

The *architecture* of such a multi-branch program involves

(1) the total number of automatically defined functions,

(2) the number of arguments (if any) possessed by each automatically defined function, and

(3) if there is more than one automatically defined function in a program, the nature of the hierarchical references (including recursive references), if any, allowed among the automatically defined functions.

Architecture-altering operations enable genetic programming to automatically determine the number of automatically defined functions, the number of arguments that each possesses, and the nature of the hierarchical references, if any, among such automatically defined functions.

Additional information on genetic programming can be found in books such as Banzhaf, Nordin, Keller, and Francone, 1998; books in the series on genetic programming from Kluwer Academic Publishers such as Langdon, 1998; Ryan, 1999; Wong and Leung, 2000; in edited collections of papers such as the *Advances in Genetic Programming* series of books from the MIT Press (Kinnear, 1994; Angeline and Kinnear, 1996; Spector et al., 1999); in the proceedings of the Genetic Programming Conference held between 1996 and 1998 (Koza et al., 1996, 1997, 1998); in the proceedings of the annual Genetic and Evolutionary Computation Conference (combining the annual Genetic Programming Conference and the International Conference on Genetic Algorithms) held starting in 1999 (Banzhaf et al., 1999; Whitley et al., 2000); in the proceedings of the annual Euro-GP conferences held starting in 1998 (Banzhaf et al., 1998; Poli et al., 1999; Poli et al., 2000); at web sites such as www.genetic-programming.org; and in the Genetic Programming and Evolvable Machines journal (from Kluwer Academic Publishers).

3 AUTOMATIC SYNTHESIS OF ANALOG ELECTRICAL CIRCUITS

Genetic programming is capable of automatically creating both the topology and sizing (component values) for analog electrical circuits composed of transistors, capacitors, resistors, and other components merely by specifying the circuit's behavior.

This automatic synthesis of circuits is performed by genetic programming even though there is no general mathematical method (prior to genetic programming) for creating (synthesizing) both the topology and sizing (component values) of analog electrical circuits from the circuit's desired (or observed) behavior (Aaserud and Nielsen, 1995; Koza et al., 1999).

For purposes of illustration, we discuss

- a lowpass filter circuits using a fitness measure based on the frequency-domain behavior of circuits, and
- a computational circuit employing transistors and using a fitness measure based on the time-domain behavior of circuits.

3.1.1 Lowpass Filter Circuit

A simple *filter* is a one-input, one-output electronic circuit that receives a signal as its input and passes the frequency components of the incoming signal that lie in a specified range (called the *passband*) while suppressing the frequency components that lie in all other frequency ranges (the *stopband*).

A *lowpass filter* passes all frequencies below a certain specified frequency, but stops all higher frequencies.

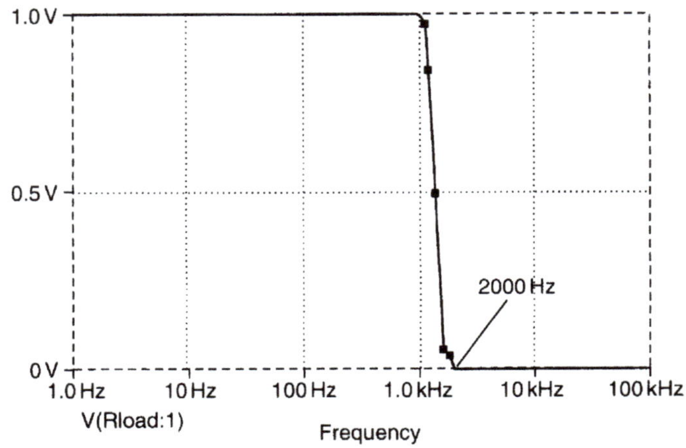

Figure 4.1. Frequency domain behavior of a lowpass filter.

Figure 4.1 shows the frequency domain behavior of an illustrative lowpass filter in which the boundary of the passband is at 1,000 Hz and the boundary of the stopband is 2,000 Hz. The horizontal axis represents the frequency of the incoming signal and ranges over five decades of frequencies between 1 and 100,000 Hz on a logarithmic scale. The vertical axis represents the peak voltage of the output and ranges between 0 to 1 V on a linear scale. This figure shows that when the input to the circuit consists of a sinusoidal signal with any frequency from 1 to 1,000 Hz, the output is a sinusoidal signal with an amplitude of a full 1 V. This figure also shows that when the input to the circuit consists of a sinusoidal signal with any frequency from 2,000 to 100,000 Hz, the amplitude of the output is essentially 0 V. The region between 1,000 and 2,000 Hz is a transition region where the voltage varies between 1 V (at 1,000 Hz) and essentially 0 V (at 2,000 Hz).

Genetic programming is capable of automatically creating both the topology and sizing (component values) for lowpass filters (and other filter circuits, such as high-pass filters, bandpass filters, bandstop filters, and filters with multiple passbands and stopbands).

A filter circuit may be evolved using a fitness measure based on frequency domain behavior. In particular, the fitness of an individual circuit is the sum, over 101 values of frequency between 1 and 100,000 Hz (equally space on a logarithmic scale), of the absolute value of the difference between the individual circuit's output voltage and the ideal voltage for an ideal lowpass filter for that frequency (i.e., the voltages shown in Figure 4.1).

For example, one run of genetic programming synthesized the lowpass filter circuit of Figure 4.2.

The evolved circuit of Figure 4.2 is what is now called a cascade (ladder) of identical π sections (Koza et al., 1999, chapter 25). The evolved circuit has the recognizable topology of the circuit for which George Campbell of American Telephone and Telegraph received U.S. patent 1,227,113 in 1917. Claim 2 of Campbell's

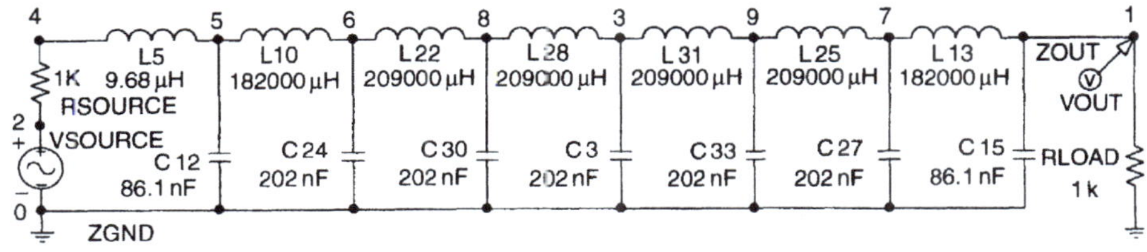

Figure 4.2. Lowpass filter created by genetic programming that infringes on Campbell's patent

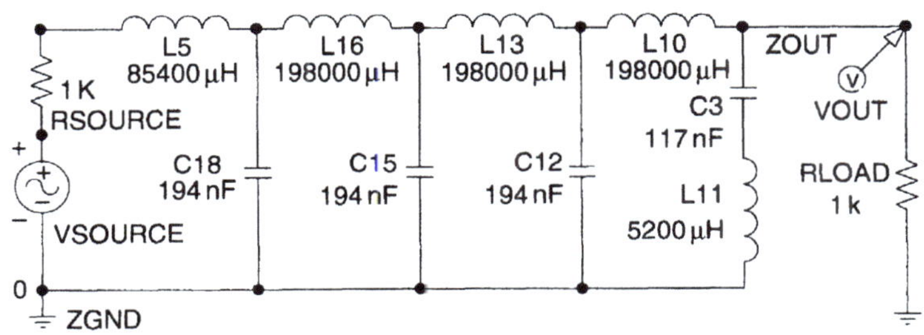

Figure 4.3. Lowpass filter created by genetic programming that infringes on Zobel's patent.

patent covered,

> An electric wave filter consisting of a connecting line of negligible attenuation composed of a plurality of sections, each section including a capacity element and an inductance element, one of said elements of each section being in series with the line and the other in shunt across the line, said capacity and inductance elements having precomputed values dependent upon the upper limiting frequency and the lower limiting frequency of a range of frequencies it is desired to transmit without attenuation, the values of said capacity and inductance elements being so proportioned that the structure transmits with practically negligible attenuation sinusoidal currents of all frequencies lying between said two limiting frequencies, while attenuating and approximately extinguishing currents of neighboring frequencies lying outside of said limiting frequencies.

In addition to possessing the topology of the Campbell filter, the numerical values of all the components in the evolved circuit closely approximate the numerical values taught in Campbell's 1917 patent.

Another run of genetic programming synthesized the lowpass filter circuit of Figure 4.3. As before, this circuit was evolved using the previously described fitness measure based on frequency domain behavior.

This evolved circuit differs from the Campbell filter in that its final section consists of both a capacitor and inductor. This filter is an improvement over the Campbell filter because its final section confers certain performance advantages on the circuit. This circuit is equivalent to what is called a cascade of three symmetric T-sections and an M-derived half section (Koza et al., 1999, chapter 25). Otto Zobel of American Telephone and Telegraph Company invented and received a patent for an "M-derived half section" used in conjunction with one or more "constant K" sections. Again, the numerical values of all the components in this evolved circuit closely approximate the numerical values taught in Zobel's 1925 patent.

Seven circuits created using genetic programming infringe on previously issued patents (Koza et al., 1999). Others duplicate the functionality of previously patented inventions in novel ways.

In both of the foregoing examples, genetic programming automatically created both the topology and sizing (component values) of the entire filter circuit by using a fitness measure expressed in terms of the signal observed at the single output point (the probe point labeled VOUT in the figures).

3.1.2 Squaring Computational Circuit

Because filters discriminate on incoming signals based on frequency, the lowpass filter circuit was automatically synthesized using a fitness measure based on the behavior of the circuit in the frequency domain. However, for many circuits, it is appropriate to synthesize the circuit using a fitness measure based on the behavior of the circuit in the time domain.

An analog electrical circuit whose output is a well-known mathematical function (e.g., square, square root) is called a *computational circuit.*

Figure 4.4 shows a squaring circuit composed of transistors, capacitors, and resistors that was automatically synthesized using a fitness measure based on the behavior of the circuit in the time domain (Mydlowec and Koza, 2000).

This circuit was evolved using a fitness measure based on time-varying input signals. In particular, fitness was the sum, taken at certain sampled times for four different time-varying input signals, of the absolute value of the difference between the individual circuit's output voltage and the desired output voltage (i.e., the square of the voltage of the input signal at the particular sampled time).

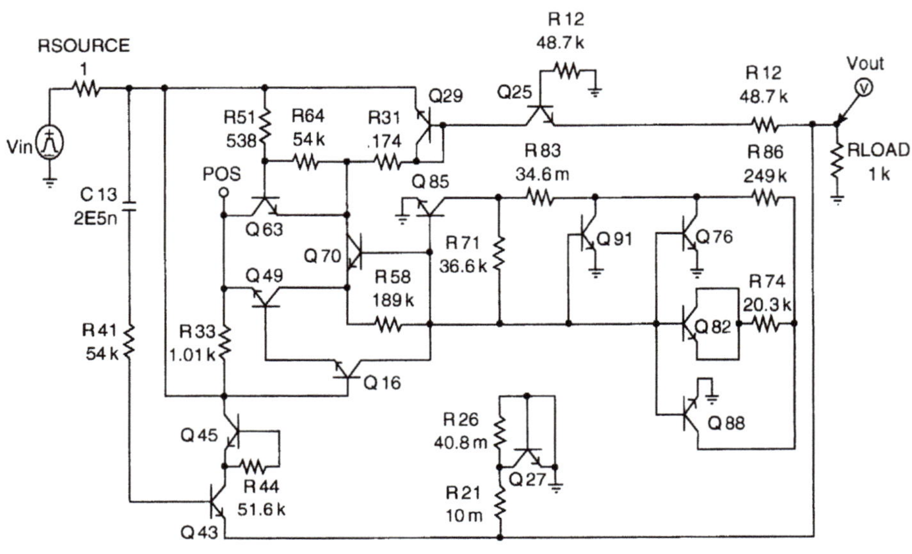

Figure 4.4. Squaring circuit created by genetic programming.

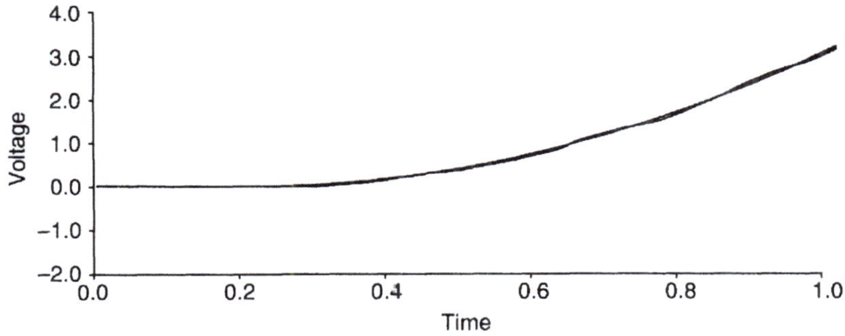

Figure 4.5. Output for rising ramp input for squaring circuit.

The four input signals were structured to provide a representative mixture of input values. All of the input signals produce outputs that are well within the range of voltages that can be handled by transistors (i.e., below 4 V). For example, one of the input signals is a rising ramp whose value remains at 0 up to 0.2 s and then rises to 2 V between 0.2 and 1.0 s. Figure 4.5 shows the output voltage produced by the evolved circuit for the rising ramp input superimposed on the (virtually indistinguishable) correct output voltage for the squaring function. As can be seen, as soon as the input signal becomes non-zero, the output is a parabolic-shaped curve representing the square of the incoming voltage.

4 AUTOMATIC SYNTHESIS OF CONTROLLERS

Genetic programming is capable of automatically creating both the topology and sizing (tuning) for controllers composed of time-domain blocks (such as integrators, differentiators, multipliers, adders, delays, gains, leads, and lags) merely by specifying the controller's effect on the to-be-controlled plant (Keane et al., 2000; Koza et al., 1999a,b, 2000a–d; Yu et al., 2000). This automatic synthesis of controllers from data is performed by genetic programming even though there is no general mathematical method for creating both the topology and sizing for controllers from a high-level statement of the design goals for the controller.

In the PID type of controller, the controller's output is the sum of proportional (P), integrative (I), and derivative (D) terms based on the difference between the plant's output and the reference signal. Albert Callender and Allan Stevenson of Imperial Chemical Limited of Northwich, England received U.S. Patent 2,175,985 in 1939 for the PI and PID controller.

Claim 1 of Callender and Stevenson (1939) covers what is now called the PI controller,

> A system for the automatic control of a variable characteristic comprising means proportionally responsive to deviations of the characteristic from a desired value, compensating means for adjusting the value of the characteristic, and electrical means associated with and actuated by responsive variations in said responsive means, for operating the compensating means

> to correct such deviations in conformity with the sum of the extent of the deviation and the summation of the deviation.

Claim 3 of Callender and Stevenson (1939) covers what is now called the PID controller,

> A system as set forth in claim 1 in which said operation is additionally controlled in conformity with the rate of such deviation.

The vast majority of automatic controllers used by industry are of the PI or PID type. However, it is generally recognized by leading practitioners in the field of control that PI and PID controllers are not ideal (Astrom and Hagglund, 1995; Boyd and Barratt, 1991).

There is no preexisting general-purpose analytic method (prior to genetic programming) for automatically creating both the topology and tuning of a controller for arbitrary linear and non-linear plants that can simultaneously optimize prespecified performance metrics. The performance metrics used in the field of control include, among others,

- minimizing the time required to bring the plant output to the desired value (as measured by, say, the integral of the time-weighted absolute error),
- satisfying time-domain constraints (involving, say, overshoot and disturbance rejection),
- satisfying frequency domain constraints (e.g., bandwidth), and
- satisfying additional constraints, such as limiting the magnitude of the control variable or the plant's internal state variables.

We employ a problem involving control of a two-lag plant (described by Dorf and Bishop 1998, p. 707) to illustrate the automatic synthesis of controllers by means of genetic programming. The problem entails synthesizing the design of both the topology and parameter values for a controller for a two-lag plant such that plant output reaches the level of the reference signal so as to minimize the integral of the time-weighted absolute error, such that the overshoot in response to a step input is less than 2%, and such that the controller is robust in the face of significant variation in the plant's internal gain, K, and the plant's time constant, τ.

Genetic programming routinely creates PI and PID controllers infringing on the 1942 of Callender and Stevenson patent during intermediate generations of runs of genetic programming on controller problems. However, the PID controller is not the best possible controller for this (and many) problems.

Figure 4.6 shows the block diagram for the best-of-run controller evolved during one run of this problem. In this figure, $R(s)$ is the reference signal; $Y(s)$ is the plant output; and $U(s)$ is the controller's output (control variable). This evolved controller is 2.42 times better than the Dorf and Bishop (1998) controller as measured by the criterion used by Dorf and Bishop. In addition, this evolved controller has only 56% of the rise time in response to the reference input, has only 32% of the settling time, and is 8.97 times better in terms of suppressing the effects of disturbance at the plant input.

This genetically evolved controller differs from a conventional PID controller in that it employs a second derivative processing block. Specifically, after applying standard manipulations to the block diagram of this evolved controller, the transfer function for

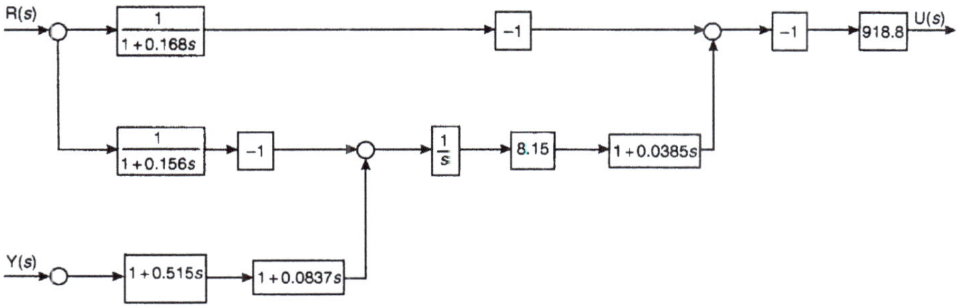

Figure 4.6. Evolved controller that infringes on Jones' patent.

the best-of-run controller can be expressed as a transfer function for a pre-filter and a transfer function for a compensator. The transfer function for the pre-filter, $G_{p32}(s)$, for the best-of-run individual from generation 32 is

$$G_{p32}(s) = \frac{1(1 + .1262s)(1 + .2029s)}{(1 + .03851s)(1 + .05146)(1 + .08375)(1 + .1561s)(1 + .1680s)}$$

The transfer function for the compensator, $G_{c32}(s)$, is

$$G_{c32}(s) = \frac{7487(1 + .03851s)(1 + .05146s)(1 + .08375s)}{s}$$

$$= \frac{7487.05 + 1300.63s + 71.2511s^2 + 1.2426s^3}{s}$$

The s^3 term (in conjunction with the s in the denominator) indicates a second derivative. Thus, the compensator consists of a second derivative in addition to proportional, integrative, and derivative functions. As it happens, Harry Jones of The Brown Instrument Company of Philadelphia received U.S. Patent 2,282,726 for this kind of controller topology in 1942.

Claim 38 of the Jones patent (Jones, 1942) states,

> In a control system, an electrical network, means to adjust said network in response to changes in a variable condition to be controlled, control means responsive to network adjustments to control said condition, reset means including a reactance in said network adapted following an adjustment of said network by said first means to initiate an additional network adjustment in the same sense, and rate control means included in said network adapted to control the effect of the first mentioned adjustment in accordance with the second or higher derivative of the magnitude of the condition with respect to time.

Note that the human user of genetic programming did not preordain, prior to the run (i.e., as part of the preparatory steps for genetic programming), that a second derivative should be used in the controller (or, from that matter, even that a P, I, or D block should be used). Genetic programming automatically discovered that the second derivative element (along with the P, I, and D elements) were useful in producing a good controller for this particular problem. That is, necessity was the mother of invention.

Similarly, the human who initiated this run of genetic programming did not preordain any particular topological arrangement of proportional, integrative, derivative, second derivative, or other functions within the automatically created controller. Instead, genetic programming automatically created a controller for the given plant without the benefit of user-supplied information concerning the total number of processing blocks to be employed in the controller, the type of each processing block, the topological interconnections between the blocks, the values of parameters for the blocks, or the existence of internal feedback (none in this instance) within the controller.

5 AUTOMATIC SYNTHESIS OF ANTENNAS

An antenna is a device for receiving or transmitting electromagnetic waves. An antenna may receive an electromagnetic wave and transform it into a signal on a transmission line. Alternately, an antenna may transform a signal from a transmission line into an electromagnetic wave that is then propagated in free space.

Maxwell's equations govern the electromagnetic waves generated and received by antennas. The behavior and characteristics of many antennas can be determined by simulation. For example, the *Numerical Electromagnetics Code* (NEC) is a method-of-moments (MoM) simulator for wire antennas that was developed at the Lawrence Livermore National Laboratory (Burke, 1992).

The task of analyzing the characteristics of a given antenna is difficult. The task of synthesizing the design of an antenna with specified characteristics typically calls for considerable creativity on the part of the antenna engineer (Balanis, 1982; Stutzman and Thiele, 1998; Linden, 1997).

Genetic programming is capable of discovering both the topological and numerical aspects of a satisfactory antenna design from a high-level specification of the antenna's behavior. In one particular problem (Comisky et al., 2000), genetic programming automatically discovered the design for a satisfactory antenna composed of wires for maximizing gain in a preferred direction over a specified range of frequencies, having a reasonable value of voltage standing wave ratio when the antenna is fed by a transmission line with a specified characteristic impedance, and fitting into a specified bounding rectangle. The design that genetic programming discovered included

(1) the number of directors in the antenna,

(2) the number of reflectors,

(3) the fact that the driven element, the directors, and the reflector are all single straight wires,

(4) the fact that the driven element, the directors, and the reflector are all arranged in parallel,

(5) the fact that the energy source (via the transmission line) is connected only to the driven element—i.e., the directors and reflectors are parasitically coupled.

The last three of the above characteristics discovered by genetic programming are the defining characteristics of an inventive design conceived in the early years of the field of antenna design (Uda, 1926, 1927; Yagi, 1928). Figure 4.7 shows the antenna created by genetic programming. It is an example of what is now called a Yagi-Uda antenna. It is approximately the same length as the conventional Yagi-Uda antenna

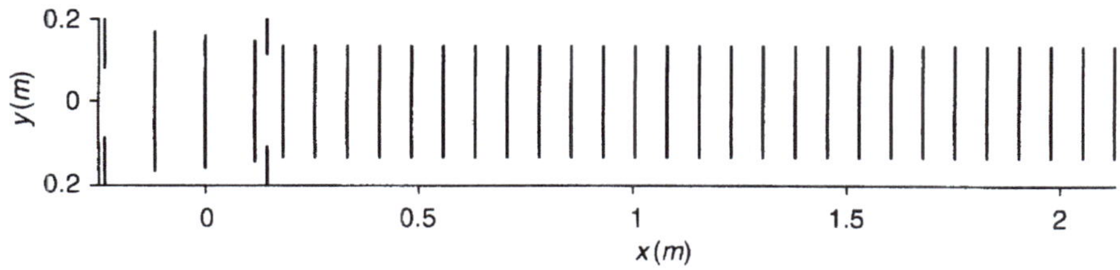

Figure 4.7. Antenna design created by genetic programming.

that a human designer might develop in order to satisfy this problem's requirements (concerning gain).

6 AUTOMATIC SYNTHESIS OF METABOLIC PATHWAYS

A living cell can be viewed as a dynamical system in which a large number of different substances react continuously and non-linearly with one another. In order to understand the behavior of a continuous non-linear dynamical system with numerous interacting parts, it is usually insufficient to study behavior of each part in isolation. Instead, the behavior must usually be analyzed as a whole (Tomita et al., 1999; Voit, 2000).

Biochemists and others have historically determined the topology and sizing of networks of chemical reactions, such as metabolic pathways, through meticulous study of particular networks of interest. However, vast amounts of time-domain data are now becoming available concerning the concentration of biologically important chemicals in living organisms (McAdams and Shapiro, 1995; Loomis and Sternberg, 1995; Arkin et al., 1997; Yuh et al., 1998; Laing et al., 1998; D'haeseleer et al., 1999). Such data include both gene expression data (obtained from microarrays) and data on the concentration of substances participating in metabolic pathways.

The question arises as to whether it is possible to start with observed time-domain concentrations of final product substance(s) and automatically create both the topology of the network of chemical reactions and the sizing of the network. In other words, is it possible to automate the process of reverse engineering a network of chemical reactions from data?

Intuitively, it might seem difficult or impossible to automatically infer both the topology and numerical parameters for a complex network from observed data. However, such intuition may be misleading.

Our approach to the problem of automatically creating both the topology and sizing of a network of chemical reactions involves

(1) establishing a representation for chemical networks involving symbolic expressions (S-expressions) and program trees that are composed of functions and terminals and that can be progressively bred (and improved) by genetic programming,

(2) converting each individual program tree in the population into an analog electrical circuit representing a network of chemical reactions,

(3) obtaining the behavior of the individual network of chemical reactions by simulating the corresponding electrical circuit,

(4) defining a fitness measure that measures how well the behavior of an individual network matches the observed time-domain data concerning concentrations of product substances, and

(5) using the fitness measure to enable genetic programming to breed a population of improving program trees.

6.1 Phospholipid Cycle

The best-of-run individual (Figure 4.8) appears in generation 225. Its fitness is almost zero (0.054). This closely matches the observed data for all data points. In addition to having the same topology as the correct metabolic pathway, the rate constants of three of the four reactions of this network match the correct rates (to three significant digits). The fourth rate is within less than 2% of the correct rate (i.e., the rate of EC 3.1.3.21 is 1.17 compared with 1.19 for the correct network).

Figure 4.9 shows the electrical circuit for the best-of-run individual from generation 225.

In the best-of-run network, the rate of the two-substrate, one-product reaction catalyzed by Triacylglycerol lipase (EC 3.1.1.3) (found at the very bottom of Figures 4.8 and 4.9) that produces the final product diacyl-glycerol (`C00165`) is given by

$$\frac{d[C00165]}{dt} = 1.45[C00162][INT\_2][\text{EC3.1.1.3}].$$

Note that genetic programming has correctly determined that the reaction that produces the network's final product diacyl-glycerol (`C00165`) has two substrates and one product; it has correctly identified enzyme EC3.1.1.3 as the catalyst for this final reaction; it has correctly determined the rate of this final reaction as 1.45; and it has correctly identified the externally supplied substance, fatty acid (C00162), as one of the two substrates for this final reaction.

Of course, genetic programming has no way of knowing that biochemists call the intermediate substance (`INT_2`) by the name Monoacyl-glycerol (`C01885`). It has,

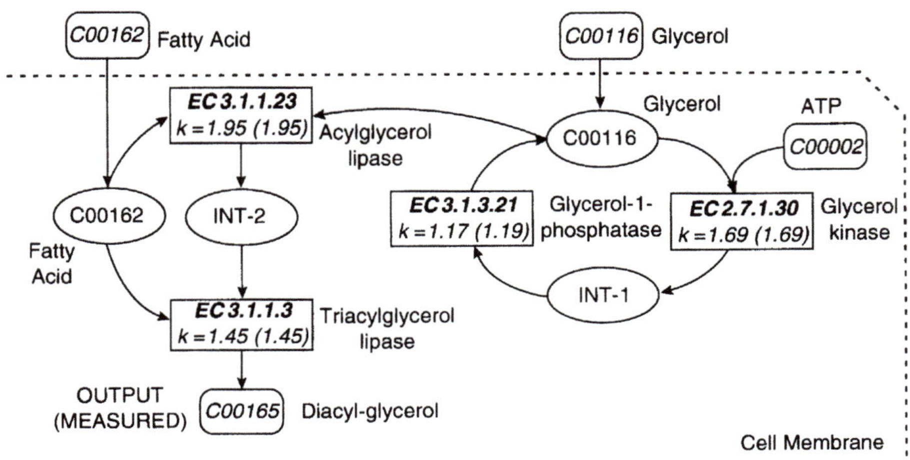

Figure 4.8. Network of chemical reactions for the best-of-run individual from generation 225.

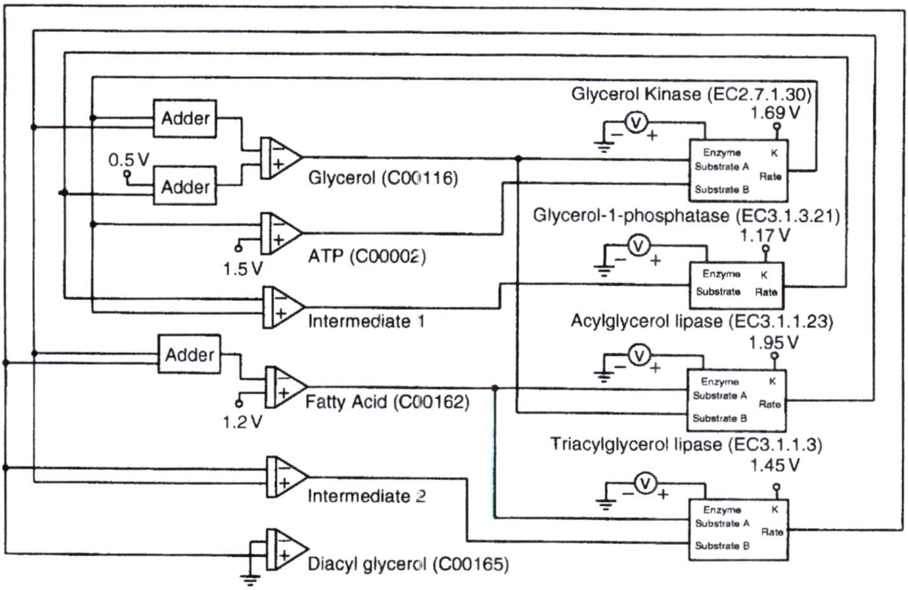

Figure 4.9. Electrical circuit for the best-of-run individual from generation 225.

however, correctly determined that an intermediate substance is needed as one of the two substrates of the network's final reaction and that this intermediate substance should, in turn, be produced by a particular other reaction (described next).

The rate of the two-substrate, one-product reaction catalyzed by Acylglycerol lipase (EC3.1.1.23) that produces intermediate substance `INT_2` is

$$\frac{d[INT\_2]}{dt} = 1.95[C00162][C00116][\text{EC3.1.1.23}] - 1.45[C00162][INT\_2][\text{EC3.1.1.3}].$$

Again, genetic programming has correctly determined that the reaction that produces the intermediate substance (`INT_2`) has two substrates and one product; it has correctly identified enzyme EC3.1.1.23 as the catalyst for this reaction; it has correctly determined the rate of this reaction as 1.95; it has correctly identified two externally supplied substance, fatty acid (C00162) and glycerol (`C00116`), as the two substrates for this reaction.

The rate of the two-substrate, one-product reaction catalyzed by Glycerol kinase (EC2.7.1.30) that produces intermediate substance `INT_1` in the internal loop is

$$\frac{d[INT\_1]}{dt} = 1.69[C00116][C00002][\text{EC2.7.1.30}] - 1.17[INT\_1][\text{EC3.1.3.21}].$$

Note that the numerical rate constant of 1.17 in the above equation is within less than 2% of the correct rate of 1.19.

Here again, genetic programming has correctly determined that the reaction that produces the intermediate substance (`INT_1`) has two substrates and one product; it has correctly identified enzyme EC2.7.1.30 as the catalyst for this reaction; it has almost correctly determined the rate of this reaction to be 1.17 (whereas the correct rate is 1.19); it has correctly identified two externally supplied substance, glycerol (`C00116`) and the cofactor ATP (`C00002`), as the two substrates for this reaction.

Genetic programming has no way of knowing that biochemists call the intermediate substance (`INT_1`) by the name sn-Glycerol-3-Phosphate (`C00093`). Genetic programming has, however, correctly determined that an intermediate substance is needed as the single substrate of the reaction catalyzed by Glycerol-1-phosphatase (EC3.1.3.21) and that this intermediate substance should, in turn, be produced by the reaction catalyzed by Glycerol kinase (EC2.7.1.30).

The rate of supply and consumption of cofactor ATP (`C00002`) is

$$\frac{d[ATP]}{dt} = 1.5 - 1.69[C00116][C00002][\mathrm{EC2.7.1.30}]$$

The rate of supply and consumption of fatty acid (`C00162`) is

$$\begin{aligned}\frac{d[C00162]}{dt} &= 1.2 - 1.95[C00162][C00116][\mathrm{EC3.1.1.23}]\\ &\quad - 1.45[C00162][INT\_2][\mathrm{EC3.1.1.3}].\end{aligned}$$

The rate of supply, consumption, and production of glycerol (`C00116`) is

$$\begin{aligned}\frac{d[C00116]}{dt} &= 0.5 + 1.17[INT\_1][\mathrm{EC3.1.3.21}]\\ &\quad - 1.69[C00116][C00002][\mathrm{EC2.7.1.30}]\\ &\quad - 1.95[C00162][C00116][\mathrm{EC3.1.1.23}]\end{aligned}$$

Again, note that the numerical rate constant of 1.17 in the above equation is slightly different from the correct rate.

Notice the internal feedback loop in which `C00116` is both consumed and produced.

In summary, driven only by the time-domain concentration values of the final product `C00165` (diacyl-glycerol), genetic programming created the entire metabolic pathway, including

- topological features such as the internal feedback loop,
- topological features such as a bifurcation point where one substance is distributed to two different reactions,
- topological features such as an accumulation point where one substance is accumulated from two sources, and
- numerical rates (sizing) for all reactions.

Notice that genetic programming created the entire metabolic pathway, including topological features (such as the internal feedback loop, the bifurcation point, and the accumulation point) and all numerical rate parameter values (sizing) of the metabolic pathway. Genetic programming also determined that two intermediate substances (`INT_1` and `INT_2`) would be used. Genetic programming did this using only the time-domain concentration values of the final product `C00165` (diacyl-glycerol).

Both the topology and sizing of the metabolic pathway were created by using 270 time-domain values of the final product. This example (and the one below) demonstrate

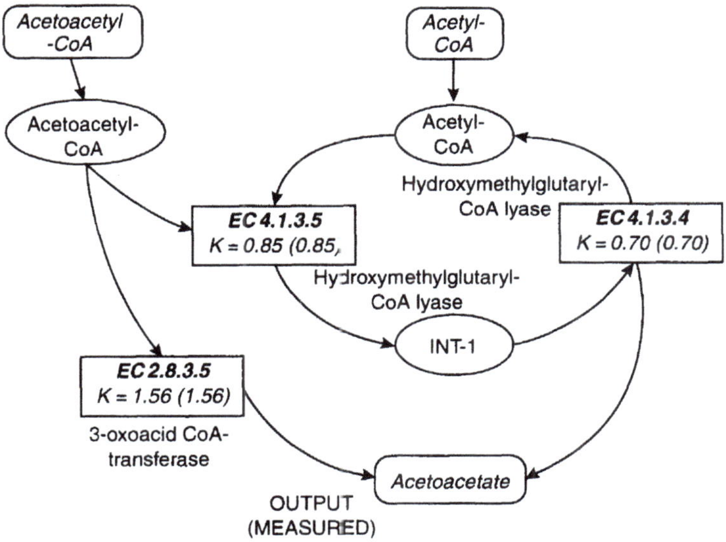

Figure 4.10. Best network of generation 97.

the principle that it is possible to reverse engineer a metabolic pathway from observed data concerning the concentration values of its final output.

For additional details, see Koza et al., 2000.

6.2 Synthesis and Degradation of Ketone Bodies

We proceed in the same way to automatically create a metabolic pathway for the synthesis and degradation of ketone bodies.

The best-of-run network appears in generation 97 (Figure 4.10). It has a fitness of 0.000 and scores 270 hits. This individual has the same topology as the correct metabolic pathway and the same rates (to three significant digits) for each of the three reactions.

In summary, driven only by the time-domain concentration values of Acetoacetate (the final product), the evolutionary process of genetic programming created the entire metabolic pathway, including

- topological features such as the internal feedback loop,
- topological features such as a bifurcation point where one substance is distributed to two different reactions,
- topological features such as an accumulation point where one substance is accumulated from two sources, and
- numerical rates (sizing) for all reactions.

7 CONCLUSIONS

This chapter has demonstrated that a biologically motivated algorithm (genetic programming) is capable of automatically synthesizing both the topology of complex graphical structures and optimal or near-optimal numerical values for all elements of the structure possessing parameters.

REFERENCES

Aaserud, O. and Nielsen, I. Ring (1995) Trends in current analog design: A panel debate. *Analog Integrated Circuits and Signal Processing*, **7**(1), 5–9.

Angeline, Peter J. and Kinnear, Kenneth E. Jr. (Eds.) (1996) *Advances in Genetic Programming 2*. The MIT Press, Cambridge, MA.

Arkin, Adam, Shen, Peidong and Ross, John (1997) A test case of correlation metric construction of a reaction pathway from measurements. *Science*, **277**, 1275–1279.

Astrom, Karl J. and Hagglund, Tore (1995) *PID Controllers: Theory, Design, and Tuning*, 2nd edition. Instrument Society of America, Research Triangle Park, NC.

Balanis, Constantine A. (1982) *Antenna Theory: Analysis and Design*. John Wiley, New York, NY.

Banzhaf, Wolfgang, Daida, Jason, Eiben, A.E., Garzon, Max H., Honavar, Vasant, Jakiela, Mark and Smith, Robert E. (Eds.) (1999) *GECCO-99: Proceedings of the Genetic and Evolutionary Computation Conference, July 13–17, 1999, Orlando, Florida USA*. Morgan Kaufmann, San Francisco, CA.

Banzhaf, Wolfgang, Nordin, Peter, Keller, Robert E. and Francone, Frank D. (1998) *Genetic Programming—An Introduction*. Morgan Kaufmann and Heidelberg: dpunkt, San Francisco, CA.

Banzhaf, Wolfgang, Poli, Riccardo, Schoenauer, Marc and Fogarty, Terence C. (1998). *Genetic Programming: First European Workshop. EuroGP'98. Paris, France, April 1998 Proceedings. Paris, France. April 1998. Lecture Notes in Computer Science*. Vol. 1391. Springer-Verlag, Berlin, Germany.

Bennett, Forrest H. III, Koza, John R., Shipman, James and Stiffelman, Oscar (1999) Building a parallel computer system for $18,000 that performs a half peta-flop per day. In: Banzhaf, Wolfgang, Daida, Jason, Eiben, A.E., Garzon, Max H., Honavar, Vasant, Jakiela, Mark and Smith, Robert E. (Eds.) (1999) *GECCO-99: Proceedings of the Genetic and Evolutionary Computation Conference, July 13–17, 1999, Orlando, Florida USA*. Morgan Kaufmann, San Francisco, CA, pp. 1484–1490.

Boyd, S.P. and Barratt, C.H. (1991) *Linear Controller Design: Limits of Performance*. Prentice Hall, Englewood Cliffs, NJ.

Burke, Gerald J. (1992) *Numerical Electromagnetics Code—NEC-4: Method of Moments—User's Manual*. Lawrence Livermore National Laboratory report UCRL-MA-109338. Lawrence Livermore National Laboratory, Livermore, CA.

Callender, Albert and Stevenson, Allan Brown (1939) *Automatic Control of Variable Physical Characteristics*. United States Patent 2,175,985. Filed February 17, 1936 in United States. Filed February 13, 1935 in Great Britain. Issued October 10, 1939 in United States.

Campbell, George A. (1917) *Electric Wave Filter*. Filed July 15, 1915. U.S. Patent 1,227,113. Issued May 22.

Comisky, William, Yu, Jessen, and Koza, John (2000) Automatic synthesis of a wire antenna using genetic programming. *Late Breaking Papers at the 2000 Genetic and Evolutionary Computation Conference, Las Vegas, Nevada*, pp. 179–186.

D'haeseleer, Patrik, Wen, Xiling, Fuhrman, Stefanie and Somogyi, Roland (1999) Linear modeling of mRNA expression levels during CNS development and injury. In Altman, Russ B. Dunker, A. Keith, Hunter, Lawrence, Klein, Teri E. and Lauderdale, Kevin (Eds.), *Pacific Symposium on Biocomputing '99*. World Scientific, Singapore pp. 41–52.

Dorf, Richard C. and Bishop, Robert H. (1998) *Modern Control Systems*, 8th edition. Addison-Wesley, Menlo Park, CA.

Holland, John H. *Adaptation in Natural and Artificial Systems: An Introductory Analysis with Applications to Biology, Control, and Artificial Intelligence*. University of Michigan Press, Ann Arbor, MI 1975. 2nd edition. The MIT Press, Cambridge, MA.

Jones, Harry S. (1942) *Control Apparatus*. United States Patent 2,282,726. Filed October 25, 1939. Issued May 12.

Keane, Martin A., Yu, Jessen and Koza, John R. (2000) Automatic synthesis of both the topology and tuning of a common parameterized controller for two families of plants using genetic programming. In: Whitley, Darrell, Goldberg, David, Cantu-Paz, Erick, Spector, Lee, Parmee, Ian and Beyer, Hans-Georg (Eds.), *GECCO-2000: Proceedings of the Genetic and Evolutionary Computation Conference, July 10–12, 2000, Las Vegas, Nevada*. Morgan Kaufmann Publishers, San Francisco, pp. 496–504.

Kinnear, Kenneth E. Jr. (Ed.) (1994) *Advances in Genetic Programming*. MIT Press, Cambridge, MA.

Koza, John R. (1992) *Genetic Programming: On the Programming of Computers by Means of Natural Selection*. MIT Press, Cambridge, MA.

Koza, John R. (1994a) *Genetic Programming II: Automatic Discovery of Reusable Programs*. Cambridge, MA: MIT Press.

Koza, John R. (1994b) *Genetic Programming II Videotape: The Next Generation*. MIT Press, Cambridge, MA.

Koza, John R., Banzhaf, Wolfgang, Chellapilla, Kumar, Deb, Kalyanmoy, Dorigo, Marco, Fogel, David B., Garzon, Max H., Goldberg, David E., Iba, Hitoshi and Riolo, Rick (Eds.) (1998) *Genetic Programming 1998: Proceedings of the Third Annual Conference*. Morgan Kaufmann, San Francisco, CA.

Koza, John R., Bennett III, Forrest H., Andre, David and Keane, Martin A. (1999a) *Genetic Programming III: Darwinian Invention and Problem Solving*. Morgan Kaufmann, San Francisco, CA.

Koza, John R., Bennett III, Forrest H., Andre, David, Keane, Martin A. and Brave Scott (1999b) *Genetic Programming III Videotape: Human-Competitive Machine Intelligence*. Morgan Kaufmann, San Francisco, CA.

Koza, John R., Deb, Kalyanmoy, Dorigo, Marco, Fogel, David B., Garzon, Max, Iba, Hitoshi and Riolo, Rick L. (Eds.), *Genetic Programming 1997: Proceedings of the Second Annual Conference, July 13–16, 1997, Stanford University*. Morgan Kaufmann, San Francisco, CA.

Koza, John R., Goldberg, David E., Fogel, David B. and Riolo, Rick L. (Eds) (1996) *Genetic Programming 1996: Proceedings of the First Annual Conference, July 28–31, 1996, Stanford University*. MIT Press, Cambridge, MA.

Koza, John R., Keane, Martin A., Bennett, Forrest H. III, Yu, Jessen, Mydlowec, William and Stiffelman, Oscar (1999c) Automatic creation of both the topology and parameters for a robust controller by means of genetic programming. *Proceedings of the 1999 IEEE International Symposium on Intelligent Control, Intelligent Systems, and Semiotics*. IEEE, Piscataway, NJ, pp. 344–352.

Koza, John R., Keane, Martin A., Yu, Jessen, Bennett, Forrest H. III and Mydlowec, William (2000a) Automatic creation of human-competitive programs and controllers by means of genetic programming. *Genetic Programming and Evolvable Machines*, (1), 121–164.

Koza, John R., Keane, Martin A., Yu, Jessen, Bennett, Forrest H. III, Mydlowec, William and Stiffelman, Oscar (1999d) Automatic synthesis of both the topology and parameters for a robust controller for a non-minimal phase plant and a three-lag plant by means of genetic programming. *Proceedings of 1999 IEEE Conference on Decision and Control*, pp. 5292–5300.

Koza, John R., Keane, Martin A., Yu, Jessen, Mydlowec, William and Bennett, Forrest H. III (2000b) Automatic synthesis of both the topology and parameters for a controller for a three-lag plant with a five-second delay using genetic programming. In: Cagnoni, Stafano et al. (Eds) *Real-World Applications of Evolutionary Computing. EvoWorkshops 2000. EvoIASP, Evo SCONDI, EvoTel, EvoSTIM, EvoRob, and EvoFlight, Edinburgh, Scotland, UK, April 2000 Proceedings. Lecture Notes in Computer Science*, Vol. 1803. Springer-Verlag, Berlin, Germany, pp. 168–177. ISBN 3-540-67353-9.

Koza, John R., Keane, Martin A., Yu, Jessen, Mydlowec, William and Bennett, Forrest H III. (2000c) Automatic synthesis of both the control law and parameters for a controller for a three-lag plant with five-second delay using genetic programming and simulation techniques. In: *Proceedings of the 2000 American Control Conference, Chicago, Illinois, June 28–30, 2000*. American Automatic Control Council, Evanston, IL, pp. 453–459.

Koza, John R., Mydlowec, William, Lanza, Guido, Yu, Jessen and Keane, Martin A. (2000d) *Reverse Engineering and Automatic Synthesis of Metabolic Pathways from Observed Data Using Genetic Programming*. Stanford Medical Informatics Technical Report SMI-2000-0851.

Koza, John R. and Rice, James P. (1992) *Genetic Programming: The Movie*. MIT Press, Cambridge, MA.

Koza, John R., Yu, Jessen, Keane, Martin A. and Mydlowec, William (2000e) Evolution of a controller with a free variable using genetic programming. In: Poli, Riccardo, Banzhaf, Wolfgang, Langdon, William B., Miller, Julian, Nordin, Peter, and Fogarty, Terence C. (Eds), *Genetic Programming: European Conference, EuroGP 2000, Edinburgh, Scotland, UK, April 2000, Proceedings. Lecture Notes in Computer Science*, Vol. 1802. Springer-Verlag, Berlin, Germany, pp. 91–105. ISBN 3-540-67339-3.

Laing, Shoudan, Fuhrman, Stefanie and Somogyi, Roland (1998) REVEAL: A general reverse engineering algorithm for inference of genetic network architecture. In: Altman, Russ B. Dunker, A. Keith, Hunter, Lawrence and Klein, Teri E. (Eds), *Pacific Symposium on Biocomputing '98*. World Scientific, Singapore, pp. 18–29.

Langdon, William B. (1998) *Genetic Programming and Data Structures: Genetic Programming + Data Structures = Automatic Programming!* Amsterdam: Kluwer.

Linden, Derek S. (1997) *Automated Design and Optimization of Wire Antennas Using Genetic Algorithms*. Ph.D. thesis. Department of Electrical Engineering and Computer Science. Massachusetts Institute of Technology.

Loomis, William F. and Sternberg, Paul W. (1995) Genetic networks. *Science*, **269**, 649.

McAdams, Harley H. and Shapiro, Lucy (1995) Circuit simulation of genetic networks. *Science*, **269**, 650–656.

Mittenthal, Jay E., Ao Yuan, Bertrand Clarke, and Scheeline, Alexander (1998) Designing metabolism: Alternative connectivities for the pentose phosphate pathway. *Bulletin of Mathematical Biology*, **60**, 815–856.

Mydlowec, William and Koza, John (2000) Use of time-domain simulations in automatic synthesis of computational circuits using genetic programming. *Late Breaking Papers at the 2000 Genetic and Evolutionary Computation Conference, Las Vegas, Nevada*, pp. 187–197.

Poli, Riccardo, Nordin, Peter, Langdon, William B. and Fogarty, Terence C. (1999) *Genetic Programming: Second European Workshop. EuroGP'99. Proceedings. Lecture Notes in Computer Science*, Vol. 1598. Springer-Verlag, Berlin, Germany.

Quarles, Thomas, Newton, A.R., Pederson, D.O. and Sangiovanni-Vincentelli, A. (1994) *SPICE 3 Version 3F5 User's Manual*. Department of Electrical Engineering and Computer Science, University of California. Berkeley, CA.

Sterling, Thomas L., Salmon, John and Becker, Donald J., and Savarese, Daniel F. (1999) *How to Build a Beowulf: A Guide to Implementation and Application of PC Clusters*. MIT Press, Cambridge, MA.

Stutzman, Warren L. and Thiele, Gary A. (1998) *Antenna Theory and Design*, 2nd edition. John Wiley, New York, NY.

Tomita, Masaru, Hashimoto, Kenta, Takahashi, Kouichi, Shimizu, Thomas Simon, Matsuzaki, Yuri, Miyoshi, Fumihiko, Saito, Kanako, Tanida, Sakura, Yugi, Katsuyuki, Venter, J. Craig, Hutchison and Clyde A. III (1999) E-CELL: Software environment for whole cell simulation. *Bioinformatics*, **15**(1), 72–84.

Uda, S. (1926) Wireless beam of short electric waves. *Journal of the IEE (Japan)*, March 273–282.

Uda, S. (1927) Wireless beam of short electric waves. *Journal of the IEE (Japan)*, March, 1209–1219.

Voit, Eberhard O. (2000) *Computational Analysis of Biochemical Systems*, Cambridge University Press, Cambridge.

Webb, Edwin C. (1992) *Enzyme Nomenclature 1992: Recommendations of the Nomenclature Committee of the International Union of Biochemistry and Molecular Biology*. Academic Press, San Diego, CA.

Whitley, Darrell, Goldberg, David, Cantu-Paz, Erick, Spector, Lee, Parmee, Ian, and Beyer, Hans-Georg (Eds) (2000) *GECCO-2000: Proceedings of the Genetic and Evolutionary Computation Conference, July 10–12, 2000, Las Vegas, Nevada*. Morgan Kaufmann Publishers, San Francisco.

Wong, Man Leung and Leung, Kwong Sak (2000) *Data Mining Using Grammar Based Genetic Programming and Applications*. Kluwer Academic Publishers, Amsterdam. ISBN: 0-7923-7746-X.

Yagi, H. (1928) Beam transmission of ultra short waves. *Proceedings of the IRE*, **26**, 714–741.

Yu, Jessen, Keane, Martin A. and Koza, John R. (2000) Automatic design of both topology and tuning of a common parameterized controller for two families of plants using genetic programming. In: *Proceedings of Eleventh IEEE International Symposium on Computer-Aided Control System Design (CACSD) Conference and Ninth IEEE International Conference on Control Applications (CCA) Conference, Anchorage, Alaska, September 25–27, 2000* (in press).

Yuh, Chiou-Hwa, Bolouri, Hamid and Davidson, Eric H. (1998) Genomic cis-regulatory logic: experimental and computational analysis of a sea urchin gene. *Science*, **279**, 1896–1902.

Zobel, Otto Julius (1925) *Wave Filter*. Filed January 15, 1921. U.S. Patent 1,538,964. Issued May 26.

Chapter 5

A GENTLE INTRODUCTION TO MEMETIC ALGORITHMS

Pablo Moscato
Grupo de Engenharia de Computação em Sistemas Complexos,
Departamento de Engenharia de Computação e Automação Industrial,
Faculdade de Engenharia Eletrónica e de Computação, Universidade Estadual de Campinas,
C.P. 6101, Campinas, SP, CEP 13083-970, Brazil
E-mail: moscato@densis.fee.unicamp.br

Carlos Cotta
Departamento de Lenguajes y Ciencias de la Computación,
Escuela Técnica Superior de Ingeniería Informática, Universidad de Málaga,
Complejo Tecnológico (3.2.49), Campus de Teatinos,
29071-Málaga, Spain
E-mail: ccottap@lcc.uma.es

1 INTRODUCTION AND HISTORICAL NOTES

The generic denomination of *'Memetic Algorithms'* (MAs) is used to encompass a broad class of metaheuristics (i.e., general purpose methods aimed to guide an underlying heuristic). The method is based on a population of agents and proved to be of practical success in a variety of problem domains and in particular for the approximate solution of $\mathcal{NP}$ Optimization problems.

Unlike traditional Evolutionary Computation (EC) methods, MAs are intrinsically concerned with exploiting *all available knowledge* about the problem under study. The incorporation of problem domain knowledge is not an optional mechanism, but a fundamental feature that characterizes MAs. This functioning philosophy is perfectly illustrated by the term "memetic". Coined by Dawkins [52], the word *'meme'* denotes an analogous to the gene in the context of cultural evolution [154]. In Dawkins' words:

> Examples of memes are tunes, ideas, catch-phrases, clothes fashions, ways of making pots or of building arches. Just as genes propagate themselves in the gene pool by leaping from body to body via sperms or eggs, so memes propagate themselves in the meme pool by leaping from brain to brain via a process which, in the broad sense, can be called imitation.

This characterization of a meme suggest that in cultural evolution processes, information is not simply transmitted unaltered between individuals. In contrast, it is processed and enhanced by the communicating parts. This enhancement is accomplished in MAs by incorporating heuristics, approximation algorithms, local search techniques, specialized recombination operators, truncated exact methods, etc. In essence, most MAs can be interpreted as a search strategy in which a population of optimizing agents cooperate and compete [169]. The success of MAs can probably be explained as being a direct consequence of the *synergy* of the different search approaches they incorporate.

The most crucial and distinctive feature of MAs, the inclusion of problem knowledge mentioned above, is also supported by strong theoretical results. As Hart and Belew [88] initially stated and Wolpert and Macready [224] later popularized in the so-called *No-Free-Lunch Theorem*, a search algorithm strictly performs in accordance with the amount and quality of the problem knowledge they incorporate. This fact clearly underpins the exploitation of problem knowledge intrinsic to MAs. Given that the term *hybridization* [42] is commonly used to denote the process of incorporating problem knowledge, it is not surprising that MAs are sometimes called 'Hybrid Evolutionary Algorithms' [51] as well. One of the first algorithms to which the MA label was assigned dates from 1988 [169], and was regarded by many as a hybrid of *traditional* Genetic Algorithms (GAs) and *Simulated Annealing* (SA). Part of the initial motivation was to find a way out of the limitations of both techniques on a well-studied combinatorial optimization problem the MIN EUCLIDEAN TRAVELING SALESMAN problem (MIN ETSP). According to the authors, the original inspiration came from *computer game tournaments* [96] used to study "*the evolution of cooperation*" [6] (see also [164,171] for more recent theoretical results in this field). That approach had several features which anticipated many current algorithms in practice today. The competitive phase of the algorithm was based on the new allocation of search points in configuration phase, a process involving a "*battle*" for survival followed by the so-called "*clonation*", which has a strong similarity with the more recent "*go with the winners*" algorithms [4,182]. The cooperative phase followed by local search may be better named "*go-with-the-local-winners*" since the optimizing *agents* were arranged with a topology of a two dimensional toroidal lattice. After initial computer experiments, an insight was derived on the particular relevance that the "*spatial*" organization, when coupled with an appropriate set of rules, had for the overall performance of population search processes. A few months later, Moscato and Norman discovered that they shared similar views with other researchers [74,157] and other authors proposing "*island models*" for GAs. Spacialization is now being recognized as the "catalyzer" responsible of a variety of phenomena [163,164]. This is an important research issue, currently only understood in a rather heuristic way. However, some proper undecidability results of have been obtained for related problems [80] giving some hope to a more formal treatment.

Less than a year later, in 1989, Moscato and Norman identified several authors who were also pioneering the introduction of heuristics to improve the solutions before recombining them [73,158] (see other references and the discussion in [154]). Particularly coming from the GA field, several authors were introducing *problem-domain knowledge* in a variety of ways. In [154] the denomination of *'memetic algorithms'* was introduced for the first time. It was also suggested that *cultural evolution* can be a better working metaphor for these metaheuristics to avoid *"biologically constrained"* thinking that was restricting progress at that time.

Ten years later, albeit unfortunately under different names, MAs have become an important optimization approach, with several successes in a variety of classical

$\mathcal{NP}$-hard optimization problems. We aim to provide an updated and self-contained introduction to MAs, focusing on their technical innards and formal features, but without loosing the perspective of their practical application and open research issues.

2 MEMETIC ALGORITHMS

Before proceeding to the description of MAs, it is necessary to provide some basic concepts and definitions. Several notions introduced in the first subsection are strongly related to the field of computational complexity. Nevertheless, they may be presented in a slightly different way and pace for the sake of the subsequent development. These basic concepts will give rise to the notions of local search and population-based search, upon which MAs are founded. This latter class of search settles the scenario for *recombination*, a crucial mechanism in the functioning of MAs that will be studied to some depth. Finally, some guidelines for designing MAs will be presented.

2.1 Basic Concepts

An *algorithm* is a detailed step-by-step procedure for solving a *computational problem*. A computational problem P denotes a class of algoritmically-doable tasks, and it has an input domain set of *instances* denoted I_P. For each instance $x \in I_P$, there is an associated set $sol_P(x)$ which denotes the *feasible* solutions for problem P given instance x. The set $sol_P(x)$ is also known as the set of *acceptable* or *valid* solutions.

We are expected to deliver an algorithm that solves problem P; this means that our algorithm, given instance $x \in I_P$, must return at least one element y from a set of *answers* $ans_P(x)$ (also called *given solutions*) that satisfies the requirements of the problem. This is the first design issue to face. To be precise, depending on the kind of answers expected, computational problems can be classified into different categories; for instance:

- finding *all* solutions in $sol_P(x)$, i.e., *enumeration* problems.
- counting *how many* solutions exist in $sol_P(x)$, i.e. *counting* problems.
- determining whether the set $sol_P(x)$ *is empty or not*, i.e., *decision* problems.
- finding a solution in $sol_P(x)$ maximizing or minimizing a given function, i.e., *optimization* problems.

In this chapter, we will focus on the last possibility, that is, a problem will be considered *solved* by finding a certain feasible solution, i.e. either finding an *optimal* $y \in sol_P(x)$ or giving an indication that no such feasible solution exists.It is thus convenient in may situations to define a Boolean *feasibility* function $feasible_P(x, y)$ in order to identify whether a given solution $y \in ans_P(x)$ is acceptable for an instance $x \in I_P$ of a computational problem P, i.e., checking if $y \in sol_P(x)$.

An algorithm is said to *solve* problem P if it can fulfill this condition for any given instance $x \in I_P$.This definition is certainly too broad, so a more restrictive characterization for our problems of interest is necessary. This characterization is provided by restricting ourselves to the so-called *combinatorial optimization* problems. These constitute a special subclass of computational problems in which for each instance $x \in I_P$:

- the cardinality of $sol_P(x)$ is finite.
- each solution $y \in sol_P(x)$ has a *goodness integer value* $m_P(y, x)$, obtained by means of an associated *objective function* m_P.
- a partial order $\prec_P$ is defined over the set of goodness values returned by the objective function, allowing determining which of two goodness values is preferable.

An instance $x \in I_P$ of a combinatorial optimization problem P is solved by finding the best solution $y^* \in sol_P(x)$, i.e., finding a solution y^* such that no other solution $y \prec_P y^*$ exists if $sol_P(x)$ is not empty. It is very common to have $\prec_P$ defining a total order. In this case, the best solution is the one that maximizes (or minimizes) the objective function.

As an example of a combinatorial optimization problem consider the 0-1 MULTIPLE KNAPSACK PROBLEM (0-1 MKP). Each instance x of this problem is defined by a vector of profits $V = \{v_0, \ldots, v_{n-1}\}$, a vector of capacities $C = \{c_0, \ldots, c_{m-1}\}$, and a matrix of capacity constraints $M = \{m_{ij}\colon 0 \leq i < m, 0 \leq j < n\}$. Intuitively, the problem consists in selecting a set of objects so as to maximize the profit of this set without violating the capacity constraints. If the objects are indexed with the elements of the set $\mathbb{N}_n = \{0, 1, \ldots, n-1\}$, the answer set $ans_P(x)$ for an instance x is simply the power set of $\mathbb{N}_n$, i.e., each subset of $\mathbb{N}_n$ is a possible answer. Furthermore, the set of feasible answers $sol_P(x)$ is composed of those subsets whose incidence vector B verifies $M \cdot B \leq C'$. Finally, the objective function is defined as $m_P(y, x) = \sum_{i \in y} v_i$, i.e., the sum of profits for all selected objects, being the goal to maximize this value.

Notice that, associated with a combinatorial optimization problem, we can define its *decisional* version. To formulate the decision problem, an integer goodness value K is considered, and instead of trying to find the best solution of instance x, we ask whether x has a solution whose goodness is equal or better than K. In the above example, we could ask whether a feasible solution y exists such that its associated profit is equal or better than K.

2.2 Search Landscapes

As mentioned above, having defined the concept of combinatorial optimization problem the goal is finding at least one of the optimal solutions for a given instance. For this purpose, a search algorithm must be used. Before discussing search algorithms, three entities must be discussed. These are the *search space*, the *neighborhood relation*, and the *guiding function*. It is important to consider that, for any given computational problem, these three entities can be instantiated in several ways, giving rise to different optimization tasks.

Let us start by defining the concept of search space for a combinatorial problem P. To do so, we consider a set $\mathcal{S}_P(x)$, whose elements have the following properties:

- Each element $s \in \mathcal{S}_P(x)$ represents at least one answer in $ans_P(x)$.
- For decision problems: at least one element of $sol_P(x)$ that stands for a 'Yes' answer must be represented by one element in $\mathcal{S}_P(x)$.

- For optimization problems: at least one *optimal* element y^* of $sol_P(x)$ is represented by one element in $\mathcal{S}_P(x)$.

Each element of $\mathcal{S}_P(x)$ will be termed a *configuration*, being related to an answer in $ans_P(x)$ by a *growth function* $g : \mathcal{S}_P(x) \to ans_P(x)$. Note that the first requirement refers to $ans_P(x)$ and not to $sol_P(x)$, i.e., some configurations in the search space may correspond to infeasible solutions. Thus, the search algorithm may need being prepared to deal with this fact. If these requirements have been achieved, we say that we have a *valid representation* or *valid formulation* of the problem. For simplicity, we will just write $\mathcal{S}$ to refer to $\mathcal{S}_P(x)$ when x and P are clear from the context. People using biologically-inspired metaphors like to call $\mathcal{S}_P(x)$ the *genotype space* and $ans_P(x)$ denotes the *phenotype space*, so we appropriately refer to g as the *growth function*.

To illustrate this notion of search space, consider again the case of the 0–1 MKP. Since solutions in $ans_P(x)$ are subsets of $\mathbb{N}_n$, we can define the search space as the set of n-dimensional binary vectors. Each vector will represent the incidence vector of a certain subset, i.e., the growth function g is defined as $g(s) = g(b_0 b_1 \cdots b_{n-1}) = \{i \,|\, b_i = 1\}$. As mentioned above, many binary vectors may correspond to infeasible sets of objects. Another possibility is defining the search space as the set of permutations of elements in $\mathbb{N}_n$ [77]. In this case, the growth function may consist of applying a greedy construction algorithm [45], considering objects in the order provided by the permutation. Unlike the binary search space previously mentioned, all configurations represent feasible solutions in this case.

The role of the search space is to provide a "ground" on while the search algorithm will act, and of course, indirectly moving in the image set $ans_P(x)$. Important properties of the search space that affect the dynamics of the search algorithm are related with the accessibility relationships between the configurations. These relationships are dependent of a *neighborhood function* $\mathcal{N} : \mathcal{S} \to 2^{\mathcal{S}}$. This function assigns to each element $s \in \mathcal{S}$ a set $\mathcal{N}(s) \subseteq \mathcal{S}$ of neighboring configurations of s. The set $\mathcal{N}(s)$ is called the *neighborhood* of s and each member $s' \in \mathcal{N}(s)$ is called a *neighbor* of s.

It must be noted that the neighborhood depends on the instance, so the notation $\mathcal{N}(s)$ is a simplified form of $\mathcal{N}_P(s, x)$ since it is clear from the context. The elements of $\mathcal{N}(s)$ need not be listed explicitly. In fact, it is very usual to define them *implicitly* by referring to a set of possible *moves*, which define *transitions* between configurations. Moves are usually defined as "*local*" modifications of some part of s, where "locality" refers to the fact that the move is done on a single solution to obtain another single solution. This "locality", is one of the key ingredients of *local search*, and actually it has also given the name to the whole search paradigm.

As examples of concrete neighborhood definitions, consider the two representations of solutions for the 0–1 MKP presented above. In the first case (binary representation), moves can be defined as changing the values of a number of bits. If just one bit is modified at a time, the resulting neighborhood structure is the n-dimensional binary hypercube. In the second case (permutation representation), moves can be defined as the interchange of two positions in the permutation. Thus, two configurations are neighboring if, and only if, they differ in exactly two positions.

This definition of locality presented above is not necessarily related to "closeness" under some kind of distance relationship between configurations (except in the tautological situation in which the distance between two configurations s and s' is defined as the number of moves needed to reach s' from s). As a matter of fact, it is possible to give common examples of very complex neighborhood definitions unrelated to intuitive distance measures.

An important feature that must be considered when selecting the class of moves to be used in the search algorithm is its *"ergodicity"*, that is the ability, given any $s \in S$ to find a sequence of moves that can reach *all other* configuration $s' \in S$. In many situations, this property is self-evident and no explicit demonstration is required. It is important since even if having a valid representation (recall the definition above), it is necessary to guarantee that *a priori* at least one optimal solution is reachable from any given initial solution. Again, consider the binary representation of solutions for a 0–1 MKP instance. If moves are defined as single bit-flips, it is easily seen that any configuration s' can be reached from another configuration s in exactly h moves, where h is the Hamming distance between these configurations. This is not always the case though.

The last entity that must be defined is the *guiding function*. To do so, we require a set $\mathcal{F}$ whose elements are termed *fitness* values (typically $\mathcal{F} \equiv \mathbb{R}$), and a partial order $\prec_{\mathcal{F}}$ on $\mathcal{F}$ (typically, but not always, $\prec_{\mathcal{F}} \equiv <$). The guiding function is defined as a function $F_g : \mathcal{S} \to \mathcal{F}$ that associates to each configuration $s \in \mathcal{S}$ a value $F_g(s)$ that assesses the quality of the solution. The behavior of the search algorithm will be "controlled" by these fitness values.

Notice that for optimization problems there is an obvious direct connection between the guiding function F_g and the objective function m_P. As a matter of fact, it is very common to enforce this relationship to the point that both terms are usually considered equivalent. However, this equivalence is not necessary and, in many situations, not even desirable. For decision problems, since a solution is a 'Yes' or 'No' answer, associated guiding functions usually take the form of *distance to satisfiability*.

A typical example is the BOOLEAN SATISFIABILITY PROBLEM, i.e., determining whether a Boolean expression in conjunctive normal form is satisfiable. In this case, solutions are assignments of Boolean values to variables, and the objective function m_P is a binary function returning 1 if the solution satisfies the Boolean expression, and returning 0 otherwise. This objective function could be used as guiding function. However, a much more typical choice is to use the number of satisfied clauses in the current configuration as guiding function, i.e., $F_g(s) = \sum_i f_i(s)$, the sum over clauses indexes i of $f_i(s)$, defined as $f_i(s) = 0$ for a yet unsatisfied clause i, and $f_i(s) = 1$ if the clause i is satisfied. Hence, the goal is to maximize this number. Notice that the guiding function is in this case the objective function of the associated $\mathcal{NP}$ Optimization problem called MAX SAT.

The above differentiation between objective function and guiding function is also very important in the context of constrained optimization problems, i.e., problems for which, in general, $sol_P(x)$ is chosen to be a proper subset of $ans_P(x)$. Since the growth function establishes a mapping from $\mathcal{S}$ to $ans_P(x)$, the search algorithm might need processing both feasible solutions (whose goodness values are well-defined) and infeasible solutions (whose goodness values are ill-defined in general). In many implementations of MAs for these problems, a guiding function is defined as a weighted sum of the value of the objective function and the distance to feasibility (which accounts for the constraints). Typically, a higher weight is assigned to the constraints, so as to give preference to feasibility over optimality. Several other remedies to this problem abound, including resorting to multi-objective techniques.

The combination of a certain problem instance and the three entities defined above induces a so-called *fitness landscape* [106]. Essentially, a fitness landscape can be defined as a weighted digraph, in which the vertices are configurations of the search space $\mathcal{S}$, and the arcs connect neighboring configurations. The weights are the difference of the guiding function of the endpoint configurations. The search can thus be seen as the process of "navigating" the fitness landscape using the information provided by the guiding function. This is a very powerful metaphor; it allows interpreting in terms of well-known topographical objects such as *peaks*, *valleys*, *mesas*, etc, of great utility to visualize the search progress, and to grasp factors affecting the performance of the process. In particular, the important notion of *local* optimum is associated to this definition of fitness landscape. To be precise, a local optimum is a vertex of the fitness landscape whose guiding function value is better than the values of all its neighbors. Notice that different moves define different neighborhoods and hence different fitness landscapes, even when the same problem instance is considered. For this reason, the notion of local optimum is not intrinsic to a problem instance as it is, sometimes, erroneously considered.

2.3 Local vs. Population-Based Search

The definitions presented in the previous subsection naturally lead to the notion of *local search algorithm*. A local search algorithm starts from a configuration $s_0 \in \mathcal{S}$, generated at random or constructed by some other algorithm. Subsequently, it iterates using at each step a transition based on the neighborhood of the current configuration. Transitions leading to preferable (according to the partial order $\prec_{\mathcal{F}}$) configurations are accepted, i.e., the newly generated configuration turns to be the current configuration in the next step. Otherwise, the current configuration is kept. This process is repeated until a certain termination criterion is met. Typical criteria are the realization of a pre-specified number of iterations, not having found any improvement in the last m iterations, or even more complex mechanisms based on estimating the probability of being at a local optimum [44].

Due to these characteristics, the approach is metaphorically called *"hill climbing"*. The whole process is sketched in Figure 5.1.

The selection of the particular type of moves (also known as *mutation* in the context of GAs) to use does certainly depend on the specific characteristics of the problem and the representation chosen. There is no general advice for this, since it is a matter

```
Procedure Local-Search-Engine (current)
begin
    repeat
        new ← GenerateNeighbor(current);
        if (F_g(new) ≺_F F_g(current)) then
            current ← new;
        endif
    until TerminationCriterion();
    return current;
end
```

Figure 5.1. A local search algorithm.

of the available computer time for the whole process as well as other algorithmic decisions that include ease of coding, etc. In some cases some moves are conspicuous, for example it can be the change of the value of one single variable or the swap of the values of two different variables. Sometimes the "step" may also be composed of a chain of transitions. For instance, in relation with MAs, Radcliffe and Surry introduced the concept of *Binomial Minimal Mutation*, where the number of mutations to perform is selected according to certain binomial distribution [189]. In the context of fitness landscapes, this is equivalent to a redefinition of the neighborhood relation, considering two configurations as neighbors when there exists a chain of transitions connecting them.

Local search algorithms are thus characterized by keeping a single configuration at a time. The immediate generalization of this behavior is the simultaneous maintenance of k, ($k \geq 2$) configurations. The term *population-based* search algorithms has been coined to denote search techniques behaving this way.

The availability of several configurations at a time allows the use of new powerful mechanisms for traversing the fitness landscape in addition to the standard mutation operator. The most popular of these mechanisms, the recombination operator, will be studied in more depth in the next section. In any case, notice that the general functioning of population-based search techniques is very similar to the pseudocode depicted in Figure 5.1. As a matter of fact, a population-based algorithm can be imagined as a procedure in which we sequentially visit vertices of a hypergraph. Each vertex of the hypergraph represents a set of configurations in $\mathcal{S}_P(x)$, i.e., a population. The next vertex to be visited, i.e., the new population, can be established according to the composition of the neighborhoods of the different transition mechanisms used in the population algorithm. Despite the analogy with local search, it is widely accepted in the scientific literature to apply the denomination 'local' just to one-configuration-at-a-time search algorithms. For this reason, the term 'local' will be used with this interpretation in the remainder of the article.

2.4 Recombination

As mentioned in the previous section, local search is based on the application of a mutation operator to a single configuration. Despite the apparent simplicity of this mechanism, "mutation-based" local search has revealed itself a very powerful mechanism for obtaining good quality solutions for $\mathcal{NP}$–hard problems (e.g., see [62,199]). For this reason, some researchers have tried to provide a more theoretically-solid background to this class of search. In this line, it is worth mentioning the definition of the *Polynomial Local Search* class (PLS) by Johnson et al. [104]. Basically, this complexity class comprises a problem and an associated search landscape such that we can decide in polynomial time if we can find a better solution in the neighborhood. Unfortunately, it is very likely that no $\mathcal{NP}$–hard problem is contained in class PLS, since that would imply that $\mathcal{NP}$ = co-$\mathcal{NP}$ [226], a conjecture usually assumed to be false. This fact has justified the quest for additional search mechanisms to be used as stand-alone operators or as complements to standard mutation.

In this line, recall that population-based search allowed the definition of generalized move operators termed *recombination* operators. In essence, recombination can be defined as a process in which a set S_{par} of n configurations (informally referred to as "parents") is manipulated to create a set $S_{desc} \subseteq sol_P(x)$ of m new configurations (informally termed "descendants"). The creation of these descendants involves the identification and combination of features extracted from the parents.

At this point, it is possible to consider properties of interest that can be exhibited by recombination operators [189]. The first property, *respect*, represents the exploitative side of recombination. A recombination operator is said to be *respectful*, regarding a particular type of features of the configurations, if, and only if, it generates descendants carrying all basic features common to all parents. Notice that, if all parent configurations are identical, a respectful recombination operator is obliged to return the same configuration as a descendant. This property is termed *purity*, and can be achieved even when the recombination operator is not generally respectful.

On the other hand, *assortment* represents the exploratory side of recombination. A recombination operator is said to be *properly assorting* if, and only if, it can generate descendants carrying any combination of compatible features taken from the parents. The assortment is said to be *weak* if it is necessary to perform several recombinations within the offspring to achieve this effect.

Finally, *transmission* is a very important property that captures the intuitive rôle of recombination. An operator is said to be transmitting if every feature exhibited by the offspring is present in at least one of the parents. Thus, a transmitting recombination operator combines the information present in the parents but does not introduce new information. This latter task is usually left to the mutation operator. For this reason, a non-transmitting recombination operator is said to introduce *implicit mutation*.

The three properties above suffice to describe the abstract input/output behaviour of a recombination operator regarding some particular features. It provides a characterization of the possible descendants that can be produced by the operator. Nevertheless, there exist other aspects of the functioning of recombination that must be studied. In particular, it is interesting to consider how the construction of $\mathcal{S}_{desc}$ is approached.

First of all, a recombination operator is said to be *blind* if it has no other input than $\mathcal{S}_{par}$, i.e., it does not use any information from the problem instance. This definition is certainly very restrictive, and hence is sometimes relaxed as to allow the recombination operator use information regarding the problem constraints (so as to construct feasible descendants), and possibly the fitness values of configurations $y \in \mathcal{S}_{par}$ (so as to bias the generation of descendants to the best parents). A typical example of a blind recombination operator is the classical *Uniform crossover* [209]. This operator is defined on search spaces $\mathcal{S} \equiv \Sigma^n$, i.e., strings of n symbols taken from an alphabet Σ. The construction of the descendant is done by randomly selecting at each position one of the symbols appearing in that position in any of the parents. This random selection can be totally uniform or can be biased according to the fitness values of the parents as mentioned before. Furthermore, the selection can be done so as to enforce feasibility (e.g., consider the binary representation of solutions in the 0–1 MKP). Notice that, in this case, the resulting operator is neither respectful nor transmitting in general.

The use of blind recombination operators has been usually justified on the grounds of not introducing excessive bias in the search algorithm, thus preventing extremely fast convergence to suboptimal solutions. This is questionable though. First, notice that the behaviour of the algorithm is in fact biased by the choice of representation and the mechanics of the particular operators. Second, there exist widely known mechanisms (e.g., spatial isolation) to hinder these problems. Finally, it can be better to quickly obtain a suboptimal solution and restart the algorithm than using blind operators for a long time in pursuit of an asymptotically optimal behaviour (not even guaranteed in most cases).

Recombination operators that use problem knowledge are commonly termed *heuristic* or *hybrid*. In these operators, problem information is utilized to guide the process of constructing the descendants. This can be done in a plethora of ways for each problem, so it is difficult to provide a taxonomy of heuristic recombination operators. Nevertheless, there exist two main aspects into which problem knowledge can be injected: the selection of the parental features that will be transmitted to the descendant, and the selection of non-parental features that will be added to it. A heuristic recombination operator can focus in one of these aspects, or in both of them simultaneously.

As an example of heuristic recombination operator focused in the first aspect, *Dynastically Optimal Recombination* (DOR) [43] must be mentioned. This operator explores the dynastic potential set (i.e., possible children) of the configurations being recombined, so as to find the best member of this set (notice that, since configurations in the dynastic potential are entirely composed of features taken from any of the parents, this is a transmitting operator). This exploration is done using a subordinate A* algorithm in order to minimize the cost of traversing the dynastic potential. The goal of this operator is thus finding the best combination of parental features giving rise to a feasible child. Hence, this operator is monotonic in the sense that any child generated is at least as good as the best parent.

Examples of heuristic recombination operators concentrating on the selection of non-parental features, one can cite the *patching-by-forma-completion* operators proposed by Radcliffe and Surry [188]. These operators are based on generating an incomplete child using a non-heuristic procedure (e.g., the RAR_ω operator [187]), and then completing the child either using a local hill climbing procedure restricted to non-specified features (*locally optimal forma completion*) or a global search procedure that finds the globally best solution carrying the specified features (*globally optimal forma completion*). Notice the similarity of this latter approach with DOR.

Finally, there exist some operators trying to exploit knowledge in both of the above aspects. A distinguished example is the *Edge Assembly Crossover* (EAX) [162]. EAX is a specialized operator for the TSP (both for symmetric and asymmetric instances) in which the construction of the child comprises two-phases: the first one involves the generation of an incomplete child via the so-called E-sets (subtours composed of alternating edges from each parent); subsequently, these subtours are merged into a single feasible subtours using a greedy repair algorithm. The authors of this operator reported impressive results in terms of accuracy and speed. It has some similarities with the recombination operator proposed in [155].

A final comment must be made in relation to the computational complexity of recombination. It is clear that combining the features of several solutions is in general computationally more expensive than modifying a single solution (i.e. a mutation). Furthermore, the recombination operation will be usually invoked a large number of times. For this reason, it is convenient (and in many situations mandatory) to keep it at a low computational cost. A reasonable guideline is to consider an $O(N \log N)$ upper bound for its complexity, where N is the size of the input (the set S_{par} *and* the problem instance x). Such limit is easily affordable for blind recombination operators, being the term *crossover* a reasonable name conveying the low complexity (yet not always used in this context) of these. However, this limit can be relatively astringent in the case of heuristic recombination, mainly when epistasis (non-additive inter-feature influence on the fitness value) is involved. This admits several solutions depending upon the particular heuristic used. For example, DOR has exponential worst case behavior, but it can be made affordable by picking larger pieces of information from each parents (the larger the size of these pieces of information, the lower the number of them needed to complete the child) [46]. In any case, consider that heuristic recombination operators provide better solutions than blind recombination operators, and hence they need not be invoked the same number of times.

2.5 Designing a Memetic Algorithm

In light of the above considerations, it is possible to provide a general template for a memetic algorithm. As mentioned in Section 2.3, this template is very similar to that of a local search procedure acting on a set of $|pop| \geq 2$ configurations. This is shown in Figure 5.2.

This template requires some explanation. First of all, the GenerateInitialPopulation procedure is responsible for creating the initial set of $|pop|$ configurations. This can be done by simply generating $|pop|$ random configurations or by using a more sophisticated seeding mechanism (for instance, some constructive heuristic), by means of which high-quality configurations are injected in the initial population [130,208]. Another possibility, the Local-Search-Engine presented in Section 2.3 could be used as shown in Figure 5.3.

As to the TerminationCriterion function, it can be defined very similarly to the case of Local Search, i.e., setting a limit on the total number of iterations, reaching a maximum number of iterations without improvement, or having performed a certain number of population restarts, etc.

The GenerateNewPopulation procedure is the core of the memetic algorithm. Essentially, this procedure can be seen as a pipelined process comprising n_{op} stages. Each of these stages consists of taking $arity_{in}^{j}$ configurations from the previous stage, generating $arity_{out}^{j}$ new configurations by applying an operator op^{j}. This pipeline is restricted to have $arity_{in}^{1} = popsize$. The whole process is sketched in Figure 5.4.

This template for the GenerateNewPopulation procedure is usually instantiated in GAs by letting $n_{op} = 3$, using a selection, a recombination, and a mutation operator.

Procedure Population-Based-Search-Engine
begin
    Initialize *pop* **using** GenerateInitialPopulation();
    repeat
        newpop ← GenerateNewPopulation(*pop*);
        pop ← UpdatePopulation (*pop*, *newpop*)
        if *pop* **has converged then**
            pop ← RestartPopulation(*pop*);
        endif
    until TerminationCriterion()
end

Figure 5.2. A population-based search algorithm.

Procedure GenerateInitialPopulation
begin
  Initialize *pop* **using** EmptyPopulation();
  parfor $j \leftarrow 1$ **to** *popsize* **do**
    $i \leftarrow$ GenerateRandomConfiguration();
    $i \leftarrow$ Local-Search-Engine (i);
    InsertInPopulation individual i **to** *pop*;
  endparfor
  return *pop*
end

Figure 5.3. Injecting high-quality solutions in the initial population.

Procedure GenerateNewPopulation (*pop*)
begin
  $buffer^0 \leftarrow pop$;
  parfor $j \leftarrow 1$ **to** n_{op} **do**
    Initialize $buffer^j$ **using** EmptyPopulation();
  endparfor
  parfor $j \leftarrow 1$ **to** n_{op} **do**
    $S^j_{par} \leftarrow$ ExtractFromBuffer ($buffer^{j-1}$, $arity^j_{in}$);
    $S^j_{desc} \leftarrow$ ApplyOperator (op^j, S^j_{par});
    for $z \leftarrow 1$ **to** $arity^j_{out}$ **do**
      InsertInPopulation individual $S^j_{desc}[z]$ **to** $buffer^j$;
    endfor
  endparfor;
  return $buffer^{n_{op}}$
end

Figure 5.4. The pipelined GenerateNewPopulation procedure.

Traditionally, mutation is applied after recombination, i.e., on each child generated by the recombination operator. However, if a heuristic recombination operator is being used, it may be more convenient to apply mutation before recombination. Since the purpose of mutation is simply to introduce new features in the configuration pool, using it in advance is also possible. Furthermore, the *smart* feature combination performed by the heuristic operator would not be disturbed this way.

This situation is slightly different in MAs. In this case, it is very common to let $n_{op} = 5$, inserting a Local-Search-Engine right after applying op^2 and op^4 (respectively recombination and mutation). Due to the local optimization performed after mutation, applying the latter after recombination is not as problematic as in GAs.

The UpdatePopulation procedure is used to reconstruct the current population using the old population *pop* and the newly generated population *newpop*. Borrowing the terminology from the evolution strategy [194,202] community, there exist two main possibilities to carry on this reconstruction: the *plus* strategy and the *comma* strategy. In the former, the current population is constructed taken the best *popsize* configurations from $pop \cup newpop$. As to the latter, the best *popsize* configurations are taken just from *newpop*. In this case, it is required to have $|newpop| > popsize$, so as to put some selective pressure on the process (the bigger the $|newpop|/posize$ ratio, the stronger the pressure). Otherwise, the search would reduce to a random wandering through $\mathcal{S}$.

There are a number of studies regarding appropriate choices for the UpdatePopulation procedure (e.g., [9,78]). As a general guideline, the comma strategy is usually regarded as less prone to stagnation, being the ratio $|newpop|/posize \simeq 6$ a common choice [8]. Nevertheless, this option can be somewhat computationally expensive if the guiding function is complex and time-consuming. Another common alternative is using a plus strategy with a low value of $|newpop|$, analogous to the so-called *steady-state* replacement strategy in GAs [222]. This option usually provides a faster convergence to high-quality solutions. However, care has to be taken with premature convergence to suboptimal regions of the search space, i.e., all configurations in the population being very similar to each other, hence hindering the exploration of other regions of $\mathcal{S}$.

The above consideration about premature convergence leads to the last component of the template shown in Figure 5.2, the restarting procedure. First of all, it must be decided whether the population has degraded or has not. To do so, it is possible to use some measure of information diversity in the population such as Shannon's entropy [50]. If this measure falls below a predefined threshold, the population is considered at a degenerate state. This threshold depends upon the representation (number of values per variable, constraints, etc.) and hence must be determined in an *ad-hoc* fashion. A different possibility is using a probabilistic approach to determine with a desired confidence that the population has converged. For example, in [99] a Bayesian approach is presented for this purpose.

Once the population is considered to be at a degenerate state, the restart procedure is invoked. Again, this can be implemented in a number of ways. A very typical strategy is keeping a fraction of the current population (this fraction can be as small as one solution, the current best), and substituting the remaining configurations with newly generated (from scratch) solutions, as shown in Figure 5.5.

```
Procedure RestartPopulation (pop)
begin
    Initialize newpop using EmptyPopulation();
    #preserved ← popsize · %preserve;
    for j ← 1 to #preserved do
        i ← ExtractBestFromPopulation(pop);
        InsertInPopulation individual i to newpop;
    endfor
    parfor j ← #preserved + 1 to popsize do
        i ← GenerateRandomConfiguration();
        i ← Local-Search-Engine (i);
        InsertInPopulation individual i to newpop;
    endparfor;
    return newpop
end
```

Figure 5.5. The RestartPopulation procedure.

The procedure shown in Figure 5.5 is also known as the *random-immigrant* strategy [38]. A different possibility is activating a *strong* or *heavy* mutation operator in order to drive the population away from its current location in the search space. Both options have their advantages and disadvantages. For example, when using the random-immigrant strategy, one has to take some caution to prevent the preserved configurations to take over the population (this can be achieved by putting a low selective pressure, at least in the first iterations after the restarting). As to the heavy mutation strategy, one has to achieve a tradeoff between an excessively strong mutation that would destroy any information contained in the current population, and a not so strong mutation that would cause the population to converge again in a few iterations.

3 APPLICATIONS OF MEMETIC ALGORITHMS

This section will provide an overview of the numerous applications of MAs. This overview is far from exhaustive since new applications are being developed continuously. However, it is intended to be illustrative of the practical impact of these optimization techniques.

3.1 Traditional $\mathcal{NP}$ Optimization Problems

Traditional $\mathcal{NP}$ Optimization problems constitute one of the most typical battlefields of MAs. A remarkable history of successes has been reported with respect to the application of MAs to $\mathcal{NP}$−hard problems such as the following: GRAPH PARTITIONING [18,19,139,142,143], MIN NUMBER PARTITIONING [14], MAX INDEPENDENT SET [2, 90,200], BIN-PACKING [195], MIN GRAPH COLORING [39,41,60,64], SET COVERING [11], SINGLE MACHINE SCHEDULING WITH SETUP-TIMES AND DUE-DATES [65,119,146], PARALLEL MACHINE SCHEDULING [33,35,135,148], MIN GENERALISED ASSIGNMENT [36], MULTIDIMENSIONAL KNAPSACK [12,45,77], NONLINEAR INTEGER PROGRAMMING [210], QUADRATIC ASSIGNMENT [17,29,137,141,142], SET PARTITIONING [120], and particularly on the MIN TRAVELLING SALESMAN PROBLEM [66,67,73,75,76,97,110,138, 142,156,189].

Regarding the theory of $\mathcal{NP}$-Completeness, most of them can be cited as *"classical"* as they appeared in Karp's notorious paper [108] on the reducibility of combinatorial problems. Remarkably, in most of them the authors claim that they have developed the best heuristic for the problem at hand. This is important since these problems have been addressed with several with different approaches from the combinatorial optimization toolbox and almost all general-purpose algorithmic techniques have been tested on them.

3.2 Other Combinatorial Optimization Problems

The MA paradigm is not limited to the above mentioned classical problems. There exist additional "non-classical" combinatorial optimization problems of similar or higher complexity in whose resolution MAs have revealed themselves as outstanding techniques. As an example of these problems, one can cite *partial shape matching* [176], *Kauffman NK Landscapes* [140], *spacecraft trajectory design* [48], *frequency allocation* [109], *multiperiod network design* [69], *degree-constrained minimum spanning tree problem* [190], *uncapacitated hub location* [1], *placement problems* [98,115,201], *vehicle routing* [101,102], *transportation problems* [71,170], *task allocation* [86], *maintenance scheduling* [25]–[27], *open shop scheduling* [34,63,124], *flowshop scheduling* [30,159,160], *project scheduling* [168,177,191], *warehouse scheduling* [216], *production planning* [57,149], *timetabling* [20–24,128,151,152,179,180,192], *rostering* [53,153], and *sport games scheduling* [40].

Obviously, this list is by no means complete since its purpose is simply to document the wide applicability of the approach for combinatorial optimization.

3.3 Machine Learning and Robotics

Machine learning and robotics are two closely related fields since the different tasks involved in the control of robots are commonly approached using artificial neural networks and/or classifier systems. MAs, generally cited as "genetic hybrids" have been used in both fields, i.e., in general optimization problems related to machine learning (e.g., the training of artificial neural networks), and in robotic applications. With respect to the former, MAs have been applied to *neural network training* [100,155,212,227], *pattern recognition* [3], *pattern classification* [113,145], and *analysis of time series* [173].

As to the application of MAs to robotics, work has been done in *reactive rulebase learning in mobile agents* [47], *path planning* [172,183,225], *manipulator motion planning* [197], *time optimal control* [31], etc.

3.4 Electronics and Engineering

Electronics and engineering are also two fields in which these methods have been actively used. For example, with regard to engineering problems, work has been done in the following areas: *structure optimization* [228], *system modeling* [215], *aeronautic design* [16,186], *trim loss minimization* [174], *traffic control* [205], and *power planning* [213]. As to practical applications in the field of electronics, the following list can illustrate the numerous areas in which these techniques have been utilized: *semiconductor manufacturing* [111], *circuit design* [83,87,220], *computer aided design* [13], *multilayered periodic strip grating* [7], *analogue network synthesis* [81], and *service restoration* [5].

3.5 Molecular Optimization Problems

We have selected this particular class of computational problems, involving nonlinear optimization issues, to help the reader to identify a common trend in the literature. Unfortunately, the authors continue referring to their technique as 'genetic', although they are closer in spirit to MAs [94].

The Caltech report that gave its name to the, at that time incipient, field of MAs [154] discussed a metaheuristic which can be viewed as a hybrid of GAs and SA developed with M.G. Norman in 1988. In recent years, several papers applied hybrids of GAs with SA or other methods to a variety of molecular optimization problems [10,49,54,59,68, 82,105,107,117,123,129,131,136,147,178,203,204,211,221,230,231]. Hybrid population approaches like this can hardly be catalogued as being 'genetic', but this denomination has appeared in previous work by Deaven and Ho [55] and then cited by J. Maddox in *Nature* [132]. Other fields of application include *cluster physics* [167]. Additional work has been done in [56,92,93,184,185,221]. Other evolutionary approaches to a variety of molecular problems can be found in: [59,89,91,134,144,193,214]. Their use for design problems is particularly appealing [37,107,223]. They have also been applied in protein design [58,118], probably due to energy landscape structure associated with proteins (see the discussion in [155] and the literature review in [94]).

3.6 Other Applications

In addition to the application areas described above, MAs have been also utilized in other fields such as, for example, *medicine* [84,85,217], *economics* [122,175], *oceanography* [161], *mathematics* [196,218,219], *imaging science and speech processing* [28,114,133,198,229], etc.

For further information about MA applications we suggest querying bibliographical databases or web browsers for the keywords *'memetic algorithms'* and *'hybrid genetic algorithm'*. We have tried to be illustrative rather than exhaustive, pointing out some selected references for well-known application areas. This means that, with high probability, many important contributions may have been inadvertently left out.

4 FUTURE DIRECTIONS

The future seems promising for MAs. This is the combination of several factors. First, MAs (less frequently disguised under different names) are showing a remarkable record of efficient implementations, providing very good results in practical problems. Second, there are reasons to believe that some new attempts to do theoretical analysis can be conducted. This includes the worst-case and average-case computational complexity of recombination procedures. Third, the ubiquitous nature of distributed systems, like networks of workstations for example, plus the inherent asynchronous parallelism of MAs and the existence of web-conscious languages like Java; all together are an excellent combination to develop highly portable and extendable object-oriented frameworks allowing algorithmic reuse. These frameworks might allow the users to solve subproblems using commercial codes or well-tested software from other users who might be specialists in the other area. Our current work on the *MemePool Project* goes in that direction. Fourth, an important and pioneering group of MAs, that of *Scatter Search*, is challenging the rôle of randomization in recombination. We expect that, as a healthy reaction, we will soon see new types of powerful MAs that blend in a more appropriate way both exhaustive (either truncated or not) and systematic search methods.

4.1 Lessons Learned

In 1998, Applegate, Bixby, Cook, and Chvatal established new breakthrough results for the MIN TSP. Interestingly enough, this work supports our view that MAs will have a central rôle as a problem solving methodology. For instance, this research team solved to optimality an instance of the TSP of 13,509 cities corresponding to all U.S. cities with populations of more than 500 people.[1] The approach, according to Bixby: "*...involves ideas from polyhedral combinatorics and combinatorial optimization, integer and linear programming, computer science data structures and*

[1] See: http://www.crpc.rice.edu/CRPC/newsArchive/tsp.html

algorithms, parallel computing, software engineering, numerical analysis, graph theory, and more". The solution of this instance demanded the use of three Digital AlphaServer 4100s (with a total of 12 processors) and a cluster of 32 Pentium-II PCs. The complete calculation took approximately three months of computer time. The code has certainly more than 1,000 pages and is based on state-of-the-art techniques from a wide variety of scientific fields.

The philosophy is the same of MAs, that of a synergy of different approaches. However, due to a comment by Johnson and McGeoch ([103], p. 301), we previously suggested that the connection with MAs might be stronger since there is an analogy with *multi-parent* recombination. In a recent unpublished manuscript, *"Finding Tours in the TSP"*, by the same authors (Bixby et al.), available from their web site, this suspicion has been now confirmed. They present results on running an optimal algorithm for solving the MIN WEIGHTED HAMILTONIAN CYCLE PROBLEM in a subgraph formed by the union of 25 Chained Lin-Kernighan tours. The approach consistently finds the optimal solution to the original MIN TSP instances with up to 4461 cities. They also attempted to apply this idea to an instance with 85,900 cities (the largest instance in TSPLIB) and from that experience they convinced themselves that it also works well for such large instances.

Their approach can possibly be classified as the most complex MA ever built for a given combinatorial optimization problem. One of the current challenges is to develop simpler algorithms that achieve these impressive results.

The approach of running a local search algorithm (Chained Lin Kernighan) to produce a collection of tours, following by the dynastical-optimal recombination method the authors named *tour merging* gave a non-optimal tour of only 0.0002% excess above the proved optimal tour for the 13,509 cities instance. We take this as a clear proof of the benefits of the MA approach and that more work is needed in developing good strategies for *complete memetic algorithms*, i.e., those that systematically and synergistically use randomized and deterministic methods and can prove optimality.

4.2 Computational Complexity of Recombination Problems

An interesting new direction for theoretical research arose after the introduction of two computational complexity classes, the PMA class (for *Polynomial Merger Algorithms problems*) and its unconstrained analogue, the uPMA class. We note that the dynastical optimal recombination, as it was the tour merging procedure just discussed, can lead to solving another $\mathcal{NP}$-hard problem of a smaller size. To try to avoid intractability of the new subproblem generated, we may wish to find interesting cases with worst-case polynomial time complexity for the generated subproblem, yet helpful towards the primary goal of solving the original optimization problem. Having these ideas in mind we will discuss two recently introduced computational complexity classes.

We will define the classes PMA and uPMA by referring to three analogous algorithms to the ones that define the class of *Polynomial Local Search* problems (PLS). These definitions (specially for PMA) are particularly dependent on an algorithm called *k-merger*, that will help to formalize the notion of *recombination of a set of k given solutions*, as generally used by most MAs (as well as other population approaches). The input of a k-merger algorithm is a set S_{par} of $k \geq 2$ *feasible* solutions, so

$S_{par} \subseteq sol_P(x)$ They can be informally called "parent" solutions and, if successful, the k-merger delivers as output *at least one* feasible solution. For the uPMA class the construction of the new solution is less restricted than for PMA. In general, recombination processes can be very complex with many side restrictions involving the detection, the preservation or avoidance, and the feasible combination of *features* already present in the parent solutions.

Definition 5.1 (Definition (uPMA)). *Let x be an instance of an optimization problem P. With $\mathcal{M}_P(S_{par}, x) \subseteq sol_P(x)$ we denote the set of all possible outputs that the k-*merger *algorithm can give if it receives as input the pair (S_{par}, x) for problem P. (Note that the set $\mathcal{M}_P(S_{par}, x)$ generalizes the idea of* dynastic potential *for more than two parents* [43]*).*

A recombination problem $P/\mathcal{M}$ belongs to uPMA if there exist three polynomial-time algorithms *p-starter*, *p′-evaluator*, and *k-merger* (where p, p' and k are integer numbers such that $p' \geq p \geq k \geq 2$) that satisfy the following properties:

- Given an input x (formally a string $\in \{0,1\}^*$), the *p-starter* determines whether $x \in I_P$ and in this case produces a set of p different feasible solutions $\{y_1, y_2, \ldots, y_p\} \subseteq sol_P(x)$.
- Given an instance $x \in I_P$ and an input (formally a string $\in \{0,1\}^*$), the *p′-evaluator* determines whether this input represents a set of feasible solutions, i.e. $\{y_1, y_2, \ldots, y_{p'}\} \subset sol_P(x)$ and in that case it computes the value of the objective function associated to each one of them, i.e. $m_P(y_j, x), \forall j = 1, \ldots, p'$.
- Given an instance $x \in I_P$ and a set of k feasible solutions $S_{par} \subseteq sol_P(x)$, the *k-merger* determines whether the set S_{par} is a *k-merger optimum*, and, if it is not, it outputs at least one feasible solution $y' \in \mathcal{M}_P(S_{par}, x)$ with strictly better value of m_P.

Analogously, the PMA class is more restricted since it embodies a particular type of recombination. For uPMA the type of recombination is implicit in the way the group neighborhood $\mathcal{M}$ is defined. However, the definition for PMA is still general enough to encompasses most of the recombination procedures used in practical population-based approaches.

Definition 5.2 (PMA). *A recombination problem $P/\mathcal{M}$ belongs to PMA if there exist three polynomial-time algorithms* p-starter, p'-evaluator, *and* k-merger *(where p, p' and k are integer numbers such that $p' \geq p \geq k \geq 2$), such that the* p-starter *and* p'-evaluator *satisfy the same properties required by the uPMA class but the* k-merger *is constrained to be of a particular type, i.e.:*

- Given an instance $x \in I_P$ and a set of k feasible solutions $S_{par} \subseteq sol_P(x)$, the *k-merger* determines whether the set S_{par} is a *k-merger optimum*, and, if it is not, it does the following:
  - —For each $y \in S_{par}$, it solves n_1 polynomial-time decision problems $\{\Pi_1(y), \ldots, \Pi_{n_1}(y)\}$. Let D be a matrix of $k \times n_1$ Boolean coefficients formed by the output of all these decision problems, i.e. $D_{i,j} = \Pi_j(y_i)$.

—It creates a set of n_2 *constraints* C, such that C can be partitioned in two subsets, i.e. $C = C_{in} \cup C_{out}$. Each constraint $c \in C$ is represented by a predicate π_c such that its associated decision problem $\Pi_c(y)$ can be solved in polynomial-time for all $y \in sol_P(x)$. Any predicate π_c is a polynomial-time computable function that has as input the Boolean matrix D and the instance x. It is required that *at least one* predicate π_c^* to be a non-constant function of *at least two different elements of* S_{par}.

—It outputs at least one *offspring*, i.e., another feasible solution $y' \in \mathcal{M}_P(S_{par}, x)$ with strictly better value of m_P, (i.e. $m_P(y', x) < \max\{m_P(y_1, x), m_P(y_2, x), \ldots, m_P(y_k, x)\}$ for a minimization problem, and $m_P(y', x) > \min\{m_P(y_1, x), m_P(y_2, x), \ldots, m_P(y_k, x)\}$ for a maximization problem) subject to

$$\max_{y' \in sol_P(x)} \left[\left(\sum_{(c \in C_{in}) \wedge \Pi_c(y')} w_c \right) - \left(\sum_{(c \in C_{out}) \wedge \Pi_c(y')} w_c \right) \right] \tag{1}$$

where w_c is an integer weight associated to constraint c.

Current research is being conducted to identify problems, and their associated recombination procedures, such that membership, in either PMA or uPMA, can be proved. It is also hoped that after some initial attempts on challenging problems completeness and reductions for the classes can be properly defined.

4.3 Exploiting Fixed-parameter Tractability Results

An interesting new avenue of research can be established by appropriately linking results from the theory of fixed-parameter tractability and the development of recombination algorithms. This is an avenue that goes both ways. On one direction, efficient, (i.e., polynomial-time), fixed-parameter algorithms can be used as *"out of the box"* tools to create efficient recombination procedures. They can be used as dynastical optimal recombination procedures or k-merger algorithms. On the opposite direction, since MAs are typically designed to deal with large instances and scale pretty well with their size, using both techniques together can produce *complete MAs* allowing to extend the benefits of fixed-parameter tractability. From a software engineering perspective, the combination is perfect both from code and algorithmic reuse.

A parameterized problem can be generally viewed as a problem that has as input two components, i.e., a pair $\langle x, k \rangle$. The former is generally an instance (i.e., $x \in I_P$) of some other decision problem P and the latter is some numerical aspect of the former (generally a positive integer assumed $k \ll |x|$) that constitutes a *parameter*. For a maximization (resp. minimization) problem P, the induced language $L_P(param)$ is the parameterized language consisting of all pairs $\langle x, k \rangle$ where $x \in I_P$ and $OPT(x) \geq k$ (resp. $OPT(x) \leq k$). If there exists an algorithm solving $L_P(param)$ in time $O(f(k)|x|^\alpha)$, where $|x|$ is a measure of the instance size, $f(k)$ an *arbitrary* function depending on k only, and α a constant independent of k or n, the parameterized problem is said to be *fixed-parameter tractable* and the language recognition problem belongs to the computational complexity class FPT.

To illustrate on this topic, we can discuss one of the most emblematic FPT problems:

VERTEX COVER (FIXED PARAMETER, DECISION VERSION)

Instance: A graph $G(V, E)$.

Parameter: A positive integer k.

Question: Is there a set $V' \subseteq V$, such that for every edge $(u, v) \in E$, at least u or v is a member of V' and $|V'| \leq k$?

In general, efficient FPT algorithms are based on the techniques of *reduction to a problem kernel* and *bounded search trees*. To understand the techniques, the reader may check a method by Chen, Kanj, and Yia [32]. This method can solve the parameterized version of vertex cover in time $O(1.271^k k^2 + kn)$. However, using this method together with the speed-up method proposed by Neidermeier and Rossmanith [166], the problem can be solved in $O(1.271^k + n)$, i.e., linear in n.

The definition of the FPT class allows $f(k)$ to grow arbitrarily fast. For the parameterized version of PLANAR DOMINATING SET the best known algorithm has $f(k) = 11^k$, but for other problems, like VERTEX COVER this situation is not that dramatic. In addition, the new general method to improve FPT algorithms that run on time $O(q(k)\beta^k + p(n))$ (where p and q are polynomials and β is a small constant) has been developed by Niedermeier and Rossmanith [166]. Their method interleaves the reduction to a problem kernel and bounded search methods. If it is the case that β^k is the size of the bounded search tree and $q(k)$ is the size of the problem kernel, the new technique is able to get rid of the $q(k)$ factor allowing the algorithm to be $O(\beta^k + p(n))$. Another problem that belongs to the FPT class is:

HITTING SET FOR SIZE THREE SETS (FIXED PARAMETER, DECISION VERSION)

Instance: A collection C of subsets of size three of a finite set S.

Parameter: A positive integer k.

Question: Is there a subset $S' \subseteq S$, with $|S'| \leq k$ which allows S' contain at least one element from each subset in C?

It can be seen as a generalization of the VERTEX COVER problem for hypergraphs, where there is an hyperedge between three vertices (instead of only two as in VERTEX COVER), and the task is to find a minimal subset of vertices covering all edges. Recently, Niedermeier and Rossmanith have presented an algorithm that solves this problem in time $O(2.311^k + n)$ [165]. We leave as an exercise to the reader the task of finding recombination operators that take advantage of this algorithm. We also leave as an exercise the following, "*is it possible to use the $O(2.311^k + n)$ to construct a k-merger algorithm for the optimization problem* MIN 3-HITTING SET *that satisfies the criteria required by the PMA and uPMA definitions?*"

4.4 Addition of negative knowledge

During the past five years, the addition of problem-dependent knowledge to enhance population-based approaches received a renewed interest among MAs researchers [2, 12,19,70,82,189,204]. Earlier papers had emphasized the relevance [79,125,154,207] of this issue, as well as Lawrence Davis from his numerous writings, also supported this idea when there was still not such a consensus, and many researchers were skeptical about their central role.

However, as many other things, the addition of problem-dependent, and instance-dependent knowledge can be addressed in different ways. Today, it is reasonable to continue finding new ways to incorporate *negative knowledge* in MAs. It is also challenging to find ways of extracting (or we may better say induce), "negative knowledge" from the current population of solutions. We quote M. Minsky [150]:

> We tend to think of knowledge in positive terms – and of experts as people who know what to do. But a 'negative' way to seem competent is, simply, never to make mistakes.

This is what heuristics do best, reducing the computational effort by radically changing the search space, so certainly there is no paradox here. Most heuristics, however, have a "positive bias". For instance, in [181] we can read:

> The strategy developed by Lin [126] for the TSP is to obtain several local optima and then identify edges that are common to all of them. These are then fixed, thus reducing the time to find more local optima. This idea is developed further in [72,127].

Curiously enough, although this strategy has been around for more than three decades it is hardly accepted by some researchers regardless all empirical evidence of the benefits of MAs that exploit, by using recombination, the correlation of highly evolved solutions [15,112,142,154,155,157].

Already recognized by R. Hofmann in 1993, there is a certain curious asymmetry in this strategy. We tend to think of common edges as pieces of "positive" information to be passed on between generations. However, after a few generations of an MA, a highly evolved (yet still diverse) population, might consistently avoid certain edges. *What is the population of agents "suggesting" about these edges ?* For instance, given an edge, we can define two tours as being equivalent if they both contain the edge, but also they can be defined equivalent *if both do not*. Hofmann recognized this duality and came with the denomination of *negative edge formae* for the latter induced equivalence class, that describes the set of tours that do not share this particular edge (see [95], p. 35).

At this point it is important to distinguish knowledge from *belief*. Formally, knowledge is generally defined as a *true belief*, for instance "agent a knows ϕ if a believes ϕ and ϕ is true". So a way of adding knowledge to the agents is to actually to attempt to *prove* things the agents believe.

For instance, if we are trying to solve an Euclidean 2-dimensional instance of the MIN TSP, it is always the case that the optimal tour must not have two crossing edges. This is a true fact that can be proven with a theorem. Since 2-opt moves can efficiently remove crossing edges with the aid of associate data structures and range queries, all agents in an MAs generally have it incorporated in their local search methods. However, it is also the case that other predicates, involving more than two edges, can be proved in all circumstances for the optimal tour (i.e., a predicate that is always true for all instances under consideration), yet not efficient associated local search methods have been found. This said, we envision that given an optimization with only one objective function, we might use methods from multi-objective MAs, using as auxiliary objective functions the number of violations of over a set of properties that the optimal solution should have. Research in this direction is also under way.

4.5 Guided Memetic Algorithms

In [181] we can read:

> As is common with heuristics, one can also argue for exactly the opposite idea. Once such common features are detected, they are *forbidden* rather than fixed. This justification is the following: If we fear that the global optimum is escaping us, this could be because our heuristic is "fooled" by this tempting features. Forbidding them could finally put us on the right track towards the global optimum. This is called *denial* in [206].

We also envision that future generations of MAs will work in at least two levels and two time scales. During the short-time scale, a set of agents would be searching in the search space associated to the problem. This is what most MAs do today. However, a long-time scale would *adapt the algorithms* associated with the agents. Here we encompass all of them, the individual search strategies, the recombination operators, etc. For instance, an MA that uses *guided local search* for an adaptive strategy has been proposed in [97]. Recombination operators with different (*rebel*, *obsequent*, and *conciliator behaviors*) were introduced in [14]. We think that these new promising directions need to be investigated, providing adaptive recombination mechanisms that take information of the progress of the current population. Moreover, it is intriguing the idea of using a combination of deterministic search procedures with a stochastic MA. The combination of both would help to systematically explore sub-regions of the search space and *guarantee* a coverage that currently is only left to some random procedure that aim to diversify the search (like the so-called *heavy mutation* procedures).

4.6 Modal Logics for MAs and Belief Search

As a logical consequence of the possible directions that MAs can take, it is reasonable to affirm that more complex schemes evolving solutions, agents, as well as representations, will soon be implemented. Some theoretical computer science researchers dismiss heuristics and metaheuristics since they are not coordinated as a formal paradigm. However, their achievements are well-recognized. From [121]:

> Explaining and predicting the impressive empirical success of some of these algorithms is one of the most challenging frontiers of the theory of computation today.

This comment is even more relevant for MAs since they generally present even better results than single-agent methods. Though metaheuristics are extremely powerful in practice, we agree that one problem with the current trend of applied research is that it allows the introduction of increasingly more complex heuristics, unfortunately most of the time parameterized by *ad-hoc* numbers. Moreover, some metaheuristics, like some *ant-systems* implementations, can basically be viewed as particular types of MAs. This is the case if you allow the "ants" to use *branch-and-bound* or *local search* methods. In addition, these methods for distributed recombination of information (or beliefs) have some points in common with *blackboard systems* [61], as it has been recognized in the past yet it is hardly being mentioned in the current metaheuristics literature.

To illustrate how Belief Search can work in an MA setting, we will use an illustrative example. Let's assume that the formula $\mathbf{B}^i_a\phi$ has the following meaning *"agent i believes with strength (at least) a that ϕ is true"*, such that the strength values a are restricted to be rational numbers in [0,1]. Let's also suppose we accept as an axiom that from ψ being true we can deduce $\mathbf{B}_1^i \psi$ for all i. Now let's suppose that our agents are trying to solve a MIN TSP and that the particular instance being considered is Euclidean and two-dimensional. Let ϕ_k represent the proposition *"edge e_k is present in the optimum tour"* and let $\chi_{k,l}$ be true if edges e_k and e_l cross each other, and false otherwise. It can be proved that for such type of instances *"if edges e_k and e_l cross each other, then e_k and e_l can not both be present in the optimal tour"*. Then we can assume that this is known by all agents, and by the previous axiom we can deduce that agent 2 now believes $\mathbf{B}_1^2(\chi_{k,l} \to \neg(\phi_k \wedge \phi_l))$. Now let's suppose that agent 1 believes, with strength 0.4, that *"either edge e_k or e_l, but not both, is present in the optimal tour"*. We will represent this as $\mathbf{B}_{0.4}^1 \phi_{k,l}$. We will not enter into the discussion of how that agent reached that belief and we take it as a fact. Now let us suppose that another agent believes, at a level 0.7 that $\chi_{k,l} \to \phi_{k,l}$, then we write $\mathbf{B}_{0.7}^3(\chi_{k,l} \to \phi_{k,l})$. Note that this kind of assumption confuses our common sense, since in general we do not see any relationship between the fact that two edges cross and that we can deduce that as a consequence one of them should be present in the optimum tour. However, note that agent 3 believes in this relationship (at a 0.7 level) for *a particular* pair of edges e_k and e_l. Now, what can we say about the *distributed belief* of this group of three agents ? How can we recombine this information ?

According to $\mathbf{PL}_n^{\otimes}$, a multi-agent epistemic logic recently introduced by Boldrin and Saffiotti, the opinions shared by different a set of n agents can be *recombined* in a distributed belief. Using $\mathbf{PL}_n^{\otimes}$ we can deduce $\mathbf{D}_{0.82}\phi_{k,l}$. The distributed belief about proposition $\phi_{k,l}$ is stronger than any individual belief about it, and is even stronger than what you would get if any agent would believe the three facts.

4.7 The MemePool Project: Our Meeting Place for Algorithm and Knowledge Reuse

As we have said before, we believe that these different research directions can blend together and make an outstanding contribution when used in combination.

One interesting example of these new MAs, is a recent work by Lamma et al. [116] applied to the field of diagnosing digital circuits. In their approach, they differentiate between genes and "*memes*". The latter group codes for the agents beliefs and assumptions. Using a logic-based technique, they modify the memes according on how the present beliefs are contradicted by integrity constraints that express observations and laws. Each agent keeps a population of chromosomes and finds a solution to the belief revision problem by means of a genetic algorithm. A *Lamarckian* operator is used to modify a chromosome using belief revision directed mutations, oriented by tracing logical derivations. As a consequence, a chromosome will satisfy a larger number of constraints. The evolution provided by the Darwinian operators, allow agents to improve the chromosomes by gaining on the experience of other agents. Central to this approach is the Lamarckian operator appropriately called *Learn*. It takes a chromosome and produces a revised chromosome as output. To achieve that, it eliminates some derivation paths that reach to contradictions.

Surprisingly enough (and here we remark the first possibility of using the theory of *fixed-parameter tractability*), the learning is achieved by finding a *hitting set* which is not necessarily minimal. The authors make clear this point by saying that: *"a hitting set generated from these support sets is not necessarily a contradiction removal set and therefore is not a solution to the belief revision problem.*" The authors might not be aware of the $O(2.311^k + n)$ exact algorithm for MIN 3-HITTING SET. They might be able to use it, but that is anecdotal at the moment. What it is important to remark is that algorithms like this one might be used *out-of-the-box* if a proper, world-wide based, algorithmic framework were created.

On the other hand, we remarked how results of logic programming and belief revision may help the current status of metaheuristics. The current situation where everybody comes with new names for the same basic techniques, and most contributions are just the addition of new parameters to guide the search, is a futile research direction. It may be possible, that belief search guided MAs can be a valid tool to help systematize the scenario. In particular, if the discussion is based on which multi-agent logic performs better rather than which parameters work better for specific problems or instances. Towards that end, we hope to convince researchers in logic programming to start to address these issues, challenging them with the task of guiding MAs for large-scale combinatorial optimization. The MemePool Project[2] seems to be a good scenario for such an interdisciplinary and international enterprise. Whether it will work or not relies more on our ability to work cooperatively more than anything else.

BIBLIOGRAPHY

[1] H.S. Abdinnour (1998) A hybrid heuristic for the uncapacitated hub location problem. *European Journal of Operational Research*, **106**(2–3), 489–99.

[2] C.C. Aggarwal, J.B. Orlin and R.P. Tai (1997) Optimized crossover for the independent set problem. *Operations Research*, **45**(2), 226–234.

[3] J. Aguilar and A. Colmenares (1998) Resolution of pattern recognition problems using a hybrid genetic/random neural network learning algorithm. *Pattern Analysis and Applications*, **1**(1), 52–61.

[4] D. Aldous and U. Vazirani (1994) "Go with the winners" algorithms. In: *Proceedings of the 35th Annual Symposium on Foundations of Computer Science*. IEEE, Los Alamitos, CA, pp. 492–501.

[5] A. Augugliaro, L. Dusonchet and E. Riva-Sanseverino (1998) Service restoration in compensated distribution networks using a hybrid genetic algorithm. *Electric Power Systems Research*, **46**(1), 59–66.

[6] R. Axelrod and W.D. Hamilton (1981) The evolution of cooperation. *Science*, **211**(4489), 1390–1396.

[7] K. Aygun, D.S. Weile and E. Michielssen (1997) Design of multilayered periodic strip gratings by genetic algorithms. *Microwave and Optical Technology Letters*, **14**(2), 81–85.

[8] T. Bäck and F. Hoffmeister (1991) Adaptive search by evolutionary algorithms. In: W. Ebeling, M. Peschel and W. Weidlich (eds.), *Models of Selforganization in Complex Systems*, number 64 in Mathematical Research, Akademie-Verlag, pp. 17–21.

[9] Th. Bäck (1996) *Evolutionary Algorithms in Theory and Practice*. Oxford University Press, New York.

[2] http://www.densis.fee.unicamp.br/~moscato/memepool.html

[10] M.J. Bayley, G. Jones, P. Willett and M.P. Williamson (1998) Genfold: A genetic algorithm for folding protein structures using NMR restraints. *Protein Science*, **7**(2), 491–499.

[11] J. Beasley and P.C. Chu (1996) A genetic algorithm for the set covering problem. *European Journal of Operational Research*, **94**(2), 393–404.

[12] J. Beasley and P.C. Chu (1998) A genetic algorithm for the multidimensional knapsack problem. *Journal of Heuristics*, **4**, 63–86.

[13] B. Becker and R. Drechsler (1994) Ofdd based minimization of fixed polarity Reed-Muller expressions using hybrid genetic algorithms. In: *Proceedings IEEE International Conference on Computer Design: VLSI in Computers and Processor*. IEEE, Los Alamitos, CA, pp. 106–110.

[14] R. Berretta and P. Moscato (1999) The number partitioning problem: An open challenge for evolutionary computation? In: D. Corne, M. Dorigo and F. Glover (eds.), *New Ideas in Optimization*. McGraw-Hill, pp. 261–278.

[15] K.D. Boese (1995) Cost versus Distance in the Traveling Salesman Problem. Technical Report TR-950018, UCLA CS Department.

[16] A.H.W. Bos (1998) Aircraft conceptual design by genetic/gradient-guided optimization. *Engineering Applications of Artificial Intelligence*, **11**(3), 377–382.

[17] D. Brown, C. Huntley and A. Spillane (1989) A parallel genetic heuristic for the quadratic assignment problem. In: J. Schaffer (ed.), *Proceedings of the Third International Conference on Genetic Algorithms*. Morgan Kaufmann, pp. 406–415.

[18] T.N. Bui and B.R. Moon (1996) Genetic algorithm and graph partitioning. *IEEE Transactions on Computers*, **45**(7), 841–855.

[19] T.N. Bui and B.R. Moon (1998) GRCA: A hybrid genetic algorithm for circuit ratio-cut partitioning. *IEEE Transactions on Computer-Aided Design of Integrated Circuits and Systems*, **17**(3), 193–204.

[20] E.K. Burke, D.G. Elliman and R.F. Weare (1995) A hybrid genetic algorithm for highly constrained timetabling problems. In: *Proceedings of the Sixth International Conference on Genetic Algorithms*. Morgan Kaufmann, San Francisco, CA, pp. 605–610.

[21] E.K. Burke, K.S. Jackson, J.H. Kingston and R.F. Weare (1997) Automated timetabling: The state of the art. *The Computer Journal*, **40**(9), 565–571.

[22] E.K. Burke and J.P. Newall (1997) A phased evolutionary approach for the timetable problem: An initial study. In: *Proceedings of the ICONIP/ANZIIS/ANNES '97 Conference*. Springer-Verlag, Dunedin, New Zealand, pp. 1038–1041.

[23] E.K. Burke, J.P. Newall and R.F. Weare (1996) A memetic algorithm for university exam timetabling. In: E.K. Burke and P. Ross (eds.), *The Practice and Theory of Automated Timetabling*, volume 1153 of *Lecture Notes in Computer Science*. Springer-Verlag, pp. 241–250.

[24] E.K. Burke, J.P. Newall and R.F. Weare (1998) Initialisation strategies and diversity in evolutionary timetabling. *Evolutionary Computation*, **6**(1), 81–103.

[25] E.K. Burke and A.J. Smith (1997) A memetic algorithm for the maintenance scheduling problem. In: *Proceedings of the ICONIP/ANZIIS/ANNES '97 Conference*. Springer-Verlag, Dunedin, New Zealand, pp. 469–472.

[26] E.K. Burke and A.J. Smith (1999) A memetic algorithm to schedule grid maintenance. In: *Proceedings of the International Conference on Computational Intelligence for Modelling Control and Automation, Vienna: Evolutionary Computation and Fuzzy Logic for Intelligent Control, Knowledge Acquisition and Information Retrieval*. IOS Press 1999, pp. 122–127.

[27] E.K. Burke and A.J. Smith (1999) A multi-stage approach for the thermal generator maintenance scheduling problem. In: *Proceedings of the 1999 Congress on Evolutionary Computation*. IEEE, Washington D.C., pp. 1085–1092.

[28] S. Cadieux, N. Tanizaki and T. Okamura (1997) Time efficient and robust 3-D brain image centering and realignment using hybrid genetic algorithm. In: *Proceedings of the 36th SICE Annual Conference*. IEEE, pp. 1279–1284.

[29] J. Carrizo, F.G. Tinetti and P. Moscato (1992) A computational ecology for the quadratic assignment problem. In: *Proceedings of the 21st Meeting on Informatics and Operations Research*. Buenos Aires, SADIO.

[30] S. Cavalieri and P. Gaiardelli (1998) Hybrid genetic algorithms for a multiple-objective scheduling problem. *Journal of Intelligent Manufacturing*, **9**(4), 361–367.

[31] N. Chaiyaratana and A.M.S. Zalzala (1999) Hybridisation of neural networks and genetic algorithms for time-optimal control. In: *Proceedings of the 1999 Congress on Evolutionary Computation*. IEEE, Washington D.C., pp. 389–396.

[32] Jianer Chen, Iyad A. Kanj and Weijia Jia (1999) Vertex cover: further observations and further improvements. In: *Proceedings of the 25th International Worksh. Graph-Theoretic Concepts in Computer Science*, number 1665 in *Lecture Notes in Computer Science*. Springer-Verlag, pp. 313–324.

[33] R. Cheng and M. Gen (1996) Parallel machine scheduling problems using memetic algorithms. In: *1996 IEEE International Conference on Systems, Man and Cybernetics. Information Intelligence and Systems*, Vol. 4. IEEE, New York, NY, pp. 2665–2670.

[34] R. Cheng, M. Gen and Y. Tsujimura (1999) A tutorial survey of job-shop scheduling problems using genetic algorithms. ii. hybrid genetic search strategies. *Computers & Industrial Engineering*, **37**(1–2), 51–55.

[35] R.W. Cheng and M. Gen (1997) Parallel machine scheduling problems using memetic algorithms. *Computers & Industrial Engineering*, **33**(3–4), 761–764.

[36] P.C. Chu and J. Beasley (1997) A genetic algorithm for the generalised assignment problem. *Computers & Operations Research*, **24**, 17–23.

[37] D.E. Clark and D.R. Westhead (1996) Evolutionary algorithms in computer-aided molecular design. *Journal of Computer-aided Molecular Design*, **10**(4), 337–358.

[38] H.G. Cobb and J.J. Grefenstette (1993) Genetic algorithms for tracking changing environments. In: S. Forrest (ed.), *Proceedings of the Fifth International*

Conference on Genetic Algorithms. Morgan Kaufmann, San Mateo, CA, pp. 529–530.

[39] P.E. Coll, G.A. Durán and P. Moscato (1999) On worst-case and comparative analysis as design principles for efficient recombination operators: A graph coloring case study. In: D. Corne, M. Dorigo and F. Glover (eds.), *New Ideas in Optimization*. McGraw-Hill, pp. 279–294.

[40] D. Costa (1995) An evolutionary tabu search algorithm and the NHL scheduling problem. *INFOR*, **33**(3), 161–178.

[41] D. Costa, N. Dubuis and A. Hertz (1995) Embedding of a sequential procedure within an evolutionary algorithm for coloring problems in graphs. *Journal of Heuristics*, **1**(1), 105–128.

[42] C. Cotta (1998) A study of hybridisation techniques and their application to the design of evolutionary algorithms. *AI Communications*, **11**(3–4), 223–224.

[43] C. Cotta, E. Alba and J.M. Troya (1998) Utilising dynastically optimal form a recombination in hybrid genetic algorithms. In: A.E. Eiben, Th. Bäck, M. Schoenauer and H.-P. Schwefel (eds.), *Parallel Problem Solving From Nature V*, volume 1498 of *Lecture Notes in Computer Science*. Springer-Verlag, Berlin, pp. 305–314.

[44] C. Cotta, E. Alba and J.M. Troya (1999) Stochastic reverse hillclimbing and iterated local search. In: *Proceedings of the 1999 Congress on Evolutionary Computation*. IEEE, Washington D.C., pp. 1558–1565.

[45] C. Cotta and J.M. Troya (1998) A hybrid genetic algorithm for the 0–1 multiple knapsack problem. In: G.D. Smith, N.C. Steele and R.F. Albrecht (eds.), *Artificial Neural Nets and Genetic Algorithms 3*. Springer-Verlag, Wien New York, pp. 251–255.

[46] C. Cotta and J.M. Troya (2000) On the influence of the representation granularity in heuristic forma recombination. In: J. Carroll, E. Damiani, H. Haddad and D. Oppenheim (eds.), *ACM Symposium on Applied Computing 2000*. ACM Press, pp. 433–439.

[47] C. Cotta and J.M. Troya (2000) Using a hybrid evolutionary A* approach for learning reactive behaviors. In: S. Cagnoni et al. (eds.), *Real-World Applications of Evolutionary Computation*, volume 1803 of *Lecture Notes in Computer Science*. Springer-Verlag Edinburgh, pp. 347–356.

[48] T. Crain, R. Bishop, W. Fowler and K. Rock (1999) Optimal interplanetary trajectory design via hybrid genetic algorithm/recursive quadratic program search. In: *Ninth AAS/AIAA Space Flight Mechanics Meeting*. Breckenridge CO, pp. 99–133.

[49] T. Dandekar and P. Argos (1996) Identifying the tertiary fold of small proteins with different topologies from sequence and secondary structure using the genetic algorithm and extended criteria specific for strand regions. *Journal of Molecular Biology*, **256**(3), 645–660.

[50] Y. Davidor and O. Ben-Kiki (1992) The interplay among the genetic algorithm operators: Information theory tools used in a holistic way. In: R. Männer and B. Manderick (eds.), *Parallel Problem Solving From Nature II*. Elsevier Science Publishers B.V., Amsterdam, pp. 75–84.

[51] L. Davis (1991) *Handbook of Genetic Algorithms*. Van Nostrand Reinhold Computer Library, New York.

[52] R. Dawkins (1976) *The Selfish Gene*. Clarendon Press, Oxford.

[53] P. de Causmaecker, G. van den Berghe and E.K. Burke (1999) Using tabu search as a local heuristic in a memetic algorithm for the nurse rostering problem. In: *Proceedings of the Thirteenth Conference on Quantitative Methods for Decision Making*, pages abstract only, poster presentation. Brussels, Belgium.

[54] P.A. de Souza, R. Garg and V.K. Garg (1998) Automation of the analysis of Mossbauer spectra. *Hyperfine Interactions*, **112**(1–4), 275–278.

[55] D.M. Deaven and K.O. Ho (1995) Molecular-geometry optimization with a genetic algorithm. *Physical Review Letters*, **75**(2), 288–291.

[56] D.M. Deaven, N. Tit, J.R. Morris and K.M. Ho (1996) Structural optimization of Lennard-Jones clusters by a genetic algorithm. *Chemical Physics Letters*, **256**(1–2), 195–200.

[57] N. Dellaert and J. Jeunet (2000) Solving large unconstrained multilevel lot-sizing problems using a hybrid genetic algorithm. *International Journal of Production Research*, **38**(5), 1083–1099.

[58] J.R. Desjarlais and T.M. Handel (1995) New strategies in protein design. *Current Opinion in Biotechnology*, **6**(4), 460–466.

[59] R. Doll and M.A. VanHove (1996) Global optimization in LEED structure determination using genetic algorithms. *Surface Science*, **355**(1–3), L393–L398.

[60] R. Dorne and J.K. Hao (1998) A new genetic local search algorithm for graph coloring. In: A.E. Eiben, Th. Bäck, M. Schoenauer and H.-P. Schwefel (eds.), *Parallel Problem Solving From Nature V*, volume 1498 of *Lecture Notes in Computer Science*. Springer-Verlag, Berlin, pp. 745–754.

[61] R. Englemore and T. Morgan (eds.) (1988) *Blackboard Systems*. Addison-Wesley.

[62] C. Ersoy and S.S. Panwar (1993) Topological design of interconnected LAN/MAN networks. *IEEE Journal on Selected Areas in Communications*, **11**(8), 1172–1182.

[63] J. Fang and Y. Xi (1997) A rolling horizon job shop rescheduling strategy in the dynamic environment. *International Journal of Advanced Manufacturing Technology*, **13**(3), 227–232.

[64] C. Fleurent and J.A. Ferland (1997) Genetic and hybrid algorithms for graph coloring. *Annals of Operations Research*, **63**, 437–461.

[65] P.M. França, A.S. Mendes and P. Moscato (1999) Memetic algorithms to minimize tardiness on a single machine with sequence-dependent setup times. In: *Proceedings of the 5th International Conference of the Decision Sciences Institute, Athens, Greece*, pp. 1708–1710.

[66] B. Freisleben and P. Merz (1996) A genetic local search algorithm for solving symmetric and asymmetric traveling salesman problems. In: *Proceedings of the 1996 IEEE International Conference on Evolutionary Computation, Nagoya, Japan*. IEEE Press, pp. 616–621.

[67] B. Freisleben and P. Merz (1996) New genetic local search operators for the traveling salesman problem. In: H.-M. Voigt, W. Ebeling, I. Rechenberg and H.-P. Schwefel (eds.), *Parallel Problem Solving from Nature IV*, volume 1141 of *Lecture Notes in Computer Science*. Springer, pp. 890–900.

[68] R.T. Fu, K. Esfarjani, Y. Hashi, J. Wu, X. Sun and Y. Kawazoe (1997) Surface reconstruction of Si (001) by genetic algorithm and simulated annealing method. *Science Reports of The Research Institutes Tohoku University Series A-Physics Chemistry and Metallurgy*, **44**(1), 77–81.

[69] B.L. Garcia, P. Mahey and L.J. LeBlanc (1998) Iterative improvement methods for a multiperiod network design problem. *European Journal of Operational Research*, **110**(1), 150–165.

[70] M. Gen and R. Cheng (1997) *Genetic Algorithms and Engineering Design*. Wiley Series in Engineering Design and Automation. John Wiley & Sons (Sd).

[71] M. Gen, K. Ida and L. Yinzhen (1998) Bicriteria transportation problem by hybrid genetic algorithm. *Computers & Industrial Engineering*, **35**(1–2), 363–366.

[72] A.J. Goldstein and A.B. Lesk (1975) Common feature techniques for discrete optimization. *Comp. Sci. Tech. Report 27*, Bell. Tel. Labs, March.

[73] M. Gorges-Schleuter (1989) ASPARAGOS: An asynchronous parallel genetic optimization strategy. In: J. David Schaffer (ed.), *Proceedings of the Third International Conference on Genetic Algorithms*. Morgan Kaufmann Publishers, pp. 422–427.

[74] M. Gorges-Schleuter (1991) Explicit parallelism of genetic algorithms through population structures. In: H.-P. Schwefel and R. Männer (eds.), *Parallel Problem Solving from Nature*. Springer-Verlag, pp. 150–159.

[75] M. Gorges-Schleuter (1991) *Genetic Algorithms and Population Structures—A Massively Parallel Algorithm*. PhD thesis, University of Dortmund, Germany.

[76] M. Gorges-Schleuter (1997) Asparagos96 and the Traveling Salesman Problem. In: T. Baeck, Z. Michalewicz and X. Yao (eds.), *Proceedings of the 1997 IEEE International Conference on Evolutionary Computation. Indianapolis, USA*. IEEE Press, pp. 171–174.

[77] J. Gottlieb (2000) Permutation-based evolutionary algorithms for multidimensional knapsack problems. In: J. Carroll, E. Damiani, H. Haddad and D. Oppenheim (eds.), *ACM Symposium on Applied Computing 2000*. ACM Press, pp. 408–414.

[78] J. Gottlieb and T. Kruse (2000) Selection in evolutionary algorithms for the traveling salesman problem. In: J. Carroll, E. Damiani, H. Haddad and D. Oppenheim (eds.), *ACM Symposium on Applied Computing 2000*. ACM Press, pp. 415–421.

[79] J. J. Grefenstette (1987) Incorporating Problem Specific Knowledge into Genetic Algorithms. In: L. Davis (ed.), *Genetic Algorithms and Simulated Annealing, Research Notes in Artificial Intelligence*. Morgan Kaufmann Publishers, pp. 42–60.

[80] P. Grim (1997) The undecidability of the spatialized prisoner's dilemma. *Theory and Decision*, **42**(1), 53–80.

[81] J.B. Grimbleby (1999) Hybrid genetic algorithms for analogue network synthesis. In: *Proceedings of the 1999 Congress on Evolutionary Computation*, IEEE, Washington D.C., pp. 1781–1787.

[82] J.R. Gunn (1997) Sampling protein conformations using segment libraries and a genetic algorithm. *Journal of Chemical Physics*, **106**(10), 4270–4281.

[83] M. Guotian and L. Changhong (1999) Optimal design of the broadband stepped impedance transformer based on the hybrid genetic algorithm. *Journal of Xidian University*, **26**(1), 8–12.

[84] O.C.L. Haas, K.J. Burnham and J.A. Mills (1998) Optimization of beam orientation in radiotherapy using planar geometry. *Physics in Medicine and Biology*, **43**(8), 2179–2193.

[85] O.C.L. Haas, K.J. Burnham, J.A. Mills, C.R. Reeves and M.H. Fisher (1996) Hybrid genetic algorithms applied to beam orientation in radiotherapy. In: *Fourth European Congress on Intelligent Techniques and Soft Computing Proceedings*, Vol. 3. Verlag Mainz, Aachen, Germany, pp. 2050–2055.

[86] A.B. Hadj-Alouane, J.C. Bean and K.G. Murty (1999) A hybrid genetic/optimization algorithm for a task allocation problem. *Journal of Scheduling*, **2**(4).

[87] S.P. Harris and E.C. Ifeachor (1998) Automatic design of frequency sampling filters by hybrid genetic algorithm techniques. *IEEE Transactions on Signal Processing*, **46**(12), 3304–3314.

[88] W.E. Hart and R.K. Belew (1991) Optimizing an arbitrary function is hard for the genetic algorithm. In: R.K. Belew and L.B. Booker (eds.), *Proceedings of the Fourth International Conference on Genetic Algorithms*. Morgan Kaufmann, San Mateo CA, pp. 190–195.

[89] B. Hartke (1993) Global geometry optimization of clusters using genetic algorithms. *Journal of Physical Chemistry*, **97**(39), 9973–9976.

[90] M. Hifi (1997) A genetic algorithm-based heuristic for solving the weighted maximum independent set and some equivalent problems. *Journal of the Operational Research Society*, **48**(6), 612–622.

[91] R. Hirsch and C.C. Mullergoymann (1995) Fitting of diffusion-coefficients in a 3-compartment sustained-release drug formulation using a genetic algorithm. *International Journal of Pharmaceutics*, **120**(2), 229–234.

[92] K.M. Ho, A.A. Shvartsburg, B.C. Pan, Z.Y. Lu, C.Z. Wang, J.G. Wacker, J.L. Fye and M.F. Jarrold (1998) Structures of medium-sized silicon clusters. *Nature*, **392**(6676), 582–585.

[93] S. Hobday and R. Smith (1997) Optimisation of carbon cluster geometry using a genetic algorithm. *Journal of The Chemical Society-Faraday Transactions*, **93**(22), 3919–3926.

[94] R.J.W. Hodgson (2000) Memetic algorithms and the molecular geometry optimization problem. In: *Proceedings of the 2000 Congress on Evolutionary Computation*. IEEE Service Center, Piscataway, NJ, pp. 625–632.

[95] Reimar Hofmann (1993) Examinations on the algebra of genetic algorithms. Master's thesis, Technische Universität München, Institut fü Informatik.

[96] D.R. Hofstadter (1983) Computer tournaments of the prisoners-dilemma suggest how cooperation evolves. *Scientific American*, **248**(5), 16–23.

[97] D. Holstein and P. Moscato (1999) Memetic algorithms using guided local search: A case study. In: D. Corne, M. Dorigo and F. Glover (eds.), *New Ideas in Optimization*. McGraw-Hill, pp. 235–244.

[98] E. Hopper and B. Turton (1999) A genetic algorithm for a 2d industrial packing problem. *Computers & Industrial Engineering*, **37**(1–2), 375–378.

[99] M. Hulin (1997) An optimal stop criterion for genetic algorithms: A bayesian approach. In: Th. Bäck (ed.), *Proceedings of the Seventh International Conference on Genetic Algorithms*. Morgan Kaufmann, San Mateo, CA, pp. 135–143.

[100] T. Ichimura and Y. Kuriyama (1998) Learning of neural networks with parallel hybrid ga using a royal road function. In: *1998 IEEE International Joint Conference on Neural Networks*, Vol. 2, IEEE, New York, NY, pp. 1131–1136.

[101] J. Berger, M. Salois and R. Begin (1998) A hybrid genetic algorithm for the vehicle routing problem with time windows. In: R.E. Mercer and E. Neufeld (eds.), *Advances in Artificial Intelligence. 12th Biennial Conference of the Canadian Society for Computational Studies of Intelligence*. Springer-Verlag, Berlin, pp. 114–127.

[102] W.R. Jih and Y.J. Hsu (1999) Dynamic vehicle routing using hybrid genetic algorithms. In: *Proceedings of the 1999 Congress on Evolutionary Computation*. IEEE, Washington D.C., pp. 453–458.

[103] D.S. Johnson and L.A. McGeoch (1997) The traveling salesman problem: A case study. In: E.H.L. Aarts and J.K. Lenstra (eds.), *Local Search in Combinatorial Optimization*. Wiley, Chichester, pp. 215–310.

[104] D.S. Johnson, C.H. Papadimitriou and M. Yannakakis (1988) How easy is local search? *Journal of Computers and System Sciences*, **37**, 79–100.

[105] G. Jones, P. Willett, R.C. Glen, A.R. Leach and R. Taylor (1997) Development and validation of a genetic algorithm for flexible docking. *Journal of Molecular Biology*, **267**(3), 727–748.

[106] T.C. Jones (1995) *Evolutionary Algorithms, Fitness Landscapes and Search*. PhD thesis, University of New Mexico.

[107] B.M. Kariuki, H. Serrano-Gonzalez, R.L. Johnston and K.D.M. Harris (1997) The application of a genetic algorithm for solving crystal structures from powder diffraction data. *Chemical Physics Letters*, **280**(3–4), 189–195.

[108] R.M. Karp (1972) Reducibility among combinatorial problems. In: R.E. Miller and J.W. Thatcher (eds.), *Complexity of Computer Computations*. Plenum, New York NY, pp. 85–103.

[109] I.E. Kassotakis, M.E. Markaki and A.V. Vasilakos (2000) A hybrid genetic approach for channel reuse in multiple access telecommunication networks. *IEEE Journal on Selected Areas in Communications*, **18**(2), 234–243.

[110] K. Katayama, H. Hirabayashi and H. Narihisa (1998) Performance analysis for crossover operators of genetic algorithm. *Transactions of the Institute of Electronics, Information and Communication Engineers*, **J81D-I**(6), 639–650.

[111] T.S. Kim and G.S. May (1999) Intelligent control of via formation by photosensitive BCB for MCM-L/D applications. *IEEE Transactions on Semiconductor Manufacturing*, **12**, 503–515.

[112] S. Kirkpatrick and G. Toulouse (1985) Configuration space analysis of traveling salesman problems. *J. Physique*, **46**, 1277–1292.

[113] K. Krishna and M. Narasimha-Murty (1999) Genetic k-means algorithm. *IEEE Transactions on Systems, Man and Cybernetics, Part B (Cybernetics)*, **29**(3), 433–439.

[114] K. Krishna, K.R. Ramakrishnan and M.A.L. Thathachar (1997) Vector quantization using genetic k-means algorithm for image compression. In: *1997 International Conference on Information, Communications and Signal Processing*, Vol. 3, IEEE, New York, NY, pp. 1585–1587.

[115] R.M. Krzanowski and J. Raper (1999) Hybrid genetic algorithm for transmitter location in wireless networks. *Computers, Environment and Urban Systems*, **23**(5), 359–382.

[116] E. Lamma, L.M. Pereira and F. Riguzzi (2000) Multi-agent logic aided lamarckian learning. Technical Report DEIS-LIA-00-004, Dipartimento di Elettronica, Informatica e Sistemistica, University of Bologna (Italy), LIA Series no. 44 (submitted for publication).

[117] E. Landree, C. Collazo-Davila and L.D. Marks (1997) Multi-solution genetic algorithm approach to surface structure determination using direct methods. *Acta Crystallographica Section B—Structural Science*, **53**, 916–922.

[118] G.A. Lazar, J.R. Desjarlais and T.M. Handel (1997) De novo design of the hydrophobic core of ubiquitin. *Protein Science*, **6**(6), 1167–1178.

[119] C.Y. Lee (1994) Genetic algorithms for single machine job scheduling with common due date and symmetric penalties. *Journal of the Operations Research Society of Japan*, **37**(2), 83–95.

[120] D. Levine (1996) A parallel genetic algorithm for the set partitioning problem. In: I.H. Osman and J.P. Kelly (eds.), *Meta-Heuristics: Theory & Applications*. Kluwer Academic Publishers, pp. 23–35.

[121] H.R. Lewis and C.H. Papadimitriou (1998) *Elements of the Theory of Computation*. Prentice-Hall, Inc., Upper Saddle River, New Jersey.

[122] F. Li, R. Morgan and D. Williams (1996) Economic environmental dispatch made easy with hybrid genetic algorithms. In: *Proceedings of the International Conference on Electrical Engineering*, Vol. 2. Int. Acad. Publishers, Beijing, China, pp. 965–969.

[123] L.P. Li, T.A. Darden, S.J. Freedman, B.C. Furie, B. Furie, J.D. Baleja, H. Smith, R.G. Hiskey and L.G. Pedersen (1997) Refinement of the NMR solution structure of the gamma-carboxyglutamic acid domain of coagulation factor IX using molecular dynamics simulation with initial Ca2+ positions determined by a genetic algorithm. *Biochemistry*, **36**(8), 2132–2138.

[124] C.F. Liaw (2000) A hybrid genetic algorithm for the open shop scheduling problem. *European Journal of Oprational Research*, **124**(1), 28–42.

[125] G.E. Liepins and M.R. Hilliard (1987) Greedy Genetics. In: J.J. Grefenstette (ed.), *Proceedings of the Second International Conference on Genetic Algorithms and their Applications*. Lawrence Erlbaum Associates, Cambridge, MA, pp. 90–99.

[126] S. Lin (1965) Computer solutions of the traveling salesman problem. *Bell System Technical Journal*, **10**, 2245–2269.

[127] S. Lin and B. Kernighan (1973) An Effective Heuristic Algorithm for the Traveling Salesman Problem. *Operations Research*, **21**, 498–516.

[128] S.E. Ling (1992) Integrating genetic algorithms with a prolog assignment program as a hybrid solution for a polytechnic timetable problem. In: *Parallel Problem Solving from Nature II*. Elsevier Science Publisher B.V., pp. 321–329.

[129] D.M. Lorber and B.K. Shoichet (1998) Flexible ligand docking using conformational ensembles. *Protein Science*, **7**(4), 938–950.

[130] S.J. Louis, X. Yin and Z.Y. Yuan (1999) Multiple vehicle routing with time windows using genetic algorithms. In: *Proceedings of the 1999 Congress on Evolutionary Computation*. Washington D.C., pp. 1804–1808. IEEE Neural Network Council—Evolutionary Programming Society—Institution of Electrical Engineers.

[131] A.L. MacKay (1995) Generalized crystallography. *THEOCHEM-Journal of Molecular Structure*, **336**(2–3), 293–303.

[132] J. Maddox (1995) Genetics helping molecular-dynamics. *Nature*, **376**(6537), 209–209.

[133] K.E. Mathias and L.D. Whitley (1994) Noisy function evaluation and the delta coding algorithm. In: *Proceedings of the SPIE—The International Society for Optical Engineering*, pp. 53–64.

[134] A.C.W. May and M.S. Johnson (1994) Protein-structure comparisons using a combination of a genetic algorithm, dynamic-programming and least-squares minimization. *Protein Engineering*, **7**(4), 475–485.

[135] A.S. Mendes, F.M. Muller, P.M. França and P. Moscato (1999) Comparing metaheuristic approaches for parallel machine scheduling problems with sequence-dependent setup times. In: *Proceedings of the 15th International Conference on CAD/CAM Robotics & Factories of the Future*, Aguas de Lindoia, Brazil.

[136] L.D. Merkle, G.B. Lamont, G.H. Jr. Gates and R. Pachter (1996) Hybrid genetic algorithms for minimization of a polypeptide specific energy model. In: *Proceedings of 1996 IEEE International Conference on Evolutionary Computation*. IEEE, New York, NY, pp. 396–400.

[137] P. Merz and B. Freisleben (1997) A Genetic Local Search Approach to the Quadratic Assignment Problem. In: T. Bäck (ed.), *Proceedings of the Seventh International Conference on Genetic Algorithms*. Morgan Kaufmann, San Mateo, CA, pp. 465–472.

[138] P. Merz and B. Freisleben (1997) Genetic local search for the TSP: new results. In: *Proceedings of the 1997 IEEE International Conference on Evolutionary Computation*. IEEE Press, pp. 159–164.

[139] P. Merz and B. Freisleben (1998) Memetic algorithms and the fitness landscape of the graph bi-partitioning problem. In: A.E. Eiben, T. Back, M. Schoenauer and H.-P. Schwefel (eds.), *Parallel Problem Solving from Nature V*, volume 1498 of *Lecture Notes in Computer Science*. Springer-Verlag, pp. 765–774.

[140] P. Merz and B. Freisleben (1998) On the effectiveness of evolutionary search in high-dimensional *NK*-landscapes. In: *Proceedings of the 1998 IEEE International Conference on Evolutionary Computation*. IEEE Press, pp. 741–745.

[141] P. Merz and B. Freisleben (1999) A comparion of Memetic Algorithms, Tabu Search, and ant colonies for the quadratic assignment problem. In: *Proceedings of the 1999 Congress on Evolutionary Computation. Washington D.C.* IEEE Service Center, Piscataway, NJ, pp. 2063–2070.

[142] P. Merz and B. Freisleben (1999) Fitness landscapes and memetic algorithm design. In: D. Corne, M. Dorigo and F. Glover (eds.), *New Ideas in Optimization*. McGraw-Hill, pp. 245–260.

[143] P. Merz and B. Freisleben (2000) Fitness landscapes, memetic algorithms, and greedy operators for graph bipartitioning. *Evolutionary Computation*, **8**(1), 61–91.

[144] J.C. Meza, R.S. Judson, T.R. Faulkner and A.M. Treasurywala (1996) A comparison of a direct search method and a genetic algorithm for conformational searching. *Journal of Computational Chemistry*, **17**(9), 1142–1151.

[145] M. Mignotte, C. Collet, P. Pérez and P. Bouthemy (2000) Hybrid genetic optimization and statistical model based approach for the classification of shadow shapes in sonar imagery. *IEEE Transactions on Pattern Analysis and Machine Intelligence*, **22**(2), 129–141.

[146] D.M. Miller, H.C. Chen, J. Matson and Q. Liu (1999) A hybrid genetic algorithm for the single machine scheduling problem. *Journal of Heuristics*, **5**(4), 437–454.

[147] S.T. Miller, J.M. Hogle and D.J. Filman (1996) A genetic algorithm for the *ab initio* phasing of icosahedral viruses. *Acta Crystallographica Section D—Biological Crystallography*, **52**, 235–251.

[148] L. Min and W. Cheng (1998) Identical parallel machine scheduling problem for minimizing the makespan using genetic algorithm combined with simulated annealing. *Chinese Journal of Electronics*, **7**(4), 317–321.

[149] X.G. Ming and K.L. Mak (2000) A hybrid hopfield network-genetic algorithm approach to optimal process plan selection. *International Journal of Production Research*, **38**(8), 1823–1839.

[150] M. Minsky (1994) Negative expertise. *International Journal of Expert Systems*, **7**(1), 13–19.

[151] A. Monfroglio (1996) Hybrid genetic algorithms for timetabling. *International Journal of Intelligent Systems*, **11**(8), 477–523.

[152] A. Monfroglio (1996) Timetabling through constrained heuristic search and genetic algorithms. *Software—Practice and Experience*, **26**(3), 251–279.

[153] A. Montfroglio (1996) Hybrid genetic algorithms for a rostering problem. *Software—Practice and Experience*, **26**(7), 851–862.

[154] P. Moscato (1989) On evolution, search, optimization, genetic algorithms and martial arts: towards memetic algorithms. Technical Report Caltech Concurrent Computation Program, Report 826, California Institute of Technology, Pasadena, California, USA.

[155] P. Moscato (1993) An introduction to population approaches for optimization and hierarchical objective functions: the role of tabu search. *Annals of Operations Research*, **41**(1–4), 85–121.

[156] P. Moscato and M.G. Norman (1992) A Memetic Approach for the Traveling Salesman Problem Implementation of a Computational Ecology for Combinatorial Optimization on Message-Passing Systems. In: M. Valero, E. Onate, M. Jane, J.L. Larriba and B. Suarez (eds.), *Parallel Computing and Transputer Applications*. IOS Press, Amsterdam, pp. 177–186.

[157] H. Mühlenbein (1991) Evolution in time and space—the parallel genetic algorithm. In: Gregory J.E. Rawlins (ed.), *Foundations of Genetic Algorithms*. Morgan Kaufmann Publishers, pp. 316–337.

[158] H. Mühlenbein M. Gorges-Schleuter and O. Krämer (1988) Evolution Algorithms in Combinatorial Optimization. *Parallel Computing*, **7**, 65–88.

[159] T. Murata and H. Ishibuchi (1994) Performance evaluation of genetic algorithms for flowshop scheduling problems. In: *Proceedings of the First IEEE Conference on Evolutionary Computation*, Vol. 2. IEEE, New York, NY, pp. 812–817.

[160] T. Murata, H. Ishibuchi and H. Tanaka (1996) Genetic algorithms for flowshop scheduling problems. *Computers & Industrial Engineering*, **30**(4), 1061–1071.

[161] M. Musil, M.J. Wilmut and N.R. Chapman (1999) A hybrid simplex genetic algorithm for estimating geoacoustic parameters using matched-field inversion. *IEEE Journal of Oceanic Engineering*, **24**(3), 358–369.

[162] Y. Nagata and Sh. Kobayashi (1997) Edge assembly crossover: a high-power genetic algorithm for the traveling salesman problem. In: Th. Bäck (ed.), *Proceedings of the Seventh International Conference on Genetic Algorithms, East Lansing, EUA*. Morgan Kaufmann, San Mateo, CA, pp. 450–457.

[163] M. Nakamaru, H. Matsuda and Y. Iwasa (1998) The evolution of social interaction in lattice models. *Sociological Theory and Methods*, **12**(2), 149–162.

[164] M. Nakamaru, H. Nogami and Y. Iwasa (1998) Score-dependent fertility model for the evolution of cooperation in a lattice. *Journal of Theoretical Biology*, **194**(1), 101–124.

[165] R. Niedermeier and P. Rossmanith (2000) An efficient fixed parameter algorithm for 3-hitting set. Technical Report WSI-99-18, Universität Tübingen, Wilhelm-Schickard-Institut für Informatik, 1999. Technical Report, Revised version accepted in *Journal of Discrete Algorithms*, August.

[166] R. Niedermeier and P. Rossmanith (2000) A general method to speed up fixed-parameter-tractable algorithms. *Information Processing Letters*, **73**, 125–129.

[167] J.A. Niesse and H.R. Mayne (1996) Global geometry optimization of atomic clusters using a modified genetic algorithm in space-fixed coordinates. *Journal of Chemical Physics*, **105**(11), 4700–4706.

[168] A.L. Nordstrom and S. Tufekci (1994) A genetic algorithm for the talent scheduling problem. *Computers & Operations-Research*, **21**(8), 927–940.

[169] M.G. Norman and P. Moscato (1991) A competitive and cooperative approach to complex combinatorial search. Technical Report Caltech Concurrent Computation Program, Report. 790, California Institute of Technology, Pasadena, California, USA, 1989. expanded version published at the *Proceedings of the 20th Informatics and Operations Research Meeting*, Buenos Aires (20th JAIIO), August pp. 3.15–3.29.

[170] A.G.N. Novaes, J.E.S. De-Cursi and O.D. Graciolli (2000) A continuous approach to the design of physical distribution systems. *Computers & Operations Research*, **27**(9), 877–893.

[171] M.A. Nowak and K. Sigmund (1998) Evolution of indirect reciprocity by image scoring. *Nature*, **393**(6685), 573–577.

[172] P. Osmera (1995) Hybrid and distributed genetic algorithms for motion control. In: V. Chundy and E. Kurekova (eds.), *Proceedings of the Fourth International Symposium on Measurement and Control in Robotics*, pp. 297–300.

[173] R. Ostermark (1999) A neuro-genetic algorithm for heteroskedastic time-series processes: empirical tests on global asset returns. *Soft Computing*, **3**(4), 206–220.

[174] R. Ostermark (1999) Solving a nonlinear non-convex trim loss problem with a genetic hybrid algorithm. *Computers & Operations Research*, **26**(6), 623–635.

[175] R. Ostermark (1999) Solving irregular econometric and mathematical optimization problems with a genetic hybrid algorithm. *Computational Economics*, **13**(2), 103–115.

[176] E. Ozcan and C.K. Mohan (1998) Steady state memetic algorithm for partial shape matching. In: V.W. Porto, N. Saravanan and D. Waagen (eds.), *Evolutionary Programming VII*, volume 1447 of *Lecture Notes in Computer Science*. Springer, Berlin, pp. 527–236.

[177] L. Ozdamar (1999) A genetic algorithm approach to a general category project scheduling problem. *IEEE Transactions on Systems, Man and Cybernetics, Part C (Applications and Reviews)*, **29**(1), 44–59.

[178] M.N. Pacey, X.Z. Wang, S.J. Haake and E.A. Patterson (1999) The application of evolutionary and maximum entropy algorithms to photoelastic spectral analysis. *Experimental Mechanics*, **39**(4), 265–273.

[179] B. Paechter, A. Cumming, M.G. Norman and H. Luchian Extensions to a Memetic timetabling system. In: E.K. Burke and P. Ross (eds.), *The Practice and Theory of Automated Timetabling*, volume 1153 of *Lecture Notes in Computer Science*. Springer Verlag, pp. 251–265.

[180] B. Paechter, R.C. Rankin and A. Cumming (1998) Improving a lecture timetabling system for university wide use. In: E.K. Burke and M. Carter (eds.),

The Practice and Theory of Automated Timetabling II, volume 1408 of *Lecture Notes in Computer Science*. Springer Verlag, pp. 156–165.

[181] C.H. Papadimitriou and K. Steiglitz (1982) *Combinatorial Optimization: Algorithms and Complexity*. Prentice-Hall, Inc., Englewood Cliffs, New Jersey.

[182] M. Peinado and T. Lengauer (1997) Parallel "go with the winners algorithms" in the LogP Model. In: IEEE Computer Society Press (ed.), *Proceedings of the 11th International Parallel Processing Symposium*. Los Alamitos, California, pp. 656–664.

[183] D.K. Pratihar, K. Deb and A. Ghosh (1999) Fuzzy-genetic algorithms and mobile robot navigation among static obstacles. In: *Proceedings of the 1999 Congress on Evolutionary Computation*. IEEE, Washington D.C., pp. 327–334.

[184] N. Pucello, M. Rosati, G. D'Agostino, F. Pisacane, V. Rosato and M. Celino (1997) Search of molecular ground state via genetic algorithm: Implementation on a hybrid SIMD-MIMD platform. *International Journal of Modern Physics C*, **8**(2), 239–252.

[185] W.J. Pullan (1997) Structure prediction of benzene clusters using a genetic algorithm. *Journal of Chemical Information and Computer Sciences*, **37**(6), 1189–1193.

[186] D. Quagliarella and A. Vicini (1998) Hybrid genetic algorithms as tools for complex optimisation problems. In: P. Blonda, M. Castellano and A. Petrosino (eds.), *New Trends in Fuzzy Logic II. Proceedings of the Second Italian Workshop on Fuzzy Logic*. World Scientific, Singapore, pp. 300–307.

[187] N.J. Radcliffe (1994) The algebra of genetic algorithms. *Annals of Mathematics and Artificial Intelligence*, **10**, 339–384.

[188] N.J. Radcliffe and P.D. Surry (1994) Fitness Variance of Formae and Performance Prediction. In: L.D. Whitley and M.D. Vose (eds.), *Proceedings of the Third Workshop on Foundations of Genetic Algorithms*. Morgan Kaufmann, San Francisco, pp. 51–72.

[189] N.J. Radcliffe and P.D. Surry (1994) Formal Memetic Algorithms. In: T. Fogarty (ed.), *Evolutionary Computing: AISB Workshop*, volume 865 of *Lecture Notes in Computer Science*. Springer-Verlag, Berlin, pp. 1–16.

[190] G.R. Raidl and B.A. Julstron (2000) A weighted coding in a genetic algorithm for the degree-constrained minimum spanning tree problem. In: J. Carroll, E. Damiani, H. Haddad and D. Oppenheim (eds.), *ACM Symposium on Applied Computing 2000*. ACM Press, pp. 440–445.

[191] E. Ramat, G. Venturini, C. Lente and M. Slimane (1997) Solving the multiple resource constrained project scheduling problem with a hybrid genetic algorithm. In: Th. Bäck (ed.), *Proceedings of the Seventh International Conference on Genetic Algorithms*. Morgan Kaufmann, San Francisco CA, pp. 489–496.

[192] R.C. Rankin (1996) Automatic timetabling in practice. In: *Practice and Theory of Automated Timetabling. First International Conference. Selected Papers*. Springer-Verlag, Berlin, pp. 266–279.

[193] M.L. Raymer, P.C. Sanschagrin, W.F. Punch, S. Venkataraman, E.D. Goodman and L.A. Kuhn (1997) Predicting conserved water-mediated and polar ligand

interactions in proteins using a k-nearest-neighbors genetic algorithm. *Journal of Molecular Biology*, **265**(4), 445–464.

[194] I. Rechenberg (1973) *Evolutionsstrategie: Optimierung technischer Systeme nach Prinzipien der biologischen Evolution*. Frommann-Holzboog Verlag, Stuttgart.

[195] C. Reeves (1996) Hybrid genetic algorithms for bin-packing and related problems. *Annals of Operations Research*, **63**, 371–396.

[196] C. Reich (2000) Simulation if imprecise ordinary differential equations using evolutionary algorithms. In: J. Carroll, E. Damiani, H. Haddad and D. Oppenheim (eds.), *ACM Symposium on Applied Computing 2000*. ACM Press, pp. 428–432.

[197] M.A. Ridao, J. Riquelme, E.F. Camacho and M. Toro (1998) An evolutionary and local search algorithm for planning two manipulators motion. In: A.P. Del Pobil, J. Mira and M. Ali (eds.), *Tasks and Methods in Applied Artificial Intelligence*, volume 1416 of *Lecture Notes in Computer Science*, Springer-Verlag, Berlin Heidelberg, pp. 105–114.

[198] C.F. Ruff, S.W. Hughes and D.J. Hawkes (1999) Volume estimation from sparse planar images using deformable models. *Image and Vision Computing*, **17**(8), 559–565.

[199] S.M. Sait and H. Youssef (2000) *VLSI Design Automation: Theory and Practice*. McGraw-Hill Book Co. (copublished by IEEE), Europe.

[200] A. Sakamoto, X.Z. Liu and T. Shimamoto (1997) A genetic approach for maximum independent set problems. *IEICE Transactions on Fundamentals of Electronics Communications and Computer Sciences*, **E80A**(3), 551–556.

[201] V. Schnecke and O. Vornberger (1997) Hybrid genetic algorithms for constrained placement problems. *IEEE Transactions on Evolutionary Computation*, **1**(4), 266–277.

[202] H.-P. Schwefel (1984) Evolution strategies: a family of non-linear optimization techniques based on imitating some principles of natural evolution. *Annals of Operations Research*, **1**, 165–167.

[203] K. Shankland, W.I.F. David and T. Csoka (1997) Crystal structure determination from powder diffraction data by the application of a genetic algorithm. *Zeitschrift Fur Kristallographie*, **212**(8), 550–552.

[204] K. Shankland, W.I.F. David T. Csoka and L. McBride (1998) Structure solution of ibuprofen from powder diffraction data by the application of a genetic algorithm combined with prior conformational analysis. *International Journal of Pharmaceutics*, **165**(1), 117–126.

[205] D. Srinivasan, R.L. Cheu, Y.P. Poh and A.K.C. Ng (2000) Development of an intelligent technique for traffic network incident detection. *Engineering Applications of Artificial Intelligence*, **13**(3), 311–322.

[206] K. Steiglitz and P. Weiner (1968) Some improved algorithms for computer solution of the traveling salesman problem. In: *Proceedings of the Sixth Allerton Conference on Circuit and System Theory*. Urbana, Illinois, pp. 814–821.

[207] J.Y. Suh and Dirk Van Gucht (1987) Incorporating heuristic information into genetic search. In: J.J. Grefenstette (ed.), *Proceedings of the Second International Conference on Genetic Algorithms and their Applications*, Lawrence Erlbaum Associates. Cambridge, MA, pp. 100–107.

[208] P.D. Surry and N.J. Radcliffe (1995) Inoculation to initialise evolutionary search. In: T.C. Fogarty (ed.), *Evolutionary Computing: AISB Workshop*, number 1143 in *Lecture Notes in Computer Science*. Springer-Verlag, pp. 269–285.

[209] G. Syswerda (1989) Uniform crossover in genetic algorithms. In: J.D. Schaffer (ed.), *Proceedings of the Third International Conference on Genetic Algorithms*. Morgan Kaufmann, San Mateo, CA, pp. 2–9.

[210] T. Taguchi, T. Yokota and M. Gen (1998) Reliability optimal design problem with interval coefficients using hybrid genetic algorithms. *Computers & Industrial Engineering*, **35**(1–2), 373–376.

[211] K.Y. Tam and R.G. Compton (1995) GAMATCH—a genetic algorithm-based program for indexing crystal faces. *Journal of Applied Crystallography*, **28**, 640–645.

[212] A.P. Topchy, O.A. Lebedko and V.V. Miagkikh (1996) Fast learning in multilayered networks by means of hybrid evolutionary and gradient algorithms. In: *Proceedings of International Conference on Evolutionary Computation and its Applications*, pp. 390–398.

[213] A.J. Urdaneta, J.F. Gómez, E. Sorrentino, L. Flores and R. Díaz (1999) A hybrid genetic algorithm for optimal reactive power planning based upon successive linear programming. *IEEE Transactions on Power Systems*, **14**(4), 1292–1298.

[214] A.H.C. vanKampen, C.S. Strom and L.M.C Buydens (1996) Lethalization, penalty and repair functions for constraint handling in the genetic algorithm methodology. *Chemometrics and Intelligent Laboratory Systems*, **34**(1), 55–68.

[215] L. Wang and J. Yen (1999) Extracting fuzzy rules for system modeling using a hybrid of genetic algorithms and kalman filter. *Fuzzy Sets and Systems*, **101**(3), 353–362.

[216] J.P. Watson, S. Rana, L.D. Whitley and A.E. Howe (1999) The impact of approximate evaluation on the performance of search algorithms for warehouse scheduling. *Journal of Scheduling*, **2**(2), 79–98.

[217] R. Wehrens, C. Lucasius, L. Buydens and G. Kateman (1993) HIPS, A hybrid self-adapting expert system for nuclear magnetic resonance spectrum interpretation using genetic algorithms. *Analytica Chimica ACTA*, **277**(2), 313–324.

[218] P. Wei and L.X. Cheng (1999) A hybrid genetic algorithm for function optimization. *Journal of Software*, **10**(8), 819–823.

[219] X. Wei and F. Kangling (2000) A hybrid genetic algorithm for global solution of nondifferentiable nonlinear function. *Control Theory & Applications*, **17**(2), 180–183.

[220] D.S. Weile and E. Michielssen (1999) Design of doubly periodic filter and polarizer structures using a hybridized genetic algorithm. *Radio Science*, **34**(1), 51–63.

[221] R.P. White, J.A. Niesse and H.R. Mayne (1998) A study of genetic algorithm approaches to global geometry optimization of aromatic hydrocarbon microclusters. *Journal of Chemical Physics*, **108**(5), 2208–2218.

[222] D. Whitley (1987) Using reproductive evaluation to improve genetic search and heuristic discovery. In: J.J. Grefenstette (ed.), *Proceedings of the Second International Conference on Genetic Algorithms and their Applications.* Lawrence Erlbaum Associates, Cambridge, MA, pp. 108–115.

[223] P. Willett (1995) Genetic algorithms in molecular recognition and design. *Trends in Biotechnology*, **13**(12), 516–521.

[224] D.H. Wolpert and W.G. Macready (1997) No free lunch theorems for optimization. *IEEE Transactions on Evolutionary Computation*, **1**(1), 67–82.

[225] J. Xiao and L. Zhang (1997) Adaptive evolutionary planner/navigator for mobile robots. *IEEE Transactions on Evolutionary Computation*, **1**(1), 18–28.

[226] M. Yannakakis (1997) Computational complexity. In: E.H.L. Aarts and J.K. Lenstra (eds.), *Local Search in Combinatorial Optimization.* Wiley, Chichester, pp. 19–55.

[227] X. Yao (1993) Evolutionary artificial neural networks. *International Journal of Neural Systems*, **4**(3), 203–222.

[228] I.C. Yeh (1999) Hybrid genetic algorithms for optimization of truss structures. *Computer Aided Civil and Infrastructure Engineering*, **14**(3), 199–206.

[229] M. Yoneyama, H. Komori and S. Nakamura (1999) Estimation of impulse response of vocal tract using hybrid genetic algorithm—a case of only glottal source. *Journal of the Acoustical Society of Japan*, **55**(12), 821–830.

[230] C.R. Zacharias, M.R. Lemes and A.D. Pino (1998) Combining genetic algorithm and simulated annealing: a molecular geometry optimization study. *THEOCHEM—Journal of Molecular Structure*, **430**(29–39).

[231] M. Zwick, B. Lovell and J. Marsh (1996) Global optimization studies on the 1-d phase problem. *International Journal of General Systems*, **25**(1), 47–59.

Chapter 6

VARIABLE NEIGHBORHOOD SEARCH

Pierre Hansen
GERAD and Ecole des Hautes Etudes Commerciales
3000 ch. de la Cote-Sainte-Catherine
Montréal H3T 2A7,
Canada
E-mail: pierreh@crt.umontreal.ca

Nenad Mladenović
Mathematical Institute,
Serbian Academy of Science
Kneza Mihajla 35
11000 Belgrade,
Yugoslavia
E-mail: nenad@mi.sanu.ac.yu

Abstract Variable neighborhood search (VNS) is a recent metaheuristic for solving combinatorial and global optimization problems whose basic idea is systematic change of neighborhood within a local search. In this survey paper we present basic rules of VNS and some of its extensions. Moreover, applications are briefly summarized. They comprise heuristic solution of a variety of optimization problems, ways to accelerate exact algorithms and to analyze heuristic solution processes, as well as computer-assisted discovery of conjectures in graph theory.

Résumé La Recherche à voisinage variable (RVV) est une métaheuristique récente pour la résolution de problèmes d'optimisation combinatoire et globale, dont l'idée de base est le changement systématique de voisinage au sein d'une recherche locale. Dans ce chaptre, nous présentons les règles de base de RVV et de certaines de ses extensions. De plus, des applications sont brièvement résumées. Elles comprennent la résolution approchée d'un ensemble de problèmes d'optimisation, des manières d'accélerer les algorithmes exacts et d'analyser le processus de résolution des heuristiques ainsi qu'un système automatique de découverte de conjectures en théorie des graphes.

1 INTRODUCTION

An optimization problem may be formulated as follows:

$$\min\{f(x)|x \in X, X \subseteq S\}. \tag{1}$$

S, X, x and f are *solution space*, *feasible set*, *feasible solution* and real valued function, respectively. If S is a finite but large set a *combinatorial optimization* problem is defined.

If $S = R^n$, we talk about *continuous optimization.* Most optimization problems are NP-hard and heuristic (suboptimal) solution methods are needed to solve them (at least for large instances or as an initial solution for some exact procedure).

Metaheuristics, or general frameworks for building heuristics to solve problem (1), are usually based upon a basic idea, or analogy. Then, they are developed, extended in various directions and possibly hybridised. The resulting heuristics often get complicated, and use many parameters. This may enhance their efficiency but obscures the reasons of their success.

Variable Neighborhood Search (VNS for short), a metaheuristic proposed just a few years ago [110,112], is based upon a simple principle: systematic change of neighborhood within the search. Its development has been rapid, with several dozen papers already published or to appear. Many extensions have been made, mainly to allow solution of large problem instances. In most of them, an effort has been made to keep the simplicity of the basic scheme.

In this paper, we survey these developments. The basic rules of VNS methods are recalled in the next section. Extensions are considered in Section 3 and Hybrids in Section 4. Applications are reviewed in Section 5, devoted to heuristic solution of combinatorial and global optimization problems, and Sections 6–8, which cover innovative uses, i.e., tools for analysis of the solution process of standard heuristics, acceleration of column generation procedures, and computer-assisted discovery in graph theory.

Desirable properties of metaheuristics are listed in Section 9 together with brief conclusions.

2 BASIC SCHEMES

Let us denote with $\mathcal{N}_k, (k = 1, \dots, k_{\max})$, a finite set of pre-selected neighborhood structures, and with $\mathcal{N}_k(x)$ the set of solutions in the kth neighborhood of x. (Most local search heuristics use only one neighborhood structure, i.e., $k_{\max} = 1$.) Neighborhoods $\mathcal{N}_k$ may be induced from one or more metric (or quasi-metric) functions introduced into a solution space S. An *optimal solution* x_{opt} (or global minimum) is a feasible solution where a minimum of (1) is reached. We call $x' \in X$ a *local minimum* of (1) with respect to $\mathcal{N}_k$ (w.r.t. $\mathcal{N}_k$ for short), if there is no solution $x \in \mathcal{N}_k(x') \subseteq X$ such that $f(x) < f(x')$. Metaheuristics (based on local search procedures) try to continue the search by other means after finding the first local minimum. VNS is based on three simple facts:

Fact 1 *A local minimum w.r.t. one neighborhood structure is not necessary so with another;*

Fact 2 *A global minimum is a local minimum w.r.t. all possible neighborhood structures.*

Fact 3 *For many problems local minima w.r.t. one or several $\mathcal{N}_k$ are relatively close to each other.*

This last observation, which is empirical, implies that a local optimum often provides some information about the global one. This may for instance be several variables with the same value in both. However, it is usually not known which ones are such. An organized study of the neighborhood of this local optimum is therefore in order, until a better one is found.

In order to solve (1) by using several neighborhoods, facts 1–3 can be used in three different ways: (i) deterministic; (ii) stochastic; (iii) both deterministic and stochastic.

(i) The **Variable neighborhood descent** (VND) method is obtained if change of neighborhoods is performed in a deterministic way and its steps are presented on Figure 6.1.

Most local search heuristics use in their descents a single or sometimes two neighborhoods ($k'_{\max} \leq 2$). Note that the final solution should be a local minimum w.r.t. all $k'_{\max}$ neighborhoods, and thus chances to reach a global one are larger than by using a single structure. Beside this *sequential* order of neighborhood structures in VND above, one can develop a *nested* strategy. Assume e.g. that $k'_{\max} = 3$; then a possible nested strategy is: perform VND from Figure 6.1 for the first two neighborhoods, in each point x' that belongs to the third ($x' \in \mathcal{N}_3(x)$). Such an approach is applied in [9,16,81].

(ii) The **Reduced VNS** (RVNS) method is obtained if random points are selected from $\mathcal{N}_k(x)$, without being followed by descent, and its steps are presented on Figure 6.2.

RVNS is useful for very large instances for which local search is costly. It is observed that the best value for the parameter $k_{\max}$ is often 2. In addition, the maximum number of iterations between two improvements is usually used as stopping condition. RVNS is akin to a Monte-Carlo method, but more systematic (see [114] where results obtained by RVNS were 30% better than those of the Monte-Carlo method in solving a continuous min–max problem). When applied to the p-Median problem, RVNS gave equally good solutions as the *Fast Interchange* heuristic of [136] in 20–40 times less time [84].

Initialization. Select the set of neighborhood structures $\mathcal{N}'_k$, for $k = 1, \ldots, k'_{\max}$, that will be used in the descent; find an initial solution x (or apply the rules to a given x);

Repeat the following sequence until no improvement is obtained:

(1) Set $k \leftarrow 1$;

(2) Repeat the following steps until $k = k'_{\max}$:

(a) Exploration of neighborhood. Find the best neighbor x' of x ($x' \in \mathcal{N}'_k(x)$);

(b) Move or not.

If the solution thus obtained x' is better than x, set $x \leftarrow x'$ and $k \leftarrow 1$; otherwise, set $k \leftarrow k + 1$;

Figure 6.1. Steps of the basic VND.

Initialization. Select the set of neighborhood structures $\mathcal{N}_k$, for $k = 1, \ldots, k_{max}$, that will be used in the search; find an initial solution x; choose a stopping condition;

Repeat the following sequence until the stopping condition is met:

(1) Set $k \leftarrow 1$;

(2) Repeat the following steps until $k = k_{\max}$:

(a) Shaking. Generate a point x' at random from the kth neighborhood of x ($x' \in \mathcal{N}_k(x)$);

(b) Move or not. If this point is better than the incumbent, move there ($x \leftarrow x'$), and continue the search with $\mathcal{N}_1$ ($k \leftarrow 1$); otherwise, set $k \leftarrow k + 1$;

Figure 6.2. Steps of the Reduced VNS.

Initialization. Select the set of neighborhood structures $\mathcal{N}_k$, for $k = 1, \ldots, k_{max}$, that will be used in the search; find an initial solution x; choose a stopping condition;

Repeat the following sequence until the stopping condition is met:

(1) Set $k \leftarrow 1$;

(2) Repeat the following steps until $k = k_{\max}$:

(a) Shaking. Generate a point x' at random from the kth neighborhood of x ($x' \in \mathcal{N}_k(x)$);

(b) Local search. Apply some local search method with x' as initial solution; denote with x'' the so obtained local optimum;

(c) Move or not. If this local optimum is better than the incumbent, move there ($x \leftarrow x''$), and continue the search with $\mathcal{N}_1$ ($k \leftarrow 1$); otherwise, set $k \leftarrow k + 1$;

Figure 6.3. Steps of the basic VNS.

(iii) The **Basic VNS** (VNS) method [112] combines deterministic and stochastic changes of neighborhood. Its steps are given on Figure 6.3.

The stopping condition may be e.g. maximum CPU time allowed, maximum number of iterations, or maximum number of iterations between two improvements. Often successive neighborhoods $\mathcal{N}_k$ will be nested. Observe that point x' is generated at random in step 2a in order to avoid cycling, which might occur if any deterministic rule was used. Note also that the Local search step (2b) may be replaced by VND. Using this VNS/VND approach led to the most successful applications recently reported (see e.g. [3,16,24–29,81,124,125]).

3 EXTENSIONS

Several easy ways to extend the basic VNS are now discussed. The basic VNS is a descent, first improvement method with randomization. Without much additional effort it could be transformed into a descent-ascent method: in Step 2c set also $x \leftarrow x''$ with some probability even if the solution is worse than the incumbent (or best solution found so far). It could also be changed into a best improvement method: make a move to the best neighborhood k^* among all $k_{\max}$ of them. Other variants of the basic VNS could be to find solution x' in Step 2a as the best among b (a parameter) randomly generated solutions from the kth neighborhood, or to introduce $k_{\min}$ and k_{step}, two parameters that control the change of neighborhood process: in the previous algorithm instead of $k \leftarrow 1$ set $k \leftarrow k_{\min}$ and instead of $k \leftarrow k + 1$ set $k \leftarrow k + k_{\text{step}}$.

While the basic VNS is clearly useful for approximate solution of many combinatorial and global optimization problems, it remains difficult or long to solve very large instances. As often, size of problems considered is limited in practice by the tools available to solve them more than by the needs of potential users of these tools. Hence, improvements appear to be highly desirable. Moreover, when heuristics are applied to really large instances their strengths and weaknesses become clearly apparent. Three improvements of the basic VNS for solving large instances are now considered.

Initialization. Select the set of neighborhood structures $\mathcal{N}_k$, for $k = 1, \ldots, k_{max}$, that will be used in the search; find an initial solution x; choose a stopping condition;

Repeat the following sequence until the stopping condition is met:

(1) Set $k \leftarrow 1$;

(2) Repeat the following steps until $k = k_{\max}$:

(a) Shaking. Generate a point x' at random from the kth neighborhood of x ($x' \in \mathcal{N}_k(x)$); in other words, let y be a set of k solution attributes present in x' but not in x ($y = x' \setminus x$).

(b) Local search. Find the local optimum in the space of y either by inspection or by some heuristic; denote the best solution found with y' and with x'' the corresponding solution in the whole space S ($x'' = (x' \setminus y) \cup y'$);

(c) Move or not. If the solution thus obtained is better than the incumbent, move there ($x \leftarrow x''$), and continue the search with $\mathcal{N}_1$ ($k \leftarrow 1$); otherwise, set $k \leftarrow k+1$;

Figure 6.4. Steps of the basic VNDS.

(iv) The **Variable Neighborhood Decomposition Search** (VNDS) method [84] extends the basic VNS into a two-level VNS scheme based upon decomposition of the problem. Its steps are presented on Figure 6.4.

Note that the only difference between the basic VNS and VNDS is in step 2b: instead of applying some local search method in the whole solution space $\mathcal{S}$ (starting from $x' \in \mathcal{N}_k(x)$), in VNDS we solve at each iteration a subproblem in some subspace $V_k \subseteq \mathcal{N}_k(x)$ with $x' \in V_k$. When the local search used in this step is also VNS, the two-level VNS-scheme arises.

VNDS can be viewed as embedding the classical successive approximation scheme (which has been used in combinatorial optimization at least since the sixties, see, e.g., [66]) in the VNS framework. Other simpler applications of this technique, where the size of the subproblems to be optimized at the lower level is fixed, are Large neighborhood search [128] and POPMUSIC [131].

(v) The **Skewed VNS** (SVNS) method [74], a second extension, addresses the problem of exploring valleys far from the incumbent solution. Indeed, once the best solution in a large region has been found it is necessary to go quite far to obtain an improved one. Solutions drawn at random in far-away neighborhoods may differ substantially from the incumbent and VNS can then degenerate, to some extent, into the Multistart heuristic (in which descents are made iteratively from solutions generated at random and which is known not to be very efficient). So some compensation for distance from the incumbent must be made and a scheme called Skewed VNS is proposed for that purpose. Its steps are presented in Figure 6.5.

SVNS makes use of a function $\rho(x, x'')$ to measure distance between the incumbent solution x and the local optimum found x''. The distance used to define the $\mathcal{N}_k$, as in the above examples, could be used also for this purpose. The parameter α must be chosen in order to accept exploring valleys far from x when $f(x'')$ is larger than $f(x)$ but not too much (otherwise one will always leave x). A good value is to be found experimentally in each case. Moreover, in order to avoid frequent moves from x to a close solution one may take a large value for α when $\rho(x, x'')$ is small. More sophisticated choices for a function $\alpha\rho(x, x'')$ could be made through some learning process.

(vi) **Parallel VNS** (PVNS) methods are a third extension. Several ways for parallelizing VNS have recently been proposed [32,106] in solving the p-Median problem.

Initialization. Select the set of neighborhood structures $\mathcal{N}_k$, for $k = 1, \dots, k_{max}$, that will be used in the search; find an initial solution x and its value $f(x)$; set $x_{opt} \leftarrow x$, $f_{opt} \leftarrow f(x)$: choose a stopping condition and a parameter value α;

Repeat the following until the stopping condition is met:

(1) Set $k \leftarrow 1$;

(2) Repeat the following steps until $k = k_{\max}$:

(a)*Shaking*. Generate a point x' at random from the kth neighborhood of x;

(b)*Local search*. Apply some local search method with x' as initial solution; denote with x'' the so obtained local optimum;

(c) *Improve or not*. If $f(x'') < f_{opt}$ set $f_{opt} \leftarrow f(x)$ and $x_{opt} \leftarrow x''$;

(d) *Move or not*. If $f(x'') - \alpha\rho(x, x'') < f(x)$ set $x \leftarrow x''$ and $k \leftarrow 1$; otherwise set $k \leftarrow k + 1$.

Figure 6.5. Steps of the Skewed VNS.

In [106] three of them are tested: (i) parallelize local search; (ii) augment the number of solutions drawn from the current neighborhood and do local search in parallel from each of them and (iii) do the same as (ii) but updating the information about the best solution found. The second version gave the best results. It is shown in [32] that assigning different neighborhoods to each processor and interrupting their work as soon as an improved solution is found gives very good results: best known solutions have been found on several large instances taken from TSP-LIB [122]. Three Parallel VNS strategies are also suggested for solving the Traveling purchaser problem in [116].

4 HYBRIDS

As change of neighborhood in the search for good solutions to (1) is a simple and a very powerful tool, several authors have added such a feature to other metaheuristics than VNS.

In this section we review the resulting Hybrids at a general level; more details concerning specific applications are given in the next section.

(i) **VNS and Tabu search** Tabu search ([62,63,72]) is a metaheuristic that has a huge number of applications (see, e.g., [64]). It explores different types of memories in the search, i.e., *recency* based (short-term), *frequency* based, *long-term* memories etc. Usually it uses one neighborhood structure and, with respect to that structure, performs descent and ascent moves building a trajectory. In principle, there are two ways of making hybrids of VNS and TS: use TS within VNS or use VNS within TS. Recently four reports on hybrids of the first kind and two on hybrids of the second kind have been proposed.

For solving the Route-median problem, TS is used instead of Local search within VNS (step 2b of Figure 6.3) in [126] (the resulting method is called VNTS), while in [20], in solving the Nurse rostering problem, in each among several neighborhoods simple descent or TS are used alternatively. A Multi-level TS is proposed in [97] for solving the Continuous min–max problem, where each level represents a ball of different size in the vicinity of the current solution, i.e., different neighborhoods, and a tabu list is constructed for each level. Reactive variable neighborhood descent (ReVND) is proposed in [14] for solving the Vehicle routing problem with time windows. In repeating a sequence of four proposed local searches, the information on unsuccessful pairs of edges is memorized so that in the next repetition those pairs are not considered. In that way the size of each neighborhood is reduced.

Note that nested VND can easily be used in a TS framework since cycling can be avoided by controlling only one neighborhood. In solving the Multisource Weber problem, several heuristics that use nested VND within TS are proposed in [16]. Moves are considered as 'tabu' regarding the *relocation* neighborhood only, while for each solution of that neighborhood *reallocation* and *alternate* moves are used in a sequence. It is also observed in [37] that the same Tabu list can be used for several different neighborhoods used sequentially (i.e., Or-opts and interchanges) for solving problems where the solution is stored as a permutation of its attributes or variables. This makes possible a hybrid of TS and VND, i.e., VND can be used within TS.

(ii) **VNS and GRASP** GRASP is a two phase metaheuristic [48]. In the first phase solutions are constructed using a greedy randomized procedure and in the second, solutions are improved by some local search or enumerative method. A natural way of hybridizing GRASP and VNS is to use VNS in the second phase of GRASP. Such an approach has been performed in solving the Steiner tree problem in graphs [108,125], the Phylogeny problem [3], the Prize-collecting Steiner tree problem [24], the Traveling purchaser problem [116] and the Max-Cut problem [49]. The results reported show improvements of the GRASP/VNS hybrid over the pure GRASP.

(iii) **VNS and Constraint programming** In the last few years, Constraint programming (CP) has attracted high attention among experts from many areas because of its potential for solving hard real-life problems. A *constraint* is a logical relation among several variables, each taking a value in a given domain. The important feature of constraints is their declarative manner, i.e., they specify what relationship must hold without specifying a computational procedure to enforce it. One aim of CP is to solve problems by stating constraints and finding solutions satisfying all these constraints. It has been noted that local search techniques can successfully be used in CP (see, e.g., [118,119,128]). In [57], two operators have been suggested and used in a local descent (VND), within Large neighborhood search (LNS) and CP framework for solving the Vehicle routing problem with time windows (VRPTW). They are called LNS-GENI and SMART (SMAll RouTing). The idea of LNS [128], which can be seen as a general decomposition heuristic, is first to 'destroy' the current solution by removing a given number of solution attributes and then rebuilt it in the best possible way by solving smaller problems. Both phases are problem dependent, but some general rules are recommended in [128]. In [12], in order to minimize the total travel costs of VRPTW, destruction of the solution is performed in a more systematic way: it starts from $k = 2$ attributes, then 3, etc.; once the improvement in rebuilding phase is obtained, $k = 2$ is set again. This version is very close to VNDS. A simpler version of LNS (called LNS/CP/GR) is suggested in [105] for solving a *Valued Constraint satisfaction problem* (VCSP). In VCSP the penalty (a valuation) must be paid for each unsatisfied variable. Then the objective is to find values of variables such that the sum of penalties is minimum. In rebuilding the solution, authors use a greedy procedure. However, in future work, they could include VNS, which has been done recently by other authors. In [107] so-called VNS/LDS + CP heuristic is proposed and tested on the Frequency assignment problem.

5 FINDING GOOD SOLUTIONS

In this section and the three following ones, we review applications of VNS, the number of which has rapidly increased since 1997, when the first paper on VNS was published. We begin with traditional ones, i.e., finding good solutions to combinatorial and global optimization problems, that is near optimal ones or possibly optimal ones but without a proof of optimality. These applications are grouped by field and within each field papers are considered chronologically.

5.1 Traveling Salesman and Vehicle Routing Problems

We shall consider here Traveling salesman problem (TSP), Vehicle routing problem (VRP), Arc routing problem (ARP) and some of their extensions that have been solved by VNS.

Traveling salesman problem Given n cities with intercity distances, the traveling salesman problem (TSP) is to find a minimum cost tour x (i.e., a permutation of the cities which minimizes the sum of the n distances between adjacent cities in the tour). It is a classical NP-hard problem.

A heuristic for the Euclidean TSP called GENIUS was developed in [59]. It is a sophisticated insertion followed by local deletion/insertion and correction procedure. The size of the neighborhood in GENIUS depends on a parameter p (the number of cities already in the tour closest to the city that is considered for possible deletion or insertion). We immediately get a set of neighborhood structures for VNS by denoting with $\mathcal{N}_p(x)$ all tours obtained by deletion/insertion with parameter value p. Details can be found in [112] where results on the same type of test problems as reported in [59] are given. VNS gives a 0.75% average improvement over GENIUS within a similar CPU time. Moreover, improvements are obtained for all problem sizes.

Probably the most popular heuristic for solving TSP is 2-opt, where in turn two links between cities in the current tour x are removed and these cities reconnected by adding links in the only other way which gives a tour. Since 2-opt is a local search descent heuristic, it stops in a local minimum. In [78] the basic VNS rules using 2-opt (or a quicker variant, in which only the shortest edges are used) as local search routine, are applied. Average results for random Euclidean problems over 100 trials for $n = 100, \ldots, 500$ and 10 trials for $n = 600, \ldots, 1000$ are reported. Average improvements in value of 2.73% and 4.43% over the classical 2-opt heuristic within a similar computing time are obtained by these two versions respectively.

In [19] a new local search heuristic, the so-called *k-hyperopt* is proposed. Then a VND heuristic uses 2-opt, 2-hyperopt and 3-hyperopt neighborhoods in descent, while a VNS heuristic uses k-hyperopt for shaking and 2-hyperopt for a local search. The new methods are compared with Iterated local search and Multistart 2-opt. Results are reported on standard test instances. It appears that tours obtained are comparable with iterated Lin-Kernighan [103] in terms of tour quality, but CPU time is larger.

In a work in progress [58], a similar VND heuristic is applied within two different decomposition schemes, i.e., VNDS: on the one hand subproblems corresponding to k successive points on the current tour are selected and re-optimized; on the other hand the same is done but with the k closest points from a randomly chosen one. It appears that the second decomposition gives the best results.

Traveling salesman problem with back-hauls The GENIUS heuristic [59] was applied to the *TSP with back-hauls* in [60]. In this problem customers (or cities) are divided into three disjoint sets: depot, line-haul and back-haul customers. Starting from the depot, a tour must be designed such that all line-haul customers are visited before all back-haul customers. This time VNS gives a 0.40% average improvement over GENIUS with a 30% increase in computing time [112]. Improvements are obtained for all problem sizes.

Route-median problem Given a set of cities with inter-cities distances, the Route-Median problem consists in choosing a subset of cities that are included in a cycle and allocating the remaining ones each to the closest chosen city in order to minimize the length of the route with an upper bound on the sum of the distances from the cities not in the route to the cities to which they are assigned. An exact branch and cut solution method has been proposed in [101]. A metaheuristic approach that combines VNS and TS and uses several neighborhood structures is proposed in [126].

Vehicle routing problem with time windows The Vehicle routing problem (VRP) consists in designing least cost routes from one depot to a given set of customers. The routes must be designed in such a way that each customer is visited only once by exactly one vehicle. There are usually capacity constraints on the load that can be carried by a vehicle, and each customer has a known demand. The time window variant to the problem (VRPTW) imposes the additional constraint that each customer must be visited within a given time interval.

Four methods for solving VRPTW have recently been proposed that include VNS ideas as well. In [57] two operators which make use of Constraint programming and local search to explore their neighborhood are proposed. These operators are combined in a VND framework. Computational results show that this method is comparable with other heuristics, often producing better solutions in terms of distance traveled. Another heuristic [14], in its third phase, uses four neighborhoods in descent. In addition, not-improving neighborhood solutions are memorized, so that in the next pass they are not visited. The resulting method is called Reactive VND (ReVND). The obvious fact that a local minimum w.r.t. one objective is not necessary so for another, can also be a powerful tool for escaping from the local minima trap. This has been explored in [14]. In the second phase the objective is minimum number of vehicles, in the third (where ReVNS is proposed) the objective is to minimize the total travel distance, and in the last phase a new objective function is defined that considers both criteria with given weights. Results are reported on standard data sets. It appears that the proposed procedure outperforms other recent heuristics for VRPTW, and that four new best known solutions are found. In [30] the same idea is used (i.e., escape from the local minima trap by changing both the neighborhood structure and the objective function), but with different local searches and in a different way: the classical k-opt exchanges improve the solution (for $k \leq 3$) in terms of number of vehicles and then another objective is considered. Beside the two objectives mentioned above, two other functions are used as well. No parameter tuning is required and no random choice is made. The algorithm has been tested on benchmark problems with favorable results, when compared with those obtained by most recent heuristics. In the most recent paper for solving VRTPTW [12], the idea of changing both the objective function and the neighborhoods within the search is explored further, and probably the best results to date on Solomon's benchmark instances are reported. The heuristic has two phases. In the first one, five neighborhood structures (2-opt, Or-opt, relocation, interchange and cross-over) and three objective functions (number of routes, maximum of the sum-of-squares route sizes, minimum of minimal delay) are considered. All five neighborhoods are not used as in VND (not in a sequential nor nested way), i.e., a single one among them is chosen at random and explored for all three objectives. The resulting procedure uses the Simulated annealing framework. In the second phase the authors implement a modified Large neighborhood search (LNS) [128] heuristic (minimizing the total distance traveled) which, according to them, make it very close to VNS.

Vehicle routing problem with backhauls (VRPB) In this problem the vehicles are not only required to deliver goods to (linehaul) customer, but also to pick up goods at (backhaul) customers locations. In [33] Reactive tabu search (RTS) and VND are developed and compared on some VRP test instances from the literature with different percentage of linehaul customers (50%, 66% and 80%). VND uses insertion and interchange moves in descent. It appears that RTS and VND are 1.8% and 5.3% above known optimal value.

Arc routing problem In Arc routing problems the aim is to determine a least cost traversal of all edges or arcs of a graph, subject to some side constraints. In the Capacitated arc routing problem (CARP) edges have a non-negative weight and each edge with a positive weight must be traversed by exactly one of several vehicles starting and ending their trip at a depot, subject to the constraint that total weight of all edges serviced by a vehicle cannot exceed its capacity. In [61] the VNS heuristic developed in [109] for the undirected CARP was compared with a Tabu search heuristic [87]. The conclusions are as follows: (i) on small test instances ($7 \leq n \leq 27$), TS and VNS perform similarly (the deviation from optimality is identical and computing times are about 20% smaller for VNS); (ii) on larger test instances VNS is better than TS in terms of solution quality, but also faster (average deviation from optimality and computing times on the 270 instances were 0.71% and 349.81 s for TS and 0.54% and 42.50 s for VNS).

Linear ordering problem Given a squared $n \times n$ matrix D, the Linear ordering problem (LOP) consists in finding permutations of rows and columns of D such that the sum of all elements above the main diagonal is maximum. LOP has a large number of applications such as triangulation of input-output matrices, archaeological seriation, scheduling etc. In [65], a basic VNS heuristic for solving LOP is proposed. Neighborhoods are derived by k row interchanges, and local searches performed using the same neighborhood as in the TS approach from [102]. The outer loop of the basic VNS has not been used, i.e., the procedure stops the first time all $k_{\max}$ neighborhoods were explored. It is shown, on the same 49 test instances, that both TS and VNS ($k_{\max}$ is set to 10) have similar performance: TS is slightly better in terms of the solution quality (number of optimal solutions found by TS was 47 and 44 for VNS); VNS is faster (average time was 0.93 seconds for TS and 0.87 for VNS).

Traveling purchaser problem (TPP) can be seen as an extension of TSP. Given a set of m items to be purchased at n markets, the cost of item k at market j ($p_{kj}, k = 1, .., m, j = 1, .., n$) and inter-market travel costs t_{ij}, the problem is to purchase all m products by visiting a subset of the markets in a tour (starting from the source $j = 0$), such that the total travel and purchase costs are minimized. This problem includes many well-known NP-hard problems such as uncapacitated facility location, set covering and group Steiner tree problems as its special cases [121].

Several constructive heuristics for the TSP are adapted in the initial solution building (within GRASP) and several neighborhoods are used for local search (within VND) for solving TPP in [116,129]. Among many possible variants that have been investigated in the sequential implementation are two from each class of heuristic (GRASP, VNS, Hybrid GRASP/VNS and TS [135]) on 16 random test instances with 50–150 markets and items. Each heuristic was restarted 5 times and average results reported. Among 8 sequential codes the GRASP/VNS hybrid had the best performance, while among 5 parallel methods, VNS was the best in average.

5.2 Location and Clustering Problems

p-Median problem Given a set L of m potential locations for p facilities and a set U of given locations for n users, the p-Median problem (PM) is to locate simultaneously the p facilities in order to minimize the total transportation distance (or cost) from users to facilities, each user being served by the closest facility. Solutions of PM are thus characterized by 0–1 vectors x with p components among $|L|$ equal to 1 indicating where facilities are located.

There are several papers that use VNS for solving the PM problem. In the first one [77], the basic VNS is applied and extensive statistical analysis of various strategies performed. Neighborhood structures are defined by moving $1, 2, \ldots, k_{\max}$ facilities and correspond to sets of 0–1 vectors at Hamming distance $2, 4, \ldots, k_{\max}$ from x. The descent heuristic used is 1–interchange, with the efficient *Fast Interchange* (FI) computational scheme [136]. Results of a comparison of heuristics for OR-Lib and some TSP-Lib problems are reported. It appears that VNS outperforms other heuristics. In order to solve larger PM problem instances, in [84] both RVNS and VNDS are applied. Subproblems with increasing number of users (that are solved by VNS) are obtained by merging the sets of users (or market areas) associated with k ($k = 1, \ldots, p$) medians. Results on 1400, 3038 and 5934 users instances from the TSP library show VNDS improves notably upon VNS in less computing time, and gives much better results than FI, in the same time that FI takes for a single descent. Moreover, Reduced VNS (RVNS), which does not use a descent phase, gives results similar to those of FI in much less computing time. Two versions of Parallel VNS for PM are proposed in [32,106] (see Section 3).

Multisource Weber problem The multisource Weber (MW) problem (also known as continuous location-allocation problem) is the continuous counterpart of PM: instead of locating the p facilities at some locations of L, they can be located anywhere in the plane. An early application of VNS to MW is given in [17]. Several other ones are discussed at length in [16]. It appears that the choice of neighborhoods is crucial. Reassignment of customers to facilities a few at a time is a poor choice, as it entails only marginal changes in the solutions considered. Much better results are obtained when the facilities themselves are moved. As they may be located anywhere in the plane target locations are needed. An easy and efficient choice is locations of customers where there is no facility as yet. Using this neighborhood structure, several basic TS and VNS heuristics were developed and an extensive empirical study carried out to evaluate various heuristics—old, recent, and new—in a unified setting. The different methods (i.e., Genetic search, three Tabu search variants, four VNS variants etc.) were compared on the basis of equivalent CPU times. Results of this study indicate that VNS can be effectively used to obtain superior solutions. For instance on a series of 20 problems with 1060 users the average error (by comparison with best known solution) is of 0.02% only for the best VNS, while it can rise to more than 20% for some well-known heuristics of the literature. Average error for a Genetic algorithm was 1.27% and for the best TS 0.13%.

Minimum-sum-of-squares clustering problem Given a set of n points in Euclidean q-dimensional space, the minimum sum-of-squares clustering problem (MSSC) is to partition this set into classes, or clusters, such that the sum of squared distances between entities and the centroids of their clusters is minimum. Among many heuristics for MSSC, the k-Means local search heuristic [93] is the most popular. It is an interchange heuristic, where points are reassigned to another cluster than their own, one at a time, until a local optimum is reached. Another popular heuristic, called H-Means [2], selects an initial partition, computes the centroids of its clusters, then reassigns entities to the closest centroid and iterates until stability. A new local search heuristic, called J-Means is proposed in [81]: centroids are relocated at some of the given points, which do not yet coincide with one of them. Results of a comparison of k-Means, H-Means, $H + K$-Means (where H-Means and k-Means are applied in sequence) and two versions of VNS are given in [81]. VNS–1 is an extension of k-Means and gives slightly better results than $H + K$-Means; VNS–2 which extends J-Means proves to be the best heuristic.

Fuzzy clustering problem The Fuzzy clustering problem (FCP) is an important one in pattern recognition. It consists in assigning (allocating) a set of patterns (or entities) to a given number of clusters such that each of them belongs to one or more clusters with different degrees of membership. The objective is to minimize the sum of squared distances to the centroids, weighted by the degrees of membership. The fuzzy clustering problem was initially formulated in [42] as a mathematical programming problem and later generalized in [13]. The most popular heuristic for solving FCP is the so-called Fuzzy C-means (F-CM) method [23]. It alternatively finds membership matrices and centroids until there is no more improvement in the objective function value. A new descent local search heuristic called Fuzzy J-means (F-JM) is proposed in [9] where the neighborhood is defined by all possible centroid-to-entity relocations (see also [16,81]). This 'integer' solution is then moved to a continuous one by an alternate step, i.e., by finding centroids with given memberships. Fuzzy VNS rules are applied as well (F-JM is used as local search subroutine).

p-Center problem The p-Center problem consists in locating p facilities and assigning clients to them in order to minimize the maximum distance between a client and the facility to which he (or she) is allocated (i.e., the closest facility). This model is used for example in locating fire stations or ambulances, where the distance from the facilities to their farthest allocated client should be minimum. In [111] basic VNS and Tabu search (e.g., the so-called Chain substitution Tabu Search, [113]) heuristics are presented. Both methods use the 1-interchange (or vertex substitution) neighborhood structure. It is shown how this neighborhood can be used even more efficiently than for solving the p-Median problem. Based on the same computing time, comparison between the Multistart 1-interchange, the Chain interchange TS (with one and two Tabu lists) and VNS are reported on standard test problems from the literature. It appears that both TS and VNS outperform the Multistart approach and give similar results with a slight edge in favor of VNS for the larger instances.

Quadratic assignment problem The Quadratic Assignment Problem can be described as follows: Given two $n \times n$ matrices A and B, find a permutation π^* minimizing the sum of the $a_{ij} \cdot b_{\pi_i \pi_j}$. A parameter free basic VNS ($k_{\max}$ is set to n) is suggested in [130], where two new methods based on Fast Ant systems (FANT) and Genetic-Descent hybrid (GDH) are also proposed. All three methods use the same simple descent local search procedure. On a series of *structured* instances from the literature, results of good quality for all three methods are reported, i.e., the average errors with respect to the best known solutions are 0.185%, 0.167% and 0.184% for FANT, GDH and VNS respectively with time necessary for 1,000 calls to the improving procedure of FANT. Best results, i.e., 0.104% error were obtained with the previous Hybrid Ant system (HAS-QAP) of [56].

Balanced MBA student teams problem In some schools and universities, students from the same grade must sometimes be divided into several teams within a classroom in such a way that each team provides a good representation of the classroom population. A problem is to form these teams and to measure the quality of their balance. In [39] mathematical models are proposed that take into account the attributes assigned to the students. Two different ways of measuring the balance among teams are proposed: *min–sum* and *min–max* models. Two formulations are considered and both exact and heuristic solution methods are developed. The exact method consists in solving a *Set Partitioning* problem based on the enumeration of all possible teams. In order to solve large problem instances, both VNS and VNDS heuristics are also developed. In the basic VNS, interchange of ℓ ($\ell = 1, \dots, \ell_{\max}$) student pairs is used for the perturbation of an incumbent solution. The local search is performed within $\mathcal{N}_1$, and an initial solution is obtained by using a random partition. Since the gap was larger than 1% on a 65 student instance with 19 attributes and 13 teams, VNDS was tested as well. The use of this extended VNS scheme led to obtain results of better quality (0.48% larger than optimum) in moderate time (29 s for VNDS versus 872 s for the exact method).

Simple plant location problem The well-known simple plant location problem (SPLP) is to locate facilities among a given set of cities in order to minimize the sum of set-up costs and distribution cost from the facilities to a given set of users, whose demand must be satisfied. The case where all fixed-costs are equal and the number of facilities is fixed is the p-Median problem. In work in progress, VNS heuristics for the p-median are extended to the SPLP. In addition to interchange moves, opening and closing of facilities are considered and a VNDS similar to that one suggested in [84] developed. Moreover, complementary slackness conditions are exploited to find a bound on the error through solution of a reduced dual problem. In this way solution within 0.04% of optimality could be obtained for problems with up to 15,000 users (see Table 6.1).

One-dimensional bin-packing problem (BPP) Packing items into boxes or bins is a task that occurs frequently in distribution and production. BPP is the simplest bin-packing problem: given a number of items $i = 1, \dots, n$ of integer sizes t_i, what is the minimum number of bins, each having the same integer capacity c, necessary to pack all items.

In developing the basic VNS for BPP, usual neighborhood structures are used in [52]: add, drop and interchange (swap) of items between bins. By using sophisticated data structure, neighborhoods considered are significantly reduced. In an intensified shaking step an item is selected at random and its best position (w.r.t. add/drop/swap restricted neighborhood) found; to get x' that belongs to $\mathcal{N}_k(x)$, this step is repeated k times. Local search uses the same neighborhoods. In addition, since a characteristic of the BPP is existence of large plateaus (many different configurations, in terms of assignment of items to bins, correspond to the same number of bins), an auxiliary objective function is

Table 6.1. SPLP on Euclidean random instances with equal fixed costs ($\sqrt{n}/1000$); Dual is reduced by both upper and lower bounds; Sun 533 Mhz; stopping rules: number of iterations without improvement = 20 for both VNDS and VNS; $k_{max} = 20$ for VNS; $k_{max} = \min\{p, 25\}$ for VNDS

			Objective values			Time (s)			Time total		
n	Fix costs	p	CPLEX	RVNS	VNDS	CPLEX	RVNS	VNDS	Best	All	Gap
500	223	347	99794.0	107984	99794	0.1	0.2	0.3	0.6	1.5	0.0000
1000	316	391	223206.0	247790	223206	0.3	0.7	1.3	2.3	6.6	0.0000
2500	500	498	542166.5	614880	542267	2.8	3.2	34.8	40.8	66.4	0.0185
5000	707	600	1028221.0	1217007	1028255	19.5	6.9	239.6	266.0	382.1	0.0033
10000	1000	752	1914908.0	2163915	1915562	167.5	36.3	1530.4	1734.2	2212.4	0.0342
15000	1224	844	2733060.5	3134626	2733979	1071.9	61.3	5103.4	6236.6	7390.1	0.0336

introduced, i.e., maximization of the sum of squared slacks of the bins. Initial solutions for VNS are obtained by modification of an existing one (called MBS'), which already produces good solutions. When tested on 1370 benchmark instances from two sources, VNS proved capable of achieving the optimal solution for 1329 of them, and could find for 4 instances solutions better than the best known ones. According to the authors, this is a remarkable performance when set against other methods. For example, when compared with the Tabu search heuristic from [127] on 1210 hardest test instances, it appears that in 1010, 1125 and 1170 instances the best solutions are found by MBS', TS and MBS' + VNS respectively.

5.3 Graphs and Networks

Let $G = (V, E)$ be a connected (un)directed graph with vertex set V and edge set E. Solution of many optimization problems on G correspond to some subset of vertices $V' \subseteq V$ or subset of edges $E' \subseteq E$ that satisfies certain conditions. A usual way to supply a solution space with some metric, and thus make possible development of VNS heuristics, is to define a distance function as the cardinality of the (symmetric) difference between any two solutions V_1 and V_2 or E_1 and E_2, i.e., node-based or edge-based neighborhoods:

$$\rho_1(V_1, V_2) = |V_1 \setminus V_2| \quad (\text{or } \rho_1 = |V_1 \Delta V_2|) \tag{2}$$

$$\rho_2(E_1, E_2) = |E_1 \setminus E_2| \quad (\text{or } \rho_2 = |E_1 \Delta E_2|) \tag{3}$$

where Δ denotes the symmetric difference operator. In the applications of VNS that follow, ρ_1 or ρ_2 or both metrics are used for inducing different neighborhood structures $\mathcal{N}_k$:

$$V_2 \in \mathcal{N}_k(V_1) \Leftrightarrow \rho_1(V_1, V_2) = k \quad \text{and/or} \quad E_2 \in \mathcal{N}_k(E_1) \Leftrightarrow \rho_2(E_1, E_2) = k. \tag{4}$$

Oil pipeline design problem Brimberg et al. [15] consider a given set of offshore platforms and on-shore wells, producing known (or estimated) amounts of oil, to be connected to a port. Connections may take place directly between platforms, well sites and the port or may go though connection points at given locations. The configuration of the network and sizes of pipes used must be chosen in order to minimize construction costs. This problem is expressed as a mixed integer program, and solved both heuristically by Tabu search and VNS methods and exactly by a branch-and-bound method. Tests are made with data from the South Gabon oil field and randomly-generated problems. VNS gives best solutions and TS is close.

Phylogeny problem The phylogeny (or evolutionary tree) problem, in its simplest form, can be formulated as follows: given A, an $n \times m$ 0–1 matrix, construct a tree with minimum number of 'evolutionary' steps, that satisfies the following conditions: (i) each node of the tree corresponds to an n-dimensional 0–1 vector, where vectors given by rows of A should be associated to leaves; (ii) degrees of all nodes that are not leaves are equal to three; (iii) two nodes are connected by an edge if their corresponding vectors differ in only one variable. The trees that satisfies (i)–(iii) and the criterion used are called phylogeny and parsimony respectively. The leaves of an evolutionary tree represent groups of species, populations of distinct species, etc. (denoted by taxons as well), while interior nodes represent hypothetical (unknown) ancestors. In [3] a VND method that uses three neighborhoods is developed, and compared with the best

Table 6.2. Phylogeny problem: final average results on 8 test problems

Heuristic	% Deviation	# of Best values	Time (min)
GRASP	0.6	5	481.8
GRASP/VND	0.6	2	321.9
VNS	0.5	6	714.4

among three local searches used alone. It appears that VND is much faster and gives solutions with better quality. In the same paper, the authors developed and compared three metaheuristic approaches to the phylogeny problem: GRASP (with the best local search among three for the second phase, hybrid GRASP with VND and VNS. The method developed as VNS use the same three neighborhoods for Shaking as for VND ($k_{\max} = k'_{\max} = 3$) and does therefore not use the full potential of VNS. Nevertheless it gives the best results. Comparative results on eight test problems are summarized in Table 6.2 (taken from [3]).

Maximum clique problem Let $G = (V, E)$ denote a graph with vertex set $V = \{v_1, v_2, \ldots, v_n\}$ and edge set $E = \{e_1, e_2, \ldots, e_m\}$. A set $C \subseteq V$ of vertices is a *clique* if any two of them are adjacent. A clique C is *maximum* if it has the largest possible number of vertices. Finding the maximum clique is a classical problem in applied graph theory. A basic VNS heuristic that combines greedy with the simplicial vertex test in its descent step is proposed and tested on standard test problems from the literature in [85]. Despite its simplicity, the proposed heuristic outperforms most of the well-known approximate solution methods. One exception is the recent Reactive Local Search (RLS) heuristic [7] which obtains results of equal quality in a quite different way.

Steiner tree problem on graphs Given a graph $G = (V, E)$, nonnegative weights w_{ij} associated with the edges (i, j) of E and a subset $X \subseteq V$ of terminal nodes, the Steiner problem (SPG) is to find a minimum weighted subtree of G which spans all terminal nodes. The solution X and corresponding tree are called Steiner nodes and Steiner minimum tree respectively. Application can be found in many areas, such as telecommunication network design, computational biology, VLSI design, etc. (see, e.g., [90] for a survey).

In [108] and later in [125], two types of neighborhoods, i.e., *node-based* (N) (2) and *path-based* (P) [134], are successfully used for solving SPG as in VND. For example, the average % of improvement over a constructive initial solution on 454 instances was 10.24% for the N–P order, while only P local search led to a 2.59% improvement. Computing time for both heuristics was the same. This N–P procedure is used within parallel GRASP in [108] and improved by Path relinking in [125], but no attempt was made to develop a full VNS, which could be an interesting task for future work.

Degree-constrained minimum spanning tree problem (DCMST) consists in finding a minimum spanning tree such that the degree of each vertex of the tree is less than or equal to a given integer b_i, for all i. Note that for $b_i = 2$ the problem becomes the Hamiltonian path problem. In [124] three neighborhood structures based on the k-edge exchange operator, $k = 1, 2, 3$ (or k-elementary tree transformations) are used in developing local search procedures within a VND heuristic. This procedure is then embedded into a VNS as a local search, where again k-edge exchange moves are used for perturbation (shaking) with parameter $k_{\max} = 0.4(|V|-1)$. VNS stops the first time $\mathcal{N}_{k_{\max}}$ has been explored without improvement. VNS and VND are then extensively compared with known methods from the literature, i.e., a genetic search, problem space search heuristics and simulated annealing, on four different sets of randomly generated problems: CRD, SYM, STR and SHRD. VND and VNS were better and much faster on almost all test instances. Some observations from their empirical study are as follows: (i) VND alone succeeded in finding optimal solutions for all problems already solved exactly (in the STR class) in less than one second of processing time, while the best among three other heuristics, i.e., the problem space search heuristic, found suboptimal solutions within up to 300 s; (ii) on SHRD class of instances, VND reduced the average relative error of Local search (with a single neighborhood $k = 1$) from 4.79% to 0.27% with a very small increase in computing time; (iii) for the same series of instances, VNS found 23 out of 24 best solutions; (iv) for instances CRD and SYM, VNS got optimal solution on 69 out of 70 instances; (v) VNS found optimal or best known solutions for all instances of class STR; it improved the average best solution values found by GA-F [99] for 35 out of the 45 pairs of instances considered. Some of the results that compare GA-F and VNS on STR instances are given in Table 6.3 (a part of Table 5.7 from [124]).

Max-cut problem Given an undirected graph $G = (V, E)$ and weights w_{ij} on the edges $(i, j) \in E, i, j \in V$ the Max-cut problem is to find a subset of vertices S that maximizes the sum of the edge weights in the cut $(S, \bar{S})$. This problem has wide applications including VLSI design. Three heuristics for solving Max-cut are developed and compared in [49]: GRASP, basic VNS and a hybrid of GRASP and VNS (G-VNS).

Table 6.3. DCMST problem: test problems STR, for $|V| = 100$; parameter *dim* designates the dimension of the space for randomly generated data; sign '−' shows that the best known solution is improved

Problem		GA-F		VNS	
b_i	dim	% deviation	time	% deviation	time
3	3	0.17	257.55	0.00	7.58
3	4	1.10	271.85	−0.01	7.39
3	5	0.07	314.29	0.00	7.09
3	6	0.58	203.85	−0.01	11.52
3	7	0.58	233.01	0.00	8.88
4	3	0.00	245.80	0.00	5.55
4	4	0.38	240.35	0.00	6.79
4	5	0.02	254.46	0.00	7.55
4	6	0.51	233.18	0.00	8.62
4	7	0.51	232.48	0.00	9.15
5	3	0.00	220.06	0.00	4.14
5	4	0.05	213.79	0.00	6.45
5	5	0.00	219.51	0.00	7.08
5	6	0.46	206.02	0.00	7.73
5	7	0.48	207.28	0.00	8.83

Solutions are represented as binary $|V|$-vectors and neighborhood structures for VNS are induced from the Hamming distance introduced into the solution space. Local search is performed in the neighborhood with distance 1. G-VNS uses VNS in the local search phase of GRASP. Results on 4 instances from the literature are reported. Preliminary computational results indicate VNS is a robust heuristic for Max-Cut. Indeed VNS is not very sensitive to the choice of initial solution, and thus, G-VNS did not perform better than VNS for the same number of iterations.

Cable layout problem Given the plan of buildings, the location of the equipment and a structure of cable supports of fixed capacities, cables of all kind have to be routed (voltage, communication, etc.). The Cable layout problem (CLP) consists in finding the design of cables that minimize the total cost, such that numerous technical requirements are met. If the cost of a cable is a linear function of its length, then the CLP can be considered as a capacitated multi-path problem. This property is used in [31] to decompose the problem, and then a VNDS heuristic is proposed. Each sub-problem consists in solving several min-cost multi-paths problems. Graphs with up to 1,000 nodes are randomly generated with up to 4% edge density (which is close to real problems) in order to compare VNDS with a previously suggested TS based heuristic [86]. Solutions obtained by VNDS were better than TS in 15 out of 16 instances, in much less computing times. Moreover, for larger instances, it is shown (from the dual bounds obtained from Lagrangian relaxation), that the duality gap is always less than 1%.

k-Cardinality tree problem Given an undirected weighted graph $G = (V, E)$ with vertex set V, edge set E and weights $w_i \in R$ associated to V or to E, the *Minimum weighted k-Cardinality tree problem* (k-CARD for short) consists of finding a subtree of G with exactly k edges whose sum of weights is minimum. There are two versions of this problem: *vertex-weighted* and *edge-weighted*, if weights are associated to V or to E respectively. The k-CARD problem is strongly NP-hard [51]. However, if G is a tree then it is polynomially solvable [51]. In [115] two VNS methods for the edge-weighted k-CARD problem, are proposed: the basic VNS (denoted by VNS-1) and VNS that uses VND as a local search (VNS-2). In VNS-1 the solution space (i.e. the set of all trees with k edges) is supplied with a distance function based on edge difference between any two trees and all neighborhood structures are induced from it. Since the minimal weighted k-cardinality tree is a spanning tree on any subgraph of G with $k + 1$ vertices, the solution space may be reduced to the set of all spanning trees and the set of spanning trees $\mathcal{T}_k$ may be supplied with another metric, i.e., as the cardinality of the difference of their vertex sets. In developing their VND method for k-CARD, the authors use three neighborhood structures.

The first two neighborhoods are induced by *edge distance*, and the third one is induced by a *vertex distance*. In Table 6.4 comparative results of VNS heuristics, two Tabu search methods (TS-1 and TS-2 for short) and the Dynamic-Dijkstra-Path (DDP for short) constructive heuristic are given. (In [45] it is shown that DDP has the best performance when compared with several others constructive heuristics). TS-1 is an implementation of Chain interchange TS rules suggested in [113] and the second TS (denoted by TS-2) is taken from [104].

Test instances include random graphs with 500, 1000, 1500 and 2000 vertices, where degree of each node is set to 20 (see [104] for details). In Table 6.4 average results on 10 random instances for each n and k are reported. Large problem instances could not be solved by DDP and TS-2 in reasonable time and for that reason we did

Table 6.4. k-Cardinality tree problem: average results on 10 random instances. Best values are boldfaced

		Objective value					
n	k	VNS-1	VNS-2	TS-1	TS-2	DDP	t_{max} (s)
500	50	1577.10	1561.50	1600.40	1713.70	**1535.90**	78
	100	3424.80	**3395.00**	3604.80	3736.40	3397.50	93
	150	5553.30	**5525**.30	5826.10	5915.80	5541.50	117
	200	7974.60	**7959**.20	8231.50	8626.80	7980.70	148
	250	10814.80	**10795**.90	11203.30	11349.10	10829.40	170
1000	100	5863.30	**5844.30**	5912.50	—	5847.20	585
	200	11947.80	**11915.40**	12079.60	—	11922.00	655
	300	18230.20	**18220.10**	18393.60	—	18222.60	775
	400	24758.90	**24748.30**	25002.30	—	24763.90	930
	500	31572.90	**31566.80**	31886.10	—	31591.90	1138
1500	150	8209.40	**8197.20**	8260.90	—	—	3000
	300	16569.60	**16557.60**	16686.40	—	—	3500
	450	25107.80	**25039.80**	25293.04	—	—	4000
	600	33842.80	**33804.10**	34099.20	—	—	5000
	750	42820.10	**42785.40**	43118.30	—	—	6000
2000	200	23722.60	**23586.30**	23924.90	—	—	3000
	400	48188.40	**48077.80**	48867.90	—	—	3500
	600	73486.50	**73344.10**	74461.30	—	—	4000
	800	99670.40	**99535.80**	100950.40	—	—	5000
	1000	126938.20	**126804.70**	128311.60	—	—	6000

not report on them in Table 6.4. It appears that VNS-2 performs best in average for all but one sizes of problems.

Prize-collecting Steiner tree problem on graphs Given a graph $G = (V, E)$, nonnegative weights w_{ij} associated with the edges (i, j) of E and a nonnegative prizes π_i associated with the vertices i of V, the *Prize-collecting Steiner tree problem* (PCSTP), is to find a subtree of G which minimizes the sum of the weights of its edges, plus the prizes of vertices not spanned. If the subset of vertices X to be spanned is known, we have the *Steiner tree problem*. PCSTP has an important application in design telecommunication of local access networks. In [24], among other heuristics, VNS has been proposed for PCSTP. In the solution space, represented by all spanning trees with cardinality $k = 1, 2, \ldots, |V|$, the neighborhood of $T(X)$ is defined as a set of spanning trees $T(X')$ having one different vertex ($|X \setminus X'| = 1$ or $|X' \setminus X| = 1$. The kth ordered neighborhood is also defined by vertex deletions or additions. Each tree considered is always transformed into another tree by a so-called peeling algorithm, which iteratively eliminates leaves whose weights are larger than corresponding prizes. The basic VNS developed is used as a post-optimization procedure, i.e., it starts with a solution obtained by a known constructive (2-approximation primal-dual) algorithm (denoted with GW) plus iterative improvement with two types of perturbations (It. Impr.) plus local search with path-relinking (LS + PR). Summary results from [24] on three known classes of hard instances are given in Table 6.5. It appears that in each class, VNS improved the best known solution for three instances.

Table 6.5. Prize-collecting Steiner tree problem: Improvements over GW

		IT. IMPR.		LS + PR			VNS		
Series	Instances	# Opt	% Impr.	# Opt	% Impr.	Time	# Opt	Impr.	Time
JMP	26	6	1.2	18	2.3	157.8	21	2.5	94.1
C	34	2	7.8	27	8.6	956.8	30	10.5	796.6
D	35	6	4.7	17	7.0	7668.1	20	7.5	2749.2

5.4 Scheduling

Single machine scheduling problems Given a set of n jobs that has to be processed without any interruption on a single machine, which can only process one job at a time, several NP-hard versions of Single machine scheduling problem that differ in objective function are solved by VND and VNS in [10]. Since the solution can be represented as a permutation of jobs, three neighborhoods of different cardinalities are used within VND, i.e., 1-opt (or transpose), insert (or Or-opt) and interchange neighborhoods have cardinalities $n-1$, $(n-1)^2$ or $n(n-1)/2$, respectively. They are used in sequence, without return to the smallest one after improvement (as in the basic VND). It appears, from extensive computations, that VND significantly improves solutions compared with those obtained by single neighborhood local searches in small additional computing time.

Another interesting observation is that ordering 1-opt, Or-opt, interchange is better than 1-opt, interchange, Or-opt, despite the fact that the cardinality of the Or-opt neighborhood is larger than that of the interchange one.

Nurse rostering problems (NRP) consist of assigning varying shift types to hospital personnel with different skills and work regulations, over a certain planning period.

Typical shift types are Morning, Day and Night shift. There are many constraints in the NRP which are divided (in [20]) into hard (which can never be violated and expressed in the set of constraints) and soft (which are preferably not violated). The objective is to minimize the number of violated soft constraints. The commercial nurse rostering package *Plane*, which is implemented in many Belgian hospitals, makes use of Tabu search [18]. For the same problem VNS is applied in [20], where several neighborhoods are constructed in order to especially satisfy some of the constraints. It appears that VND enables the system to find schedules which are hidden for single neighborhood heuristics. Moreover, the authors conclude that "it is often more beneficial to apply simple descent heuristics with [a] variety of neighborhoods than to use sophisticated heuristics which are blind for large parts of the search space".

Multiprocessor scheduling problem with communication delays This problem consists in finding a static schedule of an arbitrary task graph onto a homogeneous multiprocessor system, such that the makespan (i.e., the time when all tasks are finished) is minimum. The task graph contains a precedence relation as well as communication delays (or data transferring time) between tasks if they are executed on different processors. The multiprocessor architecture is assumed to contain identical processors with a given distance matrix (i.e., the number of processors along the shortest path) between each two processors [35]. For solving this NP-hard problem [133], a basic

VNS heuristic is developed in [36], where a k-swap neighborhood structure is used for shaking and a reduced 1-swap for local search. It is compared with Multistart local search (MLS), Tabu search (TS) [36] and Genetic algorithms [1] (PSGA). Initial solutions for all methods are obtained by modification of the well-known constructive heuristic *critical path* (CP). Starting from it a *swap* local search is run (LS). Two types of task graphs are generated: (i) with known optimal solution on the 2-dimensional hypercube (i.e. with 4 processors) and given density as proposed in [100]; (ii) with given density 0.2, 0.4, 0.5, 0.6 and 0.8. It appears that for both types of random test instances VNS performs best.

Resource-constrained scheduling problem (RCPSP) is concerned with n non-preemptive activities and m renewable resources. Given the availability of each resource $k(R_k)$, processing time for each activity j in time units (d_j), amount of resource k needed for activity j, during d_j (r_{jk}) and for each activity j a sets of immediate predecessors (P_j) or a set of immediate successors (S_j), find for each activity j its start time s_j ($s_1 = 0$) such that the makespan of the project $T = s_n$ is minimum.

Since the RCPSP is an important and difficult NP-hard problem, a considerable amount of research on it has appeared in the literature (for a recent survey see, e.g., [21]). In [53], a VNS heuristic is developed for the RCPSP. The solution is represented as a feasible sequence of activities (that respect precedence conditions), and in order to find makespan, an additional procedure is used. The disadvantage of such a representation is that the same schedule could be obtained from several permutations. Neighborhood for a local search is defined as a sequence of feasible 1-opt (or transpose) moves, i.e., an activity is inserted between its two closest ones (to the left and to the right) in the current permutation until all left activities are from P_j or right activities from S_j. In addition, this neighborhood is reduced by a parameter α which defines the maximum number of such 1-opt moves. These moves allow an efficient updating of the objective function. In order to derive a solution from $\mathcal{N}_k$, the same sequence of moves is repeated k times. In the Shaking step, k (instead of one, as in the basic VNS) solutions are generated and the best among them is used as an initial solution for the Local search. Very good results by VNS are reported on 4 classes of problems from the literature: J30 (480 instances), J60 (480), J90 (480), J120 (600). Some average comparative results, given in Table 6.6, include GA as well, which was identified in [96] as the most effective.

Capacitated lot-sizing problem with setup times (CLSP-ST) The trend towards just-in-time manufacturing has led to a drastic reduction of setup times in many manufacturing processes. Given are a fixed costs r_{it} for producing item i in period t ($i = 1, .., n; t = 1, .., T$), variable costs (per unit production cost and per unit inventory holding cost in period t), the demand for each item in each period and the amount of each resource available. In the CLSP-ST it is required to determine the lot sizes of each item in each time period, such that the total cost is minimized and the capacity constraints satisfied.

Table 6.6. RCPSP: % deviation from critical-path lower bound

Set	# Instances	Best-known	GA	Init.	Init.+	VNS	VNS+
J60	480	10.76	11.89	12.98	12.18	11.13	10.94
J120	600	32.17	36.74	40.02	37.38	33.88	33.10

Table 6.7. CLSP-ST: Comparative aggregate results from[88]

Method	Better than TTM	Same as TTM	Worse than TTM	Not solved	Gap %		Time (s)	
					Avg.	Max	Avg.	Max
TTM	—	—	—	0	4.16	31.87	0.60	3.61
DBKZ	35	2	453	31	9.11	47.86	0.13	1.90
M-DBKZ	100	39	352	1	5.22	29.28	2.00	16.03
SM	437	42	13	0	2.77	31.40	0.81	11.66
SM+VNS	447	35	10	0	2.51	22.90	5.84	117.36

Very few papers address the CLSP-ST. Usually Lagrangian relaxation with subgradient optimization is employed to calculate a lower bound and different heuristics are developed in attempt to find good feasible solutions. In[88], beside Lagrangian relaxation, a new smoothing heuristic (SM) followed by a VNS is suggested. In the Shaking step, the k setups corresponding to k smallest additional costs are switched off (and the corresponding transshipment problem is solved), while SM is used again as a local search within VNS. Only one loop of the basic VNS is applied with $k_{\max} = 7$, i.e., the first time k reaches 7, the procedure stops. All tests were carried out on a subset of the 751 benchmark test instances. Five methods were tested, two known and three new: TTM [132], DBKZ [40], M-DBKZ (modification of DBKZ), SM and SM+VNS. It appears that, among 492 hardest problem instances, SM and SM+VNS improve 437 and 447 solutions, being 13 and 10 times worse, respectively. In Table 6.7 comparative aggregate results w.r.t. TTM on 492 hardest problem instances are reported.

5.5 Artificial Intelligence

Weighted Max-SAT problem The satisfiability problem, in clausal form, consists in determining if a given set of m clauses (all in disjunctive or all in conjunctive form) built upon n logical variables has a solution or not [55]. The maximum satisfiability problem consists in finding a solution satisfying the largest possible number of clauses. In the *weighted maximum satisfiability* problem (WMAXSAT) [123] positive weights are assigned to the clauses and a solution maximizing the sum of weights of satisfied clauses is sought. Results of comparative experiments with VNS and TS heuristics on instances having 500 variables, 4500 clauses and 3 variables per clause, in direct or complemented form, are given in Table 6.8 [74]. It appears that using a restricted neighborhood consisting of a few directions of steepest descent or mildest ascent in the Shaking step does not improve results, but using this idea in conjunction with SVNS improves notably upon results of basic VNS and also upon those of a TS heuristic.

Learning Bayesian networks Let us consider a set of random variables $V = \{x_1, x_2, \ldots, x_n\}$ and a set of parameters which together specify a joint probability distribution over the random variables. Bayesian networks (BNs), also known as Belief or Causal networks, are knowledge representation tools able to manage the dependence and independence relationships among the random variables [117]. BN is represented by a directed acyclic graph (dag). Once the BN is specified, it constitutes an efficient device to perform inference tasks. The problem is to develop automatic methods for

Table 6.8. Results for *GERAD* test problems for WMAXSAT ($n = 500$)

	VNS	VNS-low	SVNS-low	TS
Number of instances where best solution is found	6	4	23	5
% Average error in 10 trials	0.2390	0.2702	0.0404	0.0630
% Best error in 10 trials	0.0969	0.1077	0.0001	0.0457
Total number of instances	25	25	25	25

building BNs capable of learning directly from given data ($D = \{v^1, v^2, \ldots, v^m\}$, containing m instances of V), as an alternative or a complement to the method of eliciting opinions from experts. This NP-hard problem is receiving increasing attention in Artificial intelligence. In [22] the current dag (solution) is represented by a permutation θ of n variables. The quality of an ordering is measured by the so-called *scoring metric*, f, i.e., a function, $f(\theta, D)$, defined for dags and a search in the space of dags compatible with θ performed. Then the basic VNS scheme (called VNS based on ordering, VNSO for short) is applied The set of neighborhood structures is defined by *k–interchange* moves in θ and local search performed for $k = 1$. VNSO has been successfully compared with two other methods from the literature.

5.6 Continuous Optimization

Bilinear programming problem Structured global optimization problems, while having several and often many local optima, possess some particular structure which may be exploited in heuristics or exact algorithms. One such problem, of considerable generality, is the bilinear programming problem (BLP) with bilinear constraints. This problem has three sets of variables, x, y and z, with cardinalities n_1, n_2 and n_3 respectively. When all variables of y are fixed it becomes a linear program in x and z. When all variables of z are fixed it becomes a linear program in x and y. This property suggests the well-known *Alternate* heuristic:

1. *Initialization:* Choose values of variables of z (or y);
2. *LP-1*: solve the linear program in (x, y) (or in (x, z));
3. *LP-2*: For y (or z) found in the previous step, solve the linear program in (x, z) (or in (x, y));
4. If stability is not reached (within a given tolerance) return to 2.

Obviously this algorithm may be used in a Multistart framework. To apply VNS one may observe that neighborhoods $\mathcal{N}_k(x, y, z)$ of a solution (x, y, z) are easy to define. They correspond to k pivots of the linear program in (x, y) or in (x, z), for $k = 1, 2, \ldots, k_{max}$. One can then apply the basic VNS of Figure 3 in a straightforward way. The local search routine is the alternate heuristic described above. Results on randomly generated test problems are presented in Table 6.9 from [6]. It appears that VNS improves in almost all cases and sometimes very substantially upon the solution provided by the Multistart Alternate heuristic.

Table 6.9. BLP: results for 10 repetitions of ALT (MALT) and VNS: each line reports average results on 4 random test problems with same parameters

Parameters					CPU time		% Error	
n	m	n_1	n_2	n_3	MALT	VNS	MALT	VNS
50	36	30	10	10	0.7	1.0	1.09	0.00
60	36	30	20	10	1.6	2.4	10.05	0.00
70	36	30	30	10	0.9	5.8	20.38	0.00
80	36	30	40	10	1.8	2.7	51.22	0.00
90	36	30	50	10	1.8	3.7	43.77	0.00
100	36	40	50	10	3.6	11.9	18.45	0.00
110	36	50	50	10	3.7	7.3	15.90	0.00
120	36	60	50	10	10.0	22.4	39.12	0.00
130	36	70	50	10	15.0	30.0	34.56	0.00
140	36	80	50	10	8.8	13.9	21.01	0.00
150	36	90	50	10	6.8	10.0	42.87	0.00
Average					4.35	9.19	23.23	0.00

Pooling problem The pooling problem, which is fundamental to the the petroleum industry, describes a situation where products possessing different attribute qualities are mixed together in a series of pools in such a way that the attribute qualities of the blended products of the end pools must satisfy given requirements. It is well known that the pooling problem can be modeled through bilinear programming. A simple alternating procedure and a VNS heuristic are developed to solve large instances, and compared with the well-known method of successive linear programming[6]. This is an application of BBLP above.

Continuous Min–Max problem The so-called multi-level Tabu search (MLTS) heuristic has been proposed in [97] for solving a continuous min–max optimization problem (in R^n with m periodic functions) of spread spectrum radar polyphase code design problem [43]. Improvements obtained with the basic VNS in both solution quality and computing time on the same set of test problems are reported in [114] and [82]. Different neighborhoods are derived from the Euclidean distance. A random point is selected from such neighborhoods in the *Shaking* step; then the gradient (feasible direction) local search with a given step size is performed on the functions that are active (functions that have a maximum value in the current point); this step is repeated until the number of active functions in the current point is equal to n (with some tolerance). It appears that VNS outperforms MLTS on all test instances.

6 UNDERSTANDING HEURISTICS AND METAHEURISTICS

The advent of metaheuristics has led, for a large variety of problems, to the design of heuristics with a much improved empirical performance, without however that theoretical reasons for such improvements be clearly understood. First steps towards a better understanding are to study the topography of local optima, and valleys or mountains, for given classes of problems, as well as the trajectories followed by the heuristics and the resulting phases of improvement or deterioration. We next describe two new tools, i.e., *mountain* (or *valley*) profiles and *distance-to-target visualization* which have been found helpful in deriving better heuristics or variants of basic schemes for VNS. We then turn to the study of an unexpected phenomenon, i.e., that when selecting moves in a heuristic *first improvement can be better than best improvement*. Finding the reason why this is so does not lead directly to a competitive heuristic for the problem under study, i.e., the traveling salesman problem, but does pinpoint a potential defect of many heuristics.

6.1 Mountain or Valley Profiles

When applying VNS, the descent from the randomly selected solution x' can lead back to the local optimum x around which the current neighborhoods are centered, or to another local optimum x'' the value of which can be better or not than that of x. It is thus natural to study the probabilities of these three outcomes as a function of distance from x to x'. It is also worthwhile to observe whether x'' is closer to x than x' (which may be interpreted as x'' belonging to the same large valley, with local rugosities in relief) or not (which may be viewed as indicating a new large valley has been found). Mountain profiles give such information on the basis of 1000 ascents from points in each successive neighborhood. Examples of such profiles for the *weighted maximum satisfiability problem* are given in Figure 6.6, from [74].

Profiles appear to vary considerably with the quality of the local optimum x: when it is bad it suffices to go a little away from x to obtain, with high probability, a better local optimum. When it is good, or very good, one must go quite far to find a new large valley and, moreover, the probability of finding a solution better than the incumbent is then low. This illustrates a weakness of the basic VNS scheme, already mentioned above: it tends to degenerate into *Multistart* when the distance from x to x' becomes large. The solution, also presented above, is to resort to the SVNS scheme.

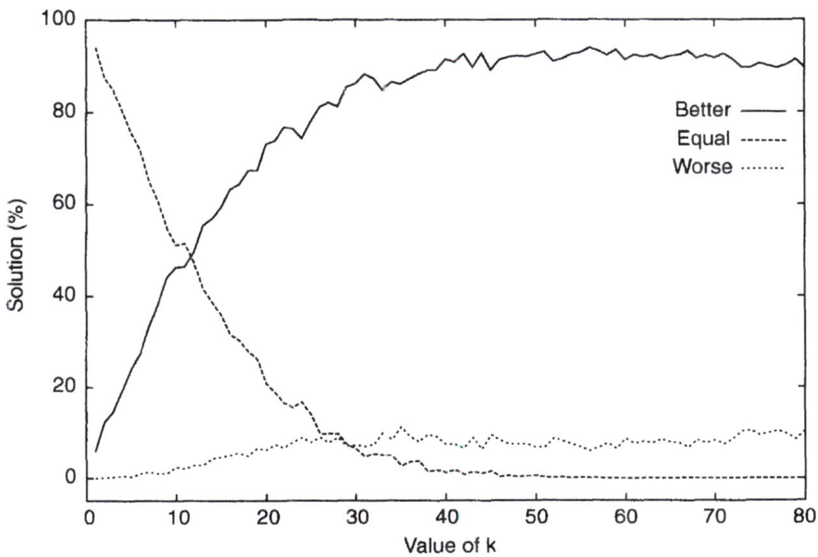

Figure 6.6. A mountain profile.

6.2 Distance-to-target Visualization

When designing heuristics for well studied classes of problems, small to medium size instances for which the optimal solution is known are often available. Such instances are used to find how often the optimal solution is reached, or what is the average percentage of error of the heuristic. But much more information can be obtained if the optimal solutions themselves are considered, and not only their values. In ongoing work on a VNS approach to the Traveling salesman problem [58] a *distance-to-target visualization* tool has been developed. This tool presents on screen the optimal solution for the training instance under study, the current solution and the symmetric difference between these two solutions. This indicates how much improvement is to be made and where. Moreover, a routine allows also the representation of the *difference* of solutions at an iteration of the heuristic and at the next. Finally, as representations of solutions for large instances may be hard to read, and many problems, in particular Euclidean ones, allow for a natural decomposition, a *focusing* routine allows representations of the above-mentioned information for chosen subproblems, e.g., in some region of the plane.

Visualizing successive steps in VND and VNS show their strengths and weaknesses, e.g., that after 2-opt is applied within VNS for the TSP much work remains to be done. Suggestions for further neighborhoods to use may then be obtained from a careful study of remaining differences between solution and target.

6.3 First versus Best Improvement

In addition to using the visualization tools described above, one can proceed to a step-by-step evaluation of the work done by a VNS or other heuristic. Moreover, variants at each of these steps can be studied. Detailed information, which would not be apparent from the global performance of the heuristic, can then be gathered. It can lead to the discovery of unexpected phenomena and the insight provided by their explanation can in turn lead to principles for building better heuristics. A typical case is the observation that when applying the 2-opt heuristic to the TSP [79], selecting at each iteration the first improvement, i.e., the first exchange of two edges which reduces the objective function value gives better results than selecting the best improvement, i.e., the exchange of two edges which reduces most the objective function (beginning from a randomly chosen solution). An explanation, corroborated by further detailed analysis, is easily obtained: if the best exchange is chosen, two edges of small but not necessarily very small length are introduced in the tour and are difficult to remove at later iterations; if a first improvement is chosen, due to ranking of edges by order of increasing lengths, a very small edge and an average length edge are introduced in the tour and the latter is easy to remove at a later iteration. Systematic study of such phenomena in a variety of heuristics could lead to many improvements.

7 IMPROVING EXACT ALGORITHMS

It is well known that many combinatorial optimization problems such as the p-median or the Multisource Weber problem [41], the clustering or partitioning problem with various objectives [44,73], the air-crew scheduling problem [38], etc. can be solved by combining column generation with integer programming. As the optimal solution of the linear relaxation is often highly degenerate, convergence may be slow and even when it is found many iterations may be needed to prove that it is indeed optimal. The dual of the column generation procedure is the outer approximation method of [95]. Slow convergence comes from the generation of many hyper-planes far from the cone vertexed at the optimal dual solution. One would therefore want to estimate this last solution and focus the generation procedure by penalizing hyperplanes far from it. If extracting the optimal dual solution is difficult, then the current dual values can be used to stabilize the generation procedure instead of focusing it.

Consider a primal problem P and its dual D, solved by column generation:

$$(P) \quad \begin{aligned} &\min c^T x \\ &\text{s.t. } Ax = b \\ &x \geqslant 0 \end{aligned}$$

$$(D) \quad \begin{aligned} &\min b^T \pi \\ &\text{s.t. } A^T \pi \leqslant c \\ &(\pi) \end{aligned}$$

In [41] a stabilization scheme is proposed which remains entirely within the linear programming column generation framework. It merges a perturbation and an exact penalty method.

A primal problem $\tilde{P}$ and its dual $\tilde{D}$ are defined as follows:

$$(\tilde{P}) \quad \begin{aligned} &\min c^T \tilde{x} - \delta_- y_- + \delta_+ y_+ \\ &\text{s.t. } A\tilde{x} - y_- + y_+ = b \\ &y_- \leq \varepsilon_- \\ &y_+ \leq \varepsilon_+ \\ &\tilde{x}, y_-, y_+ \geq 0 \end{aligned}$$

$$(\tilde{D}) \quad \begin{aligned} &\max b^T \tilde{\pi} - \varepsilon_- w_- - \varepsilon_+ w_+ \\ &\text{s.t. } A^T \tilde{\pi} \leq c \\ &-\tilde{\pi} - w_- \leq -\delta_- \\ &\tilde{\pi} - w_+ \leq \delta_+ \\ &(\tilde{\pi}), w_-, w_+ \geq 0 \end{aligned}$$

In the primal y_- and y_+ are vectors of slack and surplus variables with upper bounds ε_- and ε_+ respectively. These variables are penalized in the objective function by vectors $[-\delta_-, \delta_+]$. When applying column generation, after finding no more column with negative reduced cost, the parameters δ and ε are updated following problem-specific rules.

For this *stabilized column generation* scheme to work well, good estimates of the optimal dual variables should be obtained at the outset. Ways to find heuristically ranges for these dual variables by sensitivity analysis are sketched in [41] for a few problems. For the p-Median problem this led to exact solution of instances with over 3000 users versus 900 in the literature; for its counterpart in the plane, i.e., the Multisource Weber problem ([75,98]) it led to exact solution of instances with over 1000 users versus 30 in the literature. For the minimum sum-of-squares clustering problem combining stabilized column generation, with the ACCPM interior point method, hyperbolic 0-1 programming quadratic 0-1 programming and VNS in several steps [44] led to exact solution of problems with up to 150 entities, including the famous 150 iris example of [50].

8 COMPUTER-AIDED DISCOVERY IN GRAPH THEORY

The AutoGraphiX system Recall that a graph invariant is a variable defined on the family of all graphs (or on an appropriate sub-family) and the value of which does not depend on the numbering of vertices or edges. Numerous problems in graph theory can be viewed, or involve, optimization of invariants on a possibly constrained family of graphs. Therefore, VNS can be used for approximate solution of such problems. This led to the development of the AutoGraphiX (AGX) system and to a series of papers, next reviewed, on its principles and applications.

As explained in [29], AGX uses the basic schemes of both VNS and VND. The descent is done by first drawing a graph G at random (with a given number of vertices and edges), computing its value for the invariant under study and then examining the effect on this value of bringing some elementary change to G: removal or addition of an edge, displacement of an edge, detour, i.e., replacement of an edge between two vertices by a path of length 2 joining them through some non-adjacent vertex, and so on. Neighborhoods so defined are ranked from the smallest to the largest. The best move is determined in each of them in turn and if it improves the value of the invariant studied the corresponding change is made in G. After a local optimum has been found, neighborhoods are defined using the Hamming distance, i.e., by considering removal or addition of edges to G, and VNS is applied. The only parameter is the maximum number of edges to be removed or added.

AGX can be applied to the following problems: (a) find a graph satisfying given constraints; (b) find optimal or near optimal values for an invariant subject to constraints; (c) refute a conjecture; (d) suggest a conjecture (or sharpen one); (e) suggest a proof. For instance, three conjectures of the system Graffiti [46,47] are refuted in [29], several strengthened and one of these proved, for the particular case of trees, exploiting knowledge of moves needed to find local optima.

Automated conjecture finding Study of a set of extremal or near-extremal graphs G obtained with AGX taking the numbers n of vertices and m of edges as parameters often suggests conjectures. Moreover, there can be corroborated or refuted by performing various changes interactively. However, it would be preferable to have an entirely automated system. Three ways to automate conjecture finding are outlined in [28]: (a) a numerical method, which exploits the mathematics of Principal Component Analysis in order to find a basis of affine relations between graph invariants; (b) a geometric method which consists in finding the convex hull of the points representing extremal or near-extremal graphs in invariants space, with a 'gift-wrapping' algorithm. Hyperplanes of this convex hull correspond to conjectures; (c) a geometric approach, which consists in recognizing the families of graphs to which belong the extremal ones and using known relations between graph invariants in those families to find new relations.

Chemical graph theory. 1. Energy The π-electronic energy E is defined in chemical graph theory as the sum of absolute values of the eigenvalues of the adjacency matrix [68]. It has been extensively studied by chemists and mathematicians. As explained in [25] AGX led to the discovery of several simple, but as yet unknown relations, including the following:

$$E \geq 2\sqrt{m} \quad \text{and} \quad E \geq \frac{4m}{n}$$

which were easily proved.

Chemical graph theory. 2. Randić index The Randić index [120] and its generalization are probably the most studied invariants of Chemical graph theory. Assign to each edge of a graph G a weight equal to the inverse of the geometric mean of the degrees of its vertices. Then, the Randić index of G is the sum of all such weights. AGX was used to find extremal trees for this index [27]. Maximum values are obtained by paths and minimal ones by two kinds of caterpillars and one of modified caterpillars with an appended 4-star. This last result was proved by an original way of using linear programming. Indeed, variables are associated with the numbers of edges with given degrees of end vertices (in chemical graphs these degrees are bounded by 4). This proof technique has already been used in three other papers of Chemical graph theory.

In [5] are presented several bounds on the Randić index of chemical trees in terms of this index for general trees and of the ramification index, which is the sum for all vertices of the excess of their degree over 2. Use of AGX [76] leads to correct some of these results and obtain much strengthened and best possible versions of the others.

Chemical graph theory. 3. Polyenes with maximum HOMO–LUMO gap Using AGX, [54] a study was made of fully conjugated acyclic π systems with maximum HOMO-LUMO gap. In the simplest Hückel model, this property of the eigenvalue spectrum of the adjacency matrix of the corresponding chemical tree corresponds to reactivity. AGX gives for all even n a 'comb', i.e., a path on $n/2$ vertices to each of which is appended a single pendant edge. From this, the conjecture that the maximum gap tends to $2\sqrt{2}-2$ when n goes to infinity can be deduced.

Trees with maximum index Trees are bipartite and hence bicolorable, say in black and white. Color-constrained trees have given numbers of black and white vertices. In [34] AGX is used to study color-constrained trees with extremal index, or largest eigenvalue of the adjacency matrix. Six conjectures are obtained, five of which are proved.

Applying the numerical method for automated conjecture finding to the extremal trees found gave the conjecture

$$\alpha = \tfrac{1}{2}(m - n_1 + D - 2r)$$

where α is the stability number, n_1 the number of pendant vertices, D the diameter and r the radius of the tree. It is unlikely that a conjecture with so many invariants would be obtained without computer aid.

Trees with palindromic Hosaya polynomial The Hosaya (or distance) polynomial of a graph is defined by

$$H(G) = a_0 + a_1 x + \cdots + a_D x^D$$

where $a_o = n$, $a_1 = m$ and $a_k (k = 3, \ldots, D)$ is the number of pairs of vertices at distance k. It was conjectured in [67] and [69] that there is no tree with a palindromic Hosaya polynomial, i.e., such that $a_0 = a_D, a_1 = a_{D-1}$ and so on. AGX refuted this conjecture [26]. All counter-examples have D even. This case is easier to satisfy than that where D is odd. Indeed, in the former case, there is a free coefficient in the center, and in the latter not. Then, define the distance to the palindrome condition as

$$z_p = \sum_{k=0}^{\lfloor D/2 \rfloor} |a_k - a_{D-k}|$$

AGX gives for all $n = 10$ to $n = 50$ trees with $z_p = \lceil \frac{n}{2} \rceil$ (and higher values, due to border effects for $n \leq 10$). The conjecture

$$z_p \geq \left\lceil \frac{n}{2} \right\rceil$$

appears to be hard to prove.

Graffiti 105 The transmission of distance of a vertex of a graph G is the sum of distances from that vertex to all others. Conjecture 105 of Graffiti [47] is that in all trees the range of degrees does not exceed the range of transmission of distances. In [4] AGX is used to get hints on how to prove this conjecture and study variants of it. Minimizing range of transmission minus range of degrees always gives stars. It is then easy to prove that the conjecture holds with equality for stars only. Moreover, one gets the conjectures: (a) if T is not a star then range of transmission minus range of degree is at least $\lceil \frac{n}{2} \rceil - 2$, and (b) if T has maximum degree $D \leq \lceil \frac{n}{2} \rceil$ then range of transmission minus range of degree is not less than $n - 3D - 1$.

These results are easily proved, and, with a little more work, the extremal graphs characterized.

These examples illustrate the help the VNS-based system AGX can bring to discovery in graph theory. Many further ones are given in the cited papers.

9 CONCLUSIONS

Heuristics are an essential tool in applied optimization, and for many large and often messy problems encountered in practice the only applicable one. Their main aim is to provide, in reasonable time, near-optimal solutions to constrained or unconstrained optimization problems. Moreover, they may also be very useful within complex exact algorithms, to accelerate many steps. Finally, they are an important building block in optimization-based knowledge discovery systems. Optimization problems are ubiquitous in Operations Research and its applications to numerous fields. Artificial intelligence tends to focus more on constraint satisfaction problems. Heuristics then seek a feasible solution. Both types of problems are less different than might appear at first view as Constraint satisfaction problems can be expressed as optimization problems in a variety of ways, e.g., as minimization of a sum of artificial variables representing constraint violations.

While traditional heuristics, such as, e.g., simple descent methods, are blocked in the first local optimum found, this is not the case for heuristics built within the metaheuristics paradigms. All of them provide methods, deterministic or stochastic, for getting out of poor local optima. As such local optima often differ considerably in value from the global optimum, particularly if there are many, the practical impact of metaheuristics has been immense. In contrast to this success, the theory of metaheuristics is lagging. While good heuristics are often obtained, with some ingenuity and a lot of parameter setting, the reason(s) why they work as well as they do are largely unknown. In this respect, the situation is even worse for hybrids. So, some reflection on desirable properties of metaheuristics, which would guarantee both their practical and theoretical interest, may be in order. A tentative list of such properties is the following:

(i) *Simplicity:* the metaheuristic should be based on a simple and clear principle, which should be largely applicable;

(ii) *Precision:* steps of the metaheuristic should be formulated in precise mathematical terms, independent from the possible physical or biological analogy which was an initial source of inspiration. Meaningless statements (e.g., "kicking the function") or overly vague ones (e.g., "choosing a point based on the history of the search") should be avoided;

(iii) *Coherence:* all steps of heuristics for particular problems should follow naturally from the metaheuristic's principle;

(iv) *Efficiency:* heuristics for particular problems should provide optimal or near-optimal solutions for all or at least most realistic instances. Preferably, they should find optimal solutions for most problems of benchmarks for which such solutions are known, when available;

(v) *Effectiveness:* heuristics for particular problems should take moderate computing time to provide optimal or near-optimal solutions;

(vi) *Robustness:* performance of heuristics should be consistent over a variety of instances, i.e., not just fine-tuned to some training set and less good elsewhere;

(vii) *User-friendliness:* heuristics should be clearly expressed, easy to understand and, most important, easy to use. This implies they should have as few parameters as possible and ideally none;

(viii) *Innovation:* preferably, the metaheuristic's principle and/or the efficiency and effectiveness of the heuristics derived from it should lead to new types of applications.

Variable Neighborhood Search (VNS) is a recent metaheuristic which strives to obtain the qualities listed above. It is based on a simple and relatively unexplored principle: systematic change of neighborhood during the search. (Note that precise rules for such change are crucial; several authors have proposed on occasion to combine different types of moves, or neighborhoods, within the same heuristic, without however doing so systematically, nor that being the main idea of their heuristics.)

Reviewing the eight desirable properties of metaheuristics, it appears that VNS possesses them to a large degree. Indeed, its principle is simple and all steps of the basic and extended schemes rely upon it. Moreover, they are state in precise mathematical terms. VNS has proved efficient in solving the problems of several benchmarks with optimal or very close to optimal results, and within moderate (or at least reasonable) computing times. Moreover, its performance appears to be robust, its basic principles are easy to apply, and very easy to use, parameters being kept to a minimum and sometimes absent. Simplicity is probably its main feature. Indeed, VNS gets as good or better results than most other metaheuristics on many problems and in a much simpler way. This explains its potential for innovation already illustrated by several new type of applications: analysis of moves and their selection for the TSP, stabilized or focussed column generation and the computer-aided scientific discovery program AGX.

Metaheuristics are a fairly young research field, with many new ideas and frequent recourse to intuition rather than deduction. Attempts at organizing this field are numerous, but as the main concepts are rarely precisely defined and there are as yet very few significant theorems, no framework has gained general acceptance. Rather, each metaheuristic has its own viewpoint and ability to explain many heuristics in its own vocabulary as well as to absorb ideas from the whole field (notably under the form of hybrids). Moreover, priority claims tend to proliferate. They are often based on such vague evidence that they are hard to evaluate.

Focusing for instance, on descent methods or memory structures corresponds to different viewpoints and metaheuristics. The closest one to VNS appears to be Iterated local search (ILS) ([11,91,92]). At the price of completely forgetting VNS's main idea—systematic change of neighborhood—one can squeeze it into the ILS framework. Conversely, one could view ILS heuristics as either badly defined (as the crucial step of perturbation of local optima is often not specified or described by some vague metaphor) or particular cases of VNS, with usually a single neighborhood. This would be equally arbitrary. It appears that the babelian character of research in metaheuristics is a, hopefully temporary, mild evil. While it lasts, clear-cut successes on particular problems will be probably more important to evaluate metaheuristics than lengthy controversies. Finally, when considering the eight desirable qualities listed above, we believe that comparative efficiency should not have the dominant, sometimes exclusive role, it gets in many papers. The aim of research should be insight, not competition. In our view other qualities of heuristics and metaheuristics than efficiency can be more important in the long run, particularly, simplicity, precision, coherence and above all, innovation.

REFERENCES

[1] I. Ahmad and M. Dhodhi (1996) Multiprocessor scheduling in a genetic paradigm. *Parallel Computing* **22**, 395–406.

[2] M.R. Anderberg (1973) *Cluster Analysis for Application*. Academic Press, New York.

[3] A. Andreatta and C. Ribeiro. Heuristics for the phylogeny problem. *Journal of Heuristics* (to appear).

[4] M. Aouchiche, G. Caporossi and P. Hansen (2001) Variable neighborhood search for extremal graphs. 8. Variations on Graffiti 105, *Congressus Numeratium* (to appear).

[5] O. Araujo and J.A. de la Penã (1998) Some bounds on the connectivity index of a chemical graph. *Journal of Chemical Information and Computer Science*, **38**, 827–831.

[6] C. Audet, J. Brimberg, P. Hansen and N. Mladenović (2000) Pooling problem: alternate formulation and solution methods. *Les Cahiers du GERAD G-2000 23*.

[7] R. Battiti and M. Protasi (2001) Reactive local search for the maximum clique problem. *Algorithmica*, **29**(4), 610–637.

[8] J.E. Beasley (1985) A note on solving large *p*-median problems. *European Journal of Operational Research*, **21**, 270–273.

[9] N. Belacel, N. Mladenović and P. Hansen. Fuzzy J-Means: A new heuristic for Fuzzy clustering. *Pattern Recognition* (to appear).

[10] M. den Basten and T. Stutzle (2001) Neighborhoods revisited: Investigation into the effectiveness of Variable neighborhood descent for scheduling, MIC'2001, Porto, pp. 545–549.

[11] M. den Basten, T. Stutzle and M. Dorigo (2001) Design of iterated local search, EvoSTIM2001, 2nd EW on Scheduling and Timetabling, *Lecture Notes CS* (to appear).

[12] R. Bent and P. Van Hentenryck (2001) A two stage hybrid local search for the vehicle routing problem with time windows, Technical report, CS-01-06, Department of Computer Science, Brown University.

[13] J.C. Bezdek (1981) *Pattern Recognition with Fuzzy Objective Function Algorithm.* Plenum, New York.

[14] O. Braysy (2000) *Local Search and Variable Neighborhood Search Algorithms for the Vehicle Routing with Time Windows.* Acta Wasaensia 87, Universitas Wasaenis, Vaasa.

[15] J. Brimberg, P. Hansen, K.-W. Lih, N. Mladenović and M. Breton (2000) An oil pipeline design problem. *Les Cahiers du GERAD* G-2000-73, Montréal, Canada.

[16] J. Brimberg, P. Hansen, N. Mladenović and É. Taillard (2000) Improvements and comparison of heuristics for solving the Multisource Weber problem. *Operation Research*, **48**(3), 444–460.

[17] J. Brimberg and N. Mladenović (1996) A Variable neighborhood algorithm for solving the continuous location-allocation problem. *Studies in Location Analysis*, **10**, 1–12.

[18] E.K. Burke, De Causmaecker and G.V. Berghe (1999) A hybrid tabu search algorithm for the nurse rostering problem. *Lecture notes in AI*, **1585**, 187–194.

[19] E.K. Burke, P. Cowling and R. Keuthen (1999) Effective local and guided Variable neighborhood search methods for the asymmetric traveling salesman problem, in the Proceedings of the Evo Workshops, Springer, Lecture Notes in Computer Science, pp. 203–212.

[20] E. Burke, P. De Causmaecker, S. Petrović and G.V. Berghe (2001) Variable neighborhood search for nurse rostering problem, MIC'2001, Porto, pp. 755–760.

[21] P. Brucker, A. Drexl, R. Mohring, K. Neumann and E. Pesch (1999) Resource-constrained project scheduling: notation, classification, models and methods, *European Journal of Operational Research*, **112**, 3–41.

[22] L.M. de Campos and J.M. Puerta (2001) Stochastic local search algorithms for linear belief networks: searching in the space of orderings. In: S. Benferhat and P. Besnard (eds.), ESCQARU 2001, *Lecture Notes in AI* 2143. Springer-Verlag, Berlin Heidelberg, pp. 228–239.

[23] R.L. Canon, J.C. Bezdek and J.V. Dave (1986) Efficient implementation of the fuzzy C-means clustering algorithm. *IEEE Transactions on Pattern Recognition Machine Intelligence*, **8**(2), 248–255.

[24] S. Canuto, M. Resende and C. Ribeiro, Local search with perturbations for the prize-collecting Steiner tree problem in graphs, *Networks* (to appear).

[25] G. Caporossi, D. Cvetković, I. Gutman and P. Hansen (1999) Variable neighborhood search for extremal graphs. 2. Finding graphs with extremal energy. *Journal of Chemical Information Computer Science*, **39**, 984–996.

[26] G. Caporossi, A.A. Dobrynin, I. Gutman and P. Hansen (1999) Trees with palindromic Hosoya polynomials. *Graph Theory Notes of New York*, **37**, 10–16.

[27] G. Caporossi, I. Gutman and P. Hansen (1999) Variable neighborhood search for extremal graphs. 4. Chemical trees with extremal connectivity index. *Computers and Chemistry*, **23**, 469–477.

[28] G. Caporossi and P. Hansen (1999) Finding relations in polynomial time. In: *Proceedings of the XVI International Joint Conference on Artificial Intelligence*, pp. 780–785.

[29] G. Caporossi and P. Hansen (2000) Variable neighborhood search for extremal graphs. 1. The AutoGraphiX system. *Discrete Mathematics*, **212**, 29–44.

[30] R. Cordone, R.W. Calvo (2001) A Heuristic for the Vehicle routing problem with time windows. *Journal of Heuristics*, **7**(2), 107–129.

[31] M.-C. Costa, F.-R. Monclar and M. Zrikem (2001) Variable neighborhood search for the optimization of cable layout problem. MIC'2001, Porto, pp. 749–753.

[32] T. Crainic, M. Gendreau, P. Hansen, N. Hoeb and N. Mladenović (2001) Parallel Variable neighborhood search for the *p*-Median. MIC'2001, Porto, July 16–21, pp. 595–599.

[33] J. Crispim and J. Brandao (2001) Reactive Tabu search and variable neighborhood descent applied to the vehicle routing problem with backhauls. MIC'2001, Porto, pp. 631–636.

[34] D. Cvetković, S. Simić, G. Caporossi and P. Hansen (2001) Variable neighborhood search for extremal graphs. 3. On the largest eigenvalue of color-constrained trees. *Linear and Multi-linear Algebra* (in press).

[35] T. Davidović (2000) Scheduling heuristic for dense task graphs. *Yugoslav Journal of Operational Research*, **10**, 113–136.

[36] T. Davidović, P. Hansen and N. Mladenović (2001) Variable neighborhood search for multiprocessor scheduling with communication delays. MIC'2001, Porto, pp. 737–741.

[37] T. Davidović, P. Hansen and N. Mladenović (2001) Heuristic methods for multiprocessor scheduling with communication delays (in preparation).

[38] G. Desaulniers, J. Desrosiers, Y. Dumas, S. Marc, B. Rioux and F. Soumis (1997) Crew pairing at air France. *European Journal of Operational Research* **97**, 245–259.

[39] J. Desrosiers, N. Mladenović and D. Villeneuve (2001) Design of balanced MBA student teams. MIC'2001, Porto, July 16–21, pp. 281–285.

[40] M. Diaby, H.C. Bahl, M.H. Karwan and S. Zionts (1992) Capacitated lot-sizing and scheduling by Lagrangian relaxation. *European Journal of Operational Research*, **59**, 444–458.

[41] O. du Merle, D. Villeneuve, J. Desrosiers and P. Hansen (1992) Stabilized column generation. *Discrete Mathematics*, **194**, 229–237.

[42] J.C. Dunn (1974) A fuzzy relative of the ISODATA process and its use in detecting compact well-separated clusters. *Journal of Cybernetics*, **3**(3), 32–57.

[43] M.L. Dukić and Z. S. Dobrosavljević (1990) A method of a spread-spectrum radar polyphase code design. *IEEE Journal on Selected areas in Communications*, **8**(5), 743–749.

[44] O. du Merle, P. Hansen, B. Jaumard and N. Mladenović (2000) An interior point algorithm for Minimum sum-of-squares clustering. *SIAM Journal Scientific Computing*, **21**, 1485–1505.

[45] M. Ehrgott, J. Freitag, H. Hamacher, and F. Maffioli (1997), Heuristics for the k-cardinality tree and subgraph problems. *Asia-Pacific Journal of Operational Research*, **14**, 87–114.

[46] S. Fajtlowicz (1988) On conjectures of Graffiti. *Discrete Mathematics*, **72**, 113–118.

[47] S. Fajtlowicz (2000) Written on the wall. Version 09-2000 (regularly updated file accessible via e-mail from clarson@math.uh.edu).

[48] T. Feo and M. Resende (1995) Greedy randomized adaptive search. *Journal of Global Optimization*, **6**, 109–133.

[49] P. Festa, P. Pardalos, M. Resende and C. Ribeiro (2001), GRASP and VNS for Max-cut, *Proceedings of MIC'2001*, pp. 371–376.

[50] R.A. Fisher. The use of multiple measurements in taxonomic problems. *Annals of Eugenics* VII part II: 179–188.

[51] M. Fischetti, H. Hamacher, K. Jornsten and F. Maffioli (1994) Weighted k-cardinality trees: complexity and polyhedral structure. *Networks*, **24**, 11–21.

[52] K. Fleszar and K.S. Hindi (2001) New heuristics for one-dimensional bin-packing. *Computing and Operational Research* (to appear).

[53] K. Fleszar and K.S. Hindi (2001) Solving the resource-constrained project scheduling problem by a variable neighborhood search. *European Journal of Operational Research* (accepted s.t. revision).

[54] P. Fowler, P. Hansen, G. Caporossi and A. Sondini (2001) Polyenes with maximum HOMO-LUMO gap. (Variable Neighborhood Search for extremal graphs 7.). *Les Cahiers du GERAD*, Montréal, Canada, 2001, *Chemical Physics Letters* (to appear).

[55] M.R. Garey and D.S. Johnson (1978) *Computers and Intractability: A Guide to the Theory of NP-Completeness*. Freeman, New York.

[56] L.M. Gambardela, E. Taillard and M. Dorigo (1999) Ant colonies for the quadratic assignment problem. *Journal of Operational Research Society*, **50**, 167–176.

[57] M. Gendreau, G. Pesant and L.-M. Rousseau. Using constraint based operators with variable neighborhood search to solve the vehicle routing problem with time windows. *Journal of Heuristics* (to appear).

[58] M. Gendreau, P. Hansen, M. Labbé and N. Mladenović (2001) Variable neighborhood search for the traveling salesman problem (in preparation).

[59] M. Gendreau, A. Hertz and G. Laporte (1992) New Insertion and postoptimization procedures for the Traveling salesman problem. *Operations Research*, **40**, 1086–1094.

[60] M. Gendreau, A. Hertz and G. Laporte (1996), The traveling salesman problem with back-hauls, *Computers Operational Research*, **23**, 501–508.

[61] G. Ghiani, A. Hertz and G. Laporte (2000) Recent algorithmic advances for arc routing problems. *Les Cahiers du GERAD G-2000-40*, Montréal, Canada.

[62] F. Glover (1989) Tabu search—Part I. *ORSA J. Computing*, **1**, 190–206.

[63] F. Glover (1990) Tabu search—Part II. *ORSA J. Computing*, **2**, 4–32.

[64] F. Glover and M. Laguna (1997) *Tabu Search*. Kluwer, Boston.

[65] C.G. Gonzales and D.P. Brito (2001) A Variable neighborhood search for solving the linear ordering problem, MIC'2001, Porto, pp. 181–185.

[66] R.E. Griffith & R.A. Stewart (1961) A nonlinear programming technique for the optimization of continuous processing systems. *Management Science*, **7**, 379–392.

[67] I. Gutman (1993) A Contribution to the study of palindromic graphs. *Graph Theory Notes of New York*, XXIV:6, New York Academy of Science, pp. 51–56.

[68] I. Gutman and O.E. Polansky (1986) *Mathematical Concepts in Organic Chemistry*, Springer Berlin.

[69] I. Gutman, E. Estrada and O. Ivanciuc (1999) Some properties of the Wiener polynomial of trees. *Graph Theory Notes of New York*, XXXVI:1, New York Academy of Sciences, pp. 7–13.

[70] I. Gutman and O. Miljković (2000) Molecules with smallest connectivity index. *Match*, **41**, 57–70.

[71] I. Gutman, O. Miljković, G. Caporossi and P. Hansen (1999) Alkanes with small and large connectivity index. *Chemical Physics Letters*, **306**, 366–372.

[72] P. Hansen and B. Jaumard (1990) Algorithms for the maximum satisfiability problem. *Computing*, **44**, 279–303.

[73] P. Hansen and B. Jaumard (1997) Cluster analysis and mathematical programming. *Mathematical Programming*, **79**, 191–215.

[74] P. Hansen, B. Jaumard, N. Mladenović and A. Parreira (2000) Variable neighborhood search for Weighted maximum satisfiability problem. *Les Cahiers du GERAD* G-2000-62, Montréal, Canada.

[75] P. Hansen, S. Krau and O. du Merle. Stabilized column generation algorithm for the multisource Weber problem (in preparation).

[76] P. Hansen and H. Melot. Variable neighborhood search for extremal graphs. 6. Analyzing bounds on the connectivity index, *Les Cahiers du GERAD* (forthcoming).

[77] P. Hansen and N. Mladenović (1997) Variable neighborhood search for the p-Median. *Location Science*, **5**, 207–226.

[78] P. Hansen and N. Mladenović (1999) An introduction to variable neighborhood search. In: S. Voss et al. (eds.), *Metaheuristics, Advances and Trends in Local Search Paradigms for Optimization*. Kluwer, Dordrecht, pp. 433–458.

[79] P. Hansen and N. Mladenović (1999) First improvement may be better than best improvement: An empirical study. *Les Cahiers du GERAD* G-99-40, Montréal, Canada.

[80] P. Hansen and N. Mladenović (2001) Variable neighborhood search: Principles and applications. *European Journal of Operational Research*, **130**, 449–467.

[81] P. Hansen and N. Mladenović (2001) J-Means: A new local search heuristic for minimum sum-of-squares clustering. *Pattern Recognition*, **34**, 405–413.

[82] P. Hansen and N. Mladenović (2001) Industrial applications of the variable neighborhood search metaheuristics. In: G. Zaccour (eds.), *Decisions & Control in Management Science*. Kluwer Academic Publishers, Boston/Dordrecht/London, pp. 261–274.

[83] P. Hansen and N. Mladenović (2001) Developments of variable neighborhood search. In: C. Ribeiro, P. Hansen (eds.), *Essays and Surveys in Metaheuristics*. Kluwer Academic Publishers, Boston/Dordrecht/London, pp. 415–440.

[84] P. Hansen, N. Mladenović and D. Perez-Brito (2001) Variable neighborhood decomposition search. *Journal of Heuristics*, **7**(4) 335–350.

[85] P. Hansen, N. Mladenović and D. Urošević (2001) Variable neighborhood search for the Maximum clique. *Les Cahiers du GERAD* G-2001–25, Montréal, Canada.

[86] A. Hertz, M-C. Costa and M. Mitaz (1999) Bounds and heuristics for the shortest capacitated path problem, MIC'99, Rio de Janeiro.

[87] A. Hertz G. Laporte and M. Mittaz (2000) A Tabu search heuristic for the Capacitated arc routing problem. *Operations Research*, **48**, 129–135.

[88] K.S. Hindi, K. Fleszar and C. Charalambous (2001) An effective heuristic for the CLSP with setup times. *European Journal of Operational Research* (accepted s.t. revision).

[89] C.-T. Hsieh, C.-C. Chin and K.-M. Shen (1998) Generalized fuzzy Kohonen clustering networks, *IEICE Trans. Fundamentals*, V. E81-A (10).

[90] F.K. Hwang, D.S. Richards and P. Winter (1992) *The Steiner Tree Problem*. North-Holland, Amsterdam.

[91] T. Ibaraki, M. Kubo, T. Masuda, T. Uno and M. Yagiura (2001) Effective local search algorithms for the vehicle routing problem with general time windows, MIC'2001, Porto, pp. 293–297.

[92] S. Imahori, M. Yagiura and T. Ibaraki (2001) Local search heuristics for the rectangle packing problem with general spatial costs, MIC'2001, Porto, pp. 471–476.

[93] R.C. Jancey (1996) Multidimensional group analysis. *Australian Journal of Botany*, **14**, 127–130.

[94] Johnson D. and Trick M. (eds.), (1996) Cliques, coloring and satisfiability: Second DIMACS implementation challenge. *DIMACS Series in Discrete Mathematics and Theoretical Computer Science*, **26**.

[95] J.E. Kelley (1960) The cutting-plane method for solving convex programs. *Journal of the SIAM*, **8**, 703–712.

[96] R. Kolish and S. Hartman (1999) Heuristic algorithms for solving the resource-constrained project scheduling problem: classification and computational analysis, *Project Scheduling: Recent Models, Algorithms and Applications*, Kluwer.

[97] V. Kovačević, M. Čangalović, M. Ašić, D. Dražić and L. Ivanović (1999) Tabu search methodology in global optimization. *Computers & Mathematics with Applications*, **37**, 125–133.

[98] S. Krau (1997) *Extensions du problems de Weber*, PhD thèse, École Polytechnique de Montréal.

[99] M. Krishnamoorthy, A.T. Ernst and Y.M. Sharaiha (1998) Comparison of algorithms for the degree constrained spanning trees, Proceedings of International Conference on Optimisation Techniques and Applications (Perth, Western Australia, 1998), L. Caccetta, et al. (eds.), Curtin University of Technology, pp. 859–866.

[100] Y.-K. Kwok and I. Ahmad (1997) Efficient scheduling of arbitrary task graphs to multiprocessors using a parallel genetic algorithm. *Journal of Parallel and Distributed Computing*, **47**, 58–77.

[101] M. Labbé, G. Laporte, I. Rodriques and J. Salazar (1999) Median cycle problems. SMG Report, Université Libre de Bruxelles, Belgium. (http://smg.ulb.ac.be).

[102] M. Laguna, R. Martin and V. Campos (1999) Intensification and diversification with elite tabu search solutions for the linear ordering problem. *Computers & Operational Research*, **26**, 1217–1230.

[103] S. Lin and B.W. Kernighan (1973) An effective heuristic for the traveling salesman problem. *Operational Research*, **21**, 498–516.

[104] K. Jörnsten and A. Lokketangen (1997) Tabu search for weighted k-cardinality trees. *Asia-Pacific Journal of Operational Research*, **14**(2): 9–26.

[105] L. Lobjois, M. Lemaitre and G. Verfaille (2001) Large neighborhood search using constraint propagation and greedy reconstruction for valued CSP resolution. Research report ONERA, Toulouse, France.

[106] F.G. Lopez, B.M. Batista, J.A. Moreno Pérez and J.M. Moreno Vega (2000) The parallel variable neighborhood search for the p-median problem. Research Report, University of La Laguna, Spain. *Journal of Heuristics*. (to appear).

[107] S. Loudin and P. Boizumault (2001) VNS/LDS + CP: A hybrid method for constraint optimization in anytime contexts, MIC'2001, Porto, pp. 761–765.

[108] S.L. Martins, M.G.C. Resende, C.C. Ribeiro and P. Pardalos (2000) A parallel GRASP for the Steiner tree problem in graphs using a hybrid local search strategy. *Journal of Global Optimization*, **17**, 267–283.

[109] M. Mittaz. Problèmes de cheminements optimaux dans des réseaux avec contraintes associées aux arcs. PhD thesis, Department of Mathematics, École Polytechnique Fédérale de Lausanne, Switzerland.

[110] N. Mladenović (1995) A Variable neighborhood algorithm—a new metaheuristic for combinatorial optimization. Abstracts of papers presented at Optimization Days, Montréal, p. 112.

[111] N. Mladenović, M. Labbé and P. Hansen (2001) Solving the *p*-center problem by Tabu search and Variable neighborhood search *Networks* (to appear).

[112] N. Mladenović and P. Hansen (1997) Variable neighborhood search. *Computers Operations Research*, **24**, 1097–1100.

[113] N. Mladenović, J.P. Moreno, and J. Moreno-Vega (1996) A Chain-interchange heuristic method. *Yugoslav Journal of Operational Research*, **6**, 41–54.

[114] N. Mladenović, J. Petrović, V. Kovačević-Vujčić and M. Čangalović (2000) Solving Spread spectrum radar polyphase code design problem by Tabu search and Variable neighborhood search. SMG Report, R-00-09, Universite libre Bruxelles, Belgium. *European Journal of Operational Research* (to appear).

[115] N. Mladenović and D. Urošević (2001) Variable neighborhood search for the *k*-cardinality tree, MIC'2001, Porto, pp. 749–753.

[116] L.S. Ochi, M.B. Silva and L. Drummond (2001) Metaheuristics based on GRASP and VNS for solving Traveling purchaser problem, MIC'2001, Porto, pp. 489–494.

[117] J. Pearl (1998) *Probabilistic Reasoning in Intelligent Systems*. Morgan and Kaufman.

[118] G. Pesant and M. Gendreau (1996) A View of local search in Constraint programming, principles and practice of Constraint Programming, *Lecture Notes in Computer Science*, volume 1118. Springer-Verlag, pp. 353–366.

[119] G. Pesant and M. Gendreau (1999) A Constraint programming framework for Local search methods. *Journal of Heuristics*, **5**, 255–279.

[120] M. Randić (1975) On characterization of molecular branching. *Journal of the American Chemical Society*, **97**, 6609–6615.

[121] R. Ravi and F.S. Salman (1999) Approximation algorithms for the traveling purchaser problem and its variants in network design, *Lecture Notes in Computer Science*, volume 1643. Springer, pp. 29–40.

[122] G. Reinelt (1991) TSLIB – A Traveling salesman library. *ORSA Journal of Computing*, **3**, 376–384.

[123] M.G.C. Resende, L.S. Pitsoulis and P.M. Pardalos (1997) Approximate solution of Weighted max-sat problems using GRASP, In: Dingzhu Du et al. (eds.), *Satisfiability Problem: Theory and Applications*. DIMACS Series in Discrete Mathematics and Theoretical Computer Science 35, American Mathematical Society, Providence, Rhode Island.

[124] C. Ribeiro and C.Souza (2001) Variable neighborhood descent for the degree-constrained minimum spanning tree problem. *Discrete Applied Mathematics* (to appear).

[125] C. Ribeiro, E. Uchoa and R. Werneck (2001) A hybrid GRASP with perturbations for the Steiner problem in graphs. Technical report, Computer Science Department, Catholic University of Rio de Janeiro.

[126] I. Rodriguez, M. Moreno-Vega and J. Moreno-Perez (1999) Heuristics for Routing-median problems. SMG Report, Université Libre de Bruxelles, Belgium.

[127] A. Scholl, R. Klein and C. Jurgens (1997) BISON: A fast hybrid procedure for exactly solving the one-dimensional bin packing problem. *Computers & Operational Research*, **24**, 627–645.

[128] P. Shaw (1998) Using constraint programming and local search methods to solve vehicle routing problems. In: *Principles and Practice of Constraint Programming* (CP'98), pp. 417–431.

[129] M.B. Silva, L. Drummond and L.S. Ochi (2000) Variable neighborhood search for the Traveling purchacer problem, 27th Int. Conf. on Comp. Indust. Engineering.

[130] E. Taillard and L. Gambardella (1999) Adaptive memories for the quadratic assignment problem. *Technical report* I-87-97, IDSIA, Lugano, Switzerland.

[131] E. Taillard and S. Voss. POPMUSIC—Partial optimization metaheuristic under special intensification conditions. In: C. Ribeiro, P. Hansen (eds.), *Essays and Surveys in Metaheuristics*. Kluwer Academic Publishers, Boston/Dordrecht/London, pp. 613–630.

[132] W. Trigeiro, J. Thomas and J. McClain (1989) Capacitated lot-sizing with setup times. *Management Science*, **35**, 353–366.

[133] J. D. Ullman (1975) NP-complete scheduling problems. *Journal of Computing System Sciences* **10**(3), 384–393.

[134] M.G.A. Verhoeven, M.E.M. Severens and E.H.L. Aarts (1996) Local search for Steiner trees in graphs. In: V.J. Rayward-Smith et al. (eds.), *Modern Heuristic Search Methods*. John Wiley and Sons, pp. 117–129.

[135] S. Voss (1996) Dynamic Tabu search strategies for the traveling purchaser problem. *Annals of Operational Research* **63**, 253–275.

[136] R. Whittaker (1983) A fast algorithm for the greedy interchange for large-scale clustering and median location problems. *INFOR*, **21**, 95–108.

[137] N. Zufferey, C. Avanthay and A. Hertz. Variable neighborhood search for graph colouring (submitted).

Chapter 7

GUIDED LOCAL SEARCH

Christos Voudouris
Research Department, BTexact Technologies,
BT plc, Orion Building, mlb1/pp12,
Marthlesham Heath, Ipswich, IP5 3RE,
UK
E-mail: chris.voudouris@bt.com

Edward P. K. Tsang
Department of Computer Science,
Univeristy of Essex, Wivenhoe Park,
Colchester, CO4 3SQ, UK
E-mail: edward@essex.ac.uk

Abstract The combinatorial explosion problem prevents complete algorithms from solving many real-life combinatorial optimization problems. In many situations, heuristic search methods are needed. This chapter describes the principles of Guided Local Search (GLS) and Fast Local Search (FLS) and surveys their applications. GLS is a penalty-based meta-heuristic algorithm that sits on top of other local search algorithms, with the aim to improve their efficiency and robustness. FLS is a way of reducing the size of the neighborhood so as to improve the efficiency of local search. The chapter also provides guidance for implementing and using GLS and FLS. Four problems, representative of general application categories, are examined with detailed information provided on how to build a GLS-based method in each case.

Keywords: Heuristic Search, Meta-Heuristics, Penalty-based Methods, Guided Local Search, Tabu Search, Constraint Satisfaction

1 INTRODUCTION

Many practical problems are NP-hard in nature, which means complete, constructive search is unlikely to satisfy our computational demand. For example, suppose we have to schedule 30 jobs to 10 machines, satisfying various production constraints. The search space has 10^{30} leaf nodes. Let us assume that we use a very clever backtracking algorithm that explores only one in every 10^{10} leaf nodes. Let us generously assume that our implementation examines 10^{10} nodes per second (with today's hardware, even the most naïve backtracking algorithm should be expected to examine only 10^5 nodes per second). This still leaves us with approximately 300 years to solve the problem, in the worst case. Many real life problems cannot be realistically and reliably be solved by complete search. This motivates the development of local search, or heuristic methods.

In this paper, we describe GLS, a general meta-heuristic algorithm and its applications. GLS sits on top of other heuristic methods with the aim to improve their efficiency or robustness. GLS has been applied to a non-trivial number of problems, and found to be efficient and effective. It is relatively simple to implement and apply, with few parameters to tune.

The rest of this chapter will be divided into two parts: the first part describes the GLS and surveys its applications. The second part provides guidelines on how to implement and use GLS in practical applications.

Part I: Survey of Guided Local Search

2 BACKGROUND

Local search (LS) is the basis of most heuristic search methods. It searches in the space of candidate solutions, such as the assignment of one machine to each job in the above example. The solution representation issue is significant, though it is not the subject of our discussion here. Starting from a (possibly randomly generated) candidate solution, LS moves to a "neighbor" that is "better" than the current candidate solution according to the objective function. LS stops when all neighbors are inferior to the current solution.

LS can find good solutions very quickly. However, it can be trapped in local optima—positions in the search space that are better than all their neighbors, but not necessarily representing the best possible solution (the global optimum). To improve the effectiveness of LS, various techniques have been introduced over the years. Simulated Annealing (SA), Tabu Search (TS) and Guided Local Search (GLS) all attempt to help LS escape local optimum. This chapter focuses on GLS.

GLS is a meta-heuristic algorithm generalized and extended from a neural-network based method called GENET (Wang and Tsang, 1991, 1994; Davenport et al., 1994; Tsang et al., 1999). GLS was inspired by ideas from the area of *Search Theory* on how to distribute the searching effort (e.g. see Koopman, 1957; Stone, 1983). GENET is a weighted method for constraint satisfaction. Though originated from neural networks, GENET resembles the min-conflict heuristic repair method by Minton et al. (1992).

The principles of GLS can be summarized as follows. As a meta-heuristic method, GLS sits on top of LS algorithms. To apply GLS, one defines a set of features for the candidate solutions. When LS is trapped in local optima, certain features are selected and penalized. LS searches using the objective function as this is augmented by the accumulated penalties. The novelty of GLS is in the way that it selects features to penalize. GLS effectively distributes the search effort in the search space, favoring areas where promise is shown.

3 GUIDED LOCAL SEARCH

As mentioned earlier, GLS augments the given objective function with penalties. To apply GLS, one needs to define features for the problem. For example, in the travelling salesman problem, a feature could be "*whether the candidate tour travels immediately from city A to city B*". GLS associates a cost and a penalty to each feature. The costs can often be defined by taking the terms and their coefficients from the objective function. For example, in the travelling salesman problem, the cost of the above feature can simply be the distance between cities A and B. The penalties are initialized to 0 and will only be increased when the local search reaches a local optimum. Given an objective function g that maps every candidate solution s to a numerical value, GLS defines a function h that will be used by LS (replacing g):

$$h(s) = g(s) + \lambda \times \sum (p_i \times I_i(s)) \tag{1}$$

where s is a candidate solution, λ is a parameter to the GLS algorithm, i ranges over the features, p_i is the penalty for feature i (all p_i's are initialized to 0) and I_i is an indication of whether s exhibits feature i:

$$\begin{aligned} I_i(s) &= 1 \quad \text{if } s \text{ exhibits feature } i; \\ &= 0 \quad \text{otherwise.} \end{aligned} \tag{2}$$

Sitting on top of local search algorithms, GLS helps them to escape local optima in the following way. Whenever the local search algorithm settles in a local optimum, GLS augments the cost function by adding penalties to selected features. The novelty of GLS is mainly in the way that it selects features to penalize. The intention is to penalize "unfavorable features" or features that "matter most" when a local search settles in a local optimum. The feature that has high cost affects the overall cost more. Another factor that should be considered is the current penalty value of that feature. The utility of penalizing feature i, util_i, under a local optimum s_*, is defined as follows:

$$\text{util}_i(s_*) = I_i(s_*) \times \frac{c_i}{1 + p_i} \tag{3}$$

where c_i is the cost and p_i is the current penalty value of feature i. In other words, if a feature is not exhibited in the local optimum (indicated by I_i), then the utility of penalizing it is 0. The higher the cost of this feature (the greater c_i), the greater the utility of penalizing it. Besides, the more times that it has been penalized (the greater p_i), the lower the utility of penalizing it again. In a local optimum, the feature with the greatest util value will be *penalized*. When a feature is penalized, its penalty value is always increased by 1. The scaling of the penalty is adjusted by λ.

By taking cost and the current penalty into consideration in selecting the feature to penalize, GLS focuses its search effort on more promising areas of the search space: areas that contain candidate solutions that exhibit "good features", i.e. features involving lower cost. On the other hand, penalties help to prevent the search from directing all effort to any particular region of the search space. Following is the general GLS procedure (a more detailed pseudo-code for GLS will be given in Section 7.1):

```
Procedure GLS (L, g, λ, I, c)
/* g is an objective function; L is a LS strategy; I and c are
arrays of features and their costs; λ is a scaling number)
  1. Generate a starting candidate solution randomly or
     heuristically;
  2. Initialize all the penalty values (pi) to 0;
```

```
3. Repeat the following until a termination condition
   (e.g. a maximum number of iterations or time limit)
   has been reached:
   3.1. Perform local search (using L) according to the
        function h (which is g plus the penalty values,
        as defined in Eq. 1 above) until a local optimum
        LM has been reached;
   3.2. For each feature i which is exhibited in LM
        compute util_i = c_i/(1 + p_i);
   3.3. Penalize every feature i such that util_i is
        maximum: p_i = p_i + 1;
4. Return the best candidate solution found so far
   according to the objective function g.
End of Procedure GLS
```

Naturally the choice of the features, their costs and the setting of λ may affect the efficiency of a search. Experience shows that the features and their costs normally come directly from the objective function. In many problems, the performance of GLS is not too sensitive to the value λ. This means one does not require too much effort to apply GLS to a new problem. In certain problems, one needs expertise in selecting the features and the λ parameter. Current research aims to reduce the sensitivity of the λ parameter (Mills and Tsang, 2000b).

4 FAST LOCAL SEARCH

One factor which affects the efficiency of a local search algorithm is the size of the neighborhood. If too many neighbours are considered, then the search could be very costly. This is especially true if the search takes many steps to reach a local optima, and/or each evaluation of the objective function requires a significant amount of computation. Bentley (1992) presented the *approximate 2-Opt* method to reduce the neighborhood of 2-Opt in the TSP. We have generalised this method to a method called *Fast Local Search* (FLS). The principle is to, guided by heuristics, ignore neighbors that are unlikely to lead to improving moves in order to enhance the efficiency of a search.

The neighborhood choosen for the problem is broken down into a number of small sub-neighborhoods and an activation bit is attached to each one of them. The idea is to scan continuously the sub-neighborhoods in a given order, searching only those with the activation bit set to 1. These sub-neighborhoods are called active sub-neighborhoods. Sub-neighborhoods with the bit set to 0 are called inactive sub-neighborhoods and they are not being searched. The neighborhood search process does not restart whenever we find a better solution but it continues with the next sub-neighborhood in the given order. This order may be static or dynamic (i.e., change as a result of the moves performed).

Initially, all sub-neighborhoods are active. If a sub-neighborhood is examined and does not contain any improving moves then it becomes inactive. Otherwise, it remains active and the improving move found is performed. Depending on the move performed, a number of other sub-neighborhoods are also activated. In particular, we activate all the sub-neighborhoods where we expect other improving moves to occur as a result of the move just performed. As the solution improves the process dies out with fewer and fewer sub-neighborhoods being active until all the sub-neighborhood bits turn to 0. The solution formed up to that point is returned as an approximate local optimum.

The overall procedure could be many times faster than conventional local search. The bit setting scheme encourages chains of moves that improve specific parts of the overall solution. As the solution becomes locally better the process is settling down, examining fewer moves and saving enormous amounts of time which would otherwise be spent on examining predominantly bad moves.

Although fast local search procedures do not generally find very good solutions, when they are combined with GLS they become very powerful optimization tools. Combining GLS with FLS is straightforward. The key idea is to associate features to sub-neighborhoods. The associations to be made are such that for each feature we know which sub-neighborhoods contain moves that have an immediate effect upon the state of the feature (i.e. moves that remove the feature from the solution).

By reducing the size of the neighbourhood, one may significantly reduce the amount of computation involved in each local search iterations. The hope is to enable more local search iterations in a fixed amount of time. The danger of ignoring certain neighbours is that some improvements may be missed. The hope is that the gain out-weighs the loss.

5 GLS AND OTHER METAHEURISTICS

GLS is closely related to other heuristic and meta-heuristic methods. Recently, Choi et al. (forthcoming) show the relationship between GENET (GLS's predecesor) and the Lagrangean Method. In this section, we shall discuss the relationship between GLS and Tabu Search (TS), Genetic Algorithms (GA).

5.1 GLS and Tabu Search

GLS is closely related to Tabu Search (TS). For example, penalties in GLS can be seen as soft taboos in TS that guide LS away from local minima. There are many ways to adopt TS ideas in GLS. For example, taboo lists and aspiration ideas have been used in later versions of GLS. Penalties augment the original objective function. They help the local search to escape local optimum. However, if too many penalties are built up during the search, the local search could be misguided. Resembling the taboo lists idea, a limited number of penalties are used when GLS is applied to the quadratic assignment problem. When the list is full, old penalties will be overwritten (Voudouris and Tsang, 1998).

In our current work, aspiration (inspired by TS) is used to favor promising moves. Details of this work will be presented in another occasion (Mills and Tsang, 2000b).

5.2 GLS and Genetic Algorithms

As a meta-heuristic method, GLS can also sit on top of genetic algorithms (GA) (Holland, 1975; Goldberg, 1989). This has been demonstrated in Guided Genetic Algorithm (GGA) (Lau and Tsang, 1997 1998; Lau, 1999).

GGA is a hybrid of GA and GLS. It is designed to extend the domain of both GA and GLS. One major objective is to further improve the robustness of GLS. It can be seen as a GA with GLS to bring it out of local optimum: if no progress has been made after a specific of iterations (this number is a parameter to GGA), GLS modifies the fitness function (which is the objective function) by means of penalties, using the criteria defined in equation (3). GA will then use the modified fitness function in future generations. The penalties are also used to bias crossover and mutation in GA – genes that contribute more to the penalties are made more susceptive to changes by these two GA operators. This allows GGA to be more focussed in its search.

On the other hand, GGA can roughly be seen as a number of GLS's from different starting points running in parallel, exchanging material in a GA manner. The difference is that only one set of penalties is used in GGA whereas parallel GLS's could have used one independent set of penalties per run. Besides, learning in GGA is more selective than parallel GLS: the updating of penalties is only based on the best chromosome found at the point of penalization.

6 APPLICATIONS

GLS, FLS and their descendents have been applied to a non-trivial number of problems and achieved world-class results.

6.1 Radio Link Frequency Assignment Problem

In the *radio link frequency assignment problem* (RLFAP), the task is to assign available frequencies to communication channels satisfying constraints that prevent interference (Bouju et al., 1995). In some RLFAPs, the goal is to minimize the number of frequencies used. Bouju et al. (1995) is an early work that applied GENET to radio length frequency assignment. For the CALMA set of benchmark problems, which has been widely used, GLS+FLS reported the best results compared to all work published previously (Voudouris and Tsang, 1996). In the NATO Symposium on RLFAP in Denmark, 1998, GGA was demonstrated to improve the robustness of GLS (Lau and Tsang, 1998). In the same symposium, new and significantly improved results by GLS were reported (Voudouris and Tsang, 1998). GLS and GGA hold some of the best results in the CALMA set of benchmark problems.

6.2 Workforce Scheduling Problem

In the *workforce scheduling* problem (WSP) (Azarmi and Abdul-Hameed, 1995), the task is to assign technicians from various bases to serve the jobs, which may include customer requests and repairs, at various locations. Customer requirements and working hours restrict the times that certain jobs can be served by certain technicians. The objective is to minimize a function that takes into account the travelling cost, overtime cost and unserved jobs. In the WFS, GLS+FLS holds the best-published results in the benchmark problem available to the authors (Tsang and Voudouris, 1997).

6.3 Travelling Salesman Problem

The most significant results of GLS and FLS are probably in their application to the travelling salesman problem (TSP). The Lin-Kernighan algorithm (LK) is a specialized algorithm for TSP that has long been perceived as the champion of this problem (Lin and Kernighan, 1973; Martin and Otto, 1996). We tested GLS+FLS+2Opt against

LK (Voudouris and Tsang, 1999) in a set of benchmark problems from the public TSP library (Reinelt, 1991). Given the same amount of time (we tested 5 and 30 cpu min on a DEC Alpha 3000/600), GLS+FLS+2Opt found better results than LK in average. GLS+FLS+2Opt also out-performed Simulated Annealing (Johnson, 1990), Tabu Search (Knox, 1994) and Genetic Algorithm (Freisleben and Merz, 1996) implementations reported on the TSP. One must be cautious when interpreting such empirical results as they could be affected by many factors, including implementation issues. But given that the TSP is an extensively studied problem, it takes something special for an algorithm to out-perform the champions under any reasonable measure (*"find me the best results within a given amount of time"* must be a realistic requirement). It must be emphasized that LK is specialized for TSP but GLS and FLS are much simpler general-purpose algorithms.

GLS hybrids have also been proposed for the TSP including the combination of GLS with Memetic Algorithms (Holstein and Moscato, 1999) and also with the, dynamic-programming based, Dynasearch technique with some encouraging preliminary results reported (Congram and Potts, 1999). Finally, Padron and Balaguer (2000) have applied GLS to the related Rural Postman Problem (RPP).

6.4 Function Optimization

GLS has also been applied to general function optimization problems to illustrate that artificial features can be defined for problems in which the objective function suggests no obvious features. Results show that, as expected, GLS spreads its search effort across solution candidates depending on their quality (as measured by the objective function). Besides, GLS consistently found solutions in a landscape with many local sub-optimals (Voudouris, 1998).

6.5 Satisfiability and Max-SAT problem

Given a set of propositions in conjunctive normal form, the Satisfiability (SAT) problem is to determine whether the propositions can all be satisfied. The MAX-SAT problem is a SAT problem in which each clause is given a weight. The task is to minimize the total weight of the violated clauses. In other words, the weighted MAX-SAT problem is an optimization problem. Many researchers believe that many problems, including scheduling and planning can be formulated as SAT and MAX-SAT problems, hence these problems have received significant attention in recent years, e.g., see Gent et al. (2000).

GLSSAT, an extension of GLS, was applied to both the SAT and weighted MAX-SAT problem (Mills and Tsang, 2000a). On a set SAT problems from DIMACS, GLSSAT produced results comparable to those produced by WalkSAT (Selman and Kautz, 1993), a variation of GSAT (Selman et al., 1992), which was specifically designed for the SAT problem.

On a popular set of benchmark weighted MAX-SAT problems, GLSSAT produced better or comparable solutions, more frequently than state-of-the-art algorithms, including DLM (Shang and Wah, 1998), WalkSAT (Selman and Kautz, 1993) and GRASP (Resende and Feo, 1996).

6.6 Generalized Assignment Problem

The Generalized Assignment Problem is a generic problem in which the task is to assign agents to jobs. Each job can only be handled by one agent, and each agent has a finite resource capacity that limits the number of jobs that it can be assigned. Assigning different agents to different jobs bear different utilities. On the other hand, different agents will consume different amounts of resources when doing the same job. In a set of benchmark problems, GGA found results as good as those produced by a state-of-the-art algorithm (which was also a GA algorithm) by Chu and Beasley (1997), with improved robustness (Lau and Tsang, 1998).

6.7 Processor Configuration Problem

In the Processors Configuration Problem, one is given a set of processors, each of which with a fixed number of connections. In connecting the processors, one objective is to minimize the maximum distances between processors. Another possible objective is to minimize the average distance between pairs of processors (Chalmers, 1994). In applying GGA to the Processors Configuration Problem, representation was a key issue. To avoid generating illegal configurations, only mutation is used. GGA found configurations with shorter average communication distance than those found by other algorithms reported previously (Lau and Tsang, 1997, 1998).

6.8 Vehicle Routing Problem

In a vehicle routing problem, one is given a set of vehicles, each with its specific capacity and availability, and a set of customers to serve, each with specific weight and/or time demand on the vehicles. The vehicles are grouped in one or more depots. Both the depots and the customers are geographically distributed. The task is to serve the customers using the vehicles, satisfying time and capacity constraints. This is a practical problem. Like many practical problems, it is NP-hard.

Kilby et al. (1999, 2000) applied GLS vehicle routing problems and achieved outstanding results. As a result, their work were incorporated in *Dispatcher*, a commercial package developed by ILOG (http://www.ilog.fr/html/products/) (Backer et al., 2000).

6.9 Constrained Logic Programming

Lee and Tam (1995) and Stuckey and Tam (1998) embedded GENET in logic programming languages in order to enhance programming efficiency. In these logic programming implementations, unification is replaced by constraint satisfaction. This enhances efficiency and extends applicability of logic programming.

6.10 Other Applications of GENET and GLS

We have also applied GLS and FLS to a variety of other problems, including the Maximum Channel Assignment problem, a Bandwidth Packing problem variant, graph colouring, car sequencing problem. GLS and FLS have also been successfully applied to the 3-D Bin Packing Problem by Faroe et al. (1999). Other applications of GENET include rail traffic control (Jose and Boyce, 1997).

Part II: Practical Guidelines for Using Guided Local Search

7 IMPLEMENTING GUIDED LOCAL SEARCH

As explained in previous sections, a local search procedure for the particular problem is required. Guided Local Search is repeatedly using this procedure to optimize the augmented objective function of the problem. The augmented objective function is modified each time a local minimum is reached by increasing the penalties of one or more of the features present in the local minimum. These features are selected by using the utility function (3) described in Section 3. The pseudo-codes for implementing a Guided Local Search method and a Guided Fast Local Search method are presented and explained in the sections below.

7.1 Pseudo-code for Guided Local Search

The pseudo-code for the Guided Local Search procedure is the following:

```
procedure GuidedLocalSeach(p, g, λ, [I1,...,IM], [c1,...,cM], M)
begin
    k ← 0;
    s0  ← ConstructionMethod(p);
    /* set all penalties to 0 */
    for i ← 1 until M do
        pi ← 0;
    /* define the augmented objective function */
    h ← g + λ * ∑pi*Ii;
    while StoppingCriterion do
    begin
        sk+1 ← ImprovementMethod(sk, h);
        /* compute the utility of features */
        for i ← 1 until M do
            utili ← Ii(sk+1) * ci/(1 + pi);
        /* penalize features with maximum utility */
        for each i such that utili is maximum do
            pi ← pi + 1;
        k ← k+1;
    end
    s* ← best solution found with respect to objective
function g;
    return s*;
end
```

where p: problem, g: objective function, h: augmented objective function, λ: lambda parameter, I_i: indicator function of feature i, c_i: cost of feature i, M: number of features, p_i: penalty of feature i, *ConstructionMethod(p)*: method for constructing a initial solution for problem p, and *ImprovementMethod*(s_k, h): method for improving solution s_k according to the augmented objective function h.

7.2 Guidelines for Implementing the GLS Pseudo-code

To understand the pseudo-code, let us explain first the methods for constructing a solution and improving a solution, which are contained in there and they are both prerequisites in order to build a GLS algorithm.

7.2.1 Construction Method

As with other meta-heuristics, Guided Local Search requires a construction method to generate an initial (starting) solution for the problem. In the pseudo-code, this is denoted by *ConstructionMethod.* This method can be generating a random solution or a heuristic solution based on some known technique for constructing solutions for the particular problem. Guided Local Search is not very sensitive to the starting solution given that sufficient time is allocated to the algorithm to explore the search space of the problem.

7.2.2 Improvement Method

A method for improving the solution is also required. In the pseudo-code, this is denoted by *ImprovementMethod.* This method can be a simple local search algorithm or a more sophisticated one such as Variable Neighborhood Search (Hansen and Mladenovic, 1999), Variable Depth Search (Lin and Kernighan, 1973), Ejection Chains (Glover and Laguna, 1997) or combinations of local search methods with exact search algorithms (Pesant and Gendreau, 1999).

It is not essential for the improvement method to generate high quality local minima. Experiments with GLS and various TSP heuristics reported in (Voudouris and Tsang, 1999) have shown that high quality local minima take time to produce, resulting in less intervention by GLS in the overall allocated search time. This may sometimes lead to inferior results compared to a simple but more computationally efficient improvement method.

Note also that the improvement method is using the augmented objective function instead of the original one.

7.2.3 Indicator Functions and Feature Penalization

Given that a construction and an improvement method are available for the problem, the rest of the pseudo-code is straightforward to apply. The penalties of features are initialized to zero and they are incremented for features that maximize the utility formula, after each call to the improvement method.

The indicator functions I_i for the features rarely need to be implemented. Looking at the values of the decision variables can directly identify the features present in a local minimum. When this is not possible, data structures with constant time deletion/addition operations (e.g. based on double-linked lists) can incrementally maintain the set of features present in the working solution avoiding the need for an expensive computation when GLS reaches a local minimum.

The selection of features to penalize can be efficiently implemented by using the same loop for computing the utility formula for features present in the local minimum (the rest can be ignored) and also placing features with maximum utility in a vector. With a second loop, the features with maximum utility contained in this vector have their penalties incremented by one.

7.2.4 *The Lambda Parameter*

The lambda parameter is the only parameter to the GLS method (at least in its basic version) and in general, it is instance dependent. Fortunately, for several problems, it has been observed that good values for lambda can be found by dividing the value of the objective function in a local minimum with the average number of features present. In these problems, lambda is dynamically computed after the first local minimum and before penalties are applied to features for the first time. The user only provides an *alpha* parameter, which is relatively instance independent. The recommended formula for lambda as a function of alpha is the following:

$$\lambda = \alpha * g(\text{local minimum})/(\text{no. of features present in the local minimum}), \qquad (4)$$

where g is the objective function of the problem. Tuning alpha can result in lambda values, which work for many instances of a problem class.

Another benefit from using α is that, if tuned, it can be fixed in industrialized versions of software, resulting in ready-to-use GLS algorithms for the end-user.

7.2.5 *Augmented Objective Function and Move Evaluations*

With regard to the objective function and the augmented objective function, the program should keep track of the actual objective value in all operations relating to storing the best solution or finding a new best solution. Keeping track of the value of the augmented objective value (e.g. after adding the penalties) is not necessary since local search methods will be looking only at the differences in the augmented objective value when evaluating moves.

However, the move evaluation mechanism needs to be revised to work efficiently with the augmented objective function. Normally, the move evaluation mechanism is not directly evaluating the objective value of the new solution generated by the move. Instead, it calculates the difference Δg in the objective function. This difference should be combined with the difference in the amount of penalties. This can be easily done and has no significant impact on the time needed to evaluate a move. In particular, we have to take into account only features that change state (being deleted or added). The penalties of the features deleted are summed together. The same is done for the penalties of features added. The change in the amount of penalties due to the move is then simply given by the difference:

$$-\sum_{\text{over all features j deleted}} p_j + \sum_{\text{over all features k added}} p_k \qquad (5)$$

which then has to be multiplied by lambda and added to Δg.

Another minor improvement is to monitor the actual objective value not only for the solutions accepted by local search but also for those evaluated. Since local search is using the augmented objective function, a move that generates a new best solution may be missed. From our experience, this modification does not improve significantly the performance of the algorithm although it can be proved useful when GLS is used to find new best-known solutions to hard benchmark instances.

7.2.6 Stopping Criterion

There are many choices possible for the *StoppingCritetion*. Since GLS is not trapped in local minima, there is no obvious point when to stop the algorithm. Like in other metaheuristics, we usually resort to a measure related to the length of the search process. For example, we may choose to set a limit on the number of moves performed, the number of moves evaluated, or the CPU time spend by the algorithm. If a lower bound is known, we can utilize it as a stopping criterion by setting the excess above the lower bound that we require to be achieved. Criteria can also be combined to allow for a more flexible way to stop the GLS method.

In the next section, we look at the combination of Guided Local Search with Fast Local Search resulting in the Guided Fast Local Search method.

8 IMPLEMENTING GUIDED FAST LOCAL SEARCH

Guided Fast Local Search (GFLS) is more sophisticated than the basic GLS algorithm using a number of sub-neighborhoods, which are enabled/disabled during the search process. The main advantage of GFLS lies in its ability to provide search focus after the penalties of features are increased. This can dramatically shorten the time required by an improvement method to re-optimize the solution each time the augmented objective function is modified

Guided Fast Local Search has been described in Section 4 of this chapter. In the following sections, we provide the pseudo-code for the method and also some suggestions on how to achieve an efficient implementation. We first look at the pseudo-code for Fast Local Search, which is part of the overall Guided Fast Local Search algorithm.

8.1 Pseudo-code for Fast Local Search

The pseudo-code for Fast Local Search is the following:

```
procedure FastLocalSeach(s, h, [bit₁,...,bit_L], L)
begin
    while ∃bit, bit = 1 do
    /* i.e. while active sub-neighborhood exists */
        for  i ←1 until L do
        begin
            if bit_i =1 then
            /* search sub-neighborhood i */
            begin
                Moves ← MovesForSubneighborhood(i);
                for each move m in Moves do
                begin
                    s′ ← m(s);
                    /* s′ is the result of move m */
                    if h(s′) < h(s) then
                    /* minimization case is assumed here */
                    begin
                        /* spread activation */
                        ActivateSet ←
```

```
                        SubneighborhoodsForMove(m);
                        for each sub-neighborhood j
                        in ActivateSet do
                             bitj ← 1;
                        s ← s';
                        goto ImprovingMoveFound
                    end
                end
                biti ← 0; /* no improving move found */
            end
ImprovingMoveFound:
            continue;
        end;
    return s;
end
```

where s: solution, h: augmented objective function, L: number of sub-neighborhoods, bit_i: activation bit for sub-neighborhood i, *MovesForSubneighborhood(i)*: method which returns the set of moves contained in sub-neighborhood i, and *SubneighborhoodsForMove(m)*: method which returns the set of sub-neighborhoods to activate when move m is performed.

8.2 Guidelines for Implementing the FLS Pseudo-code

As explained in section (4), the problem's neighborhood is broken down into a number of sub-neighborhoods and an *activation bit* is attached to each one of them. The idea is to examine sub-neighborhoods in a given order, searching only those with the activation bit set to 1. The neighborhood search process does not restart whenever we find a better solution but it continues with the next sub-neighborhood in the given order. The pseudo-code given above is flexible since it does not specify which bits are initially switched on or off, something which is an input to the procedure. This allows the procedure to be focused to certain sub-neighborhoods and not the whole neighborhood, which may be a large one.

The procedure has two points that need to be customized. The first is the breaking-down of the neighborhood into sub-neighborhoods (*MovesForSubneighborhood* method in pseudo-code). The second is the mapping from moves to sub-neighborhoods for spreading activation (*SubneighborhoodsForMove* method in pseudo-code). Both these points are strongly dependent on the move operator used.

In general, the move operator depends on the solution representation. Fortunately, several problems share the same solution representation which is typically based on some well-known simple or composite combinatorial structure (e.g., selection, permutation, partition, composition, path, cyclic path, tree, graph, etc.). This allows us to use the same move operators for many different problems (e.g. 1-Opt, 2-Opt, Swaps, Insertions, etc.).

8.2.1 Breaking Down the Neighborhood into Sub-neighborhoods

The method for mapping sub-neighborhoods to moves, which is denoted in the pseudo-code by *SubneighbourhoodToMoves*, can be defined by looking at the implementation of a typical local search procedure for the problem. This implementation, at its core, will usually contain a number of nested for-loops for generating all possible move combinations. The variable in the outer-most loop in the move generation code can be used to define the sub-neighborhoods. The moves in each sub-neighborhood will be those generated by the inner loops for the particular sub-neighborhood index value at the outer-most loop.

In general, the sub-neighborhoods can be overlapping. Fast local search is usually examining limited numbers of moves compared to exhaustive neighborhood search methods and therefore duplication of moves is not a problem. Moreover, this can be desirable sometimes to give a greater range to each sub-neighborhood. Since not all sub-neighborhoods are active in the same iteration, if there is no overlapping, some of improving moves may be missed.

8.2.2 Spreading Activation When Moves are Executed

The method for spreading activation, denoted by *SubneighbourhoodsForMove*, returns a set of sub-neighborhoods to activate after a move is performed. The lower bound for this set is the sub-neighborhood where the move originated. The upper bound (although not useful) is all the sub-neighborhoods in the problem.

A way to define this method is to look at the particular move operator used. Moves will affect part of the solution directly or indirectly while leaving other parts unaffected. If a sub-neighborhood contains affected parts then it needs to be activated since an opportunity could arise there for an improving move as a result of the original move performed.

The Fast Local Search loop is settling down in a local minimum when all the bits of sub-neighborhoods turn to zero (i.e. no improving move can be found in any of the sub-neighborhoods). Fast Local Search to that respect is similar to other local search procedures. The main differences are that the method can be focused to search particular parts of the overall neighborhood and secondly, it is working in a opportunistic way looking at parts of the solution which are likely to contain improving moves rather than the whole solution. In the next section, we look at Guided Fast Local Search, which uses Fast Local Search as its improvement method.

8.3 Pseudo-code for Guided Fast Local Search

The pseudo-code for Guided Fast Local Search is given below:

```
procedure GuidedFastLocalSearch(p, g, λ, [I1,...,IM],
[c1,...,cM], M, L)
begin
    k ← 0;
    s0 ← ConstructionMethod(p);
    /* set all penalties to 0 */
    for i ←1 until M do
        pi ← 0;
    /* set all sub-neighborhoods to the active state */
    for i ←1 until L do
        biti ← 1;
    /* define the augmented objective function */
```

```
    h ← g + λ * ∑ pi*Ii;
    while StoppingCriterion do
    begin
        sk+1 ← FastLocalSearch(sk, h, [bit1, ..., bitL], L);
        /* compute the utility of features */
        for i ←1 until M do
            utili ← Ii(sk+1) * ci/(1 + pi);
        /* penalize features with maximum utility */
        for each i such that utili is maximum do
        begin
            pi ← pi + 1;
            /* activate sub-neighborhoods related
            to penalized feature i */
            ActivateSet ← SubneighborhoodsForFeature(i);
            for each sub-neighborhood j in ActivateSet do
                bitj ← 1;
        end
        k ← k + 1;
    end
    s* ← best solution found with respect to objective
function g;
    return s*;
end
```

where p: problem, g: objective function, h: augmented objective function, λ: lambda parameter, I_i: indicator function of feature i, c_i: cost of feature i, M: number of features, p_i: penalty of feature i, L: number of sub-neighborhoods, bit_i: activation bit for sub-neighborhood i, *ConstructionMethod(p)* method for constructing a initial solution for problem p, *FastLocalSearch*(s_k, h, $[bit_1, \ldots, bit_L], L$): the fast local search method as depicted in section 8.1, and *SubneighborhoodsForFeature(i)*: method which returns the set of sub-neighborhoods to activate when feature i is penalized.

8.4 Guidelines for Implementing the GFLS Pseudo-code

This pseudo-code is similar to that of Guided Local Search explained in Section 7. All differences relate to the manipulation of activation bits for the purpose of focusing Fast Local Search. These bits are initialized to 1. As a result, the first call to Fast Local Search is examining the whole neighborhood for improving moves.

Subsequent calls to Fast Local Search examine only part of the neighborhood and in particular all the sub-neighborhoods that relate to the features penalized by GLS.

8.4.1 *Identifying Sub-neighborhoods to Activate When Features are Penalized*

Identifying the sub-neighborhoods that are related to a penalized feature is the task of *SubneighborhoodsForFeature* method. The role of this method is similar to that of *SubneighborhoodsForMove* method in Fast Local Search (see Section 8.2.2).

The *SubneighborhoodsForFeature* method is usually defined based on an analysis of the move operator. After the application of penalties, we are looking for moves which remove or have the potential to remove penalized features from the solution. The sub-neighborhoods, which contain such moves, are prime candidates for activation. Specific examples will be given later in the chapter and in the context of GLS applications.

Guided Fast Local Search is much faster than basic Guided Local Search especially in large problem instances where repeatedly and exhaustively searching the whole neighborhood is computationally expensive.

9 USEFUL FEATURES FOR COMMON APPLICATIONS

Apart from the necessary specialization of the algorithms explained in previous sections, applying Guided Local Search or Guided Fast Local Search to a problem requires identifying a *suitable* set of features to guide the search process. As explained in section 3, features need to be defined in the form of indicator functions that given a solution return 1 if the feature is present in the solution or 0 otherwise.

Features provide the heuristic search expert with quite a powerful tool since any solution property can be potentially captured and used to guide local search. Usually, we are looking for solution properties, which have a direct impact on the objective function. These can be modeled as features with feature costs equal or analogous to their contribution to the objective function value. By applying penalties to features, GLS can guide the improvement method to avoid costly ("bad") properties, converging faster towards areas of the search space, which are of high quality.

Features are not necessarily specific to a particular problem and they can be used in several problems of similar structure. Real world problems, which sometimes incorporate elements from several academic problems, can benefit from using more than one feature-sets to guide local search to optimize better all different terms of a complex objective function.

Below, we provide examples of features that can be deployed in the context of various problems. The reader may find them helpful and use them in his/her own optimization application.

9.1 Routing/Scheduling Problems

In routing/scheduling problems, one is seeking to minimize the time required by a vehicle to travel between customers or for a resource to be setup from one activity to the next. Problems in this category include the Traveling Salesman Problem, Vehicle Routing Problem, Machine Scheduling with Sequence Dependent Set-up Times and others.

Travel or setup times are modeled as edges in a path or graph structure commonly used to represent the solution of these problems. The objective function (or at least part of it) is given by the sum of lengths for the edges used by the solution.

Edges are ideal GLS features. A solution either contains an edge or not. Furthermore, each edge has a cost equal to its length. We can define a feature for each possible edge and assign a cost to it equal to the edge length. GLS quickly identifies and penalizes long and costly edges guiding local search to high quality solutions, which use as much as possible the short edges available.

9.2 Assignment Problems

In assignment problems, a set of items has to be assigned to another set of items (e.g., airplanes to flights, locations to facilities, people to work etc.). Each assignment of item i to item j usually carries a cost and depending on the problem, a number of constraints required to be satisfied (e.g., capacity or compatibility constraints). The assignment of item i to item j can be seen as a solution property which is either present in the solution or not. Since each assignment also carries a cost, this is another good example of a feature to be used in a GLS implementation.

In some variations of the problem such as the Quadratic Assignment Problem, the cost function is more complicated and assignments are having an indirect impact to the cost. Even in these cases, we found the GLS can generate good results by assigning to all features the same feature costs (e.g. equal to 1 or some other arbitrary value). Although, GLS is not guiding the improvement method to good solutions (since this information is difficult to extract from the objective function), it can still diversify the search because of the penalty memory incorporated and it is capable of producing results comparable to popular heuristic methods.

9.3 Resource Allocation Problems

Assignment problems can be used to model resource allocation applications. A special but important case in resource allocation is when the resources available are not sufficient to service all requests. Usually, the objective function will contain a sum of costs for the unallocated requests, which is to be minimized. The cost incurred when a request is unallocated will reflect the importance of the request or the revenue lost in the particular scenario.

A possible feature to consider for these problems is whether a request is unallocated or not. If the request is unallocated then a cost is incurred in the objective function, which we can use as the feature cost to guide local search. The number of features in a problem is equal to the number of requests that may be left unallocated, one for each request. There may be hard constraints which state that certain requests should always be allocated a resource, there is no need to define a feature for them. Problems in this category include the Path Assignment Problem (Anderson et al., 1993), Maximum Channel Assignment Problem (Simon, 1989), Workforce Scheduling Problem (Azarmi and Abdul-Hameed, 1995) and others.

9.4 Constrained Optimization Problems

Constraints are very important in capturing processes and systems in the real world. A number of combinatorial optimization problems deals with finding a solution, which satisfies a set of constraints or, if that is not possible, minimizes the number of constraint violations (relaxations). Constraint violations may have costs (weights) associated to them, in which case the sum of constraint violation costs is to be minimized.

Local search usually considers the number of constraint violations (or their weighted sum) as the objective function even in cases where the goal is to find a solution, which satisfies all the constraints. Constraints by their nature can be easily used as features. They can be modeled by indicator functions and they also incur a cost (i.e., when violated/relaxed), which can be used as their feature cost. Problems which can benefit from this modeling include the Constraint Satisfaction and Partial Constraint

Satisfaction Problem (Tsang, 1993), the famous SAT and its MAX-SAT variant, Graph Coloring, various Frequency Assignment Problems (Murphey et al., 1999) and others.

The features proposed in past sections, we would see them used in case problems. In particular, we examine the application of GLS to the following:

— Traveling Salesman Problem (Routing/Scheduling category),
— Quadratic Assignment Problem (Assignment Problem category),
— Workforce Scheduling Problem (Resource Allocation category),
— Radio Link Frequency Assignment Problem (Constraint Optimization category).

For each case problem, we provide a short problem description along with guidelines on how to build a basic local search procedure, implement GLS and also GFLS when applicable.

10 TRAVELING SALESMAN PROBLEM (TSP)

10.1 Problem Description

There are many variations of the Traveling Salesman Problem (TSP). In here, we examine the classic symmetric TSP. The problem is defined by N cities and a symmetric distance matrix $D = [d_{ij}]$ which gives the distance between any two cities i and j. The goal in TSP is to find a tour (i.e. closed path), which visits each city exactly once and is of minimum length. A tour can be represented as a cyclic permutation π on the N cities if we interpret $\pi(i)$ to be the city visited after city $i, i = 1, \ldots, N$. The cost of a permutation is defined as:

$$g(\pi) = \sum_{i=1}^{N} d_{i\pi(i)} \tag{6}$$

and gives the cost function of the TSP.

10.2 Local Search

10.2.1 Solution Representation

The solution representation usually adopted for the TSP is that of a vector which contains the order of the cities in the tour. For example, the i-th element of the vector will contain an identifier for the i-th city to be visited. Since the solution of the TSP is a closed path there is an edge implied from the last city in the vector to the first one in order to close the tour. The solution space of the problem is that off all possible permutations of the cities as represented by the vector.

10.2.2 Construction Method

A simple construction method is to generate a random tour. If the above solution representation is adopted then all that is required is a simple procedure, which generates a random permutation of the identifiers of the cities. More advanced TSP heuristics can be used if we require a higher quality starting solution to be generated (Reinelt, 1994). This is useful in real time/online applications where a good tour may be needed very early in the search process in case the user interrupts the algorithm. If there are no concerns like that then a random tour generator suffices since the GLS meta-heuristic tends to be relatively insensitive to the starting solution and capable of finding high quality solutions even if left to run for a relatively short time.

10.2.3 Improvement Method

Most improvement methods for the TSP are based on the k-Opt moves. Using k-Opt moves, neighboring solutions can be obtained by deleting k edges from the current tour and reconnecting the resulting paths using k new edges. The k-Opt moves are the basis of the three most famous local search heuristics for the TSP, namely *2-Opt* (Croes, 1958), *3-Opt* (Lin, 1965) and *Lin-Kernighan (LK)* (Lin and Kernighan, 1973).

The reader can consider using the simple 2-Opt method, which in addition to its simplicity is very effective when combined with GLS. With 2-Opt, a neighboring solution is obtained from the current solution by deleting two edges, reversing one of the resulting paths and reconnecting the tour. In practical terms, this means reversing the order of the cities in a contiguous section of the vector or its remainder depending on which one is the shortest in length.

Computing incrementally the change in solution cost by a 2-Opt move is relatively simple. Let us assume that edges e1, e2 are removed and edges e3, e4 are added with lengths d1, d2, d3, d4 respectively. The change in cost is the following:

$$\mathrm{d3} + \mathrm{d4} - \mathrm{d1} - \mathrm{d2} \tag{7}$$

When we discuss the features used in the TSP, we will explain how this evaluation mechanism is revised to account for penalty changes in the augmented objective function.

10.3 Guided Local Search

For the TSP, a tour includes a number of edges and the solution cost (tour length) is given by the sum of the lengths of the edges in the tour (see (6)). As mentioned in Section 9.1, edges are ideal features for routing problems such as the TSP. First, a tour either includes an edge or not and second, each edge incurs a cost in the objective function equal to the edge length, as this is given by the distance matrix $D = [d_{ij}]$ of the problem. A set of features can be defined by considering all possible undirected edges e_{ij} $(i = 1, \ldots, N, j = i + 1, \ldots, N, i \neq j)$ that may appear in a tour with feature costs given by the edge lengths d_{ij}. Each edge e_{ij} connecting cities i and city j is attached a penalty p_{ij} initially set to 0 which is increased by GLS during search. When implementing the GLS algorithm for the TSP, the edge penalties can be arranged in a symmetric penalty matrix $P = [p_{ij}]$. As mentioned in section 3, penalties have to be combined with the problem's objective function to form the augmented objective function which is minimized by local search. We therefore need to consider the auxiliary distance matrix:

$$D' = D + \lambda \cdot P = [d_{ij} + \lambda \cdot p_{ij}] \tag{8}$$

Local search must use D' instead of D in move evaluations. GLS modifies P and (through that) D' whenever local search reaches a local minimum.

In order to implement this, we revise the incremental move evaluation formula (7) to take into account the edge penalties and also the lambda parameter. If p1, p2, p3, p4 are the penalties associated to edges e1, e2, e3, and e4, respectively the revised version of (7) is as follows:

$$(\mathrm{d3} + \mathrm{d4} - \mathrm{d1} - \mathrm{d2}) + \lambda * (\mathrm{p3} + \mathrm{p4} - \mathrm{p1} - \mathrm{p2}) \tag{9}$$

Similarly, we can implement GLS for higher order k-Opt moves.

The edges penalized in a local minimum are selected according to the utility function (3), which for the TSP takes the form:

$$\mathrm{Util}(\mathrm{tour}, e_{ij}) = I_{e_{ij}}(\mathrm{tour}) \cdot \frac{d_{ij}}{1 + p_{ij}}, \tag{10}$$

where

$$I_{e_{ij}}(\mathrm{tour}) = \begin{cases} 1, & e_{ij} \in \mathrm{tour} \\ 0, & e_{ij} \notin \mathrm{tour} \end{cases} \tag{11}$$

The only parameter of GLS that requires tuning is the λ parameter. Alternatively, we can tune the a parameter which is defined in Section 7.2 and it is relatively instance independent. Experimenting with a on the TSP, we found that there is an inverse relation between a and local search effectiveness. Not so effective local search heuristics such as 2-Opt require higher a values compared to more effective heuristics such as 3-Opt and LK. This is probably because the amount of penalty needed to escape from local minima decreases as the effectiveness of the heuristic increases explaining why lower values for a (and consequently for λ which is a function of a) work better with 3-Opt and LK. For 2-Opt, the following range for a generates high quality solutions for instances in the TSPLIB (Reinelt, 1991).

$$\tfrac{1}{8} \le a \le \tfrac{1}{2} \tag{12}$$

The reader may refer to (Voudouris and Tsang, 1999) for more details on the experimentation procedure and the full set of results.

10.4 Guided Fast Local Search

We can exploit the way local search works on the TSP to partition the neighborhood in sub-neighborhoods as required by Guided Fast Local Search. Each city in the problem may be seen as defining a sub-neighborhood, which contains all 2-Opt edge exchanges removing either one of the edges adjacent to the city. For a problem with N cities, the neighborhood is partitioned into N sub-neighborhoods, one for each city in the instance.

The sub-neighborhoods to be activated after a move is executed are those of the cities at the ends of the edges removed or added by the move.

Finally, the sub-neighborhoods activated after penalization are those defined by the cities at the ends of the edge(s) penalized. There is a good chance that these sub-neighborhoods will include moves that remove the one or more of the penalized edges.

11 QUADRATIC ASSIGNMENT PROBLEM (QAP)

11.1 Problem Description

The Quadratic Assignment Problem (QAP) is one of the most difficult problems in combinatorial optimization. The problem can model a variety of applications but it is mainly known for its use in facility location problems. In the following, we describe the QAP in its simplest form.

Given a set $N = \{1, 2, \ldots, n\}$ and $n \times n$ matrices $A = [a_{ij}]$ and $B = [b_{kl}]$, the QAP can be stated as follows:

$$\min_{p \in \Pi_N} \sum_{i=1}^{n} \sum_{j=1}^{n} A_{ij} \cdot B_{p(i)p(j)} \tag{13}$$

where p is a permutation of N and Π_N is the set of all possible permutations. There are several other equivalent formulations of the problem. In the facility location context, each permutation represents an assignment of n facilities to n locations. More specifically, each position i in the permutation represents a location and its contents $p(i)$ the facility assigned to that location. The matrix A is called the distance matrix and gives the distance between any two of the locations. The matrix B is called the flow matrix and gives the flow of materials between any two of the facilities. For simplicity, we only consider the Symmetric QAP case for which both the distance and flow matrices are symmetric.

11.2 Local Search

QAP solutions can be represented by permutations to satisfy the constraint that each facility is assigned to exactly one location. A move commonly used for the problem is simply to exchange the contents of two permutation positions (i.e. swap the facilities assigned to a pair of locations). A best improvement local search procedure starts with a random permutation. In each iteration, all possible moves (i.e. swaps) are evaluated and the best is selected and performed. The algorithm reaches a local minimum when there is no move, which improves further the cost of the current permutation.

An efficient update scheme can be used in the QAP which allows evaluation of moves in constant time. The scheme works only with best improvement local search. Move values of the first neighborhood search are stored and updated each time a new neighborhood search is performed to take into account changes from the move last executed, see (Taillard, 1995) for details. Move values do not need to be evaluated from scratch and thus the neighborhood can be fully searched in roughly $O(n^2)$ time instead of $O(n^3)$ required otherwise.[1] To evaluate moves in constant time, we have to examine all possible moves in each iteration and have their values updated. Because of that, the scheme can not be easily combined with Fast Local Search, which examines only a number of moves in each iteration therefore preventing benefiting substantially from GFLS in this case.

11.3 Guided Local Search

A set of features that can be used in the QAP is the set of all possible assignments of facilities to locations (i.e. location-facility pairs). This kind of feature is general and can be used in a variety of assignment problems as explained in Section 9.2. In the QAP, there are n^2 possible location-facility combinations. Because of the structure of

[1] To evaluate the change in the cost function 13 caused by a move normally requires $O(n)$ time. Since there are $O(n^2)$ moves to be evaluated, the search of the neighborhood without the update scheme requires $O(n^3)$ time.

the objective function, it is not possible to estimate easily the impact of features and assign to them appropriate feature costs. In particular, the contribution in the objective function of a facility assignment to a location depends also on the placement of the other facilities with a non-zero flow to that facility.

Experimenting with the problem, we found that if all features are assigned the same cost (e.g. 1), the algorithm is still capable of generating high quality solutions. This is due to the ability of GLS to diversify search using the penalty memory. Since features are considered of equal cost, the algorithm is distributing search efforts uniformly across the feature set. Comparative tests we conducted between GLS and the Taboo Search of (Taillard, 1991) indicate that both algorithms are performing equally well when applied to the QAPLIB instances (Burkard et al., 1997) with no clear winner across the instance set. GLS, although not using feature costs in this problem, is still very competitive to state-of-the-art techniques such as Taboo Search.

To determine λ in the QAP, one may use the formula below, which was derived experimentally:

$$\lambda = \alpha * n * (\text{mean flow}) * (\text{mean distance}) \tag{14}$$

where n is the size of the problem and the flow and distance means are computed over the distance and flow matrices respectively (including any possible 0 entries which are common in QAP instances). Experimenting with QAPLIB instances, we found that optimal performance is achieved for $a = 0.75$.

12 WORKFORCE SCHEDULING PROBLEM

12.1 Problem Description

We now look at how GLS can be applied to a real-word resource allocation problem with unallocated requests called the Workforce Scheduling problem (WSP), see (Tsang and Voudouris, 1997) for more details. The problem is to schedule a number of engineers to a set of jobs, minimizing total cost according to a function, which is to be explained below. Each job is described by a triple:

$$(Loc, Dur, Type) \tag{15}$$

where *Loc* is the location of the job (depicted by its x and y co-ordinates), *Dur* is the standard duration of the job and *Type* indicates whether this job must be done in the morning, in the afternoon, as the first job of the day, as the last job of the day, or "don't care".

Each engineer is described by a 5-tuple:

$$(Base, ST, ET, OT\_limit, Skill) \tag{16}$$

where *Base* is the x and y co-ordinates at which the engineer locates, ST and ET are this engineer's starting and ending time, *OT_limit* is his/her overtime limit, and *Skill* is a skill factor between 0 and 1 which indicates the fraction of the standard duration that this engineer needs to accomplish a job. The cost function that is to be minimized is defined as follows:

$$\textit{Total Cost} = \sum_{i=1}^{NoT} TC_i + \sum_{i=1}^{NoT} OT_i^2 + \sum_{j=1}^{NoJ} (Dur_j + Penalty) \times UF_j \tag{17}$$

where

$$\begin{aligned}
NoT &= \text{number of engineers,}\\
NoJ &= \text{number of jobs,}\\
TC_i &= \text{Travelling Cost of engineer } i,\\
OT_i &= \text{Overtime of engineer } i,\\
Dur_j &= \text{Standard duration of job } j,\\
UF_j &= 1 \text{ if job } j \text{ is unallocated; 0 otherwise,}\\
\mathit{Penalty} &= \text{constant (which is set to 60 in the tests).}
\end{aligned}$$

The traveling cost between (x_1, y_1) to (x_2, y_2) is defined as follows:

$$\mathrm{TC}((x_1, y_1), (x_2, y_2)) = \begin{cases} \dfrac{(\Delta_x/2) + \Delta_y}{8}, & \Delta_x > \Delta_y \\[2ex] \dfrac{(\Delta_y/2) + \Delta_x}{8}, & \Delta_y \geq \Delta_x \end{cases} \tag{18}$$

Here Δ_x is the absolute difference between x_1 and x_2, and Δ_y is the absolute difference between y_1 and y_2. The greater of the x and y differences is halved before summing. Engineers are required to start from and return to their base everyday. An engineer may be assigned more jobs than he/she can finish.

12.2 Local Search

12.2.1 Solution Representation

We represent a candidate solution (i.e. a possible schedule) by a permutation of the jobs. Each permutation is mapped into a schedule using the deterministic algorithm described below:

procedure **Evaluation** (input: one particular permutation of jobs)

1. For each job, order the qualified engineers in ascending order of the distances between their bases and the job (such orderings only need to be computed once and recorded for evaluating other permutations).
2. Process one job at a time, following their ordering in the input permutation. For each job x, try to allocate it to an engineer according to the ordered list of qualified engineers:
   - 2.1. to check if engineer g can do job x, make x the first job of g; if that fails to satisfy any of the constraints, make it the second job of g, and so on;
   - 2.2. if job x can be fitted into engineer g's current tour, then try to improve g's new tour (now with x in it): the improvement is done by a simple 2-opting algorithm (see section 10), modified in the way that only better tours which satisfy the relevant constraints will be accepted;
   - 2.3. if job x cannot be fitted into engineer g's current tour, then consider the next engineer in the ordered list of qualified engineers for x; the job is unallocated if it cannot fit into any engineer's current tour.
3. The cost of the input permutation, which is the cost of the schedule thus created, is returned.

12.2.2 Construction Method

The starting point of local search is generated heuristically and deterministically: the jobs are ordered by the number of qualified engineers for them. Jobs that can be served by the fewest number of qualified engineers are placed earlier in the permutation.

12.2.3 Improvement Method

Given a permutation, local search is performed in a simple way: a pair of jobs is examined at a time. Two jobs are swapped to generate a new permutation if the new permutation is evaluated (using the Evaluation procedure above) to a lower cost than the original permutation.

Note here that since the problem is also close to the Vehicle Routing Problem (VRP), one may follow a totally different approach considering VRP move operators such as insertions, swaps etc. In this case, the solution representation and construction methods need to be revised. The reader may refer to (Backer et al., 2000) for more information on the application of GLS to the VRP.

12.3 Guided Local Search

In the workforce scheduling problem, we use the feature type recommended for resource allocation problems in Section 9.3. In particular, the inability to serve jobs incurs a cost, which plays the most important part in the objective function. Therefore, we intend to bias local search to serve jobs of high importance. To do so, we define a feature for each job in the problem:

$$I_{\mathrm{job}_j}(\mathrm{schedule}) = \begin{cases} 1, & job_j \text{ is unallocated in } \textit{schedule} \\ 0, & job_j \text{ is allocated in } \textit{schedule} \end{cases} \tag{19}$$

The cost of this feature is given by $(Dur_j + Penalty)$ which is equal to the cost incurred in the cost function (17) when a job is unallocated.

The jobs penalized in a local minimum are selected according to the utility function (3) which for workforce scheduling takes the form:

$$\mathrm{Util}(\mathrm{schedule},\mathrm{job}_j) = I_{\mathrm{job}_j}(\mathrm{schedule}) \cdot \frac{(\mathrm{Dur}_j + \mathrm{Penalty})}{1 + p_j}. \tag{20}$$

WSP exhibits properties found in resource allocation problems (i.e. unallocated job costs) and also in routing problems (i.e. travel costs). In addition to the above feature type and for better performance, we may consider introducing a second feature type based on edges as suggested in section 9.1 for routing problems and explained in section 10.3 for the TSP. This feature set can help to aggressively optimize the travel costs also incorporated in the objective function (17). Furthermore, one or both features sets can be used in conjunction with a VRP based local search method.

12.4 Guided Fast Local Search

To apply Guided Fast Local Search to workforce scheduling, each job permutation position defines a separate sub-neighborhood. The activation bits are manipulated according to the general FLS algorithm of section (8). In particular:

1. all the activation bits are set to 1 (or "on") when GFLS starts;
2. the bit for job permutation position x will be switched to 0 (or "off") if every possible swap between the job at position x and the other jobs under the current permutation has been considered, but no better permutation has been found;
3. the bit for job permutation position x will be switched to 1 whenever x is involved in a swap which has been accepted.

Mapping penalized jobs to sub-neighborhoods is straightforward. We simply activate the sub-neighborhoods corresponding to the permutation positions of the penalized jobs. This essentially forces Fast Local Search to examine moves, which swap the penalized jobs.

13 RADIO LINK FREQUENCY ASSIGNMENT PROBLEM

13.1 Problem Description

The *Radio Link Frequency Assignment Problem* (RLFAP) (Tiourine et al., 1995; Murphey et al., 1999) is abstracted from the real life application of assigning frequencies to radio links. The problem belongs to the class of constraint optimization problems mentioned in Section 9.4. In brief, the interference level between the frequencies assigned to the different links has to be acceptable; otherwise communication will be distorted. The frequency assignments have to comply with certain regulations and physical characteristics of the transmitters. Moreover, the number of frequencies is to be minimized, because each frequency used in the network has to be reserved at a certain cost. In certain cases, some of the links may have pre-assigned frequencies which may be respected or preferred by the frequency assignment algorithm. In here, we examine a simplified version of the problem considering only the interference constraints. Information on the application of GLS to the full problem can be found in (Voudouris and Tsang, 1998). A definition of the simplified problem is the following.

We are given a set L of links. For each link i, a frequency f_i has to be chosen from a given domain D_i. Constraints are defined on pairs of links that limit the choice of frequencies for these pairs. For a pair of links $\{i, j\}$ these constraints are either of type

$$|f_i - f_j| > d_{ij} \tag{21}$$

or of type

$$|f_i - f_j| = d_{ij} \tag{22}$$

for a given distance $d_{ij} \geq 0$. Two links i and j involved in a constraint of type (21) are called *interfering* links, and the corresponding d_{ij} is the interfering distance. Two links bound by a constraint of type (22) are referred to as a pair of *parallel* links; every link belongs to exactly one such pair.

Some of the constraints may be violated at a certain cost. Such restrictions are called *soft*, in contrast to the hard *constraints*, which may not be violated. The constraints of type (22) are always hard. Interference costs c_{ij} for violating soft constraints of type (21) are given. An assignment of frequencies is complete if every link in L has a frequency assigned to it. We denote by C the set of all soft *interference* constraints.

The goal is to find a complete assignment that satisfies all hard constraints and is of minimum cost:

$$\min_{C} \sum c_{ij} \cdot \delta(|f_i - f_j| \leq d_{ij}) \tag{23}$$

subject to hard constraints:

$|f_i - f_j| > d_{ij}$: for all pairs of links $\{i, j\}$ involved in the hard constraints,
$|f_i - f_j| = d_{ij}$: for all pairs of parallel links $\{i, j\}$,
$f_i \in D_i$: for all links $i \in L$,

where $\delta(\cdot)$ is 1 if the condition within brackets is true and 0 otherwise.

We look next at a local search procedure for the problem.

13.2 Local Search

13.2.1 Using an Alternative Objective Function

When using heuristic search to solve a combinatorial optimization problem, it is not always necessary to use the objective function as dictated in the problem formulation. Objective functions based on the original one can be devised which result in smoother landscapes. These objective functions can sometimes generate solutions of higher quality (with respect to the original objective function) than if the original one is used.

In the RLFAP, we can define and use a simple objective function g, which is given by the sum of all constraint violation costs in the solution with all the constraints contributing equally to the sum instead of using weights as in (23). This objective function is as follows:

$$g(s) = \sum_{C \cup C^{Hard}} \delta(|f_i(s) - f_j(s)| \leq d_{ij}) \tag{24}$$

subject to hard constraints:

$$f_i(s) \in D'_i : \text{for all links } i \in L,$$

where $\delta(\ \cdot\ )$ is 1 if the condition within brackets is true and 0 otherwise, $f_i(s)$ is the frequency assigned to link i in solution s, C^{Hard} is the set of hard inequality constraints, C is the set of soft inequality constraints and D'_i is the reduced domain for link i containing only frequencies which satisfy the hard equality constraints.

A solution s with cost 0 with respect to g is satisfying all hard and soft constraints of the problem.

The motivation to use an objective function such as (24) is closely related to the rugged landscapes formed in RLFAP, if the original cost function is used. In particular, high and very low violation costs are defined for some of the soft constraints. This leads to even higher violation costs to have to be defined for hard constraints. The landscape is not smooth but full of deep local minima mainly due to the hard and soft constraints of high cost. Soft constraints of low cost are buried under these high costs.

A similar approach to replace the objective function has been used successfully by (Mills and Tsang, 2000) in the MAX-SAT problem suggesting the universal appeal of the idea in constraint optimization problems.

13.2.2 Solution Representation

An efficient solution representation for the problem takes into account the fact that each link in RLFAP is connected to exactly one other link via a hard constraint of type (22). In particular, we can define a decision variable for each pair of parallel links bound by an equality constraint (22). The domain of this variable is defined as the set of all pairs of frequencies from the original domains of the parallel links that satisfy the hard equality constraint.

13.2.3 Construction Method

A construction method can be implemented by assigning to each decision variable (assigns values to a pair of links) a random value from its domain. In large problem instances, it is beneficial to consider a domain pre-processing and reduction phase. Sophisticated techniques based on Arc-Consistency (Tsang, 1993) can be utilized during that phase to reduce domains based on the problem's hard constraints. These domains can then be used instead of the original ones for the random solution generation and also by the improvement method.

13.2.4 Improvement Method

An improvement method can based on the min-conflicts heuristic of Minton et al. (1992) for Constraint Satisfaction Problems. A 1-optimal type move is used which changes the value of one variable at a time. Starting from a random and complete assignment of values to variables, variables are examined in an arbitrary static order. Each time a variable is examined, the current value of the variable changes to the value (in the variable's domain) which yields the minimum value for the objective function. Ties are randomly resolved allowing moves which transit to solutions with equal cost. These moves are called *sideways moves* (Selman et al., 1992) and enable local search to examine plateau of solutions occurring in the landscapes of many constraint optimization problems.

13.3 Guided Local Search

The most important cost factor in the RLFAP is constraint violation costs defined for soft inequality constraints. Inequality constraints can be used to define a basic feature set for the RLFAP. Each inequality constraint is interpreted as a feature with the feature cost given by the constraint violation cost c_{ij} as defined in the problem's original cost function (23).

Hard inequality constraints are also modelled as features though the cost assigned to them is infinity. This results in their utility to be penalized to also tend to infinity. To implement this in the code, hard constraints are simply given priority over soft constraints when penalties are applied. This basically forces local search to return back to a feasible region where penalizing soft constraints can resume.

GLS is especially suited to use the alternative objective function (24) because of the definition of feature costs as described above. The application of penalties can force local search toward solutions which satisfy constraints with high violation costs, to some degree independently from the objective function used by local search while benefiting from the smoother landscape introduced by (24).

The λ parameter can be set to 1 provided that we use (24) as the objective function. The same value for λ has also been used in MAX-SAT problems in (Mills and Tsang, 2000) where the same approach is followed with respect to smoothing the landscape.

A variation of the GLS method which seems to significantly improve performance in certain RLFAP instances is to decrease penalties and not only increase them (Voudouris and Tsang, 1998). More specifically, the variant uses a circular list to retract the effects of penalty increases made earlier in the search process, in a way that very much resembles a tabu list. In particular, penalties increased are decreased after a certain number of penalty increases is performed. The scheme uses an array of size t where the t most recent features penalised are recorded. The array is treated as a circular list, adding elements in sequence in positions 1 through t and then starting over at position 1. Each time the penalty of a feature is increased (by one unit), the feature is inserted in the array and the penalty of the feature previously stored in the same position is decreased (by one unit). The rational behind the strategy is to allow GLS to return to regions of the search visited earlier in the search process, so introducing a search intensification mechanism.

13.4 Guided Fast Local Search

Best improvement local search for the RLFAP as used in the context of Tabu Search, for an example see (Hao et al., 1998), evaluates all possible 1-optimal moves over all variables before selecting and performing the best move. Given the large number of links in real world instances, greedy local search is a computationally expensive option. This is especially the case for the RLFAP where we cannot easily devise an incremental move update mechanism (such as that for the QAP) for all the problem's variations. The local search procedure described in Section 13.2 is already a faster alternative than best improvement. Using Guided Fast Local Search, things can be improved further.

To apply Guided Fast Local Search to RLFAP, each decision variable defines a subneighborhood and has a bit associated to it. Whenever a variable is examined and its value is changed (i.e., the variable's parallel links are assigned to another pair of frequencies because of an improving or sideways move) the activation bit of the variable remains to 1 otherwise it turns to 0 and the variable is excluded in future iterations of the improvement loop. Additionally, if a move is performed activation spreads to other variables which have their bits set to 1. In particular, we set to 1 the bit of variables where improving moves may occur as a result of the move just performed. These are the variables for which either one of their links is connected via a constraint to one of the links of the variable that changed value. There are five potential schemes for propagating activation after changing the value of a variable. They are the following:

1. Activate all variables connected via a constraint to the variable which changed value.
2. Activate only variables that are connected via a constraint which is violated. This resembles CSP local search methods where only variables in conflict have their neighborhood searched.
3. Activate only variables that are connected via a constraint which has become violated as a result of the move (subset of condition 2 and also 4).

4. Activate only variables that are connected via a constraint that changed state (i.e. violated → satisfied or satisfied → violated) as a result of the move (superset of condition 3).
5. Activate variables that fall under either condition 2 or 4.

Experimentation suggests that scheme 5 tends to produce better results for the real world instances of RLFAP available in the literature. Fast local search stops when all the variables are inactive or when a local minimum is detected by other means (i.e. a number of sideways moves is performed without an improving move found).

Finally, when a constraint is penalized we activate the variables connected via the constraint in an effort to find 1-Opt moves which will satisfy the constraint.

14 SUMMARY AND CONCLUSIONS

For many years, general heuristics for combinatorial optimization problems with most prominent examples the methods of Simulated Annealing and Genetic Algorithms heavily relied on randomness to generate good approximate solutions to difficult NP-Hard problems. The introduction and acceptance of Tabu Search (Glover and Laguna, 1997) by the Operations Research community initiated an important new era for heuristic methods where deterministic algorithms exploiting historical information started appearing and being used in real world applications.

Guided local search described in this chapter follows this trend. GLS heavily exploits information (not only the search history) to distribute the search effort in the various regions of the search space. Important structural properties of solutions are captured by solution features. Solutions features are assigned costs and local search is biased to spend its efforts according to these costs. Penalties on features are utilized for that purpose.

When local search settles in a local minimum, the penalties are increased for selected features present in the local minimum. By penalizing features appearing in local minima, GLS avoids the local minima visited (exploiting historical information) but also diversifies choices for the various structural properties of solutions captured by the solution features. Features of high cost are penalized more times than features of low cost: the diversification process is directed and deterministic rather than undirected and random.

In general, several penalty cycles may be required before a move is executed out of a local minimum. This should not be viewed as an undesirable situation. It is caused by the uncertainty in the information as captured by the feature costs which makes necessary the testing of the GLS decisions against the landscape of the problem.

The penalization scheme of GLS is ideally combined with FLS which limits neighbourhood search to particular parts of the overall solution leading to the GFLS algorithm. GFLS significantly reduces the computation times required to explore the area around a local minimum to find the best escape route allowing many more penalty modification cycles to be performed in a given amount of running time.

The GLS and GFLS methods are still in their early stages and future research is required to develop them further. The use of incentives implemented as negative penalties, which encourage the use of specific solution features, is one promising direction to be explored. Other interesting directions include *fuzzy features* with indicator functions returning real values in the [0, 1] interval, automated tuning of the lambda or alpha parameters, definition of effective termination criteria, alternative utility functions for selecting the features penalized and also studies about the convergence properties of GLS.

We found it relatively easy to adapt GLS and GFLS to the different problems examined in this chapter. Although local search is problem dependent, the other structures of GLS and also GFLS are problem independent. Moreover, a step by step procedure is usually followed when GLS or GFLS is applied to a new problem (i.e. implement a local search procedure, identify features, assign costs, define sub-neighborhoods, etc.) something which makes easier the use of the technique by non-specialists (e.g. software engineers).

REFERENCES

Anderson, C.A., Fraughnaugh, K., Parker, M. and Ryan, J. (1993) Path assignment for call routing: An application of tabu search. *Annals of Operations Research*, **41**, 301–312.

Azarmi, N. and Abdul-Hameed, W. (1995) Workforce scheduling with constraint logic programming. *BT Technology Journal*, **13**(1), 81–94.

Backer, B.D., Furnon, V., Shaw, P., Kilby, P. and Prosser, P. (2000) Solving vehicle routing problems using constraint programming and metaheuristics. *Journal of Heuristics*, **6**(4), 501–523.

Bentley, J.J. (1992) Fast algorithms for geometric traveling salesman problems. *ORSA Journal on Computing*, **4**, 387–411.

Bouju, A., Boyce, J.F., Dimitropoulos, C.H.D., vom Scheidt, G. and Taylor, J.G. (1995) Intelligent search for the radio link frequency assignment problem. *Proceedings of the International Conference on Digital Signal Processing*, Cyprus.

Chalmers, A.G. (1994) *A Minimum Path Parallel Processing Environment*. Research Monographs in Computer Science, Alpha Books.

Choi, K.M.F., Lee, J.H.M. and Stuckey, P.J. A Lagrangian reconstruction of GENET. *Artificial Intelligence* (to appear).

Chu, P. and Beasley, J.E. (1997) A genetic algorithm for the generalized assignment problem. *Computers and Operations Research*, **24**, 17–23.

Congram, R.K. and Potts, C.N., (1999) Dynasearch Algorithms for the Traveling Salesman Problem. Presentation at the Travelling Salesman Workshop, CORMSIS, University of Southampton.

Croes, A. (1958) A method for solving traveling-salesman problems. *Operations Research*, **5**, 791–812.

Davenport, A., Tsang, E.P.K., Wang, C.J. and Zhu, K. (1994) GENET: a connectionist architecture for solving constraint satisfaction problems by iterative improvement. *Proceedings 12th National Conference for Artificial Intelligence (AAAI)*, pp. 325–330.

Faroe, O., Pisinger, D. and Zachariasen, M (1999) Guided local search for the three-dimensional bin packing problem. Tech. Rep. 99–13, Department of Computer Science, University of Copenhagen.

Freisleben, B. and Merz, P. (1996) A genetic local search algorithm for solving symmetric and asymmetric travelling salesman problems. *Proceedings of the 1996 IEEE International Conference on Evolutionary Computation*, IEEE Press, pp. 616–621.

Gent, I.P., van Maaren, H. and Walsh, T. (2000) SAT2000, Highlights of satisfiability research in the year 2000. *Frontiers in Artificial Intelligence and Applications*, IOS Press.

Glover, F. and Laguna, M. (1997) *Tabu Search*. Kluwer Academic Publishers, Boston.

Goldberg, D.E. (1989) Genetic algorithms in search, optimization, and machine learning, Reading, MA, Addison-Wesley Pub. Co., Inc.

Hansen, P. and Mladenovic, N. (1999) An introduction to variable neighborhood search. In: S. Voss, S. Martello, I.H. Osman, and C. Roucairol (eds.), *Meta-Heuristics: Advances and Trends in Local Search Paradigms for Optimization*. Kluwer, Boston, pp. 433–458.

Hao J.-K., Dorne, R. and Galinier, P. (1998) Tabu search for frequency assignment in mobile radio Networks. *Journal of Heuristics*, **4**(1), 47–62.

Holland, J.H. (1975) *Adaptation in Natural and Artificial Systems*. University of Michigan Press, Ann Arbor, MI.

Holstein, D. and Moscato, P. (1999) Memetic algorithms using guided local search: a case study. In: D. Corne, F. Glover, and M. Dorigo (eds.), *New Ideas in Optimisation* McGraw-Hill, London, pp. 234–244.

Johnson, D. (1990) Local optimization and the traveling salesman problem. *Proceedings of the 17th Colloquium on Automata Languages and Programming*, Lecture Notes in Computer Science 443, Springer-Verlag, pp. 446–461.

Jose, R. and Boyce, J. (1997) Appication of connectionist local search to line management rail traffic control. *Proceedings of International Conf. on Practical Applications of Constraint Technology (PACT'97)*, London.

Kilby, P., Prosser, P. and Shaw, P. (1999) Guided local search for the vehicle routing problem with time windows. In: S. Voss, S. Martello, I.H. Osman and C. Roucairol (eds.), *Meta-Heuristics: Advances and Trends in Local Search Paradigms for Optimization*. Kluwer Academic Publishers, pp. 473–486.

Kilby, P., Prosser, P. and Shaw, P. (2000) A comparison of traditional and constraint-based heuristic methods on vehicle routing problems with side constraints. *Constraints*, **5**(4), 389–414.

Knox, J. (1994) Tabu search performance on the symmetric traveling salesman problem. *Computers Operations Research*, **21**(8), 867–876.

Koopman, B.O. (1957) The theory of search, part III, the optimum distribution of search effort. *Operations Research*, **5**, 613–626.

Lau, T.L. and Tsang, E.P.K. (1997) Solving the processor configuration problem with a mutation-based genetic algorithm. *International Journal on Artificial Intelligence Tools (IJAIT)*, **6**(4), 567–585.

Lau, T.L. and Tsang, E.P.K. (1998) The guided genetic algorithm and its application to the general assignment problem. *IEEE 10th International Conference on Tools with Artificial Intelligence (ICTAI'98)*, Taiwan, pp. 336–343.

Lau, T.L. and Tsang, E.P.K. (1998) Guided genetic algorithm and its application to the radio link frequency allocation problem. *Proceedings of NATO symposium on Frequency Assignment, Sharing and Conservation in Systems (AEROSPACE)*, AGARD, RTO-MP-13, Paper No. 14b.

Lau, T.L. (1999) *Guided Genetic Algorithm*. PhD Thesis, Department of Computer Science, University of Essex, Colchester, UK.

Lee, J.H.M. and Tam, V.W.L. (1995) A framework for integrating artificial neural networks and logic programming. *International Journal on Artificial Intelligence Tools*, **4**, 3–32.

Lin, S. (1965) Computer Solutions of the Traveling-Salesman Problem. *Bell Systems Technical Journal*, **44**, 2245–2269.

Lin, S. and Kernighan, B.W. (1973) An effective heuristic algorithm for the traveling salesman problem. *Operations Research*, **21**, 498–516.

Martin, O., and Otto, S.W. (1966) Combining simulated annealing with local search heuristics. In: G. Laporte and I.H. Osman (eds.), *Metaheuristics in Combinatorial Optimization, Annals of Operations Research*, Vol. 63.

Mills, P. and Tsang, E.P.K. (2000) Guided local search for solving SAT and weighted MAX-SAT problems. *Journal of Automated Reasoning*, **24**, 205–223.

Mills, P. and Tsang, E.P.K. (2000) An empirical study of extended guided local search on the quadratic assignment problem. Manuscript, submitted to Nareyek, A. (ed.), *Post-Proceedings, ECAI-2000 Workshop on Local Search for Planning and Scheduling*, Springer LNCS/LNAI book series.

Minton S., Johnston, M.D., Philips A.B. and Laird, P. (1992) Minimizing conflicts: a heuristic repair method for constraint satisfaction and scheduling problems. *Artificial Intelligence* **58**(1–3) (Special Volume on Constraint Based Reasoning), 161–205.

Murphey, R.A., Pardalos, P.M. and Resende, M.G.C. (1999) Frequency assignment problems. In: D.-Z Du and P. Pardalos (eds.), *Handbook of Combinatorial Optimization*, Vol. 4, Kluwer Academic Publishers.

Padron, V. and Balaguer, C. (2000) New Methodology to solve the RPP by means of Isolated Edge. In: A. Tuson (ed.), *2000 Cambridge Conference Tutorial Papers*, Young OR 11, UK Operational Research Society.

Pesant, G. and Gendreau, M. (1999) A constraint programming framework for local search methods. *Journal of Heuristics*, **5**(3), 255–279.

R.E. Burkard, R.E., Karisch, S.E. and Rendl F. (1997) QAPLIB—a quadratic assignment problem library. *Journal of Global Optimization*, **10**, 391–403.

Reinelt, G. (1991) A traveling salesman problem library. *ORSA Journal on Computing*, **3**, 376–384.

Reinelt, G. (1995) The traveling salesman: computational solutions for TSP applications. *Lecture Notes in Computer Science*, Vol. 840, Springer-Verlag.

Resende, M.G.C. and Feo, T.A. (1996) A GRASP for satisfiability. In: D.S. Johnson and M.A. Trick (eds.), *Cliques, coloring, and satisfiability: Second DIMACS Implementation Challenge*. DIMACS Series on Discrete Mathematics and Theoretical Computer Science, American Mathematical Society, Vol. 26, pp. 499–520.

Selman, B., Levesque, H. J. and Mitchell, D. G. (1992) A new method for solving hard satisfiability problems. *Proceedings of AAAI-92*, 440–446.

Selman, B. and Kautz, H. (1993) Domain-independent extensions to GSAT: solving large structured satisfiability problems. *Proceedings of 13th International Joint Conference on AI*, 290–295.

Shang, Y. and Wah, B.W. (1998) A discrete lagrangian-based global-search method for solving satisfiability problems. *Journal of Global Optimization*, **12**(1), 61–99.

Simon, H. U. (1989) Approximation algorithms for channel assignment in cellular radio networks. Proceedings 7th International Symposium on Fundamentals of Computation Theory, *Lecture Notes in Computer Science*, Vol. 380. Springer-Verlag, pp. 405–416.

Stone, L.D. (1983) The process of search planning: current approaches and continuing problems. *Operations Research*, **31**, 207–233.

Stuckey, P. and Tam, V. (1998) Semantics for using stochastic constraint solvers in constraint logic programming. *Journal of Functional and Logic Programming*, **2**.

Taillard, E. (1991) Robust taboo search for the QAP. *Parallel Computing* **17**, 443–455.

Taillard, E. (1995) Comparison of iterative searches for the quadratic assignment problem. *Location Science*, **3**, 87–105.

Tiourine, S., Hurkins, C. and Lenstra, J.K. (1995) An overview of algorithmic approaches to frequency assignment problems. *EUCLID CALMA Project Overview Report*, Delft University of Technology, The Netherlands.

Tsang, E.P.K. and Wang, C.J. (1992) A generic neural network approach for constraint satisfaction problems. In: J.G. Taylor (ed.), *Neural Network Applications*, Springer-Verlag, pp. 12–22.

Tsang, E.P.K. and Voudouris, C. (1997) Fast local search and guided local search and their application to British Telecom's workforce scheduling problem. *Operations Research Letters*, **20**(3), 119–127.

Tsang, E.P.K., Wang, C.J., Davenport, A., Voudouris, C. and Lau, T.L. (1999) A family of stochastic methods for constraint satisfaction and optimisation. *Proceedings of the First International Conference on The Practical Application of Constraint Technologies and Logic Programming (PACLP)*, London, pp. 359–383.

Voudouris, C. and Tsang, E.P.K. (1996) Partial constraint satisfaction problems and guided local search. *Proceedings of PACT'96*, London, pp. 337–356.

Voudouris, C. (1997) *Guided Local Search for Combinatorial Optimisation Problems.* PhD Thesis, Department of Computer Science, University of Essex, Colchester, UK.

Voudouris, C. (1998) Guided Local Search—An illustrative example in function optimisation. *BT Technology Journal*, **16**(3), 46–50.

Voudouris, C. and Tsang, E. (1998) Solving the Radio Link Frequency Assignment Problems using Guided Local Search. *Proceedings of NATO symposium on Frequency Assignment, Sharing and Conservation in Systems (AEROSPACE)*, AGARD, RTO-MP-13, Paper No. 14a.

Voudouris, C. and Tsang, E.P.K. (1999) Guided Local Search and its application to the Travelling Salesman Problem. *European Journal of Operational Research*, **113**(2), 469–499.

Wang, C.J. and Tsang, E.P.K. (1991) Solving constraint satisfaction problems using neural-networks. *Proceedings of the IEE Second International Conference on Artificial Neural Networks*, pp. 295–299.

Wang, C.J. and Tsang, E.P.K. (1994) A cascadable VLSI design for GENET. In: J.G. Delgado-Frias and W.R. Moore (eds.), *VLSI for Neural Networks and Artificial Intelligence*. Plenum Press, New York, pp. 187–196.

Chapter 8

GREEDY RANDOMIZED ADAPTIVE SEARCH PROCEDURES

Mauricio G.C. Resende
AT&T Labs Research
E-mail: mgcr@research.att.com

Celso C. Ribeiro
Catholic University of Rio de Janeiro
E-mail: celso@inf.puc-rio.br

Abstract GRASP is a multi-start metaheuristic for combinatorial problems, in which each iteration consists basically of two phases: construction and local search. The construction phase builds a feasible solution, whose neighborhood is investigated until a local minimum is found during the local search phase. The best overall solution is kept as the result. In this chapter, we first describe the basic components of GRASP. Successful implementation techniques and parameter tuning strategies are discussed and illustrated by numerical results obtained for different applications. Enhanced or alternative solution construction mechanisms and techniques to speed up the search are also described: Reactive GRASP, cost perturbations, bias functions, memory and learning, local search on partially constructed solutions, hashing, and filtering. We also discuss in detail implementation strategies of memory-based intensification and post-optimization techniques using path-relinking. Hybridizations with other metaheuristics, parallelization strategies, and applications are also reviewed.

1 INTRODUCTION

We consider in this chapter a combinatorial optimization problem, defined by a finite ground set $E = \{1, \ldots, n\}$, a set of feasible solutions $F \subseteq 2^E$, and an objective function $f : 2^E \rightarrow \mathbb{R}$. In the minimization version, we search an optimal solution $S^* \in F$ such that $f(S^*) \leq f(S), \forall S \in F$. The ground set E, the cost function f, and the set of feasible solutions F are defined for each specific problem. For instance, in the case of the traveling salesman problem, the ground set E is that of all edges connecting the cities to be visited, $f(S)$ is the sum of the costs of all edges $e \in S$, and F is formed by all egde subsets that determine a Hamiltonian cycle.

The GRASP (Greedy Randomized Adaptive Search Procedure) metaheuristic [38,39] is a multi-start or iterative process, in which each iteration consists of two phases: construction and local search. The construction phase builds a feasible solution, whose neighborhood is investigated until a local minimum is found during the local search phase. The best overall solution is kept as the result. An extensive survey of

```
procedure GRASP(Max_Iterations,Seed)
    Read_Input();
    for k = 1,...,Max_Iterations do
        Solution ← Greedy_Randomized_Construction(Seed);
        Solution ← Local_Search(Solution);
        Update_Solution(Solution,Best_Solution);
    end;
    return Best_Solution;
end GRASP.
```

Figure 8.1. Pseudo-code of the GRASP metaheuristic.

```
procedure Greedy_Randomized_Construction(Seed)
    Solution ← ∅;
    Evaluate the incremental costs of the candidate elements;
    while Solution is not a complete solution do
        Build the restricted candidate list (RCL);
        Select an element s from the RCL at random;
        Solution ← Solution ∪{s};
        Reevaluate the incremental costs;
    end;
    return Solution;
end Greedy_Randomized_Construction.
```

Figure 8.2. Pseudo-code of the construction phase.

the literature is given in [44]. The pseudo-code in Figure 8.1 illustrates the main blocks of a GRASP procedure for minimization, in which `Max_Iterations` iterations are performed and `Seed` is used as the initial seed for the pseudorandom number generator.

Figure 8.2 illustrates the construction phase with its pseudo-code. At each iteration of this phase, let the set of candidate elements be formed by all elements that can be incorporated to the partial solution under construction without destroying feasibility. The selection of the next element for incorporation is determined by the evaluation of all candidate elements according to a greedy evaluation function. This greedy function usually represents the incremental increase in the cost function due to the incorporation of this element into the solution under construction. The evaluation of the elements by this function leads to the creation of a restricted candidate list (RCL) formed by the best elements, i.e. those whose incorporation to the current partial solution results in the smallest incremental costs (this is the greedy aspect of the algorithm). The element to be incorporated into the partial solution is randomly selected from those in the RCL (this is the probabilistic aspect of the heuristic). Once the selected element is incorporated to the partial solution, the candidate list is updated and the incremental costs are reevaluated (this is the adaptive aspect of the heuristic). This strategy is similar to the semi-greedy heuristic proposed by Hart and Shogan [55], which is also a multi-start approach based on greedy randomized constructions, but without local search.

The solutions generated by a greedy randomized construction are not necessarily optimal, even with respect to simple neighborhoods. The local search phase usually

```
procedure Local_Search(Solution)
    while Solution is not locally optimal do
        Find s' ∈ N(Solution) with f(s') < f(Solution);
        Solution ← s';
    end;
    return Solution;
end Local_Search.
```

Figure 8.3. Pseudo-code of the local search phase.

improves the constructed solution. A local search algorithm works in an iterative fashion by successively replacing the current solution by a better solution in the neighborhood of the current solution. It terminates when no better solution is found in the neighborhood. The pseudo-code of a basic local search algorithm starting from the solution `Solution` constructed in the first phase and using a neighborhood N is given in Figure 8.3.

The effectiveness of a local search procedure depends on several aspects, such as the neighborhood structure, the neighborhood search technique, the fast evaluation of the cost function of the neighbors, and the starting solution itself. The construction phase plays a very important role with respect to this last aspect, building high-quality starting solutions for the local search. Simple neighborhoods are usually used. The neighborhood search may be implemented using either a *best-improving* or a *first-improving* strategy. In the case of the best-improving strategy, all neighbors are investigated and the current solution is replaced by the best neighbor. In the case of a first-improving strategy, the current solution moves to the first neighbor whose cost function value is smaller than that of the current solution. In practice, we observed on many applications that quite often both strategies lead to the same final solution, but in smaller computation times when the first-improving strategy is used. We also observed that premature convergence to a non-global local minimum is more likely to occur with a best-improving strategy.

2 CONSTRUCTION OF THE RESTRICTED CANDIDATE LIST

An especially appealing characteristic of GRASP is the ease with which it can be implemented. Few parameters need to be set and tuned. Therefore, development can focus on implementing efficient data structures to assure quick iterations. GRASP has two main parameters: one related to the stopping criterion and another to the quality of the elements in the restricted candidate list.

The stopping criterion used in the pseudo-code described in Figure 8.1 is determined by the number `Max_Iterations` of iterations. Although the probability of finding a new solution improving the currently best decreases with the number of iterations, the quality of the best solution found may only improve with the latter. Since the computation time does not vary much from iteration to iteration, the total computation time is predictable and increases linearly with the number of iterations. Consequently, the larger the number of iterations, the larger will be the computation time and the better will be the solution found.

For the construction of the RCL used in the first phase we consider, without loss of generality, a minimization problem as the one formulated in Section 1. We denote

```
procedure Greedy_Randomized_Construction(α, Seed)
    Solution ← ∅;
    Initialize the candidate set: C ← E;
    Evaluate the incremental cost c(e) for all e ∈ C;
    while C ≠ ∅ do
        c^min ← min{c(e)|e ∈ C};
        c^max ← max{c(e)|e ∈ C};
        RCL ← {e ∈ C| c(e) ≤ c^min + α(c^max − c^min)};
        Select an element s from the RCL at random;
        Solution ← Solution ∪{s};
        Update the candidate set C;
        Reevaluate the incremental costs c(e) for all e ∈ C;
    end;
    return Solution;
end Greedy_Randomized_Construction.
```

Figure 8.4. Refined pseudo-code of the construction phase.

by $c(e)$ the incremental cost associated with the incorporation of element $e \in E$ into the solution under construction. At any GRASP iteration, let $c^{\min}$ and $c^{\max}$ be, respectively, the smallest and the largest incremental costs.

The restricted candidate list RCL is made up of elements $e \in E$ with the best (i.e., the smallest) incremental costs $c(e)$. This list can be limited either by the number of elements (cardinality-based) or by their quality (value-based). In the first case, it is made up of the p elements with the best incremental costs, where p is a parameter. In this chapter, the RCL is associated with a threshold parameter $\alpha \in [0, 1]$. The restricted candidate list is formed by all "feasible" elements $e \in E$ which can be inserted into the partial solution under construction without destroying feasibility and whose quality is superior to the threshold value, i.e., $c(e) \in [c^{\min}, c^{\min} + \alpha(c^{\max} - c^{\min})]$. The case $\alpha = 0$ corresponds to a pure greedy algorithm, while $\alpha = 1$ is equivalent to a random construction. The pseudo-code in Figure 8.4 is a refinement of the greedy randomized contruction pseudo-code shown in Figure 8.2. It shows that the parameter α controls the amounts of greediness and randomness in the algorithm.

GRASP may be viewed as a repetitive sampling technique. Each iteration produces a sample solution from an unknown distribution, whose mean and variance are functions of the restrictive nature of the RCL. For example, if the RCL is restricted to a single element, then the same solution will be produced at all iterations. The variance of the distribution will be zero and the mean will be equal to the value of the greedy solution. If the RCL is allowed to have more elements, then many different solutions will be produced, implying a larger variance. Since greediness plays a smaller role in this case, the mean solution value should be worse. However, the value of the best solution found outperforms the mean value and very often is optimal. The histograms in Figure 8.5 illustrate this situation on an instance of MAXSAT with 100 variables and 850 clauses, depicting results obtained with 1000 independent constructions using the first phase of the GRASP described in [83,84]. Since this is a maximization problem, the purely greedy construction corresponds to $\alpha = 1$, whereas the random construction occurs with $\alpha = 0$. We notice that when the value of α increases from 0 to 1, the mean

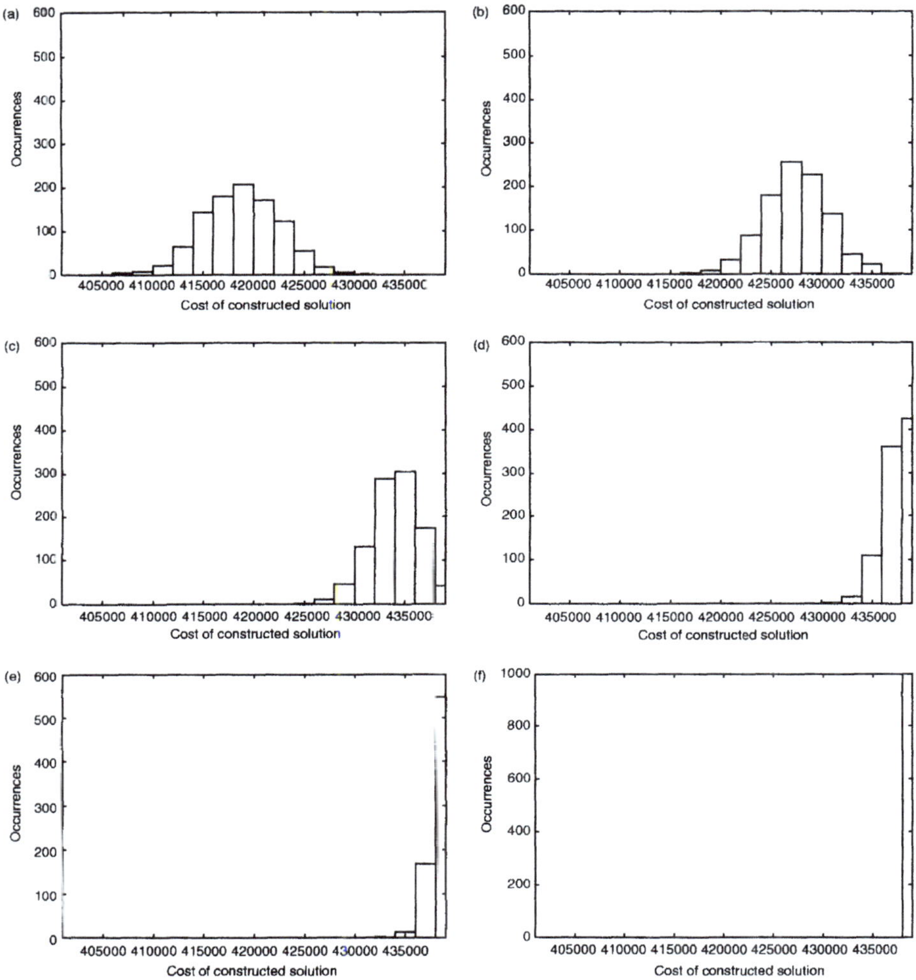

Figure 8.5. Distribution of construction phase solution values as a function of the RCL parameter α (1000 repetitions were recorded for each value of α). (a) $\alpha = 0.0$ (random construction), (b) $\alpha = 0.2$, (c) $\alpha = 0.4$, (d) $\alpha = 0.6$, (e) $\alpha = 0.8$ and (f) $\alpha = 1.0$ (greedy construction).

solution value increases towards the purely greedy solution value, while the variance approaches zero.

For each value of α, we present in Figure 8.6 histograms with the results obtained by applying local search to each of the 1000 constructed solutions. Figure 8.7 summarizes the overall results of this experiment in terms of solution diversity, solution quality, and computation time. We first observe that the larger the variance of the solution values obtained in the construction phase, the larger is the variance of the overall solution values, as shown in the top graph. The graph in the middle illustrates the relationship between the variance of the solution values and the average solution values, and how this affects the best solution found. It is unlikely that GRASP will find an optimal solution if the average solution value is low, even if there is a large variance in the

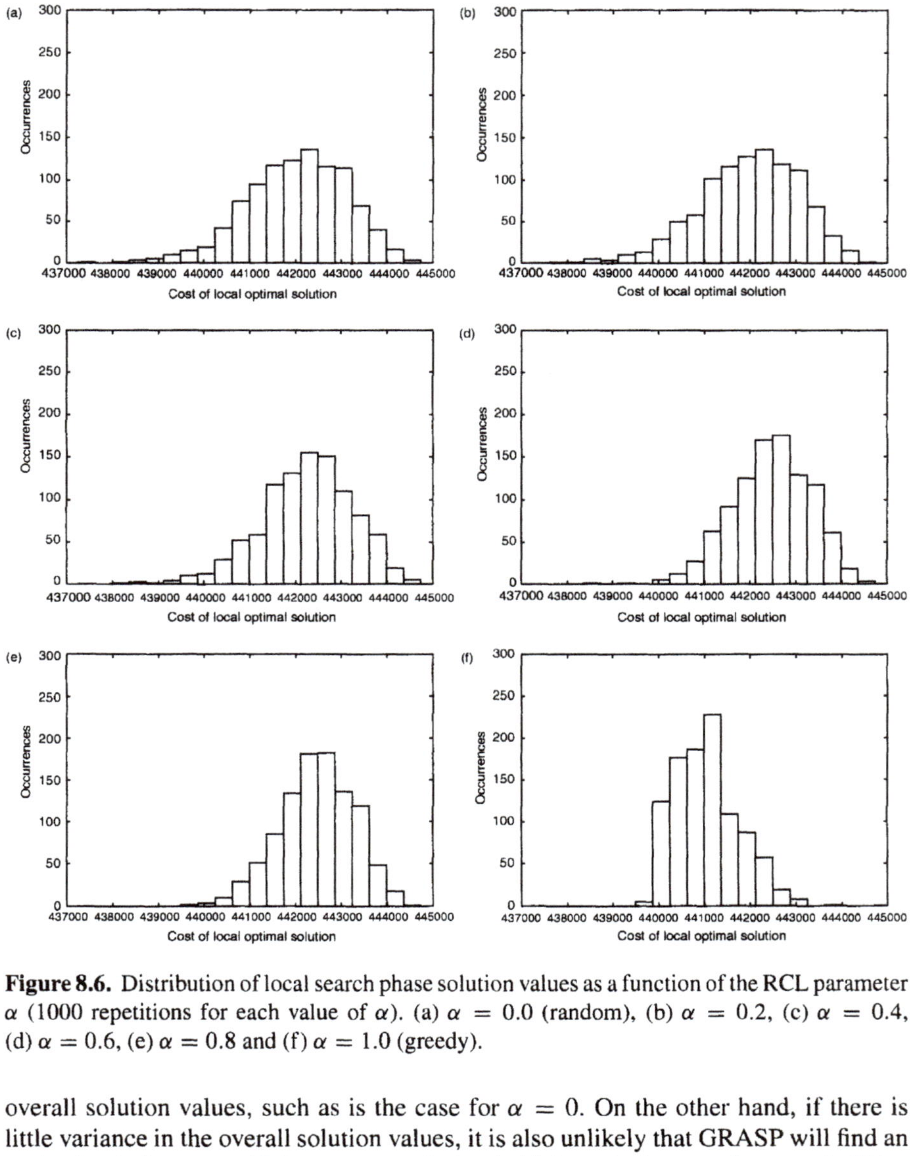

Figure 8.6. Distribution of local search phase solution values as a function of the RCL parameter α (1000 repetitions for each value of α). (a) $\alpha = 0.0$ (random), (b) $\alpha = 0.2$, (c) $\alpha = 0.4$, (d) $\alpha = 0.6$, (e) $\alpha = 0.8$ and (f) $\alpha = 1.0$ (greedy).

overall solution values, such as is the case for $\alpha = 0$. On the other hand, if there is little variance in the overall solution values, it is also unlikely that GRASP will find an optimal solution, even if the average solution is high, as is the case for $\alpha = 1$. What often leads to good solutions are relatively high average solution values in the presence of a relatively large variance, such as is the case for $\alpha = 0.8$. The middle graph also shows that the distance between the average solution value and the value of the best solution found increases as the construction phase moves from more greedy to more random. This causes the average time taken by the local search to increase, as shown in the graph in the bottom. Very often, many GRASP solutions are generated in the same amount of time required for the local optimization procedure to converge from a single random start.

These results are illustrated in Table 8.1 and Figure 8.8, for another instance of MAXSAT where 1000 iterations were run. For each value of α ranging from 0 (purely

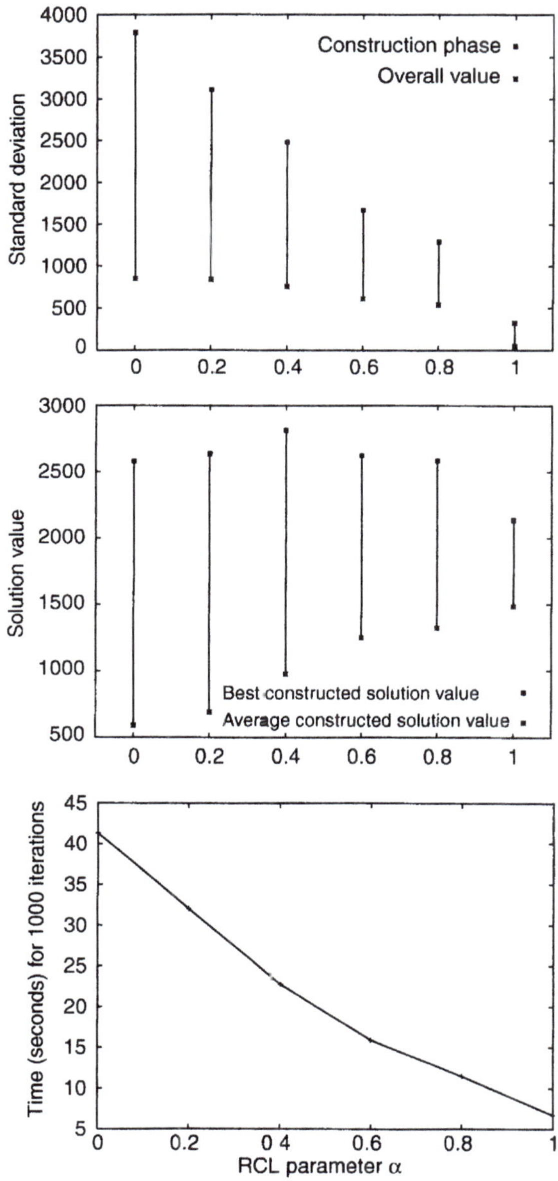

Figure 8.7. Standard deviation of the solution values found, best and average solution values found, and total processing time as a function of the RCL parameter α (1000 repetitions were recorded for each value of α).

random construction) to 1 (purely greedy construction), we give in Table 8.1 the average Hamming distance between each solution built at the end of the construction phase and the corresponding local optimum obtained after local search, the average number of moves from the first to the latter, the local search time in seconds, and the total processing time in seconds. Figure 8.8 summarizes the values observed for the total

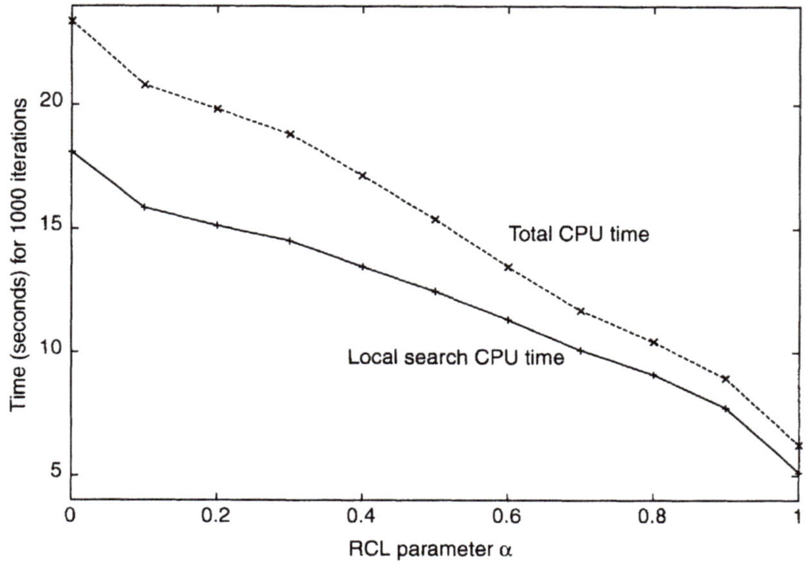

Figure 8.8. Total CPU time and local search CPU time as a function of the RCL parameter α (1000 repetitions for each value of α).

Table 8.1. Average number of moves and local search time as a function of the RCL parameter α

α	Avg. distance	Avg. moves	Local search time (s)	Total time (s)
0.0	12.487	12.373	18.083	23.378
0.1	10.787	10.709	15.842	20.801
0.2	10.242	10.166	15.127	19.830
0.3	9.777	9.721	14.511	18.806
0.4	9.003	8.957	13.489	17.139
0.5	8.241	8.189	12.494	15.375
0.6	7.389	7.341	11.338	13.482
0.7	6.452	6.436	10.098	11.720
0.8	5.667	5.643	9.094	10.441
0.9	4.697	4.691	7.753	8.941
1.0	2.733	2.733	5.118	6.235

processing time and the local search time. We notice that both time measures considerably decrease as α tends to 1, approaching the purely greedy choice. In particular, we observe that the average local search time taken by $\alpha = 0$ (purely random) is approximately 2.5 times that taken in the case $\alpha = 0.9$ (almost greedy). In this example, two to three greedily constructed solutions can be investigated in the same time needed to apply local search to one single randomly constructed solution. The appropriate choice of the value of the RCL parameter α is clearly critical and relevant to achieve a good balance between computation time and solution quality.

Prais and Ribeiro [77] have shown that using a single fixed value for the value of RCL parameter α very often hinders finding a high-quality solution, which eventually could be found if another value was used. They proposed an extension of the basic GRASP procedure, which they call *Reactive GRASP*, in which the parameter α is self-tuned and its value is periodically modifed according with the quality of the solutions obtained recently. In particular, computational experiments on the problem of traffic assignment in communication satellites [78] have shown that Reactive GRASP found better solutions than the basic algorithm for many test instances. These results motivated the study of the behavior of GRASP for different strategies for the variation of the value of the RCL parameter α:

(R) α self tuned according with the Reactive GRASP procedure;

(E) α randomly chosen from a uniform discrete probability distribution;

(H) α randomly chosen from a decreasing non-uniform discrete probability distribution; and

(F) fixed value of α, close to the purely greedy choice.

We summarize the results obtained by the experiments reported in [76,77]. These four strategies were incorporated into the GRASP procedures developed for four different optimization problems: (P-1) matrix decomposition for traffic assignment in communication satellite [78], (P-2) set covering [38], (P-3) weighted MAX-SAT [83,84], and (P-4) graph planarization [85,87]. Let $\Psi = \{\alpha_1, \ldots, \alpha_m\}$ be the set of possible values for the parameter α for the first three strategies. The strategy for choosing and self-tuning the value of α in the case of the Reactive GRASP procedure (R) is described later in Section 3. In the case of the strategy based on using the discrete uniform distribution (E), all choice probabilities are equal to $1/m$. The third case corresponds to the a hybrid strategy (H), in which we typically consider $p(\alpha = 0.1) = 0.5$, $p(\alpha = 0.2) = 0.25$, $p(\alpha = 0.3) = 0.125$, $p(\alpha = 0.4) = 0.03$, $p(\alpha = 0.5) = 0.03$, $p(\alpha = 0.6) = 0.03$, $p(\alpha = 0.7) = 0.01$, $p(\alpha = 0.8) = 0.01$, $p(\alpha = 0.9) = 0.01$, and $p(\alpha = 1.0) = 0.005$. Finally, in the last strategy (F), the value of α is fixed as recommended in the original reference where this parameter was tuned for each problem. A subset of the literature instances was considered for each class of test problems. The results reported in [77] are summarized in Table 8.2. For each problem, we first list the number of instances considered. Next, for each strategy, we give the number of times it found the best solution (hits), as well as the average CPU time (in seconds) on an IBM 9672 model R34. The number of iterations was fixed at 10,000.

Strategy (F) presented the shortest average computation times for three out of the four problem types. It was also the one with the least variability in the constructed

Table 8.2. Computational results for different strategies for the variation of parameter α

		R		*E*		*H*		*F*	
Problem	Total	Hits	Time	Hits	Time	Hits	Time	Hits	Time
P-1	36	34	579.0	35	358.2	32	612.6	24	642.8
P-2	7	7	1346.8	6	1352.0	6	668.2	5	500.7
P-3	44	22	2463.7	23	2492.6	16	1740.9	11	1625.2
P-4	37	28	6363.1	21	7292.9	24	6326.5	19	5972.0
Total	124	91		85		78		59	

solutions and, in consequence, found the best solution the fewest times. The reactive strategy (R) is the one which most often found the best solutions, however, at the cost of computation times that are longer than those of some of the other strategies. The high number of hits observed by strategy (E) also illustrates the effectiveness of strategies based on the variation of the RCL parameter.

3 ALTERNATIVE CONSTRUCTION MECHANISMS

One possible shortcoming of the standard GRASP framework is the independence of its iterations, i.e., the fact that it does not learn from the history of solutions found in previous iterations. This is so because the basic algorithm discards information about any solution encountered that does not improve the incumbent. Information gathered from good solutions can be used to implement memory-based procedures to influence the construction phase, by modifying the selection probabilities associated with each element of the RCL. In this section, we consider enhancements and alternative techniques for the construction phase of GRASP. They include Reactive GRASP, cost perturbations in place of randomized selection, bias functions, memory and learning, and local search on partially constructed solutions.

3.1 Reactive GRASP

A first possible strategy to incorporate a learning mechanism in the memoryless construction phase of the basic GRASP is the Reactive GRASP procedure introduced in Section 2. In this case, the value of the RCL parameter α is not fixed, but instead is selected at each iteration from a discrete set of possible values. This selection is guided by the solution values found along the previous iterations. One way to accomplish this is to use the rule proposed in [78]. Let $\Psi = \{\alpha_1, \ldots, \alpha_m\}$ be the set of possible values for α. The probabilities associated with the choice of each value are all initially made equal to $p_i = 1/m, i = 1, \ldots, m$. Furthermore, let z^* be the incumbent solution and let A_i be the average value of all solutions found using $\alpha = \alpha_i, i = 1, \ldots, m$. The selection probabilities are periodically reevaluated by taking $p_i = q_i / \sum_{j=1}^{m} q_j$, with $q_i = z^*/A_i$ for $i = 1, \ldots, m$. The value of q_i will be larger for values of $\alpha = \alpha_i$ leading to the best solutions on average. Larger values of q_i correspond to more suitable values for the parameter α. The probabilities associated with these more appropriate values will then increase when they are reevaluated.

The reactive approach leads to improvements over the basic GRASP in terms of robustness and solution quality, due to greater diversification and less reliance on parameter tuning. In addition to the applications in [76–78], this approach has been used in power system transmission network planning [20] and in a capacitated location problem [29].

3.2 Cost Perturbations

The idea of introducing some noise into the original costs is similar to that in the so-called "noising method" of Charon and Hudry [25,26]. It adds more flexibility into algorithm design and may be even more effective than the greedy randomized construction of the basic GRASP procedure, in circumstances where the construction algorithms are not very sensitive to randomization. This is indeed the case for the shortest-path heuristic of Takahashi and Matsuyama [95], used as one of the main building blocks of the construction phase of the hybrid GRASP procedure proposed by Ribeiro et al. [90] for the Steiner problem in graphs. Another situation where cost perturbations can be effective appears when no greedy algorithm is available for straight randomization. This happens to be the case of the hybrid GRASP developed by Canuto et al. [22] for the prize-collecting Steiner tree problem, which makes use of the primal-dual algorithm of Goemans and Williamson [52] to build initial solutions using perturbed costs.

In the case of the GRASP for the prize-collecting Steiner tree problem described in [22], a new solution is built at each iteration using node prizes updated by a perturbation function, according to the structure of the current solution. Two different prize perturbation schemes are used:

- *Perturbation by eliminations*: To enforce search diversification, the primal-dual algorithm used in the construction phase is driven to build a new solution without some of the nodes appearing in the solution constructed in the previous iteration. This is done by changing to zero the prizes of some persistent nodes, which appeared in the last solution built and remained at the end of the local search. A parameter α controls the fraction of the persistent nodes whose prizes are temporarily set to zero.
- *Perturbation by prize changes*: Another strategy to enforce the primal-dual algorithm to build different, but still good solutions, consists in introducing some noise into the node prizes, similarly to what is proposed in [25,26], so as to change the objective function. For each node i, a perturbation factor $\beta(i)$ is randomly generated in the interval $[1 - a, 1 + a]$, where a is an implementation parameter. The prize associated with node i is temporarily changed to $\overline{\pi}(i) = \pi(i) \cdot \beta(i)$, where $\pi(i)$ is its original prize.

The cost perturbation methods used in the GRASP for the minimum Steiner tree problem described in [90] incorporate learning mechanisms associated with intensification and diversification strategies, originally proposed in the context of tabu search. Let w_e denote the weight of edge e. Three distinct weight randomization methods (D, I, U) are applied. At a given GRASP iteration i, the modified weight w_e^i of each edge e is randomly selected from a uniform distribution between w_e and $r_i(e) \cdot w_e$, where the coefficient $r_i(e)$ depends on the selected weight randomization method applied at iteration i. Let $t_{i-1}(e)$ be the number of locally optimal solutions in which edge e appeared, after $i - 1$ iterations of the hybrid GRASP procedure have been performed. Clearly, $0 \leq t_{i-1}(e) \leq i - 1$. Table 8.3 displays how the coefficients $r_i(e)$ are computed by each randomization method.

Table 8.3. Maximum randomization coefficients

Method	$r_i(e)$
D	$1.25 + 0.75 \cdot t_{i-1}(e)/(i-1)$
I	$2 - 0.75 \cdot t_{i-1}(e)/(i-1)$
U	2

In method D, values of the coefficients $r_i(e)$ are larger for edges which appeared more frequently in previously found local optima. This scheme leads to a diversification strategy, since more frequently used edges are likely to be penalized with stronger augmentations. Contrarily, method I is an intensification strategy penalizing less frequent edges with larger coefficients $r_i(e)$. Finally, the third randomization method U uses a uniform penalization strategy, independent of frequency information. The original weights without any penalization are used in the first three iterations, combined with three different construction heuristics. The weight randomization methods are then cyclically applied, one at each of the remaining iterations, starting with method I, next D, then U, then I again, and so on. The alternation between diversifying (method D) and intensifying (method I) iterations characterizes a strategic oscillation approach [49]. The experimental results reported in [90] show that the strategy combining these three perturbation methods is more robust than any of them used isolated, leading to the best overall results on a quite broad mix of test instances with different characteristics. The hybrid GRASP with path-relinking using this cost perturbation strategy is among the most effective heuristics currently available for the Steiner problem in graphs.

3.3 Bias Functions

In the construction procedure of the basic GRASP, the next element to be introduced in the solution is chosen at random from the candidates in the RCL. The elements of the RCL are assigned equal probabilities of being chosen. However, any probability distribution can be used to bias the selection toward some particular candidates. Another construction mechanism was proposed by Bresina [21], where a family of such probability distributions is introduced. They are based on the rank $r(\sigma)$ assigned to each candidate element σ, according to its value of the greedy function. Several bias functions are introduced, such as:

- random bias: $\texttt{bias}(r) = 1$;
- linear bias: $\texttt{bias}(r) = 1/r$;
- log bias: $\texttt{bias}(r) = \log^{-1}(r+1)$;
- exponential bias: $\texttt{bias}(r) = e^{-r}$; and
- polynomial bias of order n: $\texttt{bias}(r) = r^{-n}$.

Let $r(\sigma)$ denote the rank of element σ and let $\texttt{bias}(r(\sigma))$ be one of the bias function defined above. Once these values have been evaluated for all elements of the RCL, the probability $\pi(\sigma)$ of selecting element σ is

$$\pi(\sigma) = \frac{\texttt{bias}(r(\sigma))}{\sum_{\sigma' \in \mathrm{RCL}} \texttt{bias}(r(\sigma'))}. \tag{8.1}$$

The evaluation of these bias functions may be restricted to the elements of the RCL. Bresina's selection procedure restricted to elements of the RCL was used in [19]. Note that the standard GRASP uses a random bias function.

3.4 Intelligent Construction: Memory and Learning

Fleurent and Glover [46] observed that the basic GRASP does not use long-term memory (information gathered in previous iterations) and proposed a long-term memory scheme to address this issue in multi-start heuristics. Long-term memory is one of the fundamentals on which tabu search relies.

Their scheme maintains a pool of elite solutions to be used in the construction phase. To become an elite solution, a solution must be either better than the best member of the pool, or better than its worst member and sufficiently different from the other solutions in the pool. For example, one can count identical solution vector components and set a threshold for rejection.

A *strongly determined variable* is one that cannot be changed without eroding the objective or changing significantly other variables. A *consistent variable* is one that receives a particular value in a large portion of the elite solution set. Let $I(e)$ be a measure of the strongly determined and consistent features of solution element $e \in E$. Then, $I(e)$ becomes larger as e appears more often in the pool of elite solutions. The intensity function $I(e)$ is used in the construction phase as follows. Recall that $c(e)$ is the greedy function, i.e., the incremental cost associated with the incorporation of element $e \in E$ into the solution under construction. Let $K(e) = F(c(e), I(e))$ be a function of the greedy and the intensification functions. For example, $K(e) = \lambda c(e) + I(e)$. The intensification scheme biases selection from the RCL to those elements $e \in E$ with a high value of $K(e)$ by setting its selection probability to be $p(e) = K(e)/\sum_{s \in \mathrm{RCL}} K(s)$.

The function $K(e)$ can vary with time by changing the value of λ, e.g., initially λ may be set to a large value that is decreased when diversification is called for. Procedures for changing the value of λ are given by Fleurent and Glover [46] and Binato et al. [19].

3.5 POP in Construction

The Proximate Optimality Principle (POP) is based on the idea that "good solutions at one level are likely to be found 'close to' good solutions at an adjacent level" [50]. Fleurent and Glover [46] provided a GRASP interpretation of this principle. They suggested that imperfections introduced during steps of GRASP construction can be "ironed-out" by applying local search during (and not only at the end of) the GRASP construction phase.

Because of efficiency considerations, a practical implementation of POP to GRASP is to apply local search during a few points in the construction phase and not during each construction iteration. In Binato et al. [19], local search is applied after 40% and 80% of the construction moves have been taken, as well as at the end of the construction phase.

4 PATH-RELINKING

Path-relinking is another enhancement to the basic GRASP procedure, leading to significant improvements in solution quality. Path-relinking was originally proposed by Glover [48] as an intensification strategy exploring trajectories connecting elite solutions obtained by tabu search or scatter search [49–51]. Starting from one or more elite solutions, paths in the solution space leading towards other elite solutions are generated and explored in the search for better solutions. This is accomplished by selecting moves that introduce attributes contained in the guiding solutions. Path-relinking may be viewed as a strategy that seeks to incorporate attributes of high quality solutions, by favoring these attributes in the selected moves.

The use of path-relinking within a GRASP procedure, as an intensification strategy applied to each locally optimal solution, was first proposed by Laguna and Martí [62]. It was followed by several extensions, improvements, and successful applications [4,22,86,90]. Two basic strategies are used:

- path-relinking is applied as a post-optimization step to all pairs of elite solutions; and
- path-relinking is applied as an intensification strategy to each local optimum obtained after the local search phase.

Applying path-relinking as an intensification strategy to each local optimum seems to be more effective than simply using it as a post-optimization step. In this context, path-relinking is applied to pairs (x_1, x_2) of solutions, where x_1 is the locally optimal solution obtained after local search and x_2 is one of a few elite solutions randomly chosen from a pool with a limited number `Max_Elite` of elite solutions found along the search. The pool is originally empty. Each locally optimal solution obtained by local search is considered as a candidate to be inserted into the pool if it is sufficiently different from every other solution currently in the pool. If the pool already has `Max_Elite` solutions and the candidate is better than the worst of them, then the former replaces the latter. If the pool is not full, the candidate is simply inserted.

The algorithm starts by computing the symmetric difference $\Delta(x_1, x_2)$ between x_1 and x_2, resulting in the set of moves which should be applied to one of them (the initial solution) to reach the other (the guiding solution). Starting from the initial solution, the best move from $\Delta(x_1, x_2)$ still not performed is applied to the current solution, until the guiding solution is attained. The best solution found along this trajectory is also considered as a candidate for insertion in the pool and the incumbent is updated. Several alternatives have been considered and combined in recent implementations:

- do not apply path-relinking at every GRASP iteration, but only periodically;
- explore two different trajectories, using first x_1, then x_2 as the initial solution;
- explore only one trajectory, starting from either x_1 or x_2; and
- do not follow the full trajectory, but instead only part of it (truncated path-relinking).

All these alternatives involve the trade-offs between computation time and solution quality. Ribeiro et al. [90] observed that exploring two different trajectories for each pair (x_1, x_2) takes approximately twice the time needed to explore only one of them, with very marginal improvements in solution quality. They have also observed that if only one trajectory is to be investigated, better solutions are found when path-relinking starts from the best among x_1 and x_2. Since the neighborhood of the initial solution is much more carefully explored than that of the guiding one, starting from the best of them gives the algorithm a better chance to investigate in more detail the neighborhood of the most promising solution. For the same reason, the best solutions are usually found closer to the initial solution than to the guiding solution, allowing pruning the relinking trajectory before the latter is reached.

Detailed computational results illustrating the trade-offs between these strategies for the problem of routing private virtual circuits in frame-relay services are reported by Resende and Ribeiro [86]. In this case, the set of moves corresponding to the symmetric difference $\Delta(x_1, x_2)$ between any pair (x_1, x_2) of solutions is the subset of private virtual circuits routed through different routes (i.e., using different edges) in x_1 and x_2. We summarize below some of these results, obtained on an SGI Challenge computer (with 28 196-MHz MIPS R10000 processors) with 7.6 Gb of memory. We considered four variants of the GRASP and path-relinking schemes previously discussed:

- `G`: This variant is a pure GRASP with no path-relinking.
- `GPRf`: This variant adds to `G` a one-way (forward) path-relinking starting from a locally optimal solution and using a randomly selected elite solution as the guiding solution.
- `GPRb`: This variant adds to `G` a one way (backwards) path-relinking starting from a randomly selected elite solution and using a locally optimal solution as the guiding solution.
- `GPRfb`: This variant combines `GPRf` and `GPRb`, performing path-relinking in both directions.

These variants are evaluated and compared in terms of their trade-offs between computation time and solution quality.

To study the effect of path-relinking on GRASP, we compared the four variants on two instances: `att` and `fr750a`, see [86] for details. Two hundred independent runs for each variant were performed for each problem. Execution was terminated when a solution of value less than or equal to a given parameter value `look4` was found. The sub-optimal values chosen for this parameter were such that the slowest variant could terminate in a reasonable amount of computation time. Empirical probability distributions for the time to target solution value are plotted in Figures 8.9 and 8.10. To plot the empirical distribution for each algorithm and each instance, we associate with the i-th smallest running time t_i a probability $p_i = (i - \frac{1}{2})/200$, and plot the points $z_i = (t_i, p_i)$, for $i = 1, \ldots, 200$. Due to the time taken by the pure GRASP procedure, we limited its plot in Figure 8.10 to 60 points.

These plots show a similar relative behavior of the four variants on the two instances. Since instance `fr750a` is harder for all variants and the associated computation times are longer, its plot is more discerning. For a given computation time, the probability of finding a solution at least as good as the target value increases from `G` to `GPRf`, from `GPRf` to `GPRfb`, and from `GPRfb` to `GPRb`. For example, there is a 9.25% probability for `GPRfb` to find a target solution value in less than 100 s, while this probability increases to 28.75% for `GPRb`. For `G`, there is a 8.33% probability of finding a target solution value within 2000 s, while for `GPRf` this probability increases to 65.25%. `GPRb` finds a target solution value in at most 129 s with 50% probability. For the same probability, this time increases to 172, 1727, and 10933 s, respectively, for variants `GPRfb`, `GPRf`, and `G`. In accordance with these results, variant `GPRb`, which does path-relinking backwards from an elite solution to a locally optimal solution, seems to be the most effective, confirming the preliminary findings reported in [90]. To further illustrate the behavior of GRASP and path-relinking, we depict in Figure 8.11 four plots representing the behavior of variant `GPRb` (GRASP with backwards path-relinking) on instance `att` with the variation of the target solution value. As before, 200 runs were performed for each target value decreasing from 126,600 to 126,000 by steps of 200. A similar behavior was observed for all other variants, with or without path-relinking, as well as for other instances and classes of test problems.

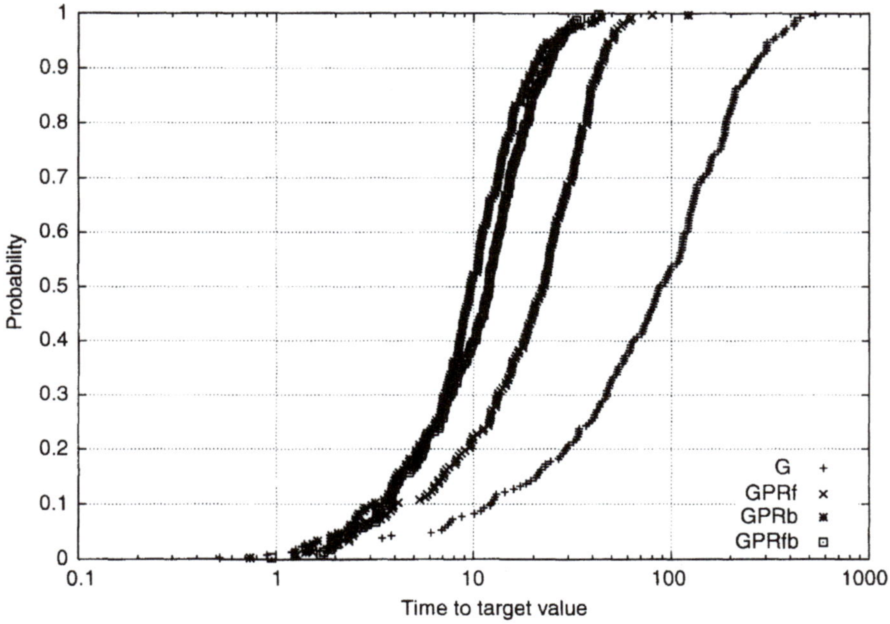

Figure 8.9. Empirical distributions of time to target solution value for GRASP, GRASP with forward path-relinking, GRASP with backwards path-relinking, and GRASP with back and forward path-relinking for instance `att`.

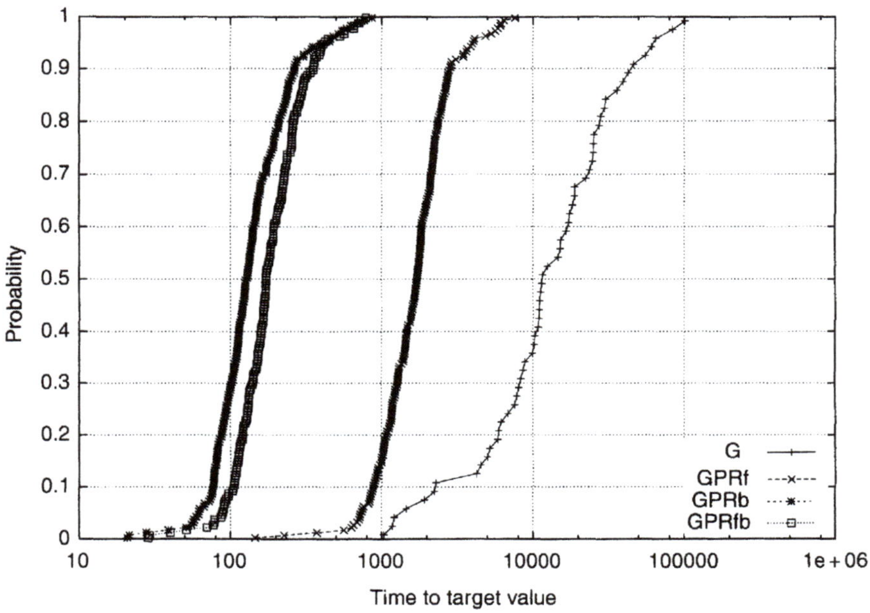

Figure 8.10. Empirical distributions of time to target solution value for GRASP, GRASP with forward path-relinking, GRASP with backwards path-relinking, and GRASP with back and forward path-relinking for instance `fr750a`.

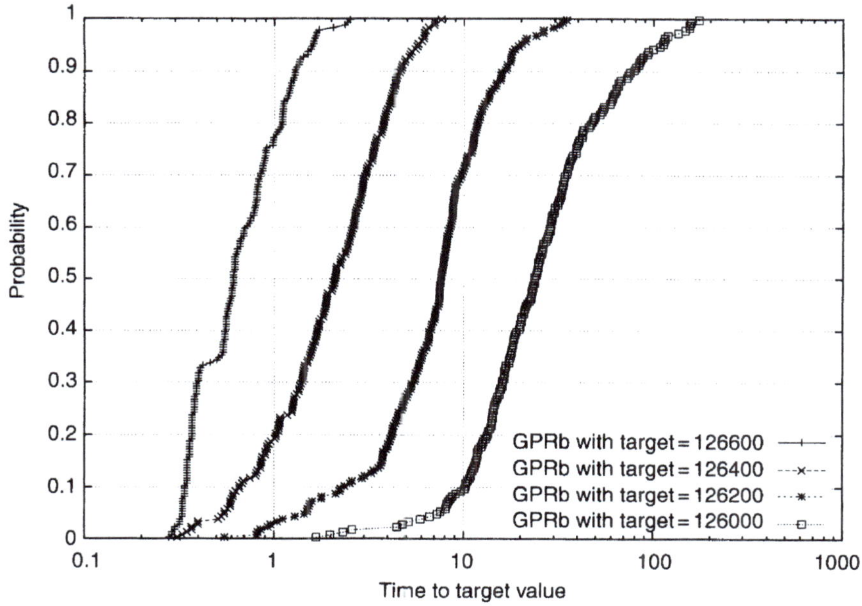

Figure 8.11. Empirical distributions of time to target solution value for GRASP with backwards path-relinking for instance `att` and different target values (`look4`).

Table 8.4. Solution values within fixed time limits over ten runs for instance `att`.

	10 s		100 s	
Variant	Best	Average	Best	Average
`GPR`	126602.883	126694.666	126227.678	126558.293
`GPRf`	126301.118	126578.323	126082.790	126228.798
`GPRb`	125960.336	126281.156	125665.785	125882.605
`GPRfb`	125961.118	126305.736	125646.460	125850.396

As a final experiment, once again we made use of the different GRASP variants for the problem of routing private virtual circuits to illustrate the effect of path-relinking in improving the solutions obtained by a pure GRASP approach, with only the construction and local search phases. This experiment was also performed using the same SGI Challenge computer (with 28,196-MHz MIPS R10000 processors) with 7.6 Gb of memory. For each of ten different seeds, we ran twice each variant for instance `att`, enforcing two different time limits: 10 and 100 s of processing time. The numerical results are reported in Table 8.4. For each variant and for each time limit, we give the average and the best solution values over the ten runs. We first note that both versions with backwards path-relinking performed systematically better, since they found better solutions for both time limits. Variants `GPRb` (GRASP with backwards path-relinking) and `GPRfb` (GRASP with path-relinking in both directions) showed similar behaviors, as it could be anticipated from the empirical probability distributions depicted in Figure 8.9. Variant `GPRb` obtained better results (in terms of both the average and the best solution values found) within the time limit of 10 s, while variant `GPRfb` performed better for the time limit of 100 s. In the first case, `GPRb` found the best solution among the two variants in seven runs, while `GPRfb` did better for only two runs. However, when the time limit was increased to 100 s, `GPRb` found the best solutions in four runs, while `GPRfb` did better for five runs.

Path-relinking is a quite effective strategy to introduce memory in GRASP, leading to very robust implementations. The results reported above can be further illustrated by those obtained with the hybrid GRASP with path-relinking algorithm for the Steiner problem in graphs described in [90], which in particular improved the best known solutions for 33 out of the 41 still open problems in series i640 of the SteinLib repository [99] on May 1, 2001.

5 EXTENSIONS

In this section, we comment on some extensions, implementation strategies, and hybrids of GRASP.

The use of hashing tables to avoid cycling in conjunction with tabu search was proposed by Woodruff and Zemel [100]. A similar approach was later explored by Ribeiro et al. [88] in their tabu search algorithm for query optimization in relational databases. In the context of GRASP implementations, hashing tables were first used by Martins et al. [66] in their multineighborhood heuristic for the Steiner problem in graphs, to avoid the application of local search to solutions already visited in previous iterations.

Filtering strategies have also been used to speed up the iterations of GRASP, see, e.g., [40,66,78]. In these cases, local search is not applied to all solutions obtained at the end of the construction phase, but instead only to some promising unvisited solutions, defined by a threshold with respect to the incumbent.

Almost all randomization effort in the basic GRASP algorithm involves the construction phase. Local search stops at the first local optimum. On the other hand, strategies such as VNS (Variable Neighborhood Search), proposed by Hansen and Mladenović [54,69], rely almost entirely on the randomization of the local search to escape from local optima. With respect to this issue, GRASP and variable neighborhood strategies may be considered as complementary and potentially capable of leading to effective hybrid methods. A first attempt in this direction was done by Martins et al. [66]. The construction phase of their hybrid heuristic for the Steiner problem in graphs follows the greedy randomized strategy of GRASP, while the local search phase makes use of two different neighborhood structures as a VND procedure [54,69]. Their heuristic was later improved by Ribeiro et al. [90], one of the key components of the new algorithm being another strategy for the exploration of different neighborhoods. Ribeiro and Souza [89] also combined GRASP with VND in a hybrid heuristic for the degree-constrained minimum spanning tree problem. Festa et al. [45] studied different variants and combinations of GRASP and VNS for the MAX-CUT problem, finding and improving the best known solutions for some open instances from the literature.

GRASP has also been used in conjunction with genetic algorithms. Basically, the greedy randomized strategy used in the construction phase of a GRASP is applied to generate the initial population for a genetic algorithm. We may cite e.g., the genetic algorithm of Ahuja et al. [3] for the quadratic assignment problem, which makes use of the GRASP proposed by Li et al. [63] to create the initial population of solutions. A similar approach was used by Armony et al. [11], with the initial population made up by both randomly generated solutions and those built by a GRASP.

The hybridization of GRASP with tabu search was first studied by Laguna and González-Velarde [61]. Delmaire et al. [29] considered two approaches. In the first, GRASP is applied as a powerful diversification strategy in the context of a tabu search procedure. The second approach is an implementation of the Reactive GRASP algorithm presented in Section 3.1, in which the local search phase is strengthened by tabu search. Results reported for the capacitated location problem show that the hybrid approaches perform better than the isolated methods previously used. Two two-stage heuristics are proposed in [1] for solving the multi-floor facility layout problem. GRASP/TS applies a GRASP to find the initial layout and tabu search to refine it.

6 PARALLEL GRASP

Even though parallelism is not yet systematically used to speed up or to improve the effectiveness of metaheuristics, parallel implementations are very robust and abound in the literature; see e.g., Cung et al. [27] for a recent survey.

Most parallel implementations of GRASP follow the *multiple-walk independent thread* strategy, based on the distribution of the iterations over the processors [6,7,40, 63,65,67,70,73,74]. In general, each search thread has to perform `Max_Iterations`/p iterations, where p and `Max_Iterations` are, respectively, the number of processors and the total number of iterations. Each processor has a copy of the sequential algorithm, a copy of the problem data, and an independent seed to generate its own pseudorandom number sequence. To avoid that the processors find the same solutions, each of them must use a different sequence of pseudorandom numbers. A single global variable is required to store the best solution found over all processors. One of the processors acts as the master, reading and distributing problem data, generating the seeds which will be used by the pseudorandom number generators at each processor, distributing the iterations, and collecting the best solution found by each processor. Since the iterations are completely independent and very little information is exchanged, linear speedups are easily obtained provided that no major load imbalance problems occur. The iterations may be evenly distributed over the processors or according with their demands, to improve load balancing.

Martins et al. [67] implemented a parallel GRASP for the Steiner problem in graphs. Parallelization is achieved by the distribution of 512 iterations over the processors, with the value of the RCL parameter α randomly chosen in the interval $[0.0, 0.3]$ at each iteration. The algorithm was implemented in C on an IBM SP-2 machine with 32 processors, using the MPI library for communication. The 60 problems from series C, D, and E of the OR-Library [18] have been used for the computational experiments. The parallel implementation obtained 45 optimal solutions over the 60 test instances. The relative deviation with respect to the optimal value was never larger than 4%. Almost-linear speedups observed for 2, 4, 8, and 16 processors with respect to the sequential implementation are illustrated in Figure 8.12.

Path-relinking may also be used in conjunction with parallel implementations of GRASP. In the case of the multiple-walk independent-thread implementation described

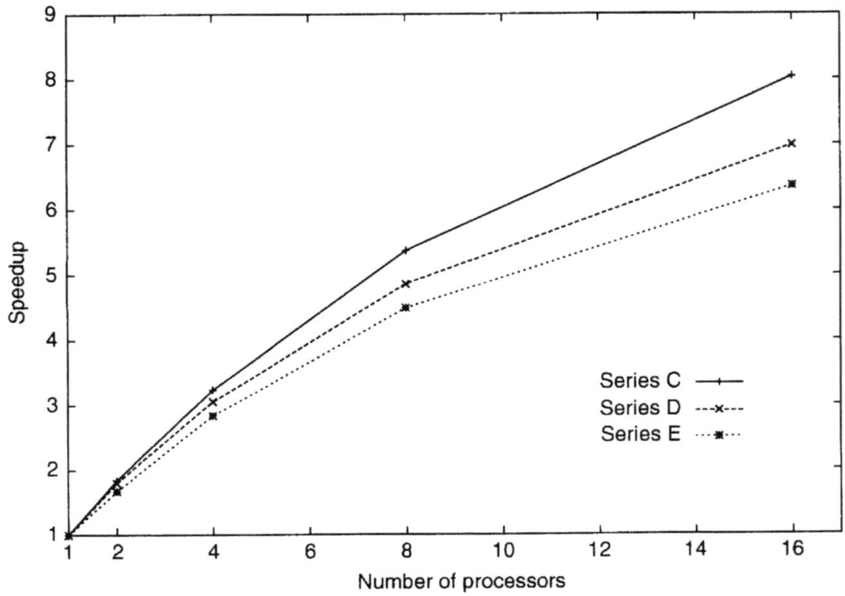

Figure 8.12. Average speedups on 2, 4, 8, and 16 processors.

by Aiex et al. [4] for the 3-index assignment problem, each processor applies path-relinking to pairs of elite solutions stored in a local pool. Computational results using MPI on an SGI Challenge computer with 28 R10000 processors showed linear speedups.

Alvim and Ribeiro [6,7] have shown that multiple-walk independent-thread approaches for the parallelization of GRASP may benefit much from load balancing techniques, whenever heterogeneous processors are used or if the parallel machine is simultaneously shared by several users. In this case, almost-linear speedups may be obtained with a heterogeneous distribution of the iterations over the p processors in $q \geq p$ packets. Each processor starts performing one packet of $\lceil \texttt{Max\_Iterations}/q \rceil$ iterations and informs the master when it finishes its packet of iterations. The master stops the execution of each slave processor when there are no more iterations to be performed and collects the best solution found. Faster or less loaded processors will perform more iterations than the others. In the case of the parallel GRASP implemented for the problem of traffic assignment described in [78], this dynamic load balancing strategy allowed reductions in the elapsed times of up to 15% with respect to the times observed for the static strategy, in which the iterations were uniformly distributed over the processors.

The efficiency of multiple-walk independent-thread parallel implementations of metaheuristics, based on running multiple copies of the same sequential algorithm, has been addressed by some authors. A given target value τ for the objective function is broadcasted to all processors which independently execute the sequential algorithm. All processors halt immediately after one of them finds a solution with value at least as good as τ. The speedup is given by the ratio between the times needed to find a solution with value at least as good as τ, using respectively the sequential algorithm and the parallel implementation with p processors. Some care is needed to ensure that no two iterations start with identical random number generator seeds. These speedups are linear for a number of metaheuristics, including simulated annealing [31,71]; iterated local search algorithms for the traveling salesman problem [33]; tabu search, provided that the search starts from a local optimum [17,94]; and WalkSAT [93] on hard random 3-SAT problems [56]. This observation can be explained if the random variable *time to find a solution within some target value* is exponentially distributed, as indicated by the following proposition [98].

Proposition 8.1. *Let $P_\rho(t)$ be the probability of not having found a given target solution value in t time units with ρ independent processes. If $P_1(t) = e^{-t/\lambda}$ with $\lambda \in \mathbf{R}^+$, corresponding to an exponential distribution, then $P_\rho(t) = e^{-\rho t/\lambda}$.*

This proposition follows from the definition of the exponential distribution. It implies that the probability $1 - e^{-\rho t/\lambda}$ of finding a solution within a given target value in time ρt with a sequential algorithm is equal to the probability of finding a solution at least as good as that in time t using ρ independent parallel processors. Hence, it is possible to achieve linear speedups in the time to find a solution within a target value by multiple independent processors. An analogous proposition can be stated for a two parameter (shifted) exponential distribution.

Proposition 8.2. *Let $P_\rho(t)$ be the probability of not having found a given target solution value in t time units with ρ independent processors. If $P_1(t) = e^{-(t-\mu)/\lambda}$ with $\lambda \in \mathbf{R}^+$ and $\mu \in \mathbf{R}^+$, corresponding to a two parameter exponential distribution, then $P_\rho(t) = e^{-\rho(t-\mu)/\lambda}$.*

Analogously, this proposition follows from the definition of the two-parameter exponential distribution. It implies that the probability of finding a solution within a given target value in time ρt with a sequential algorithm is equal to $1 - e^{-(\rho t-\mu)/\lambda}$, while the probability of finding a solution at least as good as that in time t using ρ independent parallel processors is $1 - e^{-\rho(t-\mu)/\lambda}$. If $\mu = 0$, then both probabilities are equal and correspond to the non-shifted exponential distribution. Furthermore, if $\rho\mu \ll \lambda$, then the two probabilities are approximately equal and it is possible to approximately achieve linear speedups in the time to find a solution within a target value using multiple independent processors.

Aiex et al. [5] have shown experimentally that the solution times for GRASP also have this property, showing that they fit a two-parameter exponential distribution. Figure 8.13 illustrates this result, depicting the superimposed empirical and theoretical distributions observed for one of the cases studied along the computational experiments reported by the authors, which involved 2400 runs of GRASP procedures for each of five different problems: maximum independent set [40,81], quadratic assignment [63,82], graph planarization [85,87], maximum weighted satisfiability [84], and maximum covering [79]. The same result still holds when GRASP is implemented in conjunction with a post-optimization path-relinking procedure [4].

In the case of *multiple-walk cooperative-thread* strategies, the search threads running in parallel exchange and share information collected along the trajectories they investigate. One expects not only to speed up the convergence to the best solution but, also, to find better solutions than independent-thread strategies. The most difficult aspect to be set up is the determination of the nature of the information to be shared or exchanged to improve the search, without taking too much additional memory or time to

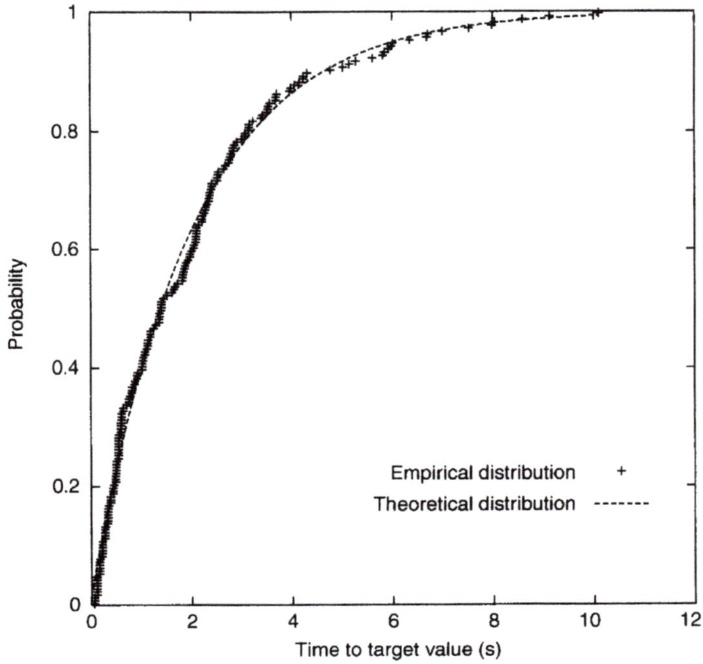

Figure 8.13. Superimposed empirical and theoretical distributions (times to target values measured in seconds on an SGI Challenge computer with 28 processors).

be collected. Cooperative-thread strategies may be implemented using path-relinking, by combining elite solutions stored in a central pool with the local optima found by each processor at the end of each GRASP iteration. Canuto et al. [22] used path-relinking to implement a parallel GRASP for the prize-collecting Steiner tree problem. Their strategy is truly cooperative, since pairs of elite solutions from a centralized unique central pool are distributed to the processors which perform path-relinking in parallel. Computational results obtained with an MPI implementation running on a cluster of 32,400-MHz Pentium II processors showed linear speedups, further illustrating the effectiveness of path-relinking procedures used in conjunction with GRASP to improve the quality of the solutions found by the latter.

7 APPLICATIONS

The first application of GRASP described in the literature concerns the set covering problem [38]. The reader is referred to Festa and Resende [44] for an annotated bibliography of GRASP and its applications. We conclude this chapter by summarizing below some references focusing the main applications of GRASP to problems in different areas:

- routing [9,12,16,24,59];
- logic [30,74,80,83];
- covering and partition [8,10,38,47,53];

- location [1,29,57,96,97];
- minimum Steiner tree [23,65–67,90];
- optimization in graphs [2,40,60,72,79,85,87];
- assignment [37,46,63,64,68,70,73,75,78];
- timetabling, scheduling, and manufacturing [13–15,19,28,32,34–36,41,42,58,91, 92,101];
- transportation [9,34,37];
- power systems [20];
- telecommunications [2,11,57,64,78,79,86];
- graph and map drawing [43,62,85,87]; and
- VLSI [8], among other areas of application.

8 CONCLUDING REMARKS

The results described in this chapter reflect successful applications of GRASP to a large number of classical combinatorial optimization problems, as well as to those that arise in real-world situations in different areas of business, science, and technology.

We underscore the simplicity of implementation of GRASP, which makes use of simple building blocks: solution construction procedures and local search methods, which often are readily available. Contrary to what occurs with other metaheuristics, such as tabu search or genetic algorithms, which use a large number of parameters in their implementations, the basic version of GRASP requires the adjustment of a single parameter.

Recent developments, presented in this chapter, show that different extensions to the basic procedure allow further improvement to the solutions found by GRASP. Among these, we highlight: reactive GRASP, which automates the adjustments of the restricted candidate list parameter; variable neighborhoods, which permit accelerated and intensified local search; and path-relinking, which beyond allowing the implementation of intensification strategies based on the memory of elite solutions, opens the way for development of very effective cooperative parallel strategies.

These and other extensions make up a set of tools that can be added to simpler heuristics to find better-quality solutions. To illustrate the effect of additional extensions on solution quality, Figure 8.14 shows some results obtained for the prize-collecting Steiner tree problem, as discussed in [22]. We consider the 40 instances of series C. The lower curve represents the results obtained exclusively with the primal-dual constructive algorithm (GW) of Goemans and Williamson [52]. The second curve shows the quality of the solutions produced with an additional local search (GW + LS), corresponding to the first iteration of GRASP. The third curve is associated with the results obtained after 500 iterations of GRASP with path-relinking (GRASP + PR). Finally, the top curve shows the results found by the complete algorithm, using variable neighborhood search as a post-optimization procedure (GRASP + PR + VNS). For a given relative deviation with respect to the optimal value, each curve indicates the number of instances for which the corresponding algorithm found a solution within that quality range. For example, we observe that the number of optimal solutions found goes from six, using only the constructive algorithm, to a total of 36, using the complete

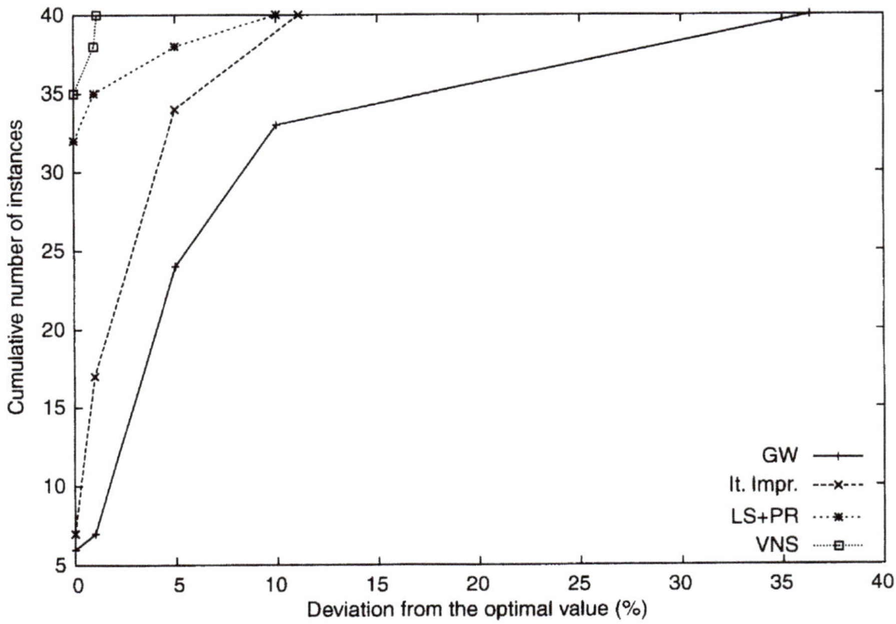

Figure 8.14. Performance of GW and successive variants of local search for Series C problems.

algorithm described in [22]. The largest relative deviation with respect to the optimal value decreases from 36.4% in the first case, to only 1.1% for the complete algorithm. It is easy to see the contribution made by each additional extension.

Parallel implementations of GRASP are quite robust and lead to linear speedups both in independent and cooperative strategies. Cooperative strategies are based on the collaboration between processors using path-relinking and a global pool of elite solutions. This allows the use of more processors to find better solutions in less computation time.

BIBLIOGRAPHY

[1] S. Abdinnour-Helm and S.W. Hadley (2000) Tabu search based heuristics for multi-floor facility layout. *International Journal of Production Research*, **38**, 365–383.

[2] J. Abello, P.M. Pardalos and M.G.C. Resende (1999) On maximum clique problems in very large graphs. In: J. Abello and J. Vitter (eds.), *External Memory Algorithms and Visualization*, volume 50 of *DIMACS Series on Discrete Mathematics and Theoretical Computer Science*. American Mathematical Society, pp. 199–130.

[3] R.K. Ahuja, J.B. Orlin and A. Tiwari (2000) A greedy genetic algorithm for the quadratic assignment problem. *Computers and Operations Research*, **27**, 917–934.

[4] R.M. Aiex, M.G.C. Resende, P.M. Pardalos and G. Toraldo (2000) GRASP with path-relinking for the three-index assignment problem. Technical report, AT&T Labs-Research.

[5] R.M. Aiex, M.G.C. Resende and C.C. Ribeiro (2002) Probability distribution of solution time in GRASP: an experimental investigation. *Journal of Heuristics*, **8**, 343–373.

[6] A.C. Alvim (1998) Parallelization strategies for the metaheuristic GRASP. Master's thesis, Department of Computer Science, Catholic University of Rio de Janeiro, Brazil (in Portuguese).

[7] A.C. Alvim and C.C. Ribeiro (1998) Load balancing for the parallelization of the GRASP metaheuristic. In: *Proceedings of the X Brazilian Symposium on Computer Architecture*, Búzios, pp. 279–282 (in Portuguese).

[8] S. Areibi and A. Vannelli (1997) A GRASP clustering technique for circuit partitioning. In: J. Gu and P.M. Pardalos (eds.), *Satisfiability Problems*, Volume 35 of *DIMACS Series on Discrete Mathematics and Theoretical Computer Science*. American Mathematical Society, pp. 711–724.

[9] M.F. Argüello, J.F. Bard and G. Yu (1997) A GRASP for aircraft routing in response to groundings and delays. *Journal of Combinatorial Optimization*, **1**, 211–228.

[10] M.F. Argüello, T.A. Feo and O. Goldschmidt (1996) Randomized methods for the number partitioning problem. *Computers and Operations Research*, **23**, 103–111.

[11] M. Armony, J.C. Klincewicz, H. Luss and M.B. Rosenwein (2000) Design of stacked self-healing rings using a genetic algorithm. *Journal of Heuristics*, **6**, 85–105.

[12] J.B. Atkinson (1998) A greedy randomised search heuristic for time-constrained vehicle scheduling and the incorporation of a learning strategy. *Journal of the Operational Research Society*, **49**, 700–708.

[13] J.F. Bard and T.A. Feo (1989) Operations sequencing in discrete parts manufacturing. *Management Science*, **35**, 249–255.

[14] J.F. Bard and T.A. Feo (1991) An algorithm for the manufacturing equipment selection problem. *IIE Transactions*. **23**, 83–92.

[15] J.F. Bard, T.A. Feo and S. Holland (1996) A GRASP for scheduling printed wiring board assembly. *IIE Transactions*, **28**, 155–165.

[16] J.F. Bard, L. Huang, P. Jaillet and M. Dror (1998) A decomposition approach to the inventory routing problem with satellite facilities. *Transportation Science*, **32**, 189–203.

[17] R. Battiti and G. Tecchiolli (1992) Parallel biased search for combinatorial optimization: Genetic algorithms and tabu. *Microprocessors and Microsystems*, **16**, 351–367.

[18] J.E. Beasley (1990) OR-Library: Distributing test problems by electronic mail. *Journal of the Operational Research Society*, **41**, 1069–1072.

[19] S. Binato, W.J. Hery, D. Loewenstern and M.G.C. Resende (2002) A GRASP for job shop scheduling. In: C.C. Ribeiro and P. Hansen (eds.), *Essays and Surveys in Metaheuristics*. Kluwer Academic Publishers, pp. 59–79.

[20] S. Binato and G.C. Oliveira (2002) A reactive GRASP for transmission network expansion planning. In: C.C. Ribeiro and P. Hansen (eds.), *Essays and Surveys in Metaheuristics*. Kluwer Academic Publishers, pp. 81–100.

[21] J.L. Bresina (1996) Heuristic-biased stochastic sampling. In: *Proceedings of the Thirteenth National Conference on Artificial Intelligence*. Portland, pp. 271–278.

[22] S.A. Canuto, M.G.C. Resende and C.C. Ribeiro (2001) Local search with perturbations for the prize-collecting Steiner tree problem in graphs. *Networks*, **38**, 50–58.

[23] S.A. Canuto, C.C. Ribeiro and M.G.C. Resende (1999) Local search with perturbations for the prize-collecting Steiner tree problem. In: *Extended Abstracts of the Third Metaheuristics International Conference*. Angra dos Reis, pp. 115–119.

[24] C. Carreto and B. Baker (2002) A GRASP interactive approach to the vehicle routing problem with backhauls. In: C.C. Ribeiro and P. Hansen (eds.), *Essays and Surveys in Metaheuristics*. Kluwer Academic Publishers, pp. 185–199.

[25] I. Charon and O. Hudry (1993) The noising method: a new method for combinatorial optimization. *Operations Research Letters*, **14**, 133–137.

[26] I. Charon and O. Hudry (2002) The noising methods: a survey. In: C.C. Ribeiro and C.C. Ribeiro (eds.), *Essays and Surveys in Metaheuristics*. Kluwer Academic Publishers, pp. 245–261.

[27] V.-D. Cung, S.L. Martins, C.C. Ribeiro and C. Roucairol (2002) Strategies for the parallel implementation of metaheuristics. In: C.C. Ribeiro and C.C. Ribeiro (eds.), *Essays and Surveys in Metaheuristics*. Kluwer Academic Publishers, pp. 263–308.

[28] P. De, J.B. Ghosj and C.E. Wells (1994) Solving a generalized model for con due date assignment and sequencing. *International Journal of Production Economics*, **34**, 179–185.

[29] H. Delmaire, J.A. Díaz, E. Fernández and M. Ortega (1999) Reactive GRASP and Tabu Search based heuristics for the single source capacitated plant location problem. *INFOR*, **37**, 194–225.

[30] A.S. Deshpande and E. Triantaphyllou (1998) A greedy randomized adaptive search procedure (GRASP) for inferring logical clauses from examples in polynomial time and some extensions. *Mathematical Computer Modelling*, **27**, 75–99.

[31] N. Dodd (1990) Slow annealing versus multiple fast annealing runs: an empirical investigation. *Parallel Computing*, **16**, 269–272.

[32] A. Drexl and F. Salewski (1997) Distribution requirements and compactness constraints in school timetabling. *European Journal of Operational Research*, **102**, 193–214.

[33] H.T. Eikelder, M. Verhoeven, T. Vossen and E. Aarts (1996) A probabilistic analysis of local search. In: I. Osman and J. Kelly (eds.), *Metaheuristics: Theory and Applications*. Kluwer Academic Publishers, pp. 605–618.

[34] T.A. Feo and J.F. Bard (1989) Flight scheduling and maintenance base planning. *Management Science*, **35**, 1415–1432.

[35] T.A. Feo and J.F. Bard (1989) The cutting path and tool selection problem in computer-aided process planning. *Journal of Manufacturing Systems*, **8**, 17–26.

[36] T.A. Feo, J.F. Bard and S. Holland (1995) Facility-wide planning and scheduling of printed wiring board assembly. *Operations Research*, **43**, 219–230.

[37] T.A. Feo and J.L. González-Velarde (1995) The intermodal trailer assignment problem: models, algorithms, and heuristics. *Transportation Science*, **29**, 330–341.

[38] T.A. Feo and M.G.C. Resende (1989) A probabilistic heuristic for a computationally difficult set covering problem. *Operations Research Letters*, **8**, 67–71.

[39] T.A. Feo and M.G.C. Resende (1995) Greedy randomized adaptive search procedures. *Journal of Global Optimization*, **6**, 109–133.

[40] T.A. Feo, M.G.C. Resende and S.H. Smith (1994) A greedy randomized adaptive search procedure for maximum independent set. *Operations Research*, **42**, 860–878.

[41] T.A. Feo, K. Sarathy and J. McGahan (1996) A GRASP for single machine scheduling with sequence dependent setup costs and linear delay penalties. *Computers and Operations Research*, **23**, 881–895.

[42] T.A. Feo, K. Venkatraman and J.F. Bard (1991) A GRASP for a difficult single machine scheduling problem. *Computers and Operations Research*, **18**, 635–643.

[43] E. Fernández and R. Martí (1999) GRASP for seam drawing in mosaicking of aerial photographic maps. *Journal of Heuristics*, **5**, 181–197.

[44] P. Festa and M.G.C. Resende (2002) GRASP: an annotated bibliography. In: C.C. Ribeiro and P. Hansen (eds.), *Essays and Surveys in Metaheuristics*. Kluwer Academic Publishers, pp. 325–367.

[45] P. Festa, M.G.C. Resende, P. Pardalos and C.C. Ribeiro (2001) GRASP and VNS for Max-Cut. In: *Extended Abstracts of the Fourth Metaheuristics International Conference*. Porto, pp. 371–376.

[46] C. Fleurent and F. Glover (1999) Improved constructive multistart strategies for the quadratic assignment problem using adaptive memory. *INFORMS Journal on Computing*, **11**, 198–204.

[47] J.B. Ghosh (1996) Computatinal aspects of the maximum diversity problem. *Operations Research Letters*, **19**, 175–181.

[48] F. Glover (1996) Tabu search and adaptive memory programming—advances, applications and challenges. In: R.S. Barr, R.V. Helgason and J.L. Kennington (eds.), *Interfaces in Computer Science and Operations Research*. Kluwer, pp. 1–75.

[49] F. Glover (2000) Multi-start and strategic oscillation methods—principles to exploit adaptive memory. In: M. Laguna and J.L. Gonzáles-Velarde (eds.), *Computing Tools for Modeling, Optimization and Simulation: Interfaces in Computer Science and Operations Research*. Kluwer, pp. 1–24.

[50] F. Glover and M. Laguna (1997) *Tabu Search*. Kluwer.

[51] F. Glover, M. Laguna and R. Martí (2000) Fundamentals of scatter search and path relinking. *Control and Cybernetics*, **39**, 653–684.

[52] M.X. Goemans and D.P. Williamson (1996) The primal dual method for approximation algorithms and its application to network design problems. In: D. Hochbaum (ed.), *Approximation Algorithms for NP-hard Problems*. PWS Publishing Co., pp. 144–191.

[53] P.L. Hammer and D.J. Rader, Jr. (2001) Maximally disjoint solutions of the set covering problem. *Journal of Heuristics*, **7**, 131–144.

[54] P. Hansen and N. Mladenović (2002) Developments of variable neighborhood search. In: C.C. Ribeiro and P. Hansen (eds.), *Essays and Surveys in Metaheuristics*. Kluwer Academic Publishers, pp. 415–439.

[55] J.P. Hart and A.W. Shogan (1987) Semi-greedy heuristics: an empirical study. *Operations Research Letters*, **6**, 107–114.

[56] H. Hoos and T. Stützle (1999) Towards a characterisation of the behaviour of stochastic local search algorithms for SAT. *Artificial Intelligence*, **112**, 213–232.

[57] J.G. Klincewicz (1992) Avoiding local optima in the p-hub location problem using tabu search and GRASP. *Annals of Operations Research*, **40**, 283–302.

[58] J.G. Klincewicz and A. Rajan (1994) Using GRASP to solve the component grouping problem. *Naval Research Logistics*, **41**, 893–912.

[59] G. Kontoravdis and J.F. Bard (1995) A GRASP for the vehicle routing problem with time windows. *ORSA Journal on Computing*, **7**, 10–23.

[60] M. Laguna, T.A. Feo and H.C. Elrod (1994) A greedy randomized adaptive search procedure for the two-partition problem. *Operations Research*, **42**, 677–687.

[61] M. Laguna and J.L. González-Velarde (1991) A search heuristic for just-in-time scheduling in parallel machines. *Journal of Intelligent Manufacturing*, **2**, 253–260.

[62] M. Laguna and R. Martí (1999) GRASP and path relinking for 2-layer straight line crossing minimization. *INFORMS Journal on Computing*, **11**, 44–52.

[63] Y. Li, P.M. Pardalos and M.G.C. Resende (1994) A greedy randomized adaptive search procedure for the quadratic assignment problem. In: P.M. Pardalos and H. Wolkowicz (eds.), *Quadratic Assignment and Related Problems*, volume 16 of *DIMACS Series on Discrete Mathematics and Theoretical Computer Science*. American Mathematical Society, pp. 237–261.

[64] X. Liu, P.M. Pardalos, S. Rajasekaran and M.G.C. Resende (2000) A GRASP for frequency assignment in mobile radio networks. In: B.R. Badrinath, F. Hsu, P.M. Pardalos and S. Rajasekaran (eds.), *Mobile Networks and Computing*, volume 52 of *DIMACS Series on Discrete Mathematics and Theoretical Computer Science*. American Mathematical Society, pp. 195–201.

[65] S.L. Martins, P.M. Pardalos, M.G.C. Resende and C.C. Ribeiro (1999) Greedy randomized adaptive search procedures for the steiner problem in graphs. In: P.M. Pardalos, S. Rajasekaran and J. Rolim (eds.), *Randomization Methods in*

Algorithmic Design, volume 43 of *DIMACS Series on Discrete Mathematics and Theoretical Computer Science*. American Mathematical Society, pp. 133–145.

[66] S.L. Martins, M.G.C. Resende, C.C. Ribeiro and P. Pardalos (2000) A parallel GRASP for the Steiner tree problem in graphs using a hybrid local search strategy. *Journal of Global Optimization*, **17**, 267–283.

[67] S.L. Martins, C.C. Ribeiro and M.C. Souza (1998) A parallel GRASP for the Steiner problem in graphs. In: A. Ferreira and J. Rolim (eds.), *Proceedings of IRREGULAR'98—5th International Symposium on Solving Irregularly Structured Problems in Parallel*, volume 1457 of *Lecture Notes in Computer Science*. Springer-Verlag, pp. 285–297.

[68] T. Mavridou, P.M. Pardalos, L.S. Pitsoulis and M.G.C. Resende (1998) A GRASP for the biquadratic assignment problem. *European Journal of Operational Research*, **105**, 613–621.

[69] N. Mladenović and P. Hansen (1997) Variable neighborhood search. *Computers and Operations Research*, **24**, 1097–1100.

[70] R.A. Murphey, P.M. Pardalos and L.S. Pitsoulis (1998) A parallel GRASP for the data association multidimensional assignment problem. In: P.M. Pardalos (ed.), *Parallel Processing of Discrete Problems*, volume 106 of *The IMA Volumes in Mathematics and Its Applications*, Springer-Verlag, pp. 159–180.

[71] L. Osborne and B. Gillett (1991) A comparison of two simulated annealing algorithms applied to the directed Steiner problem on networks. *ORSA Journal on Computing*, **3**, 213–225.

[72] P.M. Pardalos, T. Qian and M.G.C. Resende (1999) A greedy randomized adaptive search procedure for the feedback vertex set problem. *Journal of Combinatorial Optimization*, **2**, 399–412.

[73] P.M. Pardalos, L.S. Pitsoulis and M.G.C. Resende (1995) A parallel GRASP implementation for the quadratic assignment problem. In: A. Ferreira and J. Rolim (eds.), *Parallel Algorithms for Irregularly Structured Problems—Irregular'94*. Kluwer Academic Publishers, pp. 115–133.

[74] P.M. Pardalos, L.S. Pitsoulis and M.G.C. Resende (1996) A parallel GRASP for MAX-SAT problems. *Lecture Notes in Computer Science*, **1184**, 575–585.

[75] L.S. Pitsoulis, P.M. Pardalos and D.W. Hearn (2001) Approximate solutions to the turbine balancing problem. *European Journal of Operational Research*, **130**, 147–155.

[76] M. Prais and C.C. Ribeiro (1999) Parameter variation in GRASP implementations. In: *Extended Abstracts of the Third Metaheuristics International Conference*. Angra dos Reis, pp. 375–380.

[77] M. Prais and C.C. Ribeiro (2000) Parameter variation in GRASP procedures. *Investigación Operativa*, **9**, 1–20.

[78] M. Prais and C.C. Ribeiro (2000) Reactive GRASP: an application to a matrix decomposition problem in TDMA traffic assignment. *INFORMS Journal on Computing*, **12**, 164–176.

[79] M.G.C. Resende (1998) Computing approximate solutions of the maximum covering problem using GRASP. *Journal of Heuristics*, **4**, 161–171.

[80] M.G.C. Resende and T.A. Feo (1996) A GRASP for satisfiability. In: D.S. Johnson and M.A. Trick (eds.), *Cliques, Coloring, and Satisfiability: The Second DIMACS Implementation Challenge*, volume 26 of *DIMACS Series on Discrete Mathematics and Theoretical Computer Science*. American Mathematical Society, pp. 499–520.

[81] M.G.C. Resende, T.A. Feo and S.H. Smith (1998) Algorithm 787: Fortran subroutines for approximate solution of maximum independent set problems using GRASP. *ACM Transactions on Mathematical Software*, **24**, 386–394.

[82] M.G.C. Resende, P.M. Pardalos and Y. Li (1996) Algorithm 754: Fortran subroutines for approximate solution of dense quadratic assignment problems using GRASP. *ACM Transactions on Mathematical Software*, **22**, 104–118.

[83] M.G.C. Resende, L.S. Pitsoulis and P.M. Pardalos (1997) Approximate solution of weighted MAX-SAT problems using GRASP. In: J. Gu and P.M. Pardalos (eds.), *Satisfiability Problems*, volume 35 of *DIMACS Series on Discrete Mathematics and Theoretical Computer Science*. American Mathematical Society, pp. 393–405.

[84] M.G.C. Resende, L.S. Pitsoulis and P.M. Pardalos (2000) Fortran subroutines for computing approximate solutions of MAX-SAT problems using GRASP. *Discrete Applied Mathematics*, **100**, 95–113.

[85] M.G.C. Resende and C.C. Ribeiro (1997) A GRASP for graph planarization. *Networks*, **29**, 173–189.

[86] M.G.C. Resende and C.C. Ribeiro (2001) A GRASP with path-relinking for private virtual circuit routing. Technical report, AT&T Labs Research.

[87] C.C. Ribeiro and M.G.C. Resende (1999) Algorithm 797: Fortran subroutines for approximate solution of graph planarization problems using GRASP. *ACM Transactions on Mathematical Software*, **25**, 342–352.

[88] C.C. Ribeiro, C.D. Ribeiro and R.S. Lanzelotte (1997) Query optimization in distributed relational databases. *Journal of Heuristics*, **3**, 5–23.

[89] C.C. Ribeiro and M.C. Souza (2002) Variable neighborhood search for the degree constrained minimum spanning tree problem. *Discrete Applied Mathematics*, **118**, 43–54.

[90] C.C. Ribeiro, E. Uchoa and R.F. Werneck (2002) A hybrid GRASP with perturbations for the Steiner problem in graphs. *INFORMS Journal on Computing*, **14**, 228–246.

[91] R.Z. Ríos-Mercado and J.F. Bard (1998) Heuristics for the flow line problem with setup costs. *European Journal of Operational Research*, 76–98.

[92] R.Z. Ríos-Mercado and J.F. Bard (1999) An enhanced TSP-based heuristic for makespan minimization in a flow shop with setup costs. *Journal of Heuristics*, **5**, 57–74.

[93] B. Selman, H. Kautz and B. Cohen (1994) Noise strategies for improving local search. In: *Proceedings of the Twelfth National Conference on Artificial Intelligence*. Seattle, MIT Press, pp. 337–343.

[94] E. Taillard (1991) Robust taboo search for the quadratic assignment problem. *Parallel Computing*, **7**, 443–455.

[95] H. Takahashi and A. Matsuyama (1980) An approximate solution for the Steiner problem in graphs. *Mathematica Japonica*, **24**, 573–577.

[96] T.L. Urban (1998) Solution procedures for the dynamic facility layout problem. *Annals of Operations Research*, 323–342.

[97] T.L. Urban, W.-C. Chiang and R.A. Russel (2000) The integrated machine allocation and layout problem. *International Journal of Production Research*, 2913–2930.

[98] M.G.A. Verhoeven and E.H.L. Aarts (1995) Parallel local search. *Journal of Heuristics*, **1**, 43–65.

[99] S. Voss, A. Martin and T. Koch (2001) Steinlib testdata library. Online document at http://elib.zib.de/steinlib/steinlib.html, last visited on May 1.

[100] D.L. Woodruff and E. Zemel (1993) Hashing vectors for tabu search. *Annals of Operations Research*, **41**, 123–137.

[101] J. Yen, M. Carlsson, M. Chang, J.M. Garcia and H. Nguyen (2000) Constraint solving for inkjet print mask design. *Journal of Imaging Science and Technology*, **44**, 391–397.

Chapter 9

THE ANT COLONY OPTIMIZATION METAHEURISTIC: ALGORITHMS, APPLICATIONS, AND ADVANCES

Marco Dorigo
IRIDIA, Université Libre de Bruxelles, Belgium
E-mail: mdorigo@ulb.ac.be
URL: http://iridia.ulb.ac.be/~mdorigo

Thomas Stützle
Intellectics Group, TU Darmstadt, Germany
E-mail: stuetzle@informatik.tu-darmstadt.de
URL: http://www.intellektik.informatik.tu-darmstadt.de/~tom

1 INTRODUCTION

Ant Colony Optimization (ACO) [32,33] is a recent metaheuristic approach for solving hard combinatorial optimization problems. The inspiring source of ACO is the pheromone trail laying and following behavior of real ants which use pheromones as a communication medium. In analogy to the biological example, ACO is based on the indirect communication of a colony of simple agents, called (artificial) ants, mediated by (artificial) pheromone trails. The pheromone trails in ACO serve as a distributed, numerical information which the ants use to probabilistically construct solutions to the problem being solved and which the ants adapt during the algorithm's execution to reflect their search experience.

The first example of such an algorithm is Ant System (AS) [30,36–38], which was proposed using as example application, the well known Traveling Salesman Problem (TSP) [60,76]. Despite encouraging initial results, AS could not compete with state-of-the-art algorithms for the TSP. Nevertheless, it had the important role of stimulating further research both on algorithmic variants, which obtain much better computational performance, and on applications to a large variety of different problems. In fact, there exists now a considerable number of applications obtaining world class performance on problems including the quadratic assignment, vehicle routing, sequential ordering, scheduling, routing in Internet-like networks, and so on [22,26,46,47,67,85]. Motivated by this success, the ACO metaheuristic has been proposed [32,33] as a common framework for the existing applications and algorithmic variants. Algorithms which follow the ACO metaheuristic will be called in the following ACO algorithms.

Current applications of ACO algorithms fall into the two important problem classes of static and dynamic combinatorial optimization problems. Static problems are those whose topology and costs do not change while the problems are being solved. This is the case, e.g., for the classic TSP, in which city locations and intercity distances do not change during the algorithm's run-time. In contrast, in dynamic problems the topology and costs can change while solutions are built. An example of such a problem is routing in telecommunications networks [26], in which traffic patterns change all the time. The ACO algorithms for solving these two classes of problems are very similar from a high-level perspective, but they differ significantly in implementation details. The ACO metaheuristic captures these differences and is general enough to comprise the ideas common to both application types.

The (artificial) ants in ACO implement a randomized construction heuristic which makes probabilistic decisions as a function of artificial pheromone trails and possibly available heuristic information based on the input data of the problem to be solved. As such, ACO can be interpreted as an extension of traditional construction heuristics which are readily available for many combinatorial optimization problems. Yet, an important difference with construction heuristics is the adaptation of the pheromone trails during algorithm execution to take into account the cumulated search experience.

The rest of this chapter is organized as follows. In Section 2, we briefly overview construction heuristics and local search algorithms. In Section 3, we define the type of problem to which the ACO metaheuristic applies, the ants' behavior, and the ACO metaheuristic itself. Section 4 outlines the inspiring biological analogy and describes the historical developments leading to ACO. In Section 5, we illustrate how the ACO metaheuristic can be applied to different types of problems and we give an overview of its successful applications. Section 6 discusses several issues arising in the application of the ACO metaheuristic; Section 7 reports on recent developments and in Section 8 we conclude indicating future research directions.

2 TRADITIONAL APPROXIMATION APPROACHES

Many important combinatorial optimization problems are hard to solve. The notion of problem hardness is captured by the theory of computational complexity [49,74] and for many important problems it is well known that they are $\mathcal{NP}$-hard, that is, the time needed to solve an instance in the worst case grows exponentially with instance size. Often, approximate algorithms are the only feasible way to obtain near optimal solutions at relatively low computational cost.

Most approximate algorithms are either *construction* algorithms or *local search* algorithms.[1] These two types of methods are significantly different, because construction algorithms work on partial solutions trying to extend these in the best possible way to complete problem solutions, while local search methods move in the search space of complete solutions.

---

[1]Other approximate methods are also conceivable. For example, when stopping exact methods, like Branch & Bound, before completion [4,58] (e.g., after some given time bound, or when some guarantee on the solution quality is obtained through the use of lower and upper bounds), we can convert exact algorithms into approximate ones.

2.1 Construction Algorithms

Construction algorithms build solutions to a problem under consideration in an incremental way starting with an empty initial solution and iteratively adding appropriate solution components without backtracking until a complete solution is obtained. In the simplest case, solution components are added in random order. Often better results are obtained if a heuristic estimate of the myopic benefit of adding solution components is taken into account. *Greedy construction heuristics* add at each step a solution component which achieves the maximal myopic benefit as measured by some heuristic information. An algorithmic outline of a greedy construction heuristic is given in Figure 9.1. The function `GreedyComponent` returns the solution component e with the best heuristic estimate. Solutions returned by greedy algorithms are typically of better quality than randomly generated solutions. Yet, a disadvantage of greedy construction heuristics is that they can generate only a very limited number of different solutions. Additionally, greedy decisions in early stages of the construction process strongly constrain the available possibilities at later stages, often causing very poor moves in the final phases of the solution construction

As an example, consider a greedy construction heuristic for the traveling salesman problem. In the TSP we are given a complete weighted graph $G = (\mathcal{N}, \mathcal{A})$ with $\mathcal{N}$ being the set of vertices, representing the cities, and $\mathcal{A}$ the set of edges fully connecting the vertices $\mathcal{N}$. Each edge is assigned a value d_{ij}, which is the length of edge $(i, j) \in \mathcal{A}$. The TSP is the problem of finding a minimal length Hamiltonian circuit of the graph, where an Hamiltonian circuit is a closed tour visiting exactly once each of the $n = |\mathcal{N}|$ vertices of G. For symmetric TSPs, the distances between the cities are independent of the direction of traversing the edges, that is, $d_{ij} = d_{ji}$ for every pair of vertices. In the more general asymmetric TSP (ATSP) at least for one pair of vertices i, j we have $d_{ij} \neq d_{ji}$.

A simple rule of thumb to build a tour is to start from some initial city and to always choose to go to the closest still unvisited city before returning to the start city. This algorithm is known as the *nearest neighbor* tour construction heuristic.

Figure 9.2 shows a tour returned by the nearest neighbor heuristic on TSP instance `att532`, taken from TSPLIB,[2] with 532 cities in the US. Noteworthy in this example is that there are a few very long links in the tour, leading to a strongly suboptimal solution.

procedure *Greedy Construction Heuristic*
  s_p = *empty solution*
  while s_p not_a_complete_solution **do**
    $e =$ `GreedyComponent(`s_p`)`
    $s_p = s_p \otimes e$
  end
  return s_p
end *Greedy Construction Heuristic*

Figure 9.1. Algorithmic skeleton of a greedy construction heuristic. The addition of component e to a partial solution s_p is denoted by the operator $\otimes$.

[2]TSPLIB is a benchmark library for the TSP and related problems and is accessible via http://www.iwr.uni-heidelberg.de/iwr/comopt/software/TSPLIB95.

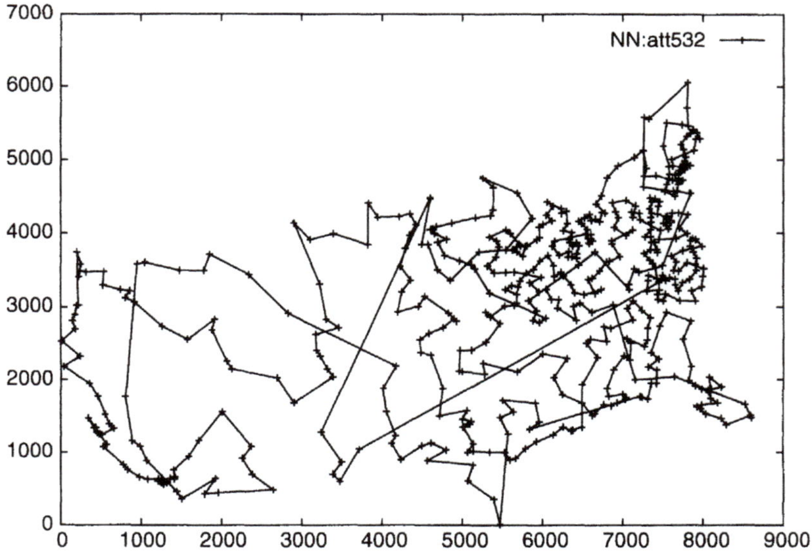

Figure 9.2. Tour returned by the nearest neighbor heuristic on TSP instance `att532` from TSPLIB.

In fact, construction algorithms are typically the fastest approximate methods, but the solutions they generate often are not of a very high quality and they are not guaranteed to be optimal with respect to small changes; the results produced by constructive heuristics can often be improved by local search algorithms.

2.2 Local Search

Local search algorithms start from a complete initial solution and try to find a better solution in an appropriately defined *neighborhood* of the current solution. In its most basic version, known as *iterative improvement*, the algorithm searches the neighborhood for an improving solution. If such a solution is found, it replaces the current solution and the local search continues. These steps are repeated until no improving neighbor solution can be found in the neighborhood of the current solution and the algorithm ends in a *local optimum*. An outline of an iterative improvement algorithm is given in Figure 9.3. The procedure `Improve` returns a better neighbor solution if one exists, otherwise it returns the current solution, in which case the algorithm stops.

The choice of an appropriate neighborhood structure is crucial for the performance of the local search algorithm and has to be done in a problem specific way. The neighborhood structure defines the set of solutions that can be reached from s in one single step of a local search algorithm. An example neighborhood for the TSP is the *k-change* neighborhood in which neighbor solutions differ by at most k edges. Figure 9.4 shows an example of a 2-change neighborhood. The 2-change algorithm systematically tests whether the current tour can be improved by replacing two edges. To fully specify a local search algorithm it is necessary to designate a particular neighborhood examination scheme that defines how the neighborhood is searched and which neighbor

```
procedure IterativeImprovement (s ∈ S)
   s' = Improve(s)
   while s' ≠ s do
      s = s'
      s' = Improve(s)
   end
   return s
end IterativeImprovement
```

Figure 9.3. Algorithmic skeleton of iterative improvement.

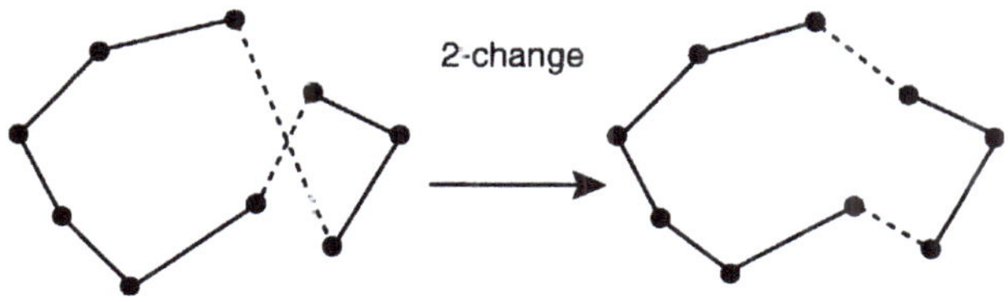

Figure 9.4. Schematic illustration of a 2-change move. The proposed move reduces the total tour length if we consider the Euclidean distance between the points.

solution replaces the current one. In the case of iterative improvement algorithms, this rule is called the *pivoting rule* [93] and examples are the *best-improvement* rule, which chooses the neighbor solution giving the largest improvement of the objective function, and the *first-improvement* rule, which uses the first improved solution found in the neighborhood to replace the current one. A common problem with local search algorithms is that they easily get trapped in local minima and that the result strongly depends on the initial solution.

3 THE ACO METAHEURISTIC

Artificial ants used in ACO are stochastic solution construction procedures that probabilistically build a solution by iteratively adding solution components to partial solutions by taking into account (i) heuristic information on the problem instance being solved, if available, and (ii) (artificial) pheromone trails which change dynamically at run-time to reflect the agents' acquired search experience.

A stochastic component in ACO allows the ants to build a wide variety of different solutions and hence explore a much larger number of solutions than greedy heuristics. At the same time, the use of heuristic information, which is readily available for many problems, can guide the ants towards the most promising solutions. More important, the ants' search experience can be used to influence, in a way reminiscent of reinforcement learning [89], the solution construction in future iterations of the algorithm. Additionally, the use of a colony of ants can give the algorithm increased robustness and in many ACO applications the collective interaction of a population of agents is needed to efficiently solve a problem.

The domain of application of ACO algorithms is vast. In principle, ACO can be applied to any discrete optimization problem for which some solution construction mechanism can be conceived. In the following of this section, we first define a generic problem representation which the ants in ACO exploit to construct solutions, then we detail the ants' behavior while constructing solutions, and finally we define the ACO metaheuristic.

3.1 Problem Representation

Let us consider the minimization problem[3] $(\mathcal{S}, f, \Omega)$, where $\mathcal{S}$ is the *set of candidate solutions*, f is the *objective function* which assigns to each candidate solution $s \in \mathcal{S}$ an objective function (cost) value $f(s, t)$,[4] and Ω is a set of constraints. The goal is to find a *globally optimal* solution $s_{opt} \in \mathcal{S}$, that is, a minimum cost solution that satisfies the constraints Ω.

The problem representation of a combinatorial optimization problem $(\mathcal{S}, f, \Omega)$ which is exploited by the ants can be characterized as follows:

- A finite set $\mathcal{C} = \{c_1, c_2, \ldots, c_{N_C}\}$ of *components* is given.
- The *states* of the problem are defined in terms of sequences $x = \langle c_i, c_j, \ldots, c_k, \ldots \rangle$ over the elements of $\mathcal{C}$. The set of all possible sequences is denoted by $\mathcal{X}$. The length of a sequence x, that is, the number of components in the sequence, is expressed by $|x|$. The maximum length of a sequence is bounded by a positive constant $n < +\infty$.
- The set of (candidate) solutions $\mathcal{S}$ is a subset of $\mathcal{X}$ (i.e., $\mathcal{S} \subseteq \mathcal{X}$).
- The finite set of *constraints* Ω defines the set of feasible states $\tilde{\mathcal{X}}$, with $\tilde{\mathcal{X}} \subseteq \mathcal{X}$.
- A non-empty set $\mathcal{S}^*$ of feasible solutions is given, with $\mathcal{S}^* \subseteq \tilde{\mathcal{X}}$ and $\mathcal{S}^* \subseteq \mathcal{S}$.
- A *cost* $f(s, t)$ is associated to each candidate solution $s \in \mathcal{S}$.
- In some cases a cost, or the estimate of a cost, $J(x_i, t)$ can be associated to states other than solutions. If x_j can be obtained by adding solution components to a state x_i then $J(x_i, t) \leq J(x_j, t)$. Note that $J(s, t) \equiv f(s, t)$.

Given this representation, artificial ants build solutions build solutions by moving on the construction graph $G = (\mathcal{C}, \mathcal{L})$, where the vertices are the components $\mathcal{C}$ and the set $\mathcal{L}$ fully connects $\mathcal{C}$ (elements of $\mathcal{L}$ are called *connections*). The problem constraints Ω are implemented in the policy followed by the artificial ants build solutions, as explained in the next section. The choice of implementing the constraints in the construction policy of the artificial ants allows a certain degree of flexibility. In fact, depending on the combinatorial optimization problem considered, it may be more reasonable to implement constraints in a hard way allowing ants to build only feasible solutions, or in a soft way, in which case ants can build infeasible solutions (that is, candidate solutions in $\mathcal{S} \setminus \mathcal{S}^*$) that will be penalized, depending on their degree of infeasibility.

3.2 Ant's Behavior

Ants can be characterized as stochastic construction procedures which build solutions moving on the construction graph $G = (\mathcal{C}, \mathcal{L})$. Ants do not move arbitrarily on G,

---

[3] The adaptation to a maximization problem is straightforward.

[4] The parameter t indicates that the the objective function can be time dependent, as it is the case in applications to dynamic problems.

but rather follow a construction policy which is a function of the problem constraints Ω. In general, ants try to build feasible solutions, but, if necessary, they can generate infeasible solutions. Components $c_i \in \mathcal{C}$ and connections $l_{ij} \in \mathcal{L}$ can have associated a *pheromone trail* τ (τ_i if associated to components, τ_{ij} if associated to connections) encoding a long-term memory about the whole ant search process that is updated by the ants themselves, and a *heuristic value* η (η_i and η_{ij}, respectively) representing a priori information about the problem instance definition or run-time information provided by a source different from the ants. In many cases η is the cost, or an estimate of the cost, of extending the current state. These values are used by the ants' heuristic rule to make probabilistic decisions on how to move on the graph.

More precisely, each ant k of the colony has the following properties:

- It exploits the graph $G = (\mathcal{C}, \mathcal{L})$ to search for feasible solutions s of minimum cost. That is, solutions s such that $\hat{f}_s = \min_s f(s, t)$.
- It has a memory $\mathcal{M}^k$ that it uses to store information about the path it followed so far. Memory can be used (i) to build feasible solutions (i.e., to implement constraints Ω), (ii) to evaluate the solution found, and (iii) to retrace the path backward to deposit pheromone.
- It can be assigned a *start state* x_s^k and one or more *termination conditions* e^k. Usually, the start state is expressed either as a unit length sequence (that is, a single component sequence), or an empty sequence.
- When in state $x_r = \langle x_{r-1}, i \rangle$ it tries to move to any vertex j in its *feasible neighborhood* $\mathcal{N}_i^k$, that is, to a state $\langle x_r, j \rangle \in \tilde{\mathcal{X}}$. If this is not possible, then the ant might be allowed to move to a vertex j in its infeasible neighborhood $\mathcal{IN}_i^k$, generating in this way an infeasible state (that is, a state $\langle x_r, j \rangle \in \mathcal{X} \setminus \tilde{\mathcal{X}}$.
- It selects the move by applying a probabilistic decision rule. Its probabilistic decision rule is a function of (i) locally available pheromone trails and heuristic values, (ii) the ant's private memory storing its past history, and (iii) the problem constraints.
- The construction procedure of ant k stops when at least one of the termination conditions e^k is satisfied. Examples of termination conditions are when a solution is completed, or when, if building infeasible solutions is not allowed, there are no feasible states reachable from the ant current state.
- When adding a component c_j to the current solution it can update the pheromone trail associated to it or to the corresponding connection. This is called *online step-by-step pheromone update*.
- Once built a solution, it can retrace the same path backward and update the pheromone trails of the used components or connections. This is called *online delayed pheromone update*.

It is important to note that ants move concurrently and independently and that each ant is complex enough to find a (probably poor) solution to the problem under consideration. Typically, good quality solutions emerge as the result of the collective interaction among the ants which is obtained via indirect communication mediated by the information ants read/write in the variables storing pheromone trail values. In a way, this is a distributed learning process in which the single agents, the ants, are not adaptive themselves but, on the contrary, they adaptively modify the way the problem is represented and perceived by other ants.

3.3 The Metaheuristic

Informally, the behavior of ants in an ACO algorithm can be summarized as follows. A colony of ants concurrently and asynchronously move through adjacent states of the problem by building paths on G. They move by applying a stochastic local decision policy that makes use of pheromone trails and heuristic information. By moving, ants incrementally build solutions to the optimization problem. Once an ant has built a solution, or while the solution is being built, the ant evaluates the (partial) solution and deposits pheromone trails on the components or connections it used. This pheromone information will direct the search of future ants.

Besides ants' activity, an ACO algorithm includes two additional procedures: *pheromone trail evaporation* and *daemon actions* (this last component being optional). Pheromone evaporation is the process by means of which the pheromone deposited by previous ants decreases over time. From a practical point of view, pheromone evaporation is needed to avoid a too rapid convergence of the algorithm towards a sub-optimal region. It implements a useful form of *forgetting*, favoring the exploration of new areas of the search space. Daemon actions can be used to implement centralized actions which cannot be performed by single ants. Examples are the activation of a local optimization procedure, or the collection of global information that can be used to decide whether it is useful or not to deposit additional pheromone to bias the search process from a non-local perspective. As a practical example, the daemon can observe the path found by each ant in the colony and choose to deposit extra pheromone on the components used by the ant that built the best solution. Pheromone updates performed by the daemon are called *off-line pheromone updates*.

In Figure 9.5 the ACO metaheuristic behavior is described in pseudo-code. The main procedure of the ACO metaheuristic manages, via the **ScheduleActivities** construct, the scheduling of the three above discussed components of ACO algorithms: (i) management of ants' activity, (ii) pheromone evaporation, and (iii) daemon actions. The **ScheduleActivities** construct does not specify how these three activities are scheduled and synchronized. In other words, it does not say whether they should be executed in a completely parallel and independent way, or if some kind of synchronization

```
procedure ACO metaheuristic
    ScheduleActivities
        ManageAntsActivity()
        EvaporatePheromone()
        DaemonActions() {Optional}
    end ScheduleActivities
end ACO metaheuristic
```

Figure 9.5. The ACO metaheuristic in pseudo-code. Comments are enclosed in braces. The procedure `DaemonActions()` is optional and refers to centralized actions executed by a daemon possessing global knowledge.

among them is necessary. The designer is therefore free to specify the way these three procedures should interact.

4 HISTORY OF ACO ALGORITHMS

The first ACO algorithm proposed was Ant System (AS). AS was applied to some rather small instances of the traveling salesman problem (TSP) with up to 75 cities. It was able to reach the performance of other general-purpose heuristics like evolutionary computation [30,38]. Despite these initial encouraging results, AS did not prove to be competitive with state-of-the-art algorithms specifically designed for the TSP when attacking large instances. Therefore, a substantial amount of recent research has focused on ACO algorithms which show better performance than AS when applied, for example, to the TSP. In the following of this section we first briefly introduce the biological metaphor on which AS and ACO are inspired, and then we present a brief history of the developments that have led from the original AS to the most recent ACO algorithms. In fact, these more recent algorithms are direct extensions of AS which add advanced features to improve the algorithm performance.

4.1 Biological Analogy

In many ant species, individual ants may deposit a pheromone (a particular chemical that ants can smell) on the ground while walking [23,52]. By depositing pheromone they create a trail that is used, e.g., to mark the path from the nest to food sources and back. In fact, by sensing pheromone trails foragers can follow the path to food discovered by other ants. Also, they are capable of exploiting pheromone trails to choose the shortest among the available paths leading to the food.

Deneubourg and colleagues [23,52] used a double bridge connecting a nest of ants and a food source to study pheromone trail laying and following behavior in controlled experimental conditions.[5] They ran a number of experiments in which they varied the ratio between the length of the two branches of the bridge. The most interesting, for our purposes, of these experiments is the one in which one branch was longer than the other. In this experiment, at the start the ants were left free to move between the nest and the food source and the percentage of ants that chose one or the other of the two branches was observed over time. The outcome was that, although in the initial phase random oscillations could occur, in most experiments all the ants ended up using the shorter branch.

This result can be explained as follows. When a trial starts there is no pheromone on the two branches. Hence, the ants do not have a preference and they select with the same probability either of the two branches. Therefore, it can be expected that, on average, half of the ants choose the short branch and the other half the long branch, although stochastic oscillations may occasionally favor one branch over the other. However, because one branch is shorter than the other, the ants choosing the short branch are the first to reach the food and to start their travel back to the nest.[6] But then, when they must

[5]The experiment described was originally executed using a laboratory colony of Argentine ants (*Iridomyrmex humilis*). It is known that these ants deposit pheromone both when leaving and when returning to the nest [52].

[6]In the ACO literature this is often called *differential path length* effect.

make a decision between the short and the long branch, the higher level of pheromone on the short branch biases their decision in its favor.[7] Therefore, pheromone starts to accumulate faster on the short branch which will eventually be used by the great majority of the ants.

It should be clear by now how real ants have inspired AS and later algorithms: the double bridge was substituted by a graph, and pheromone trails by artificial pheromone trails. Also, because we wanted artificial ants to solve problems more complicated than those solved by real ants, we gave artificial ants some extra capacities, like a memory (used to implement constraints and to allow the ants to retrace their path back to the nest without errors) and the capacity for depositing a quantity of pheromone proportional to the quality of the solution produced (a similar behavior is observed also in some real ants species in which the quantity of pheromone deposited while returning to the nest from a food source is proportional to the quality of the food source found [3]).

In the next section we will see how, starting from AS, new algorithms have been proposed that, although retaining some of the original biological inspiration, are less and less biologically inspired and more and more motivated by the need of making ACO algorithms competitive with or indeed better than other state-of-the-art algorithms. Nevertheless, many aspects of the original Ant System remain: the need for a colony, the role of autocatalysis, the cooperative behavior mediated by artificial pheromone trails, the probabilistic construction of solutions biased by artificial pheromone trails and local heuristic information, the pheromone updating guided by solution quality, and the evaporation of pheromone trail, are present in all ACO algorithms. It is interesting to note that there is one well known algorithm that, although making use in some way of the ant foraging metaphor, cannot be considered an instance of the Ant Colony Optimization metaheuristic. This is HAS-QAP, proposed in [48], where pheromone trails are not used to guide the solution construction phase; on the contrary, they are used to guide modifications of complete solutions in a local search style. This algorithm belongs nevertheless to *ant algorithms*, a new class of algorithms inspired by a number of different behaviors of social insects. Ant algorithms are receiving increasing attention in the scientific community (see, e.g., [8,9,11,31]) as a promising novel approach to distributed control and optimization.

4.2 Historical Development

As we said, AS was the first example of an ACO algorithm to be proposed in the literature. In fact, AS was originally a set of three algorithms called *ant-cycle*, *ant-density*, and *ant-quantity*. These three algorithms were proposed in Dorigo's doctoral dissertation [30] and first appeared in a technical report [36,37] that was published a few years later in the IEEE Transactions on Systems, Man, and Cybernetics [38]. Another early publication is [17].

While in *ant-density* and *ant-quantity* the ants updated the pheromone directly after a move from a city to an adjacent one, in *ant-cycle* the pheromone update was only done after all the ants had constructed the tours and the amount of pheromone deposited by each ant was set to be a function of the tour quality. Because *ant-cycle* performed better than the other two variants, it was later called simply Ant System (and in fact, it

[7] A process like this, in which a decision taken at time t increases the probability of making the same decision at time $T > t$ is said to be an *autocatalytic* process. Autocatalytic processes exploit *positive feedback*.

is the algorithm that we will present in the following subsection), while the other two algorithms were no longer studied.

The major merit of AS, whose computational results were promising but not competitive with other more established approaches, was to stimulate a number of researchers, mostly in Europe, to develop extensions and improvements of its basic ideas so as to produce better performing, and often state-of-the-art, algorithms. It is following the successes of this collective undertaking that recently Dorigo and Di Caro [32] made the synthesis effort that led to the definition of the ACO metaheuristic presented in this chapter (see also [33]). In other words, the ACO metaheuristic was defined *a posteriori* with the goal of providing a common characterization of a new class of algorithms and a reference framework for the design of new instances of ACO algorithms.

4.2.1 The First ACO Algorithm: Ant System and the TSP

The traveling salesman problem (TSP) is a paradigmatic $\mathcal{NP}$-hard combinatorial optimization problem which has attracted an enormous amount of research effort [57,60,76]. The TSP is a very important problem also in the context of Ant Colony Optimization because it is the problem to which the original AS was first applied [30,36,38], and it has later often been used as a benchmark to test new ideas and algorithmic variants.

In AS each ant is initially put on a randomly chosen city and has a memory which stores the partial solution it has constructed so far (initially the memory contains only the start city). Starting from its start city, an ant iteratively moves from city to city. When at a city i, an ant k chooses to go to an as yet unvisited city j with a probability given by

$$p_{ij}^{k}(t) = \frac{[\tau_{ij}(t)]^{\alpha} \cdot [\eta_{ij}]^{\beta}}{\sum_{l \in \mathcal{N}_i^k} [\tau_{il}(t)]^{\alpha} \cdot [\eta_{il}]^{\beta}} \quad \text{if } j \in \mathcal{N}_i^k \tag{9.1}$$

where $\eta_{ij} = 1/d_{ij}$ is *a priori* available heuristic information, α and β are two parameters which determine the relative influence of pheromone trail and heuristic information, and $\mathcal{N}_i^k$ is the feasible neighborhood of ant k, that is, the set of cities which ant k has not yet visited. Parameters α and β have the following influence on the algorithm behavior. If $\alpha = 0$, the selection probabilities are proportional to $[\eta_{ij}]^{\beta}$ and the closest cities will more likely be selected: in this case AS corresponds to a classical stochastic greedy algorithm (with multiple starting points since ants are initially randomly distributed on the cities). If $\beta = 0$, only pheromone amplification is at work: this will lead to the rapid emergence of a *stagnation* situation with the corresponding generation of tours which, in general, are strongly suboptimal [30]. (Search stagnation is defined in [38] as the situation where all the ants follow the same path and construct the same solution.)

The solution construction ends after each ant has completed a tour, that is, after each ant has constructed a sequence of length n. Next, the pheromone trails are updated. In AS this is done by first lowering the pheromone trails by a constant factor (this is pheromone evaporation) and then allowing each ant to deposit pheromone on the edges that belong to its tour:

$$\forall (i,j) \quad \tau_{ij}(t+1) = (1-\rho) \cdot \tau_{ij}(t) + \sum_{k=1}^{m} \Delta\tau_{ij}^{k}(t) \tag{9.2}$$

where $0 < \rho \leq 1$ is the pheromone trail evaporation rate and m is the number of ants. The parameter ρ is used to avoid unlimited accumulation of the pheromone trails and enables the algorithm to "forget" previous "bad" decisions. On edges which are not chosen by the ants, the associated pheromone strength will decrease exponentially with the number of iterations. $\Delta\tau_{ij}^k(t)$ is the amount of pheromone ant k deposits on the edges; it is defined as

$$\Delta\tau_{ij}^k(t) = \begin{cases} 1/L^k(t) & \text{if edge } (i, j) \text{ is used by ant } k \\ 0 & \text{otherwise} \end{cases} \tag{9.3}$$

where $L^k(t)$ is the length of the kth ant's tour. By Equation 9.3, the shorter the ant's tour is, the more pheromone is received by edges belonging to the tour.[8] In general, edges which are used by many ants and which are contained in shorter tours will receive more pheromone and therefore are also more likely to be chosen in future iterations of the algorithm.

4.2.2 Ant System and its Extensions

As previously stated, AS was not competitive with state-of-the-art algorithms for TSP. Researchers then started to extend it to try to improve its performance.

A first improvement, called the *elitist strategy*, was introduced in [30,38]. It consists in giving the best tour since the start of the algorithm (called T^{gb}, where gb stands for global-best) a strong additional weight. In practice, each time the pheromone trails are updated, those belonging to the edges of the global-best tour get an additional amount of pheromone. For these edges Equation 9.3 becomes:

$$\Delta\tau_{ij}^{gb}(t) = \begin{cases} e/L^{gb}(t) & \text{if edge } (i, j) \in T^{gb} \\ 0 & \text{otherwise} \end{cases} \tag{9.4}$$

The edges of T^{gb} are therefore reinforced with a quantity of pheromone given by e/L^{gb}, where L^{gb} is the length of T^{gb} and e is a positive integer. Note that this type of pheromone update is a first example of daemon action as described in Section 3.3.

Other improvements were the rank-based version of Ant System (AS$_{\text{rank}}$), $\mathcal{MAX}$–$\mathcal{MIN}$ Ant System ($\mathcal{MM}$AS), and Ant Colony System (ACS). AS$_{\text{rank}}$ [14] is in a sense an extension of the elitist strategy: it sorts the ants according to the lengths of the tours they generated and, after each tour construction phase, only the $(w - 1)$ best ants and the global-best ant are allowed to deposit pheromone. The rth best ant of the colony contributes to the pheromone update with a weight given by $\max\{0, w - r\}$ while the global-best tour reinforces the pheromone trails with weight w. Equation 9.2 becomes therefore:

$$\forall (i, j) \quad \tau_{ij}(t + 1) = (1 - \rho) \cdot \tau_{ij}(t) + \sum_{r=1}^{w-1} (w - r) \cdot \Delta\tau_{ij}^r(t) + w \cdot \Delta\tau_{ij}^{gb}(t) \tag{9.5}$$

where $\Delta\tau_{ij}^r(t) = 1/L^r(t)$ and $\Delta\tau_{ij}^{gb}(t) = 1/L^{gb}$.

ACS [34,35,44] improves over AS by increasing the importance of exploitation of information collected by previous ants with respect to exploration of the search

---

[8]Note that when applied to symmetric TSPs the edges are considered to be bidirectional and edges (i, j) and (j, i) are both updated. This is different for the ATSP, where edges are directed; an ant crossing edge (i, j) will update only this edge and not the edge (j, i).

space.[9] This is achieved via two mechanisms. First, a strong elitist strategy is used to update pheromone trails. Second, ants choose the next city to move to using a so-called *pseudo-random proportional* rule [35]: with probability $q_0, 0 \leq q_0 \leq 1$, they move to the city j for which the product between pheromone trail and heuristic information is maximum, that is, $j = \arg\max_{j \in \mathcal{N}_i^k} \{\tau_{ij}(t) \cdot \eta_{ij}^{\beta}\}$, while with probability $1 - q_0$ they operate a biased exploration in which the probability $p_{ij}^k(t)$ is the same as in AS (see Equation 9.1). The value q_0 is a parameter: when it is set to a value close to 1, as it is the case in most ACS applications, exploitation is favored over exploration. Obviously, when $q_0 = 0$ the probabilistic decision rule becomes the same as in AS.

As mentioned earlier, pheromone updates are performed using a strong elitist strategy: only the ant that has produced the best solution is allowed to update pheromone trails, according to a pheromone trail update rule similar to that used in AS:

$$\tau_{ij}(t+1) = (1-\rho) \cdot \tau_{ij}(t) + \rho \cdot \Delta\tau_{ij}^{\text{best}}(t) \tag{9.6}$$

The best ant can be the *iteration-best* ant, that is, the best in the current iteration, or the *global-best* ant, that is, the ant that made the best tour from the start of the trial.

Finally, ACS differs from previous ACO algorithms also because ants update the pheromone trails while building solutions (as in *ant-quantity* and in *ant-density*). In practice, ACS ants "eat" some of the pheromone trail on the edges they visit. This has the effect of decreasing the probability that the same path is used by all the ants (i.e., it favors exploration, counterbalancing this way the other two above-mentioned modifications that strongly favor exploitation of the collected knowledge about the problem). ACS has been improved also by the addition of local search routines that take the solution generated by ants to their local optimum just before the pheromone update.

$\mathcal{MM}$AS [84,87,88] introduces upper and lower bounds to the values of the pheromone trails, as well as a different initialization of their values. In practice, in $\mathcal{MM}$AS the allowed range of the pheromone trail strength is limited to the interval $[\tau_{\min}, \tau_{\max}]$, that is, $\tau_{\min} \leq \tau_{ij} \leq \tau_{\max}\ \forall\tau_{ij}$, and the pheromone trails are initialized to the upper trail limit, which causes a higher exploration at the start of the algorithm. Also, as in ACS, in $\mathcal{MM}$AS only the best ant (the global-best or the iteration-best ant) is allowed to add pheromone after each algorithm iteration. In fact, in $\mathcal{MM}$AS the iteration-best ant and the global-best ant can be used alternatingly in the pheromone update. Computational results have shown that best results are obtained when pheromone updates are performed using the global-best solution with increasing frequency during the algorithm execution. Similarly to ACS, also $\mathcal{MM}$AS often exploits local search to improve its performance.

4.2.3 Applications to Dynamic Problems

The application of ACO algorithms to dynamic problems, that is, problems whose characteristics change while being solved, is the most recent major development in the field. The first such application [79] concerned routing in circuit-switched networks

[9] ACS was an offspring of Ant-Q [43], an algorithm intended to create a link between reinforcement learning [89] and Ant Colony Optimization. Computational experiments have shown that some aspects of Ant-Q, in particular the pheromone update rule, could be strongly simplified without affecting performance. It is for this reason that Ant-Q was abandoned in favor of the simpler and equally good ACS.

(e.g., classical telephone networks). The proposed algorithm, called ABC, was demonstrated on a simulated version of the British Telecom network. The main merit of ABC was to stimulate the interest of ACO researchers in dynamic problems. In fact, only rather limited comparisons were made between ABC and state-of-the-art algorithms, so that it is not possible to judge on the quality of the results obtained.

A very successful application of ACO to dynamic problems is the AntNet algorithm, proposed by Di Caro and Dorigo [24–26,28] and discussed in Section 5. AntNet was applied to routing in packet-switched networks (e.g., the Internet). It contains a number of innovations with respect to AS and it has been shown experimentally to outperform a whole set of state-of-the-art algorithms on numerous benchmark problems.

5 EXAMPLES OF APPLICATIONS

The versatility and the practical use of the ACO metaheuristic for the solution of combinatorial optimization problems is best illustrated via example applications to a number of different problems.

The ACO application to the TSP has already been presented in the previous section. Here, we additionally discuss applications to three $\mathcal{NP}$-hard optimization problems, the single machine total weighted tardiness problem (SMTWTP), the generalized assignment problems (GAP), and the set covering problem (SCP). We have chosen these problems to make the application examples as comprehensive as possible with respect to different ways of representing solutions. While the TSP and the SMTWTP are permutation problems, that is, solutions are represented as permutations of solution components, solutions in the GAP are assignments of tasks to agents and in the SCP a solution is represented as a subset of the available solution components.

Applications of ACO to dynamic problems focus mainly on routing in data networks. As an example, in the following we present the AntNet algorithm [26].

Example 9.1. *The single machine total weighted tardiness scheduling problem (SMTWTP)*

In the SMTWTP n jobs have to be processed sequentially without interruption on a single machine. Each job has an associated processing time p_j, a weight w_j, and a due date d_j and all jobs are available for processing at time zero. The tardiness of job j is defined as $T_j = \max\{0, C_j - d_j\}$, where C_j is its completion time in the current job sequence. The goal in the SMTWTP is to find a job sequence which minimizes the sum of the weighted tardiness given by $\sum_{i=1}^{n} w_i \cdot T_i$.

For the ACO application to the SMTWTP, the set of components C is the set of all jobs. As in the TSP case, the states of the problem are all possible partial sequences. In the SMTWTP case we do not have explicit costs associated with the connections because the objective function contribution of each job depends on the partial solution constructed so far.

The SMTWTP was attacked in [22] using ACS (ACS-SMTWTP). In ACS-SMTWTP, each ant starts with an empty sequence and then iteratively appends an unscheduled job to the partial sequence constructed so far. Each ant chooses the next job using the *pseudo-random-proportional* action choice rule, where at each step the feasible neighborhood $\mathcal{N}_i^k$ of ant k is the list of as yet unscheduled jobs. Pheromone trails are defined as follows: τ_{ij} refers to the desirability of scheduling job j at position i. This definition of the pheromone trails is, in fact, used in most ACO application to scheduling problems [2,22,67,82]. Concerning the heuristic information, in [22] the use of three priority rules allowed to define three different types of heuristic information for the SMTWTP. The investigated priority rules were: (i) the earliest due date rule, which puts the jobs in non-decreasing order of the due dates d_j, (ii) the modified due date rule which puts the jobs in non-decreasing order of the modified due dates given by $mdd_j = \max\{C + p_j, d_j\}$ [2], where C is the sum of the processing times of the already sequenced jobs, and (iii) the apparent urgency rule which puts the jobs in non-decreasing order of the apparent urgency [72], given by $au_j = (w_j/p_j) \cdot \exp(-(\max\{d_j - C_j, 0\})/k\bar{p})$, where k is a parameter of the priority rule. In each case, the heuristic information was defined as $\eta_{ij} = 1/h_j$, where h_j is either d_j, mdd_j, or au_j, depending on the priority rule used.

The global and the local pheromone updates are carried out as in the standard ACS described in Section 4.2, where in the global pheromone update, T^{gb} is the total weighted tardiness of the global best solution.

In [22], ACS-SMTWTP was combined with a powerful local search algorithm. The final ACS algorithm was tested on a benchmark set available from ORLIB at `http://www.ms.ic.ac.uk/info.html`. Within the computation time limits given,[10] ACS reached a very good performance and could find in each single run the optimal or best known solutions on all instances of the benchmark set. For more details on the computational results we refer to [22].

Example 9.2. *The generalized assignment problem (GAP)*

In the GAP a set of tasks $\mathcal{I}, i = 1, \ldots, n$, has to be assigned to a set of agents $\mathcal{J}, j = 1, \ldots, m$. Each agent j has only a limited capacity a_j and each task i consumes, when assigned to agent j, a quantity b_{ij} of the agent's capacity. Also, the cost d_{ij} of assigning task i to agent j is given. The objective then is to find a feasible task assignment with minimal cost.

Let y_{ij} be one if task i is assigned to agent j and zero otherwise. Then the GAP can formally be defined as

$$\min f(y) = \sum_{i=1}^{n} \sum_{j=1}^{m} d_{ij} \cdot y_{ij} \tag{9.7}$$

subject to

$$\sum_{i=1}^{n} b_{ij} \cdot y_{ij} \leq a_j \quad j = 1, \ldots, m \tag{9.8}$$

$$\sum_{j=1}^{m} y_{ij} = 1 \quad i = 1, \ldots, n \tag{9.9}$$

$$y_{ij} \in \{0, 1\} \quad i = 1 \ldots, n; \quad j = 1, \ldots, m \tag{9.10}$$

The constraints 9.8 implement the limited resource capacity of the agents, while constraints 9.9 and 9.10 impose that each task is assigned to exactly one agent and that a task cannot be split among several agents.

[10]The maximum time for the largest instances was 20 min on a 450 MHz Pentium III PC with 256 MB RAM. Programs were written in C++ and the PC was run under Red Hat Linux 6.1.

The GAP can easily be cast into the framework of the ACO metaheuristic. The problem can be represented by a graph in which the set of components comprises the set of tasks and agents, that is $\mathcal{C} = \mathcal{I} \cup \mathcal{J}$ and the set of connections fully connect the graph. Each assignment, which consists of n couplings (i, j) of tasks and agents, corresponds to an ant's walk on this graph. Such a walk has to observe the constraints 9.9 and 9.10 to obtain a valid assignment. One particular way of generating such an assignment is by an ant's walk which iteratively switches from task vertices (vertices in the set $\mathcal{J}$) to agent vertices (vertices in the set $\mathcal{I}$) without repeating any task vertex but possibly using the same agent vertex several times (that is, several tasks may be assigned to the same agent).

At each step of the construction process, an ant has to make one of the following two basic decisions: (i) it has to decide which task to assign next and (ii) it has to decide to which agent a chosen task should be assigned. Pheromone trail and heuristic information can be associated with both steps. With respect to the first step the pheromone information can be used to learn an appropriate assignment order of the tasks, that is, τ_{ij} gives the desirability of assigning task j directly after task i, while the pheromone information in the second step is associated with the desirability of assigning a task to a specific agent.

For simplicity let us consider an approach in which the tasks are assigned in a random order. Then, at each step a task has to be assigned to an agent. Intuitively, it is better to assign tasks to agents such that small assignment costs are incurred and that the agent needs only a relatively small amount of its available resource to perform the task. Hence, one possible heuristic information is $\eta_{ij} = a_j / d_{ij} b_{ij}$ and a probabilistic selection of the assignments can follow the AS probabilistic rule (Equation 9.1) or the pseudo-random proportional rule of ACS. Yet, a complication in the construction process is that the GAP involves resource capacity constraints and, in fact, for the GAP no guarantee is given that an ant will construct a feasible solution which obeys the resource constraints given by Equation 9.8. In fact, to have a bias towards generating feasible solutions, the resource constraints should be taken into account in the definition of $\mathcal{N}_i^k$, the feasible neighborhood of ant k. For the GAP, we define $\mathcal{N}_i^k$ to consist of all those agents to which the task i can be assigned without violating the agents' resource capacity. If no agent can meet the task's resource requirement, we are forced to build an infeasible solution and in this case we can simply choose $\mathcal{N}_i^k$ as the set of all agents. Infeasibilites can then be handled, for example, by assigning penalties proportional to the amount of resource violations.

A first application of $\mathcal{MAX}$–$\mathcal{MIN}$ Ant System ($\mathcal{MM}$AS) to the GAP was presented in [75]. The approach shows some particularities, like the use of a single ant and the lack of any heuristic information. The infeasibility of solutions is only treated in the pheromone update: the amount of pheromone deposited by an ant is set to a high value if the solution it generated is feasible, to a low value if it is infeasible. These values are constants independent of the solution quality. Additionally, $\mathcal{MM}$AS was coupled with a local search based on tabu search and ejection chain elements [51] and it obtained very good performance on benchmark instances available at ORLIB.

Example 9.3. *The set covering problem (SCP)*

In the set covering problem (SCP) we are given a $m \times n$ matrix $A = (a_{ji})$ in which all the matrix elements are either zero or one. Additionally, each column is given a non-negative cost b_i. We say that a column i covers a row j if $a_{ji} = 1$. The goal in the

SCP is to choose a subset of the columns of minimal weight that covers every row. Let $\mathcal{I}$ denote a subset of the columns and y_i be a binary variable which is one, if $i \in \mathcal{I}$, and zero otherwise. The SCP can be defined formally as follows.

$$\min f(y) = \sum_{i=1}^{n} b_i \cdot y_i \tag{9.11}$$

subject to

$$\sum_{i=1}^{n} a_{ji} \cdot y_i \geq 1 \quad j = 1, \ldots, m \tag{9.12}$$

$$y_i \in \{0, 1\} \quad i = 1, \ldots, n \tag{9.13}$$

The constraints 9.12 enforce that each row is covered by at least one column.

ACO can be applied in a very straightforward way to the SCP. The columns are chosen as the solution components and have associated a cost and a pheromone trail. The constraints say that each column can be visited by an ant at most once and that a final solution has to cover all rows. A walk of an ant over the graph representation corresponds to the iterative addition of columns to the partial solution obtained so far. Each ant starts with an empty solution and adds columns until a cover is completed. A pheromone trail τ_i and a heuristic information η_i are associated to each column i. A column to be added is chosen with probability

$$p_i^k(t) = \frac{[\tau_i(t)] \cdot [\eta_i]^\beta}{\sum_{l \in \mathcal{N}^k} [\tau_l(t)] \cdot [\eta_l]^\beta} \quad \text{if } i \in \mathcal{N}^k \tag{9.14}$$

where $\mathcal{N}^k$ is the feasible neighborhood of ant k which consists of all columns which cover at least one still uncovered row. The heuristic information η_i can be chosen in several different ways. For example, a simple static information could be used, taking into account only the column cost: $\eta_i = 1/b_i$. A more sophisticate approach would be to consider the total number of rows d_i covered by a column i and to set $\eta_i = d_i/b_i$. The heuristic information could also be made dependent on the partial solution y_k of an ant k. In this case, it can be defined as $\eta_i = e_i/b_i$, where e_i is the so-called *cover value*, that is, the number of additional rows covered when adding column i to the current partial solution. In other words, the heuristic information measures the unit cost of covering one additional row.

An ant ends the solution construction when all rows are covered. In a post-optimization step, an ant can remove redundant columns—columns that only cover rows which are also covered by a subset of other columns in the final solution—or apply some additional local search to improve solutions. The pheromone update can be carried out in a standard way as described in earlier sections.

When applying ACO to the SCP we have two main differences with the previously presented applications: (i) pheromone trails are associated only to components and, (ii) the length of the ant's walks (corresponding to the lengths of the sequences) may differ among the ants and, hence, the solution construction only ends when all the ants have terminated their corresponding walks.

There already exist some first applications of ACO to the SCP. In [1], ACO has been used only as a construction algorithm and the approach has only been tested on some small SCP instances. A more recent article [55] applies Ant System to the SCP and uses techniques to remove redundant columns and local search to improve solutions. Good results are obtained on a large set of benchmark instances taken from ORLIB, but the performance of Ant System could not fully reach that of the best performing algorithms for the SCP.

Example 9.4. *AntNet for network routing applications*

Given a graph representing a telecommunications network, the problem solved by AntNet is to find the minimum cost path between each pair of vertices of the graph. It is important to note that, although finding a minimum cost path on a graph is an easy problem (it can be efficiently solved by algorithms having polynomial worst case complexity [5]), it becomes extremely difficult when the costs on the edges are time-varying stochastic variables. This is the case of routing in packet-switched networks, the target application for AntNet.

Here we briefly describe a simplified version of AntNet (the interested reader should refer to [26], where the AntNet approach to routing is explained and evaluated in detail). As stated earlier, in AntNet each ant searches for a minimum cost path between a given pair of vertices of the network. To this goal, ants are launched from each network vertex towards randomly selected destination vertices. Each ant has a source vertex s and a destination vertex d, and moves from s to d hopping from one vertex to the next until vertex d is reached. When ant k is in vertex i, it chooses the next vertex j to move to according to a probabilistic decision rule which is a function of the ant's memory and of local pheromone and heuristic information (very much like to what happened, for example, in AS).

Unlike AS, where pheromone trails are associated to edges, in AntNet pheromone trails are associated to edge-destination pairs. That is, each directed edge (i, j) has $n - 1$ trail values $\tau_{ijd} \in [0, 1]$ associated, where n is the number of vertices in the graph associated to the routing problem; in general, $\tau_{ijd} \neq \tau_{jid}$. In other words, there is one trail value τ_{ijd} for each possible destination vertex d an ant located in vertex i can have. Each edge has also associated an heuristic value $\eta_{ij} \in [0, 1]$ independent of the final destination. The heuristic values can be set for example to the values $\eta_{ij} = 1 - q_{ij} / \sum_{l \in \mathcal{N}_i} q_{il}$, where q_{ij} is the length (in bits waiting to be sent) of the queue of the link connecting vertex i with its neighbor j: links with a shorter queue have a higher heuristic value.

In AntNet, as well as in most other implementations of ACO algorithms for routing problems [79,90], the daemon component (see Figure 9.5) is not present.

Ants choose their way probabilistically, using as probability a functional composition of the local pheromone trails τ_{ijd} and of the heuristic values η_{ij}. While building the path to their destinations, ants move using the same link queues as data and experience the same delays as data packets. Therefore, the time T_{sd} elapsed while moving from the source vertex s to the destination vertex d can be used as a measure of the quality of the path they built. The overall quality of a path is evaluated by an heuristic function of the trip time T_{sd} and of a local adaptive statistical model maintained in each vertex. In fact, paths need to be evaluated relative to the network status because a trip time T judged of low quality under low congestion conditions could be an excellent one under high traffic load. Once the generic ant k has completed a path, it deposits on the visited vertices an amount of pheromone $\Delta\tau^k(t)$ proportional to the quality of the path it built. To deposit pheromone, after reaching its destination vertex, the ant moves back to its source vertex along the same path but backward and using high priority queues, to allow a fast propagation of the collected information. The pheromone trail intensity of each edge l_{ij} the ant used while it was moving from s to d is increased as follows: $\tau_{ijd}(t) \leftarrow \tau_{ijd}(t) + \Delta\tau^k(t)$. After the pheromone trail on the visited edges has been updated, the pheromone value of all the outgoing connections of the same vertex i, relative to the destination d, evaporates in such a way that the pheromone values are normalized and can continue to be usable as probabilities: $\tau_{ijd}(t) \leftarrow \tau_{ijd}(t)/(1 + \Delta\tau^k(t))$, $\forall j \in \mathcal{N}_i$, where $\mathcal{N}_i$ is the set of neighbors of vertex i.

AntNet was compared with many state-of-the-art algorithms on a large set of benchmark problems under a variety of traffic conditions. It always compared favorably with competing approaches and it was shown to be very robust with respect to varying traffic conditions and parameter settings. More details on the experimental results can be found in [26].

Applications of the ACO metaheuristic

ACO has recently raised a lot of interest in the scientific community. There are now available numerous successful implementations of the ACO metaheuristic applied to a wide range of different combinatorial optimization problems. These applications comprise two main application fields.

- $\mathcal{NP}$-hard problems, for which the best known algorithms have exponential time worst case complexity. For these problems, very often ACO algorithms are coupled with extra capabilities such as problem-specific local optimizers, which take the ants' solutions to local optima.
- Shortest path problems in which the properties of the problem's graph representation change over time concurrently with the optimization process that has to adapt to the problem's dynamics. In this case, the problem's graph can be physically available (as in network problems), but its properties, like the costs of components or of connections, can change over time. In this case we conjecture that the use of ACO algorithms becomes more and more appropriate as the variation rate of the costs increases and/or the knowledge about the variation process diminishes.

These applications are summarized in Table 9.1. In some of these applications, ACO algorithms have obtained world-class performance, which is the case, for example, for quadratic assignment [63,88], sequential ordering [45,46], vehicle routing [47], scheduling [22,67] or packet-switched network routing [26].

6 DISCUSSION OF APPLICATION PRINCIPLES

Despite being a rather recent metaheuristic, ACO algorithms have already been applied to a large number of different combinatorial optimization problems. Based on this experience, we have identified some basic issues which play an important role in several of these applications. These are discussed in the following.

6.1 Pheromone Trails Definition

An initial, very important choice when applying ACO is the definition of the intended meaning of the pheromone trails. Let us explain this issue with an example. When

Table 9.1. Current applications of ACO algorithms. Applications are listed by class of problems and in chronological order

Problem name	Authors	Algorithm name	Year	Main references
Traveling salesman	Dorigo, Maniezzo & Colorni	AS	1991	[30,37,38]
	Gambardella & Dorigo	Ant-Q	1995	[43]
	Dorigo & Gambardella	ACS & ACS-3-change	1996	[34,35,44]
	Stützle & Hoos	$\mathcal{MM}$AS	1997	[84,87,88]
	Bullnheimer, Hartl & Strauss	AS_{rank}	1997	[14]
	Cordón, et al.	BWAS	2000	[19]
Quadratic assignment	Maniezzo, Colorni & Dorigo	AS-QAP	1994	[66]
	Gambardella, Taillard & Dorigo	HAS-QAP[a]	1997	[48]
	Stützle & Hoos	$\mathcal{MM}$AS-QAP	1997	[81,88]
	Maniezzo	ANTS-QAP	1998	[63]
	Maniezzo & Colorni	AS-QAP[b]	1999	[65]
Scheduling problems	Colorni, Dorigo & Maniezzo	AS-JSP	1994	[18]
	Stützle	$\mathcal{MM}$AS-FSP	1997	[82]
	Bauer et al.	ACS-SMTTP	1999	[2]
	den Besten, Stützle & Dorigo	ACS-SMTWTP	1999	[22]
	Merkle, Middendorf & Schmeck	ACO-RCPS	2000	[67]
Vehicle routing	Bullnheimer, Hartl & Strauss	AS-VRP	1997	[12,13]
	Gambardella, Taillard & Agazzi	HAS-VRP	1999	[47]

Connection-oriented network routing	Schoonderwoerd et al.	ABC	1996	[78,79]
	White, Pagurek & Oppacher	ASGA	1998	[91]
	Di Caro & Dorigo	AntNet-FS	1998	[27]
	Bonabeau et al.	ABC-smart ants	1998	[10]
Connection-less	Di Caro & Dorigo	AntNet & AntNet-FA	1997	[24,26,29]
	Heusse et al.	CAF	1998	[56]
	van der Put & Rothkrantz	ABC-backward	1998	[90]
Sequential ordering	Gambardella & Dorigo	HAS-SOP	1997	[45,46]
Graph coloring	Costa & Hertz	ANTCOL	1997	[20]
Shortest common supersequence	Michel & Middendorf	AS-SCS	1998	[69,70]
Frequency assignment	Maniezzo & Carbonaro	ANTS-FAP	1998	[64]
Generalized assignment	Ramalhinho Lourenço & Serra	$\mathcal{MM}$AS-GAP	1998	[75]
Multiple knapsack	Leguizamón & Michalewicz	AS-MKP	1999	[61]
Optical networks routing	Navarro Varela & Sinclair	ACO-VWP	1999	[73]
Redundancy allocation	Liang & Smith	ACO-RAP	1999	[62]
Constraint satisfaction	Solnon	Ant-P-solver	2000	[80]

[a] HAS-QAP is an ant algorithm which does not follow all the aspects of the ACO metaheuristic. [b] This is a variant of the original AS-QAP.

applying ACO to the TSP, the standard interpretation of a pheromone trail τ_{ij}, used in all published ACO applications to the TSP, is that it refers to the desirability of visiting city j directly after a city i. That is, it provides some information on the desirability of the relative positioning of city i and j. Yet, another possibility, not working so well in practice, would be to interpret τ_{ij} as the desirability of visiting city i as the jth city in a tour, that is, the desirability of the absolute positioning. Conversely, when applying ACO to the SMTWTP (see Section 5) better results are obtained when using the absolute position interpretation of the pheromone trails, where τ_{ij} is the desirability of putting job j on the ith position [21]. This is intuitively due to the different role that permutations have in the two problems. In the TSP, permutations are cyclic, that is, only the relative order of the solution components is important and a permutation $\pi = (1\ 2 \cdots n)$ has the same tour length as the permutation $\pi' = (n 1\ 2 \cdots n-1)$— it represents the same tour. Therefore, a relative position based pheromone trail is the appropriate choice. On the contrary, in the SMTWTP (as well as in many other scheduling problems), π and π' represent two different solutions with most probably very different costs. Hence, in the SMTWTP the absolute position based pheromone trails are a better choice. Nevertheless, it should be noted that, in principle, both choices are possible, because any solution of the search space can be generated with both representations.

The definition of the pheromone trails is crucial and a poor choice at this stage of the algorithm design will result in poor performance. Fortunately, for many problems the intuitive choice is also a very good one, as it was the case for the previous example applications. Yet, sometimes the use of the pheromones can be somewhat more involved, which is, for example, the case in the ACO application to the shortest common supersequence problem [70].

6.2 Balancing Exploration and Exploitation

Any effective metaheuristic algorithm has to achieve an appropriate balance between the exploitation of the search experience gathered so far and the exploration of unvisited or relatively unexplored search space regions. In ACO several ways exist of achieving such a balance, typically through the management of the pheromone trails. In fact, the pheromone trails induce a probability distribution over the search space and determine which parts of the search space are effectively sampled, that is, in which part of the search space the constructed solutions are located with higher frequency. Note that, depending on the distribution of the pheromone trails, the sampling distribution can vary from a uniform distribution to a degenerate distribution which assigns probability one to a single solution and zero probability to all the others. In fact, this latter situation corresponds to the stagnation of the search as explained on page 261.

The simplest way to exploit the ants' search experience is to make the pheromone update a function of the solution quality achieved by each particular ant. Yet, this bias alone is often too weak to obtain good performance, as was shown experimentally on the TSP [84,88]. Therefore, in many ACO algorithms (see Section 4) an *elitist strategy* was introduced whereby the best solutions found during the search strongly contribute to pheromone trail updating.

A stronger exploitation of the "learned" pheromone trails can also be achieved during solution construction by applying the pseudo-random proportional rule of Ant Colony System, as explained in Section 4.2.2.

Search space exploration is achieved in ACO primarily by the ants' randomized solution construction. Let us consider for a moment an ACO algorithm that does not use heuristic information (this can be easily achieved by setting $\beta = 0$). In this case, the pheromone updating activity of the ants will cause a shift from the initial uniform sampling of the search space to a sampling focused on specific search space regions. Hence, exploration of the search space will be higher in the initial iterations of the algorithm, and will decrease as the computation goes on. Obviously, attention must be paid to avoid too strong a focus on apparently good regions of the search space, which can cause the ACO algorithm to enter a stagnation situation.

There are several ways to try to avoid such stagnation situations, thus maintaining a reasonable level of exploration of the search space. For example, in ACS the ants use a local pheromone update rule during the solution construction to make the path they have taken less desirable for following ants and, thus, to diversify search. $\mathcal{MM}$AS introduces an explicit lower limit on the pheromone trail level so that a minimal level of exploration is always guaranteed. $\mathcal{MM}$AS also uses a reinitialization of the pheromone trails, which is a way of enforcing search space exploration. Experience has shown that pheromone trail reinitialization, when combined with appropriate choices for the pheromone trail update [88], can be very useful to refocus the search on a different search space region.

Finally, an important, though somewhat neglected, role in the balance of exploration and exploitation is that of the parameters α and β, which determine the relative influence of pheromone trail and heuristic information. Consider first the influence of parameter α. For $\alpha > 0$, the larger the value of α the stronger the exploitation of the search experience, for $\alpha = 0$ the pheromone trails are not taken into account at all, and for $\alpha < 0$ the most probable choices taken by the ants are those that are less desirable from the point of view of pheromone trails. Hence, varying α could be used to shift from exploration to exploitation and vice versa. The parameter β determines the influence of the heuristic information in a similar way. In fact, systematic variations of α and β could, similarly to what is done in the strategic oscillations approach [50], be part of simple and useful strategies to balance exploration and exploitation.

6.3 ACO and Local Search

In many applications to $\mathcal{NP}$-hard combinatorial optimization problems, ACO algorithms perform best when coupled with local search algorithms (which is, in fact, a particular type of daemon action of the ACO metaheuristic). Local search algorithms locally optimize the ants' solutions and these locally optimized solutions are used in the pheromone update.

The use of local search in ACO algorithms can be very interesting as the two approaches are complementary. In fact, ACO algorithms perform a rather coarse-grained search, and the solutions they produce can then be locally optimized by an adequate local search algorithm. The coupling can therefore greatly improve the quality of the solutions generated by the ants.

On the other side, generating initial solutions for local search algorithms is not an easy task. For example, it has been shown that, for most problems, repeating local searches from randomly generated initial solutions is not efficient (see, e.g., [57]). In practice, ants probabilistically combine solution components which are part of the best locally optimal solutions found so far and generate new, promising initial solutions for the local search. Experimentally, it has been found that such a combination of a probabilistic, adaptive construction heuristic with local search can yield excellent results [6,35,87].

It is important to note that when using local search a choice must be made concerning pheromone trail update: either pheromone is added to the components and/or connections of the locally optimal solution, or to the starting solutions for the local search.[11] The quasi totality of published research has used the first approach. Although it could be interesting to investigate the second approach, some recent experimental results suggest that its performance is worse.

Despite the fact that the use of local search algorithms has been shown to be crucial for achieving best performance in many ACO applications, it should be noted that ACO algorithms also show very good performance where local search algorithms cannot be applied easily. One such example are the network routing applications described in Section 5 or the shortest common supersequence problem [70].

6.4 Heuristic Information

The possibility of using heuristic information to direct the ants' probabilistic solution construction is important because it gives the possibility of exploiting problem specific knowledge. This knowledge can be available *a priori* (this is the most frequent situation in static problems) or at run-time (this is the typical situation in dynamic problems). In static problems, the heuristic information η is usually computed once at initialization time and then it remains the same throughout the whole algorithm's run. An example is the use, in the TSP applications, of the length d_{ij} of the edge connecting cities i and j to define the heuristic information $\eta_{ij} = 1/d_{ij}$. Static heuristic information has the advantage that (i) it is easy to compute, (ii) it has to be computed only once at initialization time, and (iii) in each iteration of the ACO algorithm, a table can be pre-computed with the values of $\tau_{ij}(t) \cdot \eta_{ij}^{\beta}$, which can result in a very significant saving of computation time. In the dynamic case, the heuristic information may also depend on the partial solution constructed so far and therefore has to be computed at each step of an ant's walk. This determines a higher computational cost that may be compensated by the higher accuracy of the computed heuristic values. For example, in the ACO application to the SMTWTP we found that the use of dynamic heuristic information based on the modified due date or the apparent urgency heuristics (see Section 5) resulted in a better overall performance.

Another way of computing heuristic information was introduced in the ANTS algorithm [63], where it is computed using lower bounds on the solution cost of the completion of an ant's partial solution. This method has the advantage that it facilitates the exclusion of certain choices because they lead to solutions that are worse than the best found so far. It allows therefore the combination of knowledge on the calculation of lower bounds from mathematical programming with the ACO paradigm. Nevertheless, a disadvantage is that the computation of the lower bounds can be time consuming, especially because they have to be calculated at each single construction step by each ant.

---

[11] Making an analogy with genetic algorithms, we could call the first approach "Lamarckian" and the second "Darwinian".

Finally, it should be noted that while the use of heuristic information is rather important for a generic ACO algorithm, its importance is strongly reduced if local search is used to improve solutions. This is due to the fact that local search takes into account information about the cost to improve solutions in a more direct way. Fortunately, this means that ACO algorithms can also achieve, in combination with a local search algorithm, very good performance for problems for which it is difficult to define *a priori* a very informative heuristic information.

6.5 Number of Ants

Why use a colony of ants instead of using one single ant? In fact, although a single ant is capable of generating a solution, efficiency considerations suggest that the use of a colony of ants is often a desirable choice. This is particularly true for geographically distributed problems, because the differential path length effect exploited by ants in the solution of this class of problems can only arise in presence of a colony of ants. It is also interesting to note that in routing problems ants solve many shortest path problems in parallel (one between each pair of vertices) and a colony of ants must be used for each of these problems.

On the other hand, in the case of combinatorial optimization problems the differential length effect is not exploited and the use of m ants, $m > 1$, that build r solutions each (i.e., the ACO algorithm is run for r iterations) could be equivalent to the use of one ant that generates $m \cdot r$ solutions. Nevertheless, in this case theoretical results on the convergence of some specific ACO algorithms, which will be presented in Section 7, as well as experimental evidence suggest that ACO algorithms perform better when the number m of ants is set to a value $m > 1$.

In general, the best value for m is a function of the particular ACO algorithm chosen as well as of the class of problems being attacked, and most of the times it must be set experimentally. Fortunately, ACO algorithms seem to be rather robust to the actual number of ants used.

6.6 Candidate Lists

One possible difficulty encountered by ACO algorithms is when they are applied to problems with a large neighborhood in the solution construction. In fact, an ant that visits a state with a large neighborhood has a correspondingly large number of possible moves among which to choose. Possible problems are that the solution construction is significantly slowed down and that the probability that many ants visit the same state is very small. Such a situation can occur, for example, in the ACO application to large TSPs or large SCPs.

In such situations, the above-mentioned problem can be considerably reduced by the use of candidate lists. Candidate lists comprise a small set of promising neighbors of the current state. They are created using *a priori* available knowledge on the problem, if available, or dynamically generated information. Their use allows ACO algorithms to focus on the more interesting components, strongly reducing the dimension of the search space.

As an example, consider the ACO application to the TSP. For the TSP it is known that very often optimal solutions can be found within a surprisingly small subgraph consisting of all the cities and of those edges that connect each city to only a few of its nearest neighbors. For example, for the TSPLIB instance `pr2392.tsp` with 2392 cities an optimal solution can be found within a subgraph of the 8 nearest neighbors [76]. This knowledge can be used for defining candidate lists, which was first done in the context of ACO algorithms in [44]. A candidate list includes for each city its *cl* nearest neighbors. During solution construction an ant tries to choose the city to move to only among the cities in the candidate list. Only if all these cities have already been visited, the ant can choose among the other cities.

So far, in ACO algorithms the use of candidate lists or similar approaches is still rather unexplored. Inspiration from other techniques like Tabu Search [51] or GRASP [42], where strong use of candidate lists is made, could be useful for the development of effective candidate list strategies for ACO.

7 OTHER DEVELOPMENTS

7.1 Proving Convergence

The simplest stochastic optimization algorithm is random search. Besides simplicity, random search has also the nice property that it guarantees that it will find, sooner or later, the optimal solution to your problem. Unfortunately, it is very inefficient. Stochastic optimization algorithms can be seen as ways of biasing random search so to make it more efficient. Unfortunately, once a stochastic algorithm is biased, it is no longer guaranteed to find, at some point, the optimal solution. In fact, the bias could simply rule out this possibility. It is therefore interesting to have convergence proofs that reassure you that this does not happen.

Although the problem of proving convergence to the optimal solution of a generic ACO algorithm is open (and it will most probably remain so, given the generality of the ACO metaheuristic), convergence has recently been proved for a few instances of ACO algorithms. The first of such proofs was provided by Gutjahr [53,54] who proved convergence to the optimal solution for a particular ACO algorithm he calls *Graph-based Ant System* (GBAS): the proof states that, given a small $\epsilon > 0$ and for fixed values of some algorithm parameters, after a number of cycles $t \geq t_0$ the algorithm will find the optimal solution with probability $P_t \geq 1-\epsilon$, where $t_0 = f(\epsilon)$. A limitation of this important result is that it applies to an ACO algorithm, GBAS, for which no experimental results are available (i.e., we do not know what is its performance). More recently, Stützle and Dorigo [86] have proved convergence for two of the experimentally most successful ACO algorithms: ACS and $\mathcal{MM}$AS.

7.2 Parallel Implementations

The very nature of ACO algorithms lends them to be parallelized in the data or population domains. In particular, many parallel models used in other population-based algorithms can be easily adapted to the ACO structure. Most parallelization strategies can be classified into *fine-grained* and *coarse-grained* strategies. Characteristic of fine-grained parallelization is that very few individuals are assigned to one single processor and that frequent information exchange among the processors takes place. On the contrary, in coarse grained approaches larger subpopulations or even full populations are assigned to single processors and information exchange is rather rare. We refer, for example, to [16] for an overview.

Fine-grained parallelization schemes have been investigated with parallel versions of AS for the TSP on the Connection Machine CM-2 adopting the approach of attributing a single processing unit to each ant [7]. Experimental results showed that communication overhead can be a major problem with this approach on fine grained parallel machines, since ants end up spending most of their time communicating to other ants the modifications they made to pheromone trails. Similar negative results have also been reported in [15].

As shown by several researches [7,15,59,71,83], coarse grained parallelization schemes are much more promising for ACO. When applied to ACO, coarse grained schemes run p subcolonies in parallel, where p is the number of available processors. Information among the subcolonies is exchanged at certain intervals. For example, in the Partially Asynchronous Parallel Implementation (PAPI) of Bullnheimer, Kotsis and Strauss [15], for which high speed-up was observed, the subcolonies exchange pheromone information every fixed number of iterations performed by each subcolony. Krüger, Merkle and Middendorf [59] investigated which information should be exchanged between the m subcolonies and how this information should be used to update the subcolony's trail information. Their results showed that it was better to exchange the best solutions found so far and to use them in the pheromone update than to exchange complete pheromone matrices for modifications of the pheromone matrix of a local subcolony. Middendorf, Reischle, and Schmeck [71] investigate different ways of exchanging solutions among m ant colonies. They consider an exchange of the global best solutions among all colonies and local exchanges based on a virtual neighborhood among subcolonies which corresponds to a directed ring. Their main observation was that the best solutions, with respect to computing time and solution quality, were obtained by limiting the information exchange to a local neighborhood of the colonies. In the extreme case, there is no communication among the subcolonies, resulting in parallel independent runs of an algorithm. This is the easiest way to parallelize randomized algorithms and can be very effective as has been shown by computational results presented by Stützle [83].

8 CONCLUSIONS

The field of ACO algorithms is very lively, as testified, for example, by the successful biannual workshop (ANTS—From Ant Colonies to Artificial Ants: A Series of International Workshops on Ant Algorithms; `http://iridia.ulb.ac.be/~ants/`) where researchers meet to discuss the properties of ACO and other ant algorithms, both theoretically and experimentally.

From the theory side, the most interesting ongoing work concerns the study of the relationship between ACO algorithms and other well-known stochastic optimization techniques. For example, it has been shown [68] that, when interpreting ACO algorithms as methods for searching in the space of pheromones (i.e., artificial ants are seen as procedures to update pheromone trails so to increase the probability of generating very good solutions to the combinatorial optimization problem), then AS can be interpreted as an approximate form of stochastic gradient descent [92] (a well-known algorithm which has been extensively used in machine learning) in the space of pheromones. Similarly, it has been shown [41,94] that there are strong connections between ACO algorithms and the Cross-Entropy method [77].

From the experimental side, most of the current research is in the direction of increasing the number of problems that are successfully solved by ACO algorithms, including real-word, industrial applications [39].

Currently, the great majority of problems attacked by ACO are static and well-defined combinatorial optimization problems, that is, problems for which all the necessary information is available and does not change during problem solution. For this kind of problems ACO algorithms must compete with very well established algorithms, often specialized for the given problem. Also, very often the role played by local search is extremely important to obtain good results (see for example [46]). Although rather successful on these problems, we believe that ACO algorithms will really show their greatest advantage when they are systematically applied to "ill-structured" problems for which it is not clear how to apply local search, or to highly dynamic domains with only local information available. A first step in this direction has already been made with the application to telecommunications networks routing, but much further research will be necessary. The reader interested in learning more about ACO is referred to the book "Ant Colony Optimization" by the same authors [40].

ACKNOWLEDGMENTS

We thank Joshua Knowles and Michael Sampels for their useful comments on a draft version of this paper. Marco Dorigo acknowledges support from the Belgian FNRS, of which he is a Senior Research Associate. This work was partially supported by the "Metaheuristics Network", a Research Training Network funded by the Improving Human Potential programme of the CEC, grant HPRN-CT-1999-00106. The information provided is the sole responsibility of the authors and does not reflect the Community's opinion. The Community is not responsible for any use that might be made of data appearing in this publication.

REFERENCES

[1] D.A. Alexandrov and Y.A. Kochetov (2000) The behavior of the ant colony algorithm for the set covering problem. In: K. Inderfurth, G. Schwödiauer, W. Domschke, F. Juhnke, P. Kleinschmidt, and G. Wäscher (eds.), *Operations Research Proceedings 1999*. Springer-Verlag, Berlin, Germany, pp. 255–260.

[2] A. Bauer, B. Bullnheimer, R.F. Hartl and C. Strauss (1999) An ant colony optimization approach for the single machine total tardiness problem. In: *Proceedings of the 1999 Congress on Evolutionary Computation (CEC'99)*, IEEE Press, Piscataway, NJ, pp. 1445–1450.

[3] R. Beckers, J.-L. Deneubourg and S. Goss (1993) Modulation of trail laying in the ant *Lasius niger* (hymenoptera: Formicidae) and its role in the collective selection of a food source. *Journal of Insect Behavior*, **6**(6), 751–759.

[4] R. Bellman, A.O. Esogbue and I. Nabeshima (1982) *Mathematical Aspects of Scheduling and Applications*. Pergamon Press, New York, NJ.

[5] D. Bertsekas (1998) *Network Optimization: Continuous and Discrete Models*. Athena Scientific, Belmont, MA.

[6] K.D. Boese, A.B. Kahng and S. Muddu (1994) A new adaptive multi-start technique for combinatorial global optimization. *Operations Research Letters*, **16**, 101–113.

[7] M. Bolondi and M. Bondanza (1993) Parallelizzazione di un algoritmo per la risoluzione del problema del commesso viaggiatore. Master's thesis, Dipartimento di Elettronica, Politecnico di Milano, Italy.

[8] E. Bonabeau, M. Dorigo and G. Theraulaz (1999) *Swarm Intelligence: From Natural to Artificial Systems*. Oxford University Press, New York, NJ.

[9] E. Bonabeau, M. Dorigo and G. Theraulaz (2000) Inspiration for optimization from social insect behavior. *Nature*, **406**, 39–42.

[10] E. Bonabeau, F. Henaux, S. Guérin, D. Snyers, P. Kuntz and G. Theraulaz (1998) Routing in telecommunication networks with "Smart" ant-like agents. In: *Proceedings of IATA'98, Second International Workshop on Intelligent Agents for Telecommunication Applications*, volume 1437 of *Lecture Notes in Artificial Intelligence*. Springer-Verlag, Berlin, Germany, pp. 60–72.

[11] E. Bonabeau and G. Theraulaz (2000) Swarm smarts. *Scientific American*, **282**(3), 54–61.

[12] B. Bullnheimer, R.F. Hartl and C. Strauss (1999) Applying the Ant System to the vehicle routing problem. In: S. Voß, S. Martello, I.H. Osman and C. Roucairol (eds.), *Meta-Heuristics: Advances and Trends in Local Search Paradigms for Optimization*. Kluwer Academic Publishers, Dordrecht, The Netherlands, pp. 285–296.

[13] B. Bullnheimer, R.F. Hartl and C. Strauss (1999) An improved Ant System algorithm for the vehicle routing problem. *Annals of Operations Research*, **89**, 319–328.

[14] B. Bullnheimer, R.F. Hartl and C. Strauss (1999) A new rank-based version of the Ant System: A computational study. *Central European Journal for Operations Research and Economics*, **7**(1), 25–38.

[15] B. Bullnheimer, G. Kotsis and C. Strauss (1998) Parallelization strategies for the Ant System. In: R. De Leone, A. Murli, P. Pardalos and G. Toraldo (eds.), *High Performance Algorithms and Software in Nonlinear Optimization*, volume 24 of *Applied Optimization*. Kluwer Academic Publishers, Dordrecht, The Netherlands, pp. 87–100.

[16] E. Cantú-Paz (2000) *Efficient and Accurate Parallel Genetic Algorithms*. Kluwer Academic Publishers, Boston, MA.

[17] A. Colorni, M. Dorigo and V. Maniezzo (1992) Distributed optimization by ant colonies. In: F.J. Varela and P. Bourgine (eds.), *Proceedings of the First European Conference on Artificial Life*, MIT Press, Cambridge, MA, pp. 134–142.

[18] A. Colorni, M. Dorigo, V. Maniezzo and M. Trubian (1994) Ant System for job-shop scheduling. *JORBEL—Belgian Journal of Operations Research, Statistics and Computer Science*, **34**(1), 39–53.

[19] O. Cordón, I. Fernández de Viana, F. Herrera and L. Moreno (2000) A new ACO model integrating evolutionary computation concepts: The best-worst Ant System. In: M. Dorigo, M. Middendorf and T. Stützle (eds.), *Abstract proceedings*

of ANTS2000—From Ant Colonies to Artificial Ants: A Series of International Workshops on Ant Algorithms. IRIDIA, Université Libre de Bruxelles, Belgium, pp. 22–29.

[20] D. Costa and A. Hertz (1997) Ants can colour graphs. *Journal of the Operational Research Society*, **48**, 295–305.

[21] M. den Besten (2000) Ants for the single machine total weighted tardiness problem. Master's thesis, University of Amsterdam, The Netherlands.

[22] M.L. den Besten, T. Stützle and M. Dorigo (2000) Ant colony optimization for the total weighted tardiness problem. In: M. Schoenauer, K. Deb, G. Rudolph, X. Yao, E. Lutton, J.J. Merelo and H.-P. Schwefel (eds.), *Proceedings of PPSN-VI, Sixth International Conference on Parallel Problem Solving from Nature*, volume 1917 of *Lecture Notes in Computer Science*. Springer-Verlag, Berlin, Germany, pp. 611–620.

[23] J.-L. Deneubourg, S. Aron, S. Goss and J.-M. Pasteels (1990) The self-organizing exploratory pattern of the Argentine ant. *Journal of Insect Behavior*, **3**, 159–168.

[24] G. Di Caro and M. Dorigo (1997) AntNet: A mobile agents approach to adaptive routing. Technical Report IRIDIA/97-12, IRIDIA, Université Libre de Bruxelles, Belgium.

[25] G. Di Caro and M. Dorigo (1998) Ant colonies for adaptive routing in packet-switched communications networks. In: A.E. Eiben, T. Bäck, M. Schoenauer and H.-P. Schwefel (eds.), *Proceedings of PPSN-V, Fifth International Conference on Parallel Problem Solving from Nature*, volume 1498 of *Lecture Notes in Computer Science*. Springer-Verlag, Berlin, Germany, pp. 673–682.

[26] G. Di Caro and M. Dorigo (1998) AntNet: Distributed stigmergetic control for communications networks. *Journal of Artificial Intelligence Research*, **9**, 317–365.

[27] G. Di Caro and M. Dorigo (1998) Extending AntNet for best-effort Quality-of-Service routing. Unpublished presentation at *ANTS'98—From Ant Colonies to Artificial Ants: First International Workshop on Ant Colony Optimization*, http://iridia.ulb.ac.be/ants98/ants98.html, October 15–16.

[28] G. Di Caro and M. Dorigo (1998) Mobile agents for adaptive routing. In: H. El-Rewini (ed.), *Proceedings of the 31st International Conference on System Sciences (HICSS-31)*. IEEE Computer Society Press, Los Alamitos, CA, pp. 74–83.

[29] G. Di Caro and M. Dorigo (1998) Two ant colony algorithms for best-effort routing in datagram networks. In: Y. Pan, S.G. Akl and K. Li (eds.), *Proceedings of the Tenth IASTED International Conference on Parallel and Distributed Computing and Systems (PDCS'98)*, IASTED/ACTA Press, Anaheim, CA, pp. 541–546.

[30] M. Dorigo (1992) *Optimization, Learning and Natural Algorithms* (in Italian). PhD thesis, Dipartimento di Elettronica, Politecnico di Milano, Italy, 140 pp.

[31] M. Dorigo, E. Bonabeau and G. Theraulaz (2000) Ant algorithms and stigmergy. *Future Generation Computer Systems*, **16**(8), 851–871.

[32] M. Dorigo and G. Di Caro (1999) The Ant Colony Optimization meta-heuristic. In: D. Corne, M. Dorigo and F. Glover (eds.), *New Ideas in Optimization*. McGraw Hill, London, UK, pp. 11–32.

[33] M. Dorigo, G. Di Caro and L.M. Gambardella (1999) Ant algorithms for discrete optimization. *Artificial Life*, **5**(2), 137–172.

[34] M. Dorigo and L.M. Gambardella (1997) Ant colonies for the traveling salesman problem. *BioSystems*, **43**, 73–81.

[35] M. Dorigo and L.M. Gambardella (1997) Ant Colony System: A cooperative learning approach to the traveling salesman problem. *IEEE Transactions on Evolutionary Computation*, **1**(1), 53–66.

[36] M. Dorigo, V. Maniezzo and A. Colorni (1991) The Ant System: An autocatalytic optimizing process. Technical Report 91-016 Revised, Dipartimento di Elettronica, Politecnico di Milano, Italy.

[37] M. Dorigo, V. Maniezzo and A. Colorni (1991) Positive feedback as a search strategy. Technical Report 91-016, Dipartimento di Elettronica, Politecnico di Milano, Italy.

[38] M. Dorigo, V. Maniezzo, and A. Colorni (1996) The Ant System: Optimization by a colony of cooperating agents. *IEEE Transactions on Systems, Man, and Cybernetics—Part B*, **26**(1), 29–41.

[39] M. Dorigo, M. Middendorf and T. Stützle (2000) (eds.) *Abstract proceedings of ANTS2000—From Ant Colonies to Artificial Ants: A Series of International Workshops on Ant Algorithms*. IRIDIA, Université Libre de Bruxelles, Belgium, 7–9 September.

[40] M. Dorigo and T. Stützle *Ant Colony Optimization*. MIT Press, Cambridge, MA (forthcoming).

[41] M. Dorigo, M. Zlochin, N. Meuleau and M. Birattari (2001) Updating ACO pheromones using stochastic gradient ascent and cross-entropy methods. Technical Report IRIDIA/2001-19, IRIDIA, Université Libre de Bruxelles, Belgium. *Proceedings of the 2nd European Workshop on Evolutionary Computation in Combinatorial Optimization (EvoCOP-2002)*, (to appear).

[42] T.A. Feo and M.G.C. Resende (1995) Greedy randomized adaptive search procedures. *Journal of Global Optimization*, **6**, 109–133.

[43] L.M. Gambardella and M. Dorigo (1995) Ant-Q: A reinforcement learning approach to the traveling salesman problem. In: A. Prieditis and S. Russell (eds.), *Proceedings of the Twelfth International Conference on Machine Learning (ML-95)*, Morgan Kaufmann Publishers, Palo Alto, CA, pp. 252–260.

[44] L.M. Gambardella and M. Dorigo (1996) Solving symmetric and asymmetric TSPs by ant colonies. In: *Proceedings of the 1996 IEEE International Conference on Evolutionary Computation (ICEC'96)*, IEEE Press, Piscataway, NJ, pp. 622–627.

[45] L.M. Gambardella and M. Dorigo (1997) HAS-SOP: An hybrid Ant System for the sequential ordering problem. Technical Report IDSIA-11-97, IDSIA, Lugano, Switzerland.

[46] L.M. Gambardella and M. Dorigo (2000) Ant Colony System hybridized with a new local search for the sequential ordering problem. *INFORMS Journal on Computing*, **12**(3), 237–255.

[47] L.M. Gambardella, È D. Taillard and G. Agazzi (1999) MACS-VRPTW: A multiple ant colony system for vehicle routing problems with time windows. In: D. Corne, M. Dorigo and F. Glover (eds.), *New Ideas in Optimization*. McGraw Hill, London, UK, pp. 63–76.

[48] L.M. Gambardella, È.D. Taillard and M. Dorigo (1999) Ant colonies for the quadratic assignment problem. *Journal of the Operational Research Society*, **50**(2), 167–176.

[49] M.R. Garey and D.S. Johnson (1979) *Computers and Intractability: A Guide to the Theory of $\mathcal{NP}$-Completeness*. Freeman, San Francisco, CA.

[50] F. Glover (1990) Tabu search—part II. *ORSA Journal on Computing*, **2**(1), 4–32.

[51] F. Glover and M. Laguna (1997) *Tabu Search*. Kluwer Academic Publishers, Boston, MA.

[52] S. Goss, S. Aron, J.L. Deneubourg and J.M. Pasteels (1989) Self-organized shortcuts in the Argentine ant. *Naturwissenschaften*, **76**, 579–581.

[53] W.J. Gutjahr (2000) A Graph-based Ant System and its convergence. *Future Generation Computer Systems*, **16**(8), 873–888.

[54] W.J. Gutjahr (2002) ACO algorithms with guaranteed convergence to the optimal solution. *Information Processing Letters*, (in press).

[55] R. Hadji, M. Rahoual, E. Talbi and V. Bachelet (2000) Ant colonies for the set covering problem. In: M. Dorigo, M. Middendorf and T. Stützle (eds.), *Abstract proceedings of ANTS2000—From Ant Colonies to Artificial Ants: A Series of International Workshops on Ant Algorithms*. IRIDIA, Université Libre de Bruxelles, Belgium, pp. 63–66.

[56] M. Heusse, S. Guérin, D. Snyers and P. Kuntz (1998) Adaptive agent-driven routing and load balancing in communication networks. *Advances in Complex Systems*, **1**(2), 237–254.

[57] D.S. Johnson and L.A. McGeoch (1997) The travelling salesman problem: A case study in local optimization. In: E.H.L. Aarts and J.K. Lenstra (eds.), *Local Search in Combinatorial Optimization*. John Wiley & Sons, Chichester, UK, pp. 215–310.

[58] M. Jünger, G. Reinelt and S. Thienel (1994) Provably good solutions for the traveling salesman problem. *Zeitschrift für Operations Research*, **40**, 183–217.

[59] F. Krüger, D. Merkle and M. Middendorf Studies on a parallel Ant System for the BSP model. Unpublished manuscript.

[60] E.L. Lawler, J.K. Lenstra, A.H.G. Rinnooy Kan and D.B. Shmoys (1985) *The Travelling Salesman Problem*. John Wiley & Sons, Chichester, UK.

[61] G. Leguizamón and Z. Michalewicz (1999) A new version of Ant System for subset problems. In: *Proceedings of the 1999 Congress on Evolutionary Computation (CEC'99)*. IEEE Press, Piscataway, NJ, pp. 1459–1464.

[62] Y.-C. Liang and A.E. Smith (1999) An Ant System approach to redundancy allocation. In: *Proceedings of the 1999 Congress on Evolutionary Computation (CEC'99)*. IEEE Press, Piscataway, NJ, pp. 1478–1484.

[63] V. Maniezzo (1999) Exact and approximate nondeterministic tree-search procedures for the quadratic assignment problem. *INFORMS Journal on Computing*, **11**(4), 358–369.

[64] V. Maniezzo and A. Carbonaro (2000) An ANTS heuristic for the frequency assignment problem. *Future Generation Computer Systems*, **16**(8), 927–935.

[65] V. Maniezzo and A. Colorni (1999) The Ant System applied to the quadratic assignment problem. *IEEE Transactions on Data and Knowledge Engineering*, **11**(5), 769–778.

[66] V. Maniezzo, A. Colorni and M. Dorigo (1994) The Ant System applied to the quadratic assignment problem. Technical Report IRIDIA/94-28, IRIDIA, Université Libre de Bruxelles, Belgium.

[67] D. Merkle, M. Middendorf and H. Schmeck (2000) Ant colony optimization for resource-constrained project scheduling. In: *Proceedings of the Genetic and Evolutionary Computation Conference (GECCO-2000)*, Morgan Kaufmann Publishers, San Francisco, CA, pp. 893–900.

[68] N. Meuleau and M. Dorigo (2002) Ant colony optimization and stochastic gradient descent. *Artificial Life*, (in press).

[69] R. Michel and M. Middendorf (1998) An island model based Ant System with lookahead for the shortest supersequence problem. In: A.E. Eiben, T. Bäck, M. Schoenauer and H.-P. Schwefel (eds.), *Proceedings of PPSN-V, Fifth International Conference on Parallel Problem Solving from Nature*, volume 1498 of *Lecture Notes in Computer Science*. Springer-Verlag, Berlin, Germany, pp. 692–701.

[70] R. Michel and M. Middendorf (1999) An ACO algorithm for the shortest supersequence problem. In: D. Corne, M. Dorigo and F. Glover (eds.), *New Ideas in Optimization*. McGraw Hill, London, UK, pp. 51–61.

[71] M. Middendorf, F. Reischle and H. Schmeck (2002) Multi colony ant algorithms. *Journal of Heuristics*, (in press).

[72] T.E. Morton, R.M. Rachamadugu and A. Vepsalainen (1984) Accurate myopic heuristics for tardiness scheduling. GSIA Working Paper 36-83-84, Carnegie–Mellon University, PA.

[73] G. Navarro Varela and M.C. Sinclair (1999) Ant colony optimisation for virtual-wavelength-path routing and wavelength allocation. In: *Proceedings of the 1999 Congress on Evolutionary Computation (CEC'99)*. IEEE Press, Piscataway, NJ, pp. 1809–1816.

[74] C.H. Papadimitriou (1994) *Computational Complexity*. Addison-Wesley, Reading, MA.

[75] H. Ramalhinho Lourenço and D. Serra (1998) Adaptive approach heuristics for the generalized assignment problem. Technical Report Economic Working Papers Series No. 304, Universitat Pompeu Fabra, Department of Economics and Management, Barcelona, Spain.

[76] G. Reinelt (1994) *The Traveling Salesman: Computational Solutions for TSP Applications*, volume 840 of *Lecture Notes in Computer Science*. Springer-Verlag, Berlin, Germany.

[77] R.Y. Rubinstein (2001) Combinatorial optimization via the simulated cross-entropy method. In: *Encyclopedia of Operations Research and Management Science*. Kluwer Academic Publishers, Boston, MA.

[78] R. Schoonderwoerd, O. Holland and J. Bruten (1997) Ant-like agents for load balancing in telecommunications networks. In: *Proceedings of the First International Conference on Autonomous Agents*. ACM Press, New York, NY, pp. 209–216.

[79] R. Schoonderwoerd, O. Holland, J. Bruten and L. Rothkrantz (1996) Ant-based load balancing in telecommunications networks. *Adaptive Behavior*, **5**(2), 169–207.

[80] C. Solnon (2000) Solving permutation constraint satisfaction problems with artificial ants. In: W. Horn (ed.), *Proceedings of the 14th European Conference on Artificial Intelligence*. IOS Press, Amsterdam, The Netherlands, pp. 118–122.

[81] T. Stützle (1997) $\mathcal{MAX}$–$\mathcal{MIN}$ Ant System for the quadratic assignment problem. Technical Report AIDA-97-4, FG Intellektik, FB Informatik, TU Darmstadt, Germany.

[82] T. Stützle (1998) An ant approach to the flow shop problem. In: *Proceedings of the 6th European Congress on Intelligent Techniques & Soft Computing (EUFIT'98)*, volume 3, Verlag Mainz, Wissenschaftsverlag, Aachen, Germany, pp. 1560–1564.

[83] T. Stützle (1998) Parallelization strategies for ant colony optimization. In: A.E. Eiben, T. Bäck, M. Schoenauer and H.-P. Schwefel (eds.), *Proceedings of PPSN-V, Fifth International Conference on Parallel Problem Solving from Nature*, volume 1498 of *Lecture Notes in Computer Science*. Springer Verlag, Berlin, Germany, pp. 722–731.

[84] T. Stützle (1999) *Local Search Algorithms for Combinatorial Problems: Analysis, Improvements and New Applications*. Infix, Sankt Augustin, Germany.

[85] T. Stützle and M. Dorigo (1999) ACO algorithms for the quadratic assignment problem. In: D. Corne, M. Dorigo and F. Glover (eds.), *New Ideas in Optimization*. McGraw Hill, London, UK, pp. 33–50.

[86] T. Stützle and M. Dorigo (2002) A short convergence proof for a class of ACO algorithms. *IEEE Transactions on Evolutionary Computation* (in press).

[87] T. Stützle and H.H. Hoos (1997) The $\mathcal{MAX}$–$\mathcal{MIN}$ Ant System and local search for the traveling salesman problem. In: T. Bäck, Z. Michalewicz and X. Yao (eds.), *Proceedings of the 1997 IEEE International Conference on Evolutionary Computation (ICEC'97)*, IEEE Press, Piscataway, NJ, pp. 309–314.

[88] T. Stützle and H.H. Hoos (2000) $\mathcal{MAX}$–$\mathcal{MIN}$ Ant System. *Future Generation Computer Systems*, 16(8):889–914.

[89] R.S. Sutton and A.G. Barto (1998) *Reinforcement Learning: An Introduction*. MIT Press, Cambridge, MA.

[90] R. van der Put (1998) Routing in the faxfactory using mobile agents. Technical Report R&D-SV-98-276, KPN Research, The Netherlands.

[91] T. White, B. Pagurek and F. Oppacher (1998) Connection management using adaptive mobile agents. In: H.R. Arabnia (ed.), *Proceedings of the International Conference on Parallel and Distributed Processing Techniques and Applications (PDPTA'98)*. CSREA Press, pp. 802–809.

[92] R.J. Williams (1992) Simple statistical gradient-following algorithms for connectionist reinforcement learning. *Machine Learning*, **8**(3), 229–256.

[93] M. Yannakakis (1997) Computational complexity. In: E.H.L. Aarts and J.K. Lenstra (eds.), *Local Search in Combinatorial Optimization*. John Wiley & Sons, Chichester, UK, pp. 19–55.

[94] M. Zlochin, M. Birattari, N. Meuleau and M. Dorigo (2001) Combinatorial optimization using model-based search. Technical Report IRIDIA/2001-15, IRIDIA, Université Libre de Bruxelles, Belgium. (submitted to *Annals of Operations Research*).

Chapter 10

THE THEORY AND PRACTICE OF SIMULATED ANNEALING

Darrall Henderson
Department of Mathematical Sciences
United States Military Academy
West Point, NY 10996-1786, USA
E-mail: darrall@stanfordalumni.org

Sheldon H. Jacobson
Department of Mechanical and Industrial Engineering
University of Illinois at Urbana-Champaign
1206 West Green Street, MC-244
Urbana, Illinois 61801-2906, USA
E-mail: shj@uiuc.edu

Alan W. Johnson
Department of Mathematical Sciences
United States Military Academy
West Point, NY 10996-1786, USA
E-mail: aa2895@usma.edu

Abstract Simulated annealing is a popular local search meta-heuristic used to address discrete and, to a lesser extent, continuous optimization problems. The key feature of simulated annealing is that it provides a means to escape local optima by allowing hill-climbing moves (i.e., moves which worsen the objective function value) in hopes of finding a global optimum. A brief history of simulated annealing is presented, including a review of its application to discrete and continuous optimization problems. Convergence theory for simulated annealing is reviewed, as well as recent advances in the analysis of finite time performance. Other local search algorithms are discussed in terms of their relationship to simulated annealing. The chapter also presents practical guidelines for the implementation of simulated annealing in terms of cooling schedules, neighborhood functions, and appropriate applications.

Keywords: Local Search Algorithms, Simulated Annealing, Heuristics, Meta-heuristics

1 BACKGROUND SURVEY

Simulated annealing is a local search algorithm (meta-heuristic) capable of escaping from local optima. Its ease of implementation, convergence properties and its use of hill-climbing moves to escape local optima have made it a popular technique over the past two decades. It is typically used to address discrete, and to a lesser extent, continuous optimization problems. Recent survey articles that provide a good overview of simulated annealing's theoretical development and domains of application include Eglese (1990), Fleischer (1995), Koulamas et al. (1994), and Romeo and Sangiovanni-Vincentelli (1991). Aarts and Korst (1989) and van Laarhoven and Aarts (1988) devote entire books to the subject. Aarts and Lenstra (1997) dedicate a chapter to simulated annealing in their book on local search algorithms for discrete optimization problems.

1.1 History and Motivation

Simulated annealing is so named because of its analogy to the process of physical annealing with solids, in which a crystalline solid is heated and then allowed to cool very slowly until it achieves its most regular possible crystal lattice configuration (i.e., its minimum lattice energy state), and thus is free of crystal defects. If the cooling schedule is sufficiently slow, the final configuration results in a solid with such superior structural integrity. Simulated annealing establishes the connection between this type of thermodynamic behavior and the search for global minima for a discrete optimization problem. Furthermore, it provides an algorithmic means for exploiting such a connection.

At each iteration of a simulated annealing algorithm applied to a discrete optimization problem, the objective function generates values for two solutions (the current solution and a newly selected solution) are compared. Improving solutions are always accepted, while a fraction of non-improving (inferior) solutions are accepted in the hope of escaping local optima in search of global optima. The probability of accepting non-improving solutions depends on a temperature parameter, which is typically non-increasing with each iteration of the algorithm.

The key algorithmic feature of simulated annealing is that it provides a means to escape local optima by allowing *hill-climbing* moves (i.e., moves which worsen the objective function value). As the temperature parameter is decreased to zero, hill-climbing moves occur less frequently, and the solution distribution associated with the inhomogeneous Markov chain that models the behavior of the algorithm converges to a form in which all the probability is concentrated on the set of globally optimal solutions (provided that the algorithm is convergent; otherwise the algorithm will converge to a local optimum, which may or not be globally optimal).

1.2 Definition of Terms

To describe the specific features of a simulated annealing algorithm for discrete optimization problems, several definitions are needed. Let Ω be the *solution space* (i.e., the set of all possible solutions). Let $f : \Omega \rightarrow \Re$ be an *objective function* defined on the solution space. The goal is to find a global minimum, ω^* (i.e., $\omega^* \in \Omega$ such that $f(\omega) \geq f(\omega^*)$ for all $\omega \in \Omega$). The objective function must be bounded to ensure that ω^* exists. Define $\mathsf{N}(\omega)$ to be the *neighborhood function* for $\omega \in \Omega$. Therefore, associated with every solution, $\omega \in \Omega$, are neighboring solutions, $\mathsf{N}(\omega)$, that can be reached in a single iteration of a local search algorithm.

Simulated annealing starts with an initial solution $\omega \in \Omega$. A neighboring solution $\omega' \in \mathsf{N}(\omega)$ is then generated (either randomly or using some pre-specified rule). Simulated annealing is based on the Metropolis acceptance criterion (Metropolis et al., 1953), which models how a thermodynamic system moves from the current solution (state) $\omega \in \Omega$ to a candidate solution $\omega' \in \mathsf{N}(\omega)$, in which the energy content is being minimized. The candidate solution, ω', is accepted as the current solution based on the acceptance probability

$$P\{\text{Accept } \omega' \text{ as next solution}\} = \begin{cases} \exp[-(f(\omega') - f(\omega))/t_k] & \text{if } f(\omega') - f(\omega) > 0 \\ 1 & \text{if } f(\omega') - f(\omega) \leq 0. \end{cases} \tag{1}$$

Define t_k as the temperature parameter at (outer loop) iteration k, such that

$$t_k > 0 \quad \text{for all } k \text{ and } \lim_{k \to +\infty} t_k = 0 \tag{2}$$

This acceptance probability is the basic element of the search mechanism in simulated annealing. If the temperature is reduced sufficiently slowly, then the system can reach an equilibrium (steady state) at each iteration k. Let $f(\omega)$ and $f(\omega')$ denote the energies (objective function values) associated with solutions $\omega \in \Omega$ and $\omega' \in \mathsf{N}(\omega)$, respectively. This equilibrium follows the Boltzmann distribution, which can be described as the probability of the system being in state $\omega \in \Omega$ with energy $f(\omega)$ at temperature T such that

$$\mathrm{P}\,\{\text{System is in state } \omega \text{ at temperature } T\} = \frac{\exp(-f(\omega)/t_k)}{\sum_{\omega'' \in \Omega} \exp(-f(\omega'')/t_k)}. \tag{3}$$

If the probability of generating a candidate solution ω' from the neighbors of solution $\omega \in \Omega$ is $g_k(\omega, \omega')$, where

$$\sum_{\omega' \in \mathsf{N}(\omega)} g_k(\omega, \omega') = 1, \quad \text{for all } \omega \in \Omega, \; k = 1, 2, \ldots, \tag{4}$$

then a non-negative square stochastic matrix $\mathbf{P}_k$ can be defined with transition probabilities

$$\mathbf{P}_k(\omega, \omega') = \begin{cases} g_k(\omega, \omega') \exp(-\Delta_{\omega,\omega'}/t_k) & \omega' \in \mathsf{N}(\omega), \omega' \neq \omega \\ 0 & \omega' \notin \mathsf{N}(\omega), \omega' \neq \omega \\ 1 - \sum\limits_{\substack{\omega'' \in \mathsf{N}(\omega) \\ \omega'' \neq \omega}} \mathbf{P}_k(\omega, \omega'') & \omega' = \omega \end{cases} \tag{5}$$

for all solutions $i \in \Omega$ and all iterations $k = 1, 2, \ldots$ and $\Delta_{\omega,\omega'} \equiv f(\omega') - f(\omega)$. These transition probabilities define a sequence of solutions generated from an inhomogeneous Markov chain (Romeo and Sangiovanni-Vincentelli, 1991). Note that boldface type indicates matrix/vector notation, and all vectors are row vectors.

1.3 Statement of Algorithm

Simulated annealing is outlined in pseudo-code (Eglese, 1990).

Select an initial solution $\omega \in \Omega$
Select the temperature change counter $k = 0$
Select a temperature cooling schedule, t_k
Select an initial temperature $T = t_0 \geq 0$
Select a repetition schedule, M_k that defines the number of iterations executed at each temperature, t_k
Repeat
Set repetition counter $m = 0$
    Repeat
    Generate a solution $\omega' \in N(\omega)$
    Calculate $\Delta_{\omega,\omega'} = f(\omega') - f(\omega)$
    If $\Delta_{\omega,\omega'} \leq 0$, then $\omega \leftarrow \omega'$
    If $\Delta_{\omega,\omega'} > 0$, then $\omega \leftarrow \omega'$ with probability $\exp(-\Delta_{\omega,\omega'}/t_k)$
    $m \leftarrow m + 1$
    Until $m = M_k$
$k \leftarrow k + 1$
Until stopping criterion is met

This simulated annealing formulation results in $M_0 + M_1 + \cdots + M_k$ total iterations being executed, where k corresponds to the value for t_k at which the stopping criteria is met. In addition, if $M_k = 1$ for all k, then the temperature changes at each iteration.

1.4 Discrete versus Continuous Problems

The majority of the theoretical developments and application work with simulated annealing has been for discrete optimization problems. However simulated annealing has also been used as a tool to address problems in the continuous domain. There is considerable interest in using simulated annealing for global optimization over regions containing several local and global minima (due to inherent non-linearity of objective functions). Fabian (1997) studies the performance of simulated annealing methods for finding a global minimum of a function and Bohachevsky et al. (1986) presents a generalized simulated annealing algorithm for function optimization for use in statistical applications. Optimization of continuous functions involves finding a candidate location by picking a direction from the current (incumbent) solution and a step size to take in this direction, and evaluating the function at the new (candidate) location. If the function value of this candidate location is an improvement over the function value of the incumbent location, then the candidate becomes the incumbent. This migration through local minima in search of a global minimum continues until the global minimum is found or some termination criteria are reached. Belisle (1992) presents a special simulated annealing algorithm for global optimization, which uses a heuristically motivated cooling schedule. This algorithm is easy to implement and provides a reasonable alternative to existing methods.

Belisle et al. (1993) discusses convergence properties of simulated annealing algorithms applied to continuous functions and applies these results to hit-and-run algorithms used in global optimization. His convergence properties are consistent with those presented by Hajek (1988) and he provides a good contrast between convergence in probability and the stronger almost sure convergence. Zabinsky et al. (1993) extends this work to an improved hit-and-run algorithm used for global optimization.

Fleischer and Jacobson (1996) proposes cybernetic optimization by simulated annealing as a method of parallel processing that accelerated the convergence of simulated annealing to the global optima. Fleischer (1999) extends the theory of cybernetic optimization by simulated annealing into the continuous domain by applying probabilistic feedback control to the generation of candidate solutions. The probabilistic feedback control method of generating candidate solutions effectively accelerates convergence to a global optimum using parallel simulated annealing on continuous variable problems.

Locatelli (1996) presents convergence properties for a class of simulated annealing algorithms for continuous global optimization by removing the restriction that the next candidate point must be generated according to a probability distribution whose support is the whole feasible set. Siarry et al. (1997) studies simulated annealing algorithms for globally minimizing functions of many-continuous variables. Their work focuses on how high-dimensionality can be addressed using variables discretization, as well as considering the design and implementation of several complementary stopping criteria. Yang (2000) and Locatelli (2000) also provide convergence results and criteria for simulated annealing applied to continuous global optimization problems. Kiatsupaibul and Smith (2000) introduces a general purpose simulated annealing algorithm to solve mixed integer linear programs. The simulated annealing algorithm is constructed using a Markov chain sampling algorithm to generate uniformly distributed points on an arbitrary bounded region of a high dimensional integer lattice. They show that their algorithm converges in probability to a global optimum. Romeijn et al. (1999) also studies a simulated annealing algorithm that uses a reflection generator for mixed integer/continuous global optimization problems.

2 CONVERGENCE RESULTS

Convergence results for simulated annealing have typically taken one of two directions; either the algorithm has been modeled as a sequence of homogeneous Markov chains or as a single inhomogeneous Markov chain.

The homogeneous Markov chain approach (see, e.g., Aarts and van Laarhoven, 1985; Faigle and Kern, 1991; Granville et al., 1994; Johnson and Jacobson, 2002a,b; Lundy and Mees, 1986; Mitra et al., 1986; Rossier et al., 1986) assumes that each temperature t_k is held constant for a sufficient number of iterations m such that the stochastic matrix $\mathbf{P}_k$ can reach its stationary (steady state) distribution π_k. Note that in the interest of simplifying notation, the inner loop index m is suppressed. However, the index k should be interpreted as the double index k,m, where a sequence of $m = 1, 2, \ldots, \mathrm{M}_k$ simulated annealing iterations occur for each fixed k.) The existence of a stationary distribution at each iteration k follows from Theorem 10.1. (Note: to ensure that Theorem 1 is consistent with the simulated annealing algorithm depicted in Section 1.3, without loss of generality, let t_k be a function only of each outer loop iteration k, and let the respective number of inner loop iterations M_k and outer loop iterations k each approach infinity).

Theorem 10.1. *Let $\mathbf{P}_k(\omega, \omega')$ be the probability of moving from solution ω to solution ω' in one inner iteration at outer loop k, and let $\mathbf{P}_k^{(m)}(\omega, \omega')$ be the probability of going from solution ω to solution ω' in m inner loops. If the Markov chain associated with $\mathbf{P}_k^{(m)}(\omega, \omega')$ is irreducible and aperiodic with finitely many solutions, then $\lim_{m\to\infty} \mathbf{P}_k^{(m)}(\omega, \omega') = \pi_k(\omega')$ exists for all $\omega, \omega' \in \Omega$ and iterations k. Furthermore, $\pi_k(\omega')$ is the unique strictly positive solution of*

$$\pi_k(\omega') = \sum_{\omega\in\Omega} \pi_k(\omega)\mathbf{P}_k(\omega, \omega'), \quad \text{for all } \omega \in \Omega, \tag{6}$$

and

$$\sum_{\omega\in\Omega} \pi_k(\omega) = 1. \tag{7}$$

Proof. See Cinlar (1974) p. 153. ■

The key requirements for the existence of the stationary distributions and for the convergence of the sequence of $\boldsymbol{\pi}_k$ vectors include

1. transition matrix irreducibility (for every finite outer loop k, the transition matrix can assign a path of non-zero probability between any two solutions $\omega, \omega' \in \Omega$),
2. aperiodicity (starting at solution $\omega' \in \Omega$, it is possible to return to ω, ω' with period 1. See Isaacson and Madsen (1976),
3. a non-zero stationary probability distribution, as the number of outer loops k approaches infinity.

Note that all simulated annealing proofs of convergence in the literature based on homogeneous Markov chain theory, either explicitly or implicitly, use the sufficient condition of reversibility (also called detailed balance) (Ross, 1996), defined as

$$\pi_k(\omega)\mathbf{P}_k(\omega, \omega') = \pi_k(\omega')\mathbf{P}_k(\omega', \omega), \text{ for all } \omega, \omega' \in \Omega, \text{ and all iterations } k. \tag{8}$$

Reversibility is sufficient condition for a unique solution to exist for (6) and (7) at each outer loop iteration k. Ross (1997) shows that a necessary condition for reversibility is multiplicativity (i.e., for any three solutions $\omega, \omega', \omega'' \in \Omega$ such that $f(\omega) \le f(\omega') \le f(\omega'')$, and for all iterations k,

$$a_k(\Delta_{\omega,\omega''}) = a_k(\Delta_{\omega,\omega'})a_k(\Delta_{\omega',\omega''}) \tag{9}$$

where $a_k(\Delta_{\omega,\omega'})$ is the probability of accepting the transition from solution ω to solution ω' at outer loop iteration k). Reversibility is enforced by assuming conditions of symmetry on the solution generation probabilities and by either directly expressing the acceptance probability using an exponential form, or by requiring the multiplicative condition in (9).

The homogeneous Markov chain proofs of convergence in the literature (implicitly or explicitly) require the condition in (9) to hold for the acceptance function, and then address the sufficient conditions on the solution generation matrix. For example, the original homogeneous proofs of convergence (Aarts and van Laarhoven, 1985; Lundy and Mees, 1986) require the multiplicative condition for the acceptance function, and then assume that the solution generation function is symmetric and constant for all outer loop iterations k. Rossier et al. (1986) partitions the solution space into blocks composed of neighboring solutions of equal objective function value, and then requires that only the solution generation probabilities be symmetric between these blocks. Rossier et al. (1986) then expresses the acceptance function as a ratio of the stationary distribution probabilities (discussed in Section 2.1.3). Faigle and Schrader (1988) and Faigle and Kern (1991) use a graph theoretic approach to relax the solution generation function symmetry condition. However, they require that the solution acceptance probability function satisfies (9).

Granville et al. (1994) proposes a simulated annealing procedure for filtering binary images, where the acceptance function is based on the probability of the current solution, instead of the change in objective function value. The probability function that Granville et al. (1994) present for accepting a candidate solution at (outer loop) iteration k is based on the ratio of the stationary probability of the incumbent solution from iteration $k-1$, versus the stationary probability of an initial solution (which is based on a maximum likelihood estimate). The acceptance probability is

$$\xi_k = (q\pi_0(\omega)/\pi_{k-1}(\omega')^{\phi(k)} \tag{10}$$

where $q = \inf_{\omega\in\Omega}\pi(\omega)/\sup_{\omega'\in\Omega}\pi(\omega')$ (q must also be estimated), and $\phi(k)$ is a slowly increasing function. Therefore, the probability of a solution transition does not consider the objective function value of the candidate solution. Granville et al. (1994) provides a proof of asymptotic convergence of this approach, but note that the proof methodology does not show that the set of globally optimal solutions are asymptotically uniformly distributed.

Simulated annealing and the homogeneous convergence theory are based on the work of Metropolis et al. (1953), which addresses problems in equilibrium statistical mechanics (Hammersley and Handscomb, 1964). To see this relationship, consider a system in thermal equilibrium with its surroundings, in solution (state) S with energy $F(S)$. The probability density in phase space of the point representing S is proportional to

$$\exp(-F(S)/bT), \tag{11}$$

where b is the Boltzmann constant, and T is the absolute temperature of the surroundings. Therefore the proportion of time that the system spends in solution S is proportional to (11) (Hammersley and Handscomb, 1964), hence the equilibrium probability density for all $S \in \Omega$ is

$$\pi_S = \frac{\exp(-F(S)/bT)}{\int \exp(-F(S)/bT)\,dS}. \tag{12}$$

The expectation of any valid solution function $f(S)$ is thus

$$E[f] = \frac{\int f(S)\exp(-F(S)/bT)\,dS}{\int \exp(-F(S)/bT)\,dS}. \tag{13}$$

Unfortunately, for many solution functions, (13) cannot be evaluated analytically. Hammersley and Handscomb (1964) notes that one could theoretically use naive Monte Carlo techniques to estimate the value of the two integrals in (11). However, this often fails in practice since the exponential factor means that a significant portion of the integrals is concentrated in a very small region of the solution space Ω. This problem can be overcome using importance sampling (see Bratley et al., 1987; Chapter 2), by generating solutions with probability density (12). This approach would also seem to fail, because of the integral in the denominator of (12). However, Metropolis et al. (1953) solves this problem by first discretizing the solution space, such that the integrals in (12) and (13) are replaced by summations over the set of discrete solutions $\omega' \in \Omega$, and then by constructing an irreducible, aperiodic Markov chain with transition probabilities $\mathbf{P}(\omega, \omega')$ such that

$$\pi(\omega') = \sum_{\omega \in \Omega} \pi(\omega)\mathbf{P}(\omega, \omega') \quad \text{for all } \omega' \in \Omega, \tag{14}$$

where

$$\pi(\omega') = \frac{\exp(-F(\omega')/bT)}{\sum_{\omega \in \Omega} \exp(-F(\omega)/bT)} \quad \text{for all } \omega' \in \Omega. \tag{15}$$

Note that to compute the equilibrium distribution $\boldsymbol{\pi}$, the denominator of (13) (a normalizing constant) does not need to be calculated. Instead, the ratios $\pi(\omega')/\pi(\omega)$ need only be computed and a transition matrix $\mathbf{P}$ defined that satisfies (14). Hammersley and Handscomb (1964) show that Metropolis et al. (1953) accomplishes this by defining $\mathbf{P}$ as the product of symmetric solution generation probabilities $g(\omega, \omega')$, and the equilibrium ratios $\pi(\omega)'/\pi(\omega)$,

$$\mathbf{P}(\omega, \omega') = \begin{cases} g(\omega, \omega')\pi(\omega')/\pi(\omega) & \text{if } \pi(\omega')/\pi(\omega) < 1, \omega' \neq \omega \\ g(\omega, \omega') & \text{if } \pi(\omega')/\pi(\omega) \geq 1, \omega' \neq \omega \\ g(\omega, \omega') + \sum\limits_{\substack{\omega'' \in \Omega, \\ \pi(\omega'') < \pi(\omega)}} g(\omega, \omega'')(1 - (\pi(\omega'')/\pi(\omega))) & \text{if } \omega' = \omega \end{cases} \tag{16}$$

where

$$g(\omega, \omega') \geq 0, \sum_{\omega' \in \Omega} g(\omega, \omega') = 1, \text{ and } g(\omega, \omega') = g(\omega', \omega) \quad \text{for all } \omega, \omega' \in \Omega \tag{17}$$

The use of stationary probability ratios to define the solution acceptance probabilities, combined with symmetric solution generation probabilities, enable Metropolis et al. (1953) to use the reversibility condition in (8) to show that (16) and (17) satisfy (14).

Homogeneous proofs of convergence for simulated annealing become more difficult to establish when the reversibility condition is not satisfied. Note that the existence of a unique stationary distribution for each outer loop iteration k is easily shown by specifying that each transition matrix $\mathbf{P}_k$ be irreducible and aperiodic. On the other hand, it becomes very difficult to derive an explicit closed-form expression for each stationary distribution $\boldsymbol{\pi}_k$ that remains analytically tractable as the problem's solution space becomes large. One can no longer use (8) to describe each stationary distribution, because in general, the multiplicative condition is not met. Instead, one must directly solve the system of equations formed with (6) and (7). For example, Davis (1991) attempts to obtain a closed-form expression for $\boldsymbol{\pi}_k$ by using Cramer's rule and rewriting (6) and (7) as

$$\boldsymbol{\pi}_k(\mathbf{I} - \mathbf{P}_k) = 0 \tag{18}$$

and

$$\boldsymbol{\pi}_k \mathbf{e}^{\mathrm{T}} = 1, \tag{19}$$

respectively, where boldface type indicates vector/matrix notation, **I** is the identity matrix, and $\mathbf{e}^{\mathrm{T}}$ is a column vector of ones. Note that the card(Ω) × card(Ω) transition matrix $\mathbf{P}_k$ associated with (18) is of rank card(Ω)-1 (Çinlar, 1974). Therefore, by deleting any one equation from (18), and substituting (19), the result is the set of card(Ω) linearly independent equations

$$\boldsymbol{\pi}_k \underline{(\mathbf{I} - \mathbf{P}_k)} = \mathbf{e}_i \tag{20}$$

where the square matrix $\underline{(\mathbf{I} - \mathbf{P}_k)}$ is obtained by substituting the ith column of matrix $(\mathbf{I} - \mathbf{P}_k)$ with a column vector of ones. The vector $\mathbf{e}_i$ is a row vector of zeroes, except for a one in the ith position. Since $\underline{(\mathbf{I} - \mathbf{P}_k)}$ is of full rank, then its determinant (written as $\det\underline{(\mathbf{I} - \mathbf{P}_k)}$), is non-zero. Define $\underline{(\mathbf{I} - \mathbf{P}_k)}^{\omega}$ to be the same matrix as $\underline{(\mathbf{I} - \mathbf{P}_k)}$ except that the elements of the ωth row of $\underline{(\mathbf{I} - \mathbf{P}_k)}$ are replaced by the vector $\mathbf{e}_{\omega}$. Therefore, for all iterations k,

$$\pi_k(\omega) = \frac{\det\underline{(\mathbf{I} - \mathbf{P}_k)}^{\omega}}{\det\underline{(\mathbf{I} - \mathbf{P}_k)}}, \quad \text{for all } \omega \in \Omega \tag{21}$$

Davis (1991) attempts to solve (21) for each $\omega \in \Omega$ via a multivariate Taylor series expansion of each determinant, but is not able to derive a closed-form analytical expression.

Overall, the difficulty of explicitly expressing the stationary distributions for large solution spaces, combined with bounding the transition matrix condition number for large k, suggests that it is very difficult to prove asymptotic convergence of the simulated annealing algorithm by treating (5) and (6) as a linear algebra problem.

Lundy and Mees (1986) notes that for each fixed outer loop iteration k, convergence to the solution equilibrium probability distribution vector $\boldsymbol{\pi}_k$ (in terms of the Euclidean distance between $P_k^{(m)}$ and $\boldsymbol{\pi}_k$, as $m \to +\infty$) is geometric since the solution space is finite, and the convergence factor is given by the second largest eigenvalue of the transition matrix $\mathbf{P}_k$. This result is based on a standard convergence theorem for irreducible, aperiodic homogeneous Markov chains (see Çinlar, 1974). Note that a large solution space precludes practical calculation of this eigenvalue. Lundy and Mees (1986) conjectures that when the temperature t_k is near zero, the second largest eigenvalue will be close to one for problems with local optima, and thus convergence to the equilibrium distribution will be very slow (recall that the dominant eigenvalue for $\mathbf{P}_k$ is one, with algebraic multiplicity one (Isaacson and Madsen, 1976). Lundy and Mees (1986) uses its conjecture to justify why simulated annealing should be initiated with a relatively high temperature. For an overview of current methods for assessing non-asymptotic rates of convergence for general homogeneous Markov chains, see Rosenthal (1995).

The assumption of stationarity for each outer loop iteration k limits practical application of homogeneous Markov chain theory—Romeo and Sangiovanni-Vincentelli (1991) shows that if equilibrium (for a Markov chain that satisfies the reversibility condition) is reached in a finite number of steps, then it is achieved in one step. Thus,

Romeo and Sangiovanni-Vincentelli (1991) conjectures that there is essentially no hope for the most-used versions of simulated annealing to reach equilibrium in a finite number of iterations.

The second convergence approach for simulated annealing is based on inhomogeneous Markov chain theory (Anily and Federgruen, 1987; Gidas, 1985; Mitra et al., 1986). In this approach, the Markov chain need not reach a stationary distribution (e.g., the simulated annealing inner loop need not be infinitely long) for each outer loop k. On the other hand, an infinite sequence of (outer loop) iterations k must still be examined, with the condition that the temperature parameter t_k cool sufficiently slowly. The proof given by Mitra et al. (1986) is based on satisfying the inhomogeneous Markov chain conditions of weak and strong ergodicity (Isaacson and Madsen, 1976; Seneta, 1981). The proof requires four conditions:

1. The inhomogeneous simulated annealing Markov chain must be weakly ergodic (i.e., dependence on the initial solution vanishes in the limit).
2. An eigenvector $\boldsymbol{\pi}_k$ with eigenvalue one must exist such that (6) and (7) hold for every iteration k.
3. The Markov chain must be strongly ergodic (i.e., the Markov chain must be weakly ergodic and the sequence of eigenvectors $\boldsymbol{\pi}_k$ must converge to a limiting form), i.e.,

$$\sum_{k=0}^{\infty} \|\pi_k - \pi_{k+1}\| < +\infty. \tag{22}$$

4. The sequence of eigenvectors must converge to a form where all probability mass is concentrated on the set of globally optimal solutions ω^*.Therefore,

$$\lim_{k\to\infty} \pi_k = \pi^{\text{opt}}. \tag{23}$$

   where π^{opt} is the equilibrium distribution with only global optima having probabilities greater than zero. (Note that weak and strong ergodicity are equivalent for homogeneous Markov chain theory.)

Mitra et al. (1986) satisfies condition (1) (weak ergodicity) by first forming a lower bound on the probability of reaching any solution from any local minimum, and then showing that this bound does not approach zero too quickly. For example, they define the lower bound for the simulated annealing transition probabilities in (5) as

$$\mathbf{P}^{(m)}(\omega, \omega') \geq w^m \exp(-m\Delta_L / t_{km-1}) \tag{24}$$

where m is the number of transitions needed to reach any solution from any solution of non-maximal objective function value, $w > 0$ is a lower bound on the one-step solution generation probabilities, and Δ_L is the maximum one-step increase in objective function value between any two solutions. Mitra et al. (1986) shows that the Markov chain is weakly ergodic if

$$\sum_{k=k_0}^{\infty} \exp(-m\Delta_L / t_{km-1}) = +\infty \tag{25}$$

Therefore, weak ergodicity is obtained if the temperature t_k is reduced sufficiently slowly to zero such that (25) is satisfied. In general, the (infinite) sequence of temperatures $\{t_k\}, k = 1, 2, \ldots$ must satisfy

$$t_k \geq \frac{\beta}{\log(k)} \tag{26}$$

where $\lim_{k\to\infty} t_k = 0$, β is a problem-dependent constant, and k is the number of iterations. Mitra et al. (1986) shows that conditions (2), (3), and (4) are satisfied by using the homogeneous Markov chain theory developed for the transition probabilities (5), and assuming that the solution generation function is symmetric.

Romeo and Sangiovanni-Vincentelli (1991) notes that while the logarithmic cooling schedule in (26) is a sufficient condition for the convergence, there are only a few values for β which make the logarithmic rule also necessary. Furthermore, there exists a unique choice for β which makes the logarithmic rule both necessary and sufficient for the convergence of simulated annealing to the set of global optima. Hajek (1988) was the first to show that the logarithmic cooling schedule (26) is both necessary and sufficient, by developing a tight lower bound for β, namely the depth of the deepest local minimum which is not a global minimum, under a weak reversibility assumption. (Note that Hajek requires the depth of global optima to be infinitely deep.) Hajek defines a Markov chain to be weakly reversible if, for any pair of solutions $\omega, \omega' \in \Omega$ and for any non-negative real number h, ω is reachable at height h from ω' if and only if ω' is reachable at height h from ω. Note that Hajek (1988) does not attempt to satisfy the conditions of weak and strong ergodicity, but instead uses a more general probabilistic approach to develop a lower bound on the probability of escaping local, but not global optima. Connors and Kumar (1989) substantiate the necessary and sufficient conditions in Hajek (1988) using the orders of recurrence,

$$B_i \equiv \sup\left\{x \geq 0 : \sum_{k=0}^{\infty} \exp(-x/t_k)\pi_k(\omega) = +\infty\right\} \quad \text{for all } i \in \Omega. \tag{27}$$

Connors and Kumar (1989) shows that these orders of recurrence quantify the asymptotic behavior of each solution's probability in the solution distribution. The key result is that the simulated annealing inhomogeneous Markov chain converges in a Cesaro sense to the set of solutions having the largest recurrence orders. Borkar (1992) improves this convergence result by using a convergence/oscillation dichotomy result for martingales. Tsitsiklis (1989) uses bounds and estimates for singularly perturbed, approximately stationary Markov chains to develop a convergence theory that subsumes the condition of weak reversibility in Hajek (1988). (Note that Tsitsiklis (1989) defines $N(h) \subset \Omega$ as the set of all local minima (in terms of objective function value) of depth $h+1$ or more. Therefore β is the smallest h such that all local (but not global) minima have depth h or less. Tsitsiklis (1989) conjectures that without some form of reversibility, there does not exist any h such that the global optima are contained in the set of local optima.) Note that Chiang and Chow (1988, 1994), Borkar (1992), Connors and Kumar (1989), Hajek (1988), and Mitra et al. (1986) all require (either explicitly or implicitly) the multiplicative condition (9) for their proofs of convergence.

Anily and Federgruen (1987) uses perturbation analysis techniques (e.g., see Meyer, 1980) to prove convergence of a particular stochastic hill-climbing algorithm, by bounding the deviations of the sequence of stationary distributions of the particular hill-climbing algorithm against the sequence of known stationary distributions corresponding to a simulated annealing algorithm. In general, this convergence proof approach is only useful for a restrictive class of simulated annealing algorithms, since the transition matrix condition number grows exponentially as the number of iterations k becomes large.

Anily and Federgruen (1987) also presents a proof of convergence for simulated annealing with general acceptance probability functions. Using inhomogeneous Markov chain theory, it proves convergence under the following necessary and sufficient conditions:

1. The acceptance probability function must, for any iteration k, allow any hill-climbing transition to occur with positive probability.
2. The acceptance probability function must be bounded and asymptotically monotone, with limit zero for hill-climbing solution transitions.
3. In the limit, the stationary probability distribution must have zero probability mass for every non-globally optimal solution.
4. The probability of escaping from any locally (but not globally) optimal solution must not approach zero too quickly.

Anily and Federgruen (1987) uses condition (3) to relax the acceptance function multiplicative condition (9). However, in practice, condition (3) would be very difficult to check without assuming that (9) holds. Condition (4) provides the necessary condition for the rate that the probability of hill-climbing transitions approaches zero. Condition (4) is expressed quantitatively as follows: let t_k be defined by (2), and define the minimum one-step acceptance probability as

$$\underline{a}_{t_k} = \min_{\substack{\omega\in\Omega,\\ \omega'\in \mathrm{N}(\omega)}} a_{t_k}(\omega,\omega') \tag{28}$$

Define the set of local optima $L \subset \Omega$ such that $\omega \in L$ implies that $f(\omega) \leq f(\omega')$ for all $\omega' \in \mathrm{N}(\omega)$, and let

$$\bar{a}_{t_k} = \max_{\substack{\omega\in L,\\ \omega'\in \mathrm{N}(\omega)\backslash L}} a_{t_k}(\omega,\omega') \tag{29}$$

Finally, let any solution $\omega' \in \Omega$, be reachable from any solution $\omega \in \Omega$, in q transitions or less. Then if (non-globally) locally optimal solutions exist,

$$\sum_{k=1}^{\infty} (\underline{a}_{t_k})^q = +\infty \tag{30}$$

and conditions (1), (2), and (3) hold, then the simulated annealing algorithm will asymptotically converge to the set of global optima with probability one. However, if (non-globally) locally optimal solutions exist and

$$\sum_{k=1}^{\infty} \bar{a}_{t_k} < +\infty \tag{31}$$

then the probability of each solution is asymptotically dependent upon the initial solution. Therefore, the simulated annealing algorithm will not always converge to the set of global optima with probability one. Johnson and Jacobson (2002b) relaxes the sufficient conditions found in Anily and Federgruen (1987) by using a path argument between global optima and local (but not global) optima.

Yao and Li (1991) and Yao (1995) also discuss simulated annealing algorithms with general acceptance probabilities, though their primary contribution is with respect to general neighborhood generation distributions. Schuur (1997) provides a description of acceptance functions ensuring the convergence of the associated simulated annealing algorithm to the set of global optima.

The inhomogeneous proof concept is stronger than the homogeneous approach in that it provides necessary conditions for the rate of convergence, but its asymptotic nature suggests that practical implementation may not be feasible. Romeo and Sangiovanni-Vincentelli (1991) notes "there is no reason to believe that truncating the logarithmic temperature sequence would yield a good configuration, since the tail of the sequence is the essential ingredient in the proof." In addition, the logarithmic cooling schedule dictates a very slow rate of convergence. Therefore, most recent work has focused on methods of improving simulated annealing's finite-time behavior and modifying or blending the algorithm with other search methods such as genetic algorithms (Liepins and Hilliard, 1989), tabu search (Glover, 1994), or both (Fox, 1993).

3 RELATIONSHIP TO OTHER LOCAL SEARCH ALGORITHMS

The hill-climbing strategy inherent in simulated annealing has lead to the formulation of other such algorithms (e.g., threshold accepting, the noising method). Moreover, though different in how they traverse the solution space, both tabu search and genetic algorithm share with simulated annealing the objective of using local information to find global optima over solution spaces contaminated with multiple local optima.

3.1 Threshold Accepting

Questioning the very need for a randomized acceptance function, Dueck and Scheuer (1990), and independently, Moscato and Fontanari (1990) propose the threshold accepting algorithm, where the acceptance function is defined as

$$a_k(\Delta_{\omega,\omega'}) = \begin{cases} 1 & \text{if } Q_k \geq \Delta_{\omega,\omega'} \\ 0 & \text{otherwise} \end{cases}$$

with Q_k defined as the threshold value at iteration k. Q_k is typically set to be a deterministic, non-increasing step function in k. Dueck and Scheuer (1990) reports computational results that suggest dramatic improvements in traveling salesman problem solution quality and algorithm run-time over basic simulated annealing. Moscato and Fontanari (1990) reports more conservative results—they conjecture that simulated annealing's probabilistic acceptance function does not play a major role in the search for near-optimal solutions.

Althofer and Koschnick (1991) develops a convergence theory for threshold accepting based on the concept that simulated annealing belongs to the convex hull of threshold accepting. The idea presented in Althofer and Koschnick (1991) is that (for a finite Q_k threshold sequence) there can exist only finitely many threshold accepting transition matrices; but simulated annealing can have infinitely many transition matrices because of the real-valued nature of the temperature at each iteration. However, every simulated annealing transition matrix for a given problem can be represented as a convex combination of the finitely many threshold accepting transition matrices. Althofer and Koschnick (1991) is unable to prove that threshold accepting will asymptotically reach a global minimum, but it does prove the existence of threshold schedules that provide convergence to within an ε-neighborhood of the optimal solutions. Jacobson and Yücesan (2002a) proves that if the threshold value approaches zero as k approaches infinity, then the algorithm does not converge in probability to the set of globally optimal solutions.

Hu et al. (1995) modifies threshold accepting to include a non-monotonic, self-tuning threshold schedule in the hope of improving the algorithm's finite-time performance. Hu et al. (1995) allows the threshold Q_k to change dynamically (either up or down), based on the perceived likelihood of being near a local minimum. These changes are accomplished using a principle they call *dwindling expectation*—when the algorithm fails to move to neighboring solutions, the threshold Q_k is gradually increased, in the hope of eventually escaping a local optimum. Conversely, when solution transitions are successful, the threshold is reduced, in order to explore local optima. The experimental results based on two traveling salesman problems presented in Hu et al. (1995) showed that the proposed algorithm outperformed previous hill-climbing methods in terms of finding good solutions earlier in the optimization process.

Threshold accepting's advantages over simulated annealing lie in its ease of implementation and its generally faster execution time, due to the reduced computational effort in avoiding acceptance probability computations and generation of random numbers (Moscato and Fontanari, 1990). However, compared to simulated annealing, relatively few threshold accepting applications are reported in the literature (Lin et al., 1995; Scheermesser and Bryngdahl, 1995; Nissen and Paul, 1995).

3.2 Noising Method

Charon and Hudry (1993) advocates a simple descent algorithm called the *noising method.* The algorithm first perturbs the solution space by adding random noise to the problem's objective function values. The noise is gradually reduced to zero during the algorithm's execution, allowing the original problem structure to reappear. Charon and Hudry (1993) provides computational results, but does not prove that the algorithm will asymptotically converge to the set of globally optimal solutions. Charon and Hudry (2001) shows how the noising method is a generalization of simulated annealing and threshold accepting.

Storer et al. (1992) proposes an optimization strategy for sequencing problems, by integrating fast, problem-specific heuristics with local search. Its key contribution is to base the definition of the search neighborhood on a heuristic problem pair (h, p), where h is a fast, known, problem-specific heuristic and p represents the problem data. By perturbing the heuristic, the problem, or both, a neighborhood of solutions is developed. This neighborhood then forms the basis for local search. The hope is that the perturbations will cluster good solutions close together, thus making it easier to perform local search.

3.3 Tabu Search

Tabu search (Glover, 1994) is a general framework for a variety of iterative local search strategies for discrete optimization. Tabu search uses the concept of *memory* by controlling the algorithm's execution via a dynamic list of forbidden moves. This allows the tabu search algorithm to intensify or diversify its search of a given problem's solution space in an effort to avoid entrapment in local optima. Note that a criticism of simulated annealing is that it is completely memoryless (i.e., simulated annealing disregards all historical information gathered during the algorithm's execution). On the other hand, no proofs of convergence exist in the literature for the general tabu search algorithm.

Faigle and Kern (1992) proposes a particular tabu search algorithm called *probabilistic tabu search* (Glover, 1989) as a meta-heuristic to help guide simulated annealing. Probabilistic tabu search attempts to capitalize on both the asymptotic optimality of simulated annealing and the memory feature of tabu search. In probabilistic tabu search, the probabilities of generating and accepting each candidate solution are set as functions of both a temperature parameter (as in simulated annealing) and information gained in previous iterations (as for tabu search). Faigle and Kern (1992) are then able to prove asymptotic convergence of their particular tabu search algorithm by using methods developed for simulated annealing (Faigle and Kern, 1991; Faigle and Schraeder, 1988). Note that the results of Faigle and Kern (1992) build upon Glover (1989) where probabilistic tabu search was first introduced and contrasted with simulated annealing.

3.4 Genetic Algorithms

Genetic algorithms (Liepens and Hilliard, 1989) emulate the evolutionary behavior of biological systems. They generate a sequence of populations of qcandidate solutions to the underlying optimization problem by using a set of genetically inspired stochastic solution transition operators to transform each population of candidate solutions into a descendent population. The three most popular transition operators are reproduction, cross-over, and mutation (Davis, 1991). Davis and Principe (1991) and Rudolph (1994) attempt to use homogeneous finite Markov chain techniques to prove convergence of genetic algorithms (Cerf, 1998), but are unable to develop a theory comparable in scope to that of simulated annealing.

Mühlenbein (1997) presents a theoretical analysis of genetic algorithms based on population genetics. He counters the popular notion that models that mimic natural phenomenon are superior to other models. The article argues that evolutionary algorithms can be inspired by nature, but do not necessarily have to copy a natural phenomenon. He addresses the behavior of transition operators and designs new genetic operators that are not necessarily related to events in nature, yet still perform well in practice.

One criticism of simulated annealing is the slow speed at which it sometimes converges. Delport (1998) combines simulated annealing with evolutionary algorithms to improve performance in terms of speed and quality of solution. He improves this hybrid system of simulated annealing and evolutionary selection by improving the cooling schedule based on fast recognition of the thermal equilibrium in terms of selection intensity. This technique results in much faster convergence of the algorithm.

Sullivan and Jacobson (2000) links genetic algorithms with simulated annealing using generalized hill climbing algorithms (Jacobson et al., 1998). They first link genetic algorithms to ordinal hill climbing algorithms, which can then be used, through its formulation within the generalized hill climbing algorithm framework, to form a bridge with simulated annealing. Though genetic algorithms have proven to be effective for addressing intractable discrete optimization problems, and can be classified as a type of hill-climbing approach, its link with generalized hill climbing algorithms (through the ordinal hill climbing formulation) provides a means to establish well-defined relationships with other generalized hill climbing algorithms (like simulated annealing and threshold accepting). They also present two formulations of genetic algorithms that provide a first step towards developing a bridge between genetic algorithms and other local search strategies like simulated annealing.

4 PRACTICAL GUIDELINES

Implementation issues for simulated annealing can follow one of two paths—that of specifying problem-specific choices (neighborhood, objective function, and constraints), and that of specifying generic choices (generation and acceptance probability functions, and the cooling schedule) (Eglese, 1990). The principal shortcoming of simulated annealing is that it often requires extensive computer time. Implementation modifications generally strive to retain simulated annealing's asymptotic convergence character, but at reduced computer run-time. The methods discussed here are mostly heuristic.

Problem-Specific Choices

Objective Functions One problem-specific choice involves the objective function specification. Stern (1992) recommends a heuristic temperature-dependent penalty function as a substitute for the actual objective function for problems where low cost solutions have neighbors of much higher cost, or in cases of degeneracy (i.e., large neighborhoods of solutions of equal, but high costs). The original objective function surfaces, as the penalty and the temperature are gradually reduced to zero. This technique is similar to the noising method presented by Charon and Hudrey (1993), where the penalty function is described as noise and is reduced at each outer loop iteration of the algorithm. One speed-up technique is to evaluate only the difference in objective functions, $\Delta_{\omega,\omega'}$ instead of calculating both $f(\omega)$ and $f(\omega')$. Tovey (1988) suggests several methods of approximating $\Delta_{\omega,\omega'}$ by using surrogate functions (that are faster to evaluate than $\Delta_{\omega,\omega'}$, but not as accurate) probabilistically for cases when evaluation of $\Delta_{\omega,\omega'}$ is expensive; this technique is referred to as the *surrogate function swindle*.

Straub et al. (1995) improves the performance of simulated annealing on problems in chemical physics by using the classical density distribution instead of the molecular dynamics of single point particles to describe the potential energy landscape. Ma and Straub (1994) reports that using this distribution has the effect of smoothing the energy landscape by reducing both the number and depth of local minima.

Yan and Mukai (1992) considers the case when a closed-form formula for the objective function is not available. They use a probabilistic simulation (termed the stochastic ruler method) to generate a sample objective function value for an input solution, and then accept the solution if the sample objective function value falls within a predetermined bound. They also provide a proof of asymptotic convergence by extrapolating the convergence proofs for simulated annealing, and analyze the rate of convergence.

Generic Choices

Generation Probability Functions Generation probability functions are usually chosen as uniform distributions with probabilities proportional to the size of the neighborhood. The generation probability function is typically not temperature-dependent. Fox (1993) suggests that instead of *blindly* generating neighbors uniformly, adopt an *intelligent* generation mechanism that modifies the neighborhood and its probability distribution to accommodate search intensification or diversification, in the same spirit of the tabu search meta-heuristic. Fox (1993) also notes that simulated annealing convergence theory does not preclude this idea. Tovey (1988) suggests an approach with a similar effect, called the *neighborhood prejudice swindle*.

Acceptance Probability Functions The literature reports considerable experimentation with acceptance probability functions for hill-climbing transitions. The most popular is the exponential form (1). Ogbu and Smith (1990) considers replacing the basic simulated annealing acceptance function $a_k(\Delta_{\omega,\omega'})$ with a geometrically decreasing form that is independent of the change in objective function value. They adopt a *probabilistic-exhaustive* heuristic technique in which randomly chosen neighbors of a solution are examined and all solutions that are accepted are noted, but only the last solution accepted becomes the new incumbent. The hope is that this scheme will explore a broader area of the solution space of a problem. Their acceptance probability function is defined for all solutions $\omega, \omega' \in \Omega$ and for $k = 1, 2, \ldots, K$ as

$$a_k(\omega, \omega') = a_k = \begin{cases} c_1 x^{k-1} & \text{if } f(\omega') > f(\omega) \\ 1 & \text{otherwise} \end{cases}$$

where a_1 is the initial acceptance probability value, $x \in (0,1)$ is a reducing factor, and K is the number of stages (equivalent to a temperature cooling schedule). They also experiment with this method (and a neighborhood of large cardinality) on a permutation flow shop problem, and reports that its approach found comparable solutions to the basic simulated annealing algorithm in one-third the computation time.

4.1 Choice of Cooling Schedule

The simulated annealing cooling schedule is fully defined by an initial temperature, a schedule for reducing/changing the temperature, and a stopping criterion. Romeo and Sangiovanni-Vincentelli (1991) notes that an effective cooling schedule is essential to reducing the amount of time required by the algorithm to find an optimal solution. Therefore much of the literature on cooling schedules (e.g., Cardoso et al., 1994; Fox and Heine, 1993; Nourani and Andersen, 1998; and Cohn and Fielding, 1999) is devoted to this topic.

Homogeneous simulated annealing convergence theory has been used to design effective cooling schedules. Romeo and Sangiovanni-Vincentelli (1991) suggests the following procedure for designing a cooling schedule:

1. Start with an initial temperature t_0 for which a good approximation of the stationary distribution π_{t_0} is quickly reached.

2. Reduce t_0 by an amount $\delta(t)$ small enough such that π_{t_0} is a good starting point to approximate $\pi_{t_0-\delta(t)}$.
3. Fix the temperature at a constant value during the iterations needed for the solution distribution to approximate $\pi_{t_0-\delta(t)}$.

Repeat the above process of cooling and iterating until no further improvement seems possible.

Cooling schedules are grouped into two classes: *static* schedules, which must be completely specified before the algorithm begins; and *adaptive* schedules, which adjust the temperature's rate of decrease from information obtained during the algorithm's execution. Cooling schedules are almost always heuristic; they seek to balance moderate execution time with simulated annealing's dependence on asymptotic behavior.

Strenski and Kirkpatrick (1991) presents an exact (non-heuristic) characterization of finite-length annealing schedules. They consider extremely small problems that represent features (local optima and smooth/hilly topologies), and solve for solution probabilities after a finite number of iterations to gain insights into some popular assumptions and intuition behind cooling schedules. Their experiments suggest that optimal cooling schedules are *not* monotone decreasing in temperature. They also show that for the test problem (a white noise surface), geometric and linear cooling schedules perform better than inverse logarithmic cooling schedules, when sufficient computing effort is allowed. Moreover, their experiments do not show measurable performance differences between linear and geometric cooling schedules. They also observe that geometric cooling schedules are not greatly affected by excessively high initial temperatures. The results presented suggest that the even the most robust adaptive cooling schedule "produces annealing trajectories which are never in equilibrium" (Strenski and Kirkpatrick, 1991). However, they also conclude that the transition acceptance rate is not sensitive to the degree of closeness to the equilibrium distribution.

Christoph and Hoffmann (1993) also attempts to characterize optimal cooling schedules. They derive a relationship between a finite sequence of optimal temperature values (i.e., outer loops) and the number of iterations (i.e., inner loops) at each respective temperature for several small test problems to reach optimality (i.e., the minimal mean final energy). They find that this *scaling behavior* is of the form

$$x_m = a_m v^{-b_m} \tag{32}$$

where a and b are scaling coefficients, $x_m = e^{-1/t_k}$ is referred to as the temperature, v is the number of inner loop iterations at temperature x_m, and m is the number of outer loops at which the temperature x_m is reduced. The proposed approach is to solve for the coefficients a and b based on known temperature and iteration parameter values for an optimal schedule based on several replications of the algorithm using $(m \times v)$ iterations for each replication, and then use (32) to interpolate the optimal cooling schedule for intermediate iterations. They however do not make any suggestions on how to efficiently solve for the necessary optimal cooling schedules for a (typically large) problem instance.

Romeo and Sangiovanni-Vincentelli (1991) presents a theoretical framework for evaluating the performance of the simulated annealing algorithm. They discuss annealing schedules in terms of initial temperature T, the number of inner loops for each value of T, the rate that the temperature T decreases (i.e., cooling schedule) and the criteria for stopping the algorithm. They conclude that the theoretical results obtained thus far have not been able to explain why simulated annealing is so successful even when a diverse collection of static cooling schedule heuristics is used. Many heuristic methods are available in the literature to find optimal cooling schedules, but the effectiveness of these schedules can only be compared through experimentation. They conjecture that the neighborhood and the corresponding topology of the objective function are responsible for the behavior of the algorithm.

Cohn and Fielding (1999) conducts a detailed analysis of various cooling schedules and how they affect the performance of simulated annealing. Convergent simulated annealing algorithms are often too slow in practice, whereas a number of non-convergent algorithms may be preferred for good finite-time performance. They analyze various cooling schedules and present cases where repeated independent runs using a non-convergent cooling schedule provide acceptable results in practice. They provide examples of when it is both practically and theoretically justified to use a very high, fixed temperature, or even fast cooling schedules which have a small probability of reaching global minima and apply these cooling schedules to traveling salesman problems of various sizes. Fielding (2000) computationally studies fixed temperature cooling schedules for the traveling salesman problem, the quadratic assignment problem, and the graph partitioning problem, and demonstrates that a fixed temperature cooling schedule can yield superior results in locating optimal and near-optimal solutions. Orosz and Jacobson (2002a,b) present finite-time performance measures for simulated annealing with fixed temperature cooling schedules. They illustrate their measures using randomly generated instances of the traveling salesman problem.

Another approach to increasing the speed of simulated annealing is to implement a two-staged simulated annealing algorithm. In two-staged simulated annealing algorithms, a fast heuristic is used to replace simulated annealing at higher temperatures, with a traditional simulated annealing algorithm implemented at lower temperatures to improve on the fast heuristic solution. In addition to implementing an intelligent cooling schedule, finding the initial temperature t_0 to initialize the traditional simulated annealing algorithm is important to the success of the two-staged algorithm. Varanelli and Cohoon (1999) proposes a method for determining an initial temperature t_0 for two-staged simulated annealing algorithms using traditional cooling schedules. They note that if t_0 is too low at the beginning of the traditional simulated annealing phase, the algorithm can get trapped in an inferior solution, while if the initial temperature t_0 is too high, the algorithm can waste too many iterations (hence computing time) by accepting too many hill-climbing moves.

4.2 Choice of Neighborhood

A key problem-specific choice concerns the neighborhood function definition. The efficiency of simulated annealing is highly influenced by the neighborhood function used (Moscato, 1993). The choice of neighborhood serves to enforce a topology—Eglese (1990) reports that "a neighborhood structure which imposes a 'smooth' topology where the local minima are shallow is preferred to a 'bumpy' topology where there are many deep local minima." Solla et al. (1986) and Fleischer and Jacobson (1999) report similar conclusions. This also supports the result in Hajek (1988) that shows that asymptotic convergence to the set of global optima depends on the depth of the local minima.

Another factor to consider when choosing neighborhood functions is the neighborhood size. No theoretical results are available, other than the necessity of reachability (in a finite number of steps) from any solution to any other solution. Cheh et al. (1991) reports that small neighborhoods are best, while Ogbu and Smith (1990) provides evidence that larger neighborhoods result in better simulated annealing performance. Goldstein and Waterman (1988) conjectures that if the neighborhood size is small compared to the total solution space cardinality, then the Markov chain cannot move around the solution space fast enough to find the minimum in a reasonable time. On the other hand, a very large neighborhood has the algorithm merely sampling randomly from a large portion of the solution space, and thus, is unable to focus on specific areas of the solution space. It is reasonable to believe that neighborhood size is heavily problem-specific. For example, problems where the smoothness of its solution space topology is moderately insensitive to different neighborhood definitions may benefit from larger neighborhood sizes.

Fleischer (1993) and Fleischer and Jacobson (1999) use concepts from information theory to show that the neighborhood structure can affect the *information rate* or total uncertainty associated with simulated annealing. Fleischer (1993) shows that simulated annealing tends to perform better as the entropy level of the associated Markov chain increases, and thus conjectures that an entropy measure could be useful for predicting when simulated annealing would perform well on a given problem. However, efficient ways of estimating the entropy are needed to make this a practical tool.

Another issue on neighborhood function definition addresses the solution space itself. Chardaire et al. (1995) proposes a method for addressing 0–1 optimization problems, in which the solution space is progressively reduced by fixing the value of *strongly persistent* variables (which have the same value in all optimal solutions). They isolate the persistent variables during simulated annealing's execution by periodically estimating the expectation of the random variable (a vector of binary elements) that describes the current solution, and fixing the value of those elements in the random variable that meet threshold criteria.

4.3 Domains—Types of Problems with Examples

Simulated annealing has developed into a popular tool for optimization in the last decade. It has been used to address numerous discrete optimization problems as well as continuous variable problems. Several application and computational survey articles have been published on simulated annealing. Johnson et al. (1989, 1991) present a series of articles on simulated annealing applied to certain well-studied discrete optimization problems. The first in the series of articles uses the graph partitioning problem to illustrate simulated annealing and highlight the effectiveness of several modifications to the basic simulated annealing algorithm. The second in the series focuses on applying lessons learned from the first article to the graph coloring and number partitioning problems. Local optimization techniques were previously thought to be unacceptable approaches to these two problems. Johnson et al. (1991) also observes that for long run lengths, simulated annealing outperforms the traditional techniques used to solve graph coloring problems. However, simulated annealing did not compare well with traditional techniques on the number partitioning problem except for small problem instances. The third article in the series (not yet published) uses simulated annealing to approach the well-known traveling salesman problem.

Koulamas et al. (1994) focuses specifically on simulated annealing applied to applications in production/operations management and operations research. They discuss traditional problems such as single machine, flow shop and job shop scheduling, lot sizing, and traveling salesman problems as well as non-traditional problems to include graph coloring and number partitioning. They conclude that simulated annealing is an effective tool for solving many problems in operations research and that the degree of accuracy that the algorithm achieves can be controlled by the practitioner, in terms of number of iterations and neighborhood functions (i.e., an increased number of iterations (outer loops) combined with increased number of searches at each iteration (inner loops) can result in solutions with a higher probability of converging to the optimal solution). Fleischer (1995) discusses simulated annealing from a historical and evolutionary point of view in terms of solving difficult optimization problems. He summarizes on-going research and presents an application of simulated annealing to graph problems.

The simulated annealing algorithm has proved to be a good technique for solving difficult discrete optimization problems. In engineering optimization, simulated annealing has emerged as an alternative tool to address problems that are difficult to solve by conventional mathematical programming techniques. The algorithm's major disadvantage is that solving a complex system problem may be an extremely slow, albeit convergent process, using much more processor time than some conventional algorithms. Consequently, simulated annealing has not been widely embraced as an optimization algorithm for engineering problems. Attempts have been made to improve the performance of the algorithm either by reducing the annealing length or changing the generation and the acceptance mechanisms. However, these faster schemes, in general, do not inherit the property of escaping local minima. A more efficient way to reduce the processor time and make simulated annealing a more attractive alternative for engineering problems is to add parallelism (e.g., see Hamma et al., 2000). However, the implementation and efficiency of parallel simulated annealing algorithms are typically problem-dependent. Leite et al. (1999) considers the evaluation of parallel schemes for engineering problems where the solution spaces may be very complex and highly constrained, with function evaluations varying from medium to high cost. In addition, they provide guidelines for selecting appropriate schemes for engineering problems. They also present an engineering problem with relatively low fitness evaluation cost and strong time constraints to demonstrate the lower bounds of applicability of parallel schemes.

Many signal processing applications create optimization problems with multimodal and non-smooth cost functions. Gradient methods are ineffective in these situations because of multiple local minima and the requirement to calculate gradients. Chen and Luk (1999) proposes an adaptive simulated annealing algorithm as a viable optimization tool for addressing such difficult non-linear optimization problems. The adaptive simulated annealing algorithm maintains the advantages of simulated annealing, but converges faster. Chen and Luk demonstrate the effectiveness of adaptive simulated annealing with three signal processing applications: maximum likelihood joint channel and data estimation, infinite-impulse-response filter design and evaluation of minimum symbol-error-rate decision feedback equalizer. They conclude that the adaptive simulated annealing algorithm is a powerful global optimization tool for solving such signal processing problems.

Abramson et al. (1999) describes the use of simulated annealing for solving the school time tabling problem. They use the scheduling problem to highlight the performance of six different cooling schedules: the basic geometric cooling schedule, a scheme that uses multiple cooling rates, geometric reheating, enhanced geometric reheating, non-monotonic cooling, and reheating as a function of cost. The basic geometric cooling schedule found in van Laarhoven and Aarts (1987) is used as the baseline schedule for comparison purposes. Experimental results suggest that using multiple cooling rates for a given problem yields better quality solutions in less time than the solutions produced by a single cooling schedule. They conclude that the cooling scheme that uses the phase transition temperature (i.e., when sub-parts of the combinatorial optimization problem are solved) in combination with the best solution to date produces the best results.

Emden-Weinert and Proksch (1999) presents a study of a simulated annealing algorithm for the airline crew-pairing problem based on an algorithm run-cutting formulation. They found that the algorithm run-time can be decreased and solution quality can be improved by using a problem-specific initial solution, relaxing constraints, combining simulated annealing with a problem-specific local improvement heuristic, and by conducting multiple independent runs.

There is no question that simulated annealing can demand significant computational time to reach global minima. Recent attempts to use parallel computing schemes to speed up simulated annealing have provided promising results. Chu et al. (1999) presents a new, efficient, and highly general-purpose parallel optimization method based upon simulated annealing that does not depend on the structure of the optimization problem being addressed. Their algorithm was used to analyze a network of interacting genes that control embryonic development and other fundamental biological processes. They use a two-stage procedure which monitors and pools performance statistics obtained simultaneously from all processors and then mixes states at intervals to maintain a Boltzman-like distribution of costs. They demonstrate that their parallel simulated annealing approach leads to nearly optimal parallel efficiency for certain optimization problems. In particular, the approach is appropriate when the relative effort required to compute the cost function is large compared to the relative communication effort between parallel machines for pooling statistics and mixing states.

Alrefaei and Andradottir (1999) presents a modified simulated annealing algorithm with a constant temperature to address discrete optimization problems and use two approaches to estimate an optimal solution to the problem. One approach estimates the optimal solution based on the state most visited versus the state last visited, while the other approach uses the best average estimated objective function value to estimate the optimal solution. Both approaches are guaranteed to converge almost surely to the set of global optimal solutions under mild conditions. They compare performance of the modified simulated annealing algorithm to other forms of simulated annealing used to solve discrete optimization problems.

Creating effective neighborhood functions or neighborhood generation mechanisms is a critical element in designing efficient simulated annealing algorithms for discrete optimization problems. Tian et al. (1999) investigates the application of simulated annealing to discrete optimization problems with a permutation property, such as the traveling salesman problem, the flow-shop scheduling problem, and the quadratic assignment problems. They focus on the neighborhood function of the discrete optimization problem and in particular the generation mechanism for the algorithm used to address the problem. They introduce six types of perturbation scheme for generating random permutation solutions and prove that each scheme satisfies asymptotic convergence requirements. The results of the experimental evaluations on the traveling salesman problem, the flow-shop scheduling problem, and the quadratic assignment problem suggest that the efficiencies of the perturbation schemes are different for each problem type and solution space. They conclude that with the proper perturbation scheme, simulated annealing produces efficient solutions to different discrete optimization problems that possess a permutation property.

Research continues to focus on the idea of simulated annealing applied to optimization of continuous functions. Continuous global optimization is defined as the problem of finding points on a bounded subset of $\Re^n$ where some real valued function f assumes its optimal (maximal or minimal) value. Application of simulated annealing to continuous optimization generally falls into two classes. The first approach closely follows the original idea presented by Kirkpatrick et al. (1983) in that the algorithm mimics the physical annealing process. The second approach describes the annealing process with Langevin equations, where the global minimum is found by solving stochastic differential equations (see Aluffi-Pentini et al., 1985). Geman and Hwang (1986) proves that continuous optimization algorithms based on Langevin equations converge to the global optima. Dekkers and Aarts (1991) proposes a third stochastic approach to address global optimization based on simulated annealing. Their approach is very similar to the formulation of simulated annealing as applied to discrete optimization problems. They extend the mathematical formulation of simulated annealing to continuous optimization problems, and prove asymptotic convergence to the set of global optima based on the equilibrium distribution of Markov chains. They also discuss an implementation of the proposed algorithm and compares its performance with other well-known algorithms on a standard set of test functions from the literature.

5 FUTURE DIRECTIONS

5.1 Generalized Hill Climbing Algorithms

Generalized Hill Climbing algorithms (GHC) (Jacobson et al., 1998) provide a framework for modeling local search algorithms used to address intractable discrete optimization problems. All generalized hill climbing algorithms have the same basic structure, but can be tailored to a specific instance of a problem by changing the hill-climbing random variable (which is used to accept or reject inferior solutions) and neighborhood functions. Generalized hill climbing algorithms are described in pseudo-code form:

Select an initial solution $\omega \in \Omega$

Set the outer loop counter bound K and the inner loop counter bounds
 $M_k, k = 1, 2, \ldots, K$

Define a set of hill-climbing (random) variables $R_k: \Omega \times \Omega \rightarrow \Re \cup \{-\infty, +\infty\}$,
 $k = 1, 2, \ldots, K$

Set the iteration indices $k = m = 1$

Repeat while $k \leq K$

Repeat while $m \leq \mathrm{M}_k$

Generate a solution $\omega' \in \mathrm{N}(\omega)$

Calculate $\Delta_{\omega,\omega'} = f(\omega') - f(\omega)$

If $\mathrm{R}_k(\omega, \omega') \geq \Delta_{\omega,\omega'}$, then $\omega \leftarrow \omega'$

If $\mathrm{R}_k(\omega, \omega') < \Delta_{\omega,\omega'}$, then $\omega \leftarrow \omega$

$m \leftarrow m + 1$

Until $m = \mathrm{M}_k$

$m \leftarrow 1, k \leftarrow k + 1$

Until $k = K$

Note that the outer and inner loop bounds, K and $\mathrm{M}_k, k = 1, 2, \ldots, K$, respectively, may all be fixed, or K can be fixed with the $\mathrm{M}_k, k = 1, 2, \ldots, K$, defined as random variables whose values are determined by the solution at the end of each set of inner loop iterations satisfying some property (e.g., the solution is a local optima).

Generalized hill climbing algorithms can be viewed as sampling procedures over the solution space Ω. The key distinction between different generalized hill climbing algorithm formulations is in how the sampling is performed. For example, simulated annealing produces biased samples, guided by the neighborhood function, the objective function, and the temperature parameter. More specifically, simulated annealing can be described as a generalized hill climbing algorithm by setting the hill-climbing random variable, $\mathrm{R}_k(\omega, \omega') = -t_k \ln(\mu_k), \omega \in \Omega, \omega' \in \mathrm{N}(\omega), k = 1, 2, \ldots, K$, and the $\{\mu_k\}$ are independent and identically distributed U(0,1) random variables. To formulate Monte Carlo search as a generalized hill climbing algorithm, set $\mathrm{R}_k(\omega, \omega') = +\infty, \omega \in \Omega, \omega' \in \mathrm{N}(\omega), k = 1, 2, \ldots, K$. Deterministic local search accepts only neighbors of improving (lower) objective function value and can be expressed as a generalized hill climbing algorithm with $\mathrm{R}_k(\omega, \omega') = 0, \omega \in \Omega, \omega' \in \mathrm{N}(\omega), k = 1, 2, \ldots, K$. Other algorithms that can be described using the generalized hill climbing framework include threshold accepting (1990) some simple forms of tabu search (1997), and Weibull accepting. Jacobson et al. (1998), Sullivan and Jacobson (2001), and Johnson and Jacobson (2002b) provide a complete discussion of these algorithms and a description of how these algorithms fit into the generalized hill climbing algorithm framework.

5.2 Convergence versus Finite-Time Performance

The current literature focuses mainly on asymptotic convergence properties of simulated annealing algorithms (Section 2 outlines and describes several of these results); however, considerable work on finite-time behavior of simulated annealing has been presented over the past decade. Chiang and Chow (1989) and Mazza (1992) investigate the statistical properties of the first visit time to a global optima which provides insight into the time-asymptotic properties of the algorithm as the outer loop counter, $k \to +\infty$. Catoni (1996) investigates optimizing a finite-horizon cooling schedule to maximize the number of visits to a global optimum after a finite number of iterations. Desai (1999) focuses on finite-time performance by incorporating size-asymptotic information supplied by certain eigenvalues associated with the transition matrix. Desai provides some quantitative and qualitative information about the performance of simulated annealing after a finite number of steps by observing the quality of solutions related to the number of steps that the algorithm has taken.

Srichander (1995) examines the finite-time performance of simulated annealing using spectral decomposition of matrices. He proposes that an annealing schedule on the temperature is not necessary for the final solution of the simulated annealing algorithm to converge to the global minimum with probability one. Srichander shows that initiating the simulated annealing algorithm with high initial temperatures produces an inefficient algorithm in the sense that the number of function evaluations required to obtain a global minima is very large. A modified simulated annealing algorithm is presented with a low initial temperature and an iterative schedule on the size of the neighborhood sets that leads to a more efficient algorithm. The new algorithm is applied to a real-world example and performance is reported.

Fleischer and Jacobson (1999) uses a reverse approach to establish theoretical relationships between the finite-time performance of an algorithm and the characteristics of problem instances. They observe that the configuration space created by an instance of a discrete optimization problem determines the efficiency of simulated annealing when applied to that problem. The entropy of the Markov chain embodying simulated annealing is introduced as a measure that captures the entire topology of the configuration space associated with the particular instance of the discrete optimization problem. By observing the expected value of the final state in a simulated annealing algorithm as it relates to the entropy value of the underlying Markov chain, the article presents measures of performance that determine how well the simulated annealing algorithm performs in finite-time. Their computational results suggest that superior finite-time performance of a simulated annealing algorithm are associated with higher entropy measures.

5.3 Extensions

The popularity of simulated annealing has spawned several new annealing-like algorithms. Pepper et al. (2000) introduce demon algorithms and test them on the traveling salesman problem. Ohlmann and Bean (2000) introduce another variant of simulated annealing termed compressed annealing. They incorporate the concepts of pressure and volume, in addition to temperature, to address discrete optimization problems with relaxed constraints. They also introduce a primal/dual meta-heuristic by simultaneously adjusting temperature and pressure in the algorithm.

Much work continues in the area of convergence and comparing the performance of simulated annealing algorithms to other local search strategies. Jacobson and Yücesan (2002b) presents necessary and sufficient (asymptotic) convergence conditions for generalized hill climbing algorithms that include simulated annealing as a special case. They also introduce new performance measures that can be used to evaluate and compare both convergent and non-convergent generalized hill climbing algorithms with random restart local search (Jacobson, 2002). Such a comparison provides insights into both asymptotic and finite-time performance of discrete optimization algorithms. For example, they use the global visit probability to evaluate the performance of simulated annealing using random restart local search as a benchmark. These results suggest that random restart local search may outperform simulated annealing provided that a sufficiently large number of restarts are executed. Ferreira and Zerovnik (1993) develop bounds on the probability that simulated annealing obtains an optimal (or near-optimal) solution, and use these bound to obtain similar results for random restart local search and simulated annealing. Fox (1994) notes that this result is only true if both the number of accepted and rejected moves are counted. Fox (1994) also provides a clever example to illustrate this point, and notes that comparing random restart local search and simulating annealing may not be prudent. Fox (1993, 1995) presents modifications of simulated annealing that circumvent this counting issue, hence yielding superior performing simulated annealing algorithm implementations. The primary value of using simulated annealing may therefore be for finite-time executions that obtain near-optimal solutions reasonably quickly. This, in turn, suggests that one should focus on the finite-time behavior of simulated annealing rather than the asymptotic convergence results that dominate the literature.

ACKNOWLEDGEMENTS

This work is supported in part by the Air Force Office of Scientific Research (F49620-01-1-0007) and the National Science Foundation (DMI-9907980).

BIBLIOGRAPHY

Aarts, E.H.L. and Korst, J. (1989) *Simulated Annealing and Boltzmann Machines: A Stochastic Approach to Combinatorial Optimization and Neural Computing*. John Wiley & Sons, Chichester, England.

Aarts, E.H.L. and Lenstra, J.K. (1997) *Local Search in Combinatorial Optimization*. John Wiley & Sons, Chichester, England.

Aarts, E.H.L. and van Laarhoven, P.J.M. (1985) Statistical cooling: A general approach to combinatorial optimization problems. *Phillips Journal of Research*, **40**, 193–226.

Abramson, D., Krishnamoorthy, M. and Dang, H. (1999) Simulated annealing cooling schedules for the school timetabling problem. *Asia-Pacific Journal of Operational Research*, **16**, 1–22.

Alrefaei, M.H. and Andradottir, S. (1999) A simulated annealing algorithm with constant temperature for discrete stochastic optimization. *Management Science*, **45**, 748–764.

Althofer, I. and Koschnick, K.U. (1991) On the convergence of threshold accepting. *Applied Mathematics and Optimization*, **24**, 183–195.

Aluffi-Pentini, F., Parisi, V. and Zirilli, F. (1985) Global optimization and stochastic differential equations. *Journal of Optimization Theory and Applications*, **47**, 1–16.

Anily, S. and Federgruen, A. (1987) Simulated annealing methods with general acceptance probabilities. *Journal of Applied Probability*, **24**, 657–667.

Belisle, C.J.P. (1992) Convergence theorems for a class of simulated annealing algorithms on R^D. *Journal of Applied Probability*, **29**, 885–895.

Belisle, C.J.P., Romeijn, H.E. and Smith, R.L. (1993) Hit-and-run algorithms for generating multivariate distributions. *Mathematics of Operations Research*, **18**, 255–266.

Bohachevsky, I.O., Johnson, M.E. and Stein, M.L. (1986) Generalized simulated annealing for function optimization. *Technometrics*, **28**, 209–217.

Borkar, V.S. (1992) Pathwise recurrence orders and simulated annealing. *Journal of Applied Probability*, **29**, 472–476.

Bratley, P., Fox, B.L. and Schrage, L. (1987) *A Guide to Simulation*, Springer-Verlag, New York.

Cardoso, M.F., Salcedo, R.L. and de Azevedo, S.F. (1994) Nonequilibrium simulated annealing: a faster approach to combinatorial minimization. *Industrial Engineering and Chemical Research*, **33**, 1908–1918.

Catoni, O. (1996) Metropolis, simulated annealing, and iterated energy transformation algorithms: theory and experiments. *Journal of Complexity*, **12**, 595–623.

Cerf, R. (1998) Asymptotic convergence of genetic algorithms. *Advances in Applied Probability*, **30**, 521–550.

Chardaire, P., Lutton, J.L. and Sutter, A. (1995) Thermostatistical persistency: a powerful improving concept for simulated annealing algorithms. *European Journal of Operational Research*, **86**, 565–579.

Charon, I. and Hudry, O. (1993) The noising method—a new method for combinatorial optimization. *Operations Research Letters*, **14**, 133–137.

Charon, I. and Hudry, O. (2001) The noising methods—a generalization of some metaheuristics. *European Journal of Operational Research*, **135**, 86–101.

Cheh, K.M., Goldberg, J.B. and Askin, R.G. (1991) A note on the effect of neighborhood-structure in simulated annealing. *Computers and Operations Research*, **18**, 537–547.

Chen, S. and Luk, B.L. (1999) Adaptive simulated annealing for optimization in signal processing applications. *Signal Processing*, **79**, 117–128.

Chiang, T.S. and Chow, Y.S. (1988) On the convergence rate of annealing processes. *SIAM Journal on Control and Optimization*, **26**, 1455–1470.

Chiang, T.S. and Chow, Y.Y. (1989) A limit-theorem for a class of inhomogeneous markov-processes. *Annals of Probability*, **17**, 1483–1502.

Chiang, T.S. and Chow, Y.Y. (1994) The asymptotic-behavior of simulated annealing processes with absorption. *SIAM Journal on Control and Optimization*, **32**, 1247–1265.

Christoph, M. and Hoffmann, K.H. (1993) Scaling behavior of optimal simulated annealing schedules. *Journal of Physics A—Mathematical and General*, **26**, 3267–3277.

Chu, K.W., Deng, Y.F. and Reinitz, J. (1999) Parallel simulated annealing by mixing of states. *Journal of Computational Physics*, **148**, 646–662.

Çinlar, E. (1974) *Introduction to Stochastic Processes*. Prentice-Hall, Englewood Cliffs, New Jersey.

Cohn, H. and Fielding, M. (1999) Simulated annealing: searching for an optimal temperature schedule. *SIAM Journal on Optimization*, **9**, 779–802.

Connors, D.P. and Kumar, P.R. (1989) Simulated annealing type markov-chains and their order balance-equations. *SIAM Journal on Control and Optimization*, **27**, 1440–1461.

Davis, T.E. (1991) *Toward an Extrapolation of the Simulated Annealing Convergence Theory onto the Simple Genetic Algorithm* (Doctoral Dissertation), University of Florida, Gainesville, Florida.

Davis, T.E. and Principe, J.C. (1991) A simulated annealing like convergence theory for the simple genetic algorithm. In: *Fourth Conference on Genetic Algorithm*, pp. 174–181.

Dekkers, A. and Aarts, E. (1991) Global optimization and simulated annealing. *Mathematical Programming*, **50**, 367–393.

Delport, V. (1998) Parallel simulated annealing and evolutionary selection for combinatorial optimisation. *Electronics Letters*, **34**, 758–759.

Desai, M.P. (1999) Some results characterizing the finite time behaviour of the simulated annealing algorithm. *Sadhana-Academy Proceedings in Engineering Sciences*, **24**, 317–337.

Dueck, G. and Scheuer, T. (1990) Threshold accepting—a general-purpose optimization algorithm appearing superior to simulated annealing. *Journal of Computational Physics*, **90**, 161–175.

Eglese, R.W. (1990) Simulated annealing: a tool for operational research. *European Journal of Operational Research*, **46**, 271–281.

Emden-Weinert, T. and Proksch, M. (1999) Best practice simulated annealing for the airline crew scheduling problem. *Journal of Heuristics*, **5**, 419–436.

Fabian, V. (1997) Simulated annealing simulated. *Computers and Mathematics with Applications*, **33**, 81–94.

Faigle, U. and Kern, W. (1991) Note on the convergence of simulated annealing algorithms. *SIAM Journal on Control and Optimization*, **29**, 153–159.

Faigle, U. and Kern, W. (1992) Some convergence results for probabilistic tabu search. *ORSA Journal on Computing*, **4**, 32–37.

Faigle, U. and Schrader, R. (1988) On the convergence of stationary distributions in simulated annealing algorithms. *Information Processing Letters*, **27**, 189–194.

Faigle, U. and Schrader, R. (1988) Simulated annealing—a case-study. *Angewandte Informatik*, 259–263.

Fielding, M. (2000) Simulated annealing with an optimal fixed temperature. *SIAM Journal of Optimization*, **11**, 289–307.

Fleischer, M.A. (1995) *Assessing the Performance of the Simulated Annealing Algorithm Using Information Theory* (Doctoral Dissertation), Department of Operations Research, Case Western Reserve University, Clevelend, Ohio.

Fleischer, M.A. (1995) Simulated annealing: past, present, and future. In: C. Alexopoulos, K. Kang, W.R. Lilegdon and D. Goldsman (eds.), *Proceedings of the 1995 Winter Simulation Conference*, IEEE Press, pp. 155–161.

Fleischer, M.A. (1999) Generalized cybernetic optimization: solving continuous variable problems. In: S. Voss, S. Martello, C. Roucairol, H. Ibrahim, and I.H. Osman (eds.), *Meta-heuristics: Advances and Trends in Local Search Paradigms for Optimization*, Kluwer Academic Publishers, pp. 403–418.

Fleischer, M.A. and Jacobson, S.H. (1996) Cybernetic optimization by simulated annealing: an implementation of parallel processing using probabilistic feedback control. In: I.H. Osman and J.P. Kelly (eds.), *Meta-Heuristics: Theory and Applications*, Kluwer Academic Publishers, pp. 249–264.

Fleischer, M.A. and Jacobson, S.H. (1999) Information theory and the finite-time behavior of the simulated annealing algorithm: experimental results. *INFORMS Journal on Computing*, **11**, 35–43.

Fox, B.L. (1993) Integrating and accelerating tabu search, simulated annealing, and genetic algorithms. *Annals of Operations Research*, **41**, 47–67.

Fox, B.L. (1994) Random restarting versus simulated annealing. *Computers and Mathematics with Applications*, **27**, 33–35.

Fox, B.L. (1995) Faster simulated annealing. *Siam Journal of Optimzation*, **5**, 485–505.

Fox, B.L. and Heine, G.W. (1993) Simulated annealing with overrides, technical, Department of Mathematics, University of Colorado, Denver, Colorado.

Gemen, S. and Hwang, C.R. (1986) Diffusions for global optimization. *SIAM Journal on Control and Optimization*, **24**, 1031–1043.

Gidas, B. (1985) Nonstationary markov chains and convergence of the annealing algorithm. *Journal of Statistical Physics*, **39**, 73–131.

Glover, F. (1989) Tabu search—Part I. *ORSA Journal on Computing*, **1**, 190–206.

Glover, F. (1994) Tabu search for nonlinear and parametric optimization (with links to genetic algorithms). *Discrete Applied Mathematics*, **49**, 231–255.

Glover, F. and Laguna, M. (1997) *Tabu Search*, Kluwer Academic Publishers, Boston, Massachusetts.

Goldstein, L. and Waterman, M. (1988) Neighborhood size in the simulated annealing algorithm. *American Journal of Mathematical and Management Sciences*, **8**, 409–423.

Granville, V., Krivanek, M. and Rasson, J.P. (1994) Simulated annealing—a proof of convergence. *IEEE Transactions on Pattern Analysis and Machine Intelligence*, **16**, 652–656.

Hajek, B. (1988) Cooling schedules for optimal annealing. *Mathematics of Operations Research*, **13**, 311–329.

Hamma, B., Viitanen, S. and Torn, A. (2000) Parallel continuous simulated annealing for global optimization. *Optimization Methods and Software*, **13**, 95–116.

Hammersley, J.M. and Handscomb, D.C. (1964) *Monte Carlo Methods*, Methuen, John Wiley & Sons, London, New York.

Hu, T.C., Kahing, A.B. and Tsao, C.W.A. (1995) Old bachelor acceptance: a new class of non-monotone threshold accepting methods. *ORSA Journal on Computing*, **7**, 417–425.

Isaacson, D.L. and Madsen, R.W. (1976) *Markov Chains, Theory and Applications*. John Wiley & Sons, New York.

Jacobson, S.H. (2002) Analyzing the performance of local search algorithms using generalized hill climbing algorithms. pp. 441–467. (Chapter 20 in *Essays and Surveys on Metaheuristics*, P. Hansen and C.C. Ribeiro (eds.), Kluwer Academic Publishers, Norwell, Massachusetts.

Jacobson, S.H., Sullivan, K.A. and Johnson, A.W. (1998) Discrete manufacturing process design optimization using computer simulation and generalized hill climbing algorithms. *Engineering Optimization*, **31**, 247–260.

Jacobson, S.H. and Yücesan, E. (2002a) A performance measure for generalized hill climbing algorithms. Technical Report, Department of Mechanical and Industrial Engineering, University of Illinois, Urbana, Illinois.

Jacobson, S.H. and Yücesan, E. (2002b) On the effectiveness of generalized hill climbing algorithms. Technical Report, Department of Mechanical and Industrial Engineering, University of Illinois, Urbana, Illinois.

Johnson, A.W. and Jacobson, S.H. (2002a) A class of convergent generalized hill climbing algorithms. *Applied Mathematics and Computation*, **125**(2–3), 359–373.

Johnson, A.W. and Jacobson, S.H. (2002b) On the convergence of generalized hill climbing algorithms. *Discrete Applied Mathematics (To Appear)*.

Johnson, D.S., Aragon, C.R., McGeoch, L.A. and Schevon, C. (1989) Optimization by simulated annealing—an experimental evaluation; part 1, graph partitioning. *Operations Research*, **37**, 865–892.

Johnson, D.S., Aragon, C.R., McGeoch, L.A. and Schevon, C. (1991) Optimization by simulated annealing—an experimental evaluation; part 2, graph-coloring and number partitioning. *Operations Research*, **39**, 378–406.

Kiatsupaibul, S. and Smith, R.L. (2000) A general purpose simulated annealing algorithm for integer linear programming. Technical Report, Department of Industrial and Operations Engineering, University of Michigan, Ann Arbor, Michigan.

Kirkpatrick, S., Gelatt, Jr., C.D. and Vecchi, M.P. (1983) Optimization by simulated annealing. *Science*, **220**, 671–680.

Koulamas, C., Antony, S.R. and Jaen, R. (1994) A survey of simulated annealing applications to operations-research problems. *OMEGA-International Journal of Management Science*, **22**, 41–56.

Leite, J.P.B. and Topping, B.H.V. (1999) Parallel simulated annealing for structural optimization. *Computers and Structures*, **73**, 545–564.

Liepins, G.E. and Hilliard, M.R. (1989) Genetic algorithms: foundations and applications. *Annals of Operations Research*, **21**, 31–58.

Lin, C.K.Y., Haley, K.B. and Sparks, C. (1995) A comparative study of both standard and adaptive versions of threshold accepting and simulated annealing algorithms in three scheduling problems. *European Journal of Operational Research*, **83**, 330–346.

Locatelli, M. (1996) Convergence properties of simulated annealing for continuous global optimization. *Journal of Applied Probability*, **33**, 1127–1140.

Locatelli, M. (2000) Simulated annealing algorithms for continuous global optimization: convergence conditions. *Journal of Optimization Theory and Applications*, **104**, 121–133.

Lundy, M. and Mees, A. (1986) Convergence of an annealing algorithm. *Mathematical Programming*, **34**, 111–124.

Ma, J. and Straub, J.E. (1994) Simulated annealing using the classical density distribution. *Journal of Chemical Physics*, **101**, 533–541.

Mazza, C. (1992) Parallel simulated annealing. *Random Structures and Algorithms*, **3**, 139–148.

Metropolis, N., Rosenbluth, A., Rosenbluth, M., Teller, A. and Teller, E. (1953) Equation of state calculations by fast computing machines. *Journal of Chemical Physics*, **21**, 1087–1092.

Meyer, C.D. (1980) The condition of a finite markov chain and perturbation bounds for the limiting probabilities. *SIAM Journal of Algebraic and Discrete Methods*, **1**, 273–283.

Mitra, D., Romeo, F. and Sangiovanni-Vincentelli, A.L. (1986) Convergence and finite time behavior of simulated annealing. *Advances in Applied Probability*, **18**, 747–771.

Moscato, P. (1993) An introduction to population approaches for optimization and hierarchical objective functions: a discussion on the role of tabu search. *Annals of Operations Research*, **41**, 85–121.

Moscato, P. and Fontanari, J.F. (1990) Convergence and finite-time behavior of simulated annealing. *Advances in Applied Probability*, **18**, 747–771.

Muhlenbein, H. (1997) Genetic algorithms. In: E. Aarts and J.K. Lenstra (eds.), *Local Search in Combinatorial Optimization*. John Wiley & Sons, New York, New York, pp. 137–172.

Nissen, V. and Paul, H. (1995) A modification of threshold accepting and its application to the quadratic assignment problem. *OR Spektrum*, **17**, 205–210.

Nourani, Y. and Andresen, B. (1998) A comparison of simulated annealing cooling strategies. *Journal of Physics A—Mathematical and General*, **31**, 8373–8385.

Ogbu, F.A. and Smith, D.K. (1990) The application of the simulated annealing algorithm to the solution of the N/M/Cmax flowshop problem. *Computers and Operations Research*, **17**, 243–253.

Ohlmann, J.W. and Bean, J.C. (2000) Compressed annealing: simulated annealing under pressure. Technical Report, Department of Industrial and Operations Engineering, University of Michigan, Ann Arbor, Michigan.

Orosz, J.E. and Jacobson, S.H. (2002a) Finite-time performance analysis of static simulated annealing algorithms. *Computational Optimization and Applications*, **21**, 21–53.

Orosz, J.E. and Jacobson, S.H. (2002b) Analysis of static simulated annealing algorithms. *Journal of Optimization Theory and Application* (to appear).

Pepper, J.W., Golden, B.L. and Wasil, E.A. (2000) Solving the traveling salesman problem with demon algorithms and variants. Technical Report, Smith School of Business, University of Maryland, College Park, Maryland.

Romeijn, H.E., Zabinsky, Z.B., Graesser, D.L. and Noegi, S. (1999) New reflection generator for simulated annealing in mixed-integer/continuous global optimization. *Journal of Optimization Theory and Applications*, **101**, 403–427.

Romeo, F. and Sangiovanni-Vincentelli, A. (1991) A theoretical framework for simulated annealing. *Algorithmica*, **6**, 302–345.

Rosenthal, J.S. (1995) Convergence rates for markov chains. *SIAM Review*, **37**, 387–405.

Ross, S.M. (1996) *Stochastic Processes*. John Wiley & Sons, New York, New York.

Ross, S.M. (1997) *Introduction to Probability Models*. Academic Press, San Diego, California.

Rossier, Y., Troyon, M. and Liebling, T.M. (1986) Probabilistic exchange algorithms and euclidean traveling salesman problems. *OR Spektrum*, **8**, 151–164.

Rudolph, G. (1994) Convergence analysis of cononical genetic algorithms. *IEEE Transactions on Neural Networks, Special Issue on Evolutional Computing*, **5**, 96–101.

Scheermesser, T. and Bryngdahl, O. (1995) Threshold accepting for constrained half-toning. *Optics Communications*, **115**, 13–18.

Schuur, P.C. (1997) Classification of acceptance criteria for the simulated annealing algorithm. *Mathematics of Operations Research*, **22**, 266–275.

Seneta, E. (1981) *Non-Negative Matrices and Markov Chains*. Springer-Verlag, New York, New York.

Siarry, P., Berthiau, G., Durbin, F. and Haussy, J. (1997) Enhanced simulated annealing for globally minimizing functions of many-continuous variables. *ACM Transactions On Mathematical Software*, **23**, 209–228.

Solla, S.A., Sorkin, G.B. and White, S.R. (1986) Configuration space analysis for optimization problems. *NATO ASI Series, Disordered Systems and Biological Organization*, **F20**, 283–293.

Srichander, R. (1995) Efficient schedules for simulated annealing. *Engineering Optimization*, **24**, 161–176.

Stern, J.M. (1992) Simulated annealing with a temperature dependent penalty function. *ORSA Journal on Computing*, **4**, 311–319.

Storer, R.H., Wu, S.D. and Vaccari, R. (1992) New search spaces for sequencing problems with application to job shop scheduling. *Management Science*, **38**, 1495–1509.

Straub, J.E., Ma, J. and Amara, P. (1995) Simulated annealing using coarse grained classical dynamics: smouuchowski dynamics in the gaussian density approximation. *Journal of Chemical Physics*, **103**, 1574–1581.

Strenski, P.N. and Kirkpatrick, S. (1991) Analysis of finite length annealing schedules. *Algorithmica*, **6**, 346–366.

Sullivan, K.A. and Jacobson, S.H. (2000) Ordinal hill climbing algorithms for discrete manufacturing process design optimization problems. *Discrete Event Dynamical Systems*, **10**, 307–324.

Sullivan, K.A. and Jacobson, S.H. (2001) A convergence analysis of generalized hill climbing algorithms. *IEEE Transactions on Automatic Control*, **46**, 1288–1293.

Tian, P., Ma, J. and Zhang, D.M. (1999) Application of the simulated annealing algorithm to the combinatorial optimisation problem with permutation property: an investigation of generation mechanism. *European Journal of Operational Research*, **118**, 81–94.

Tovey, C.A. (1988) Simulated simulated annealing. *American Journal of Mathematical and Management Sciences*, **8**, 389–407.

Tsitsiklis, J.N. (1989) Markov chains with rare transitions and simulated annealing. *Mathematics of Operations Research*, **14**, 70–90.

van Laarhoven, P.J.M. (1988) *Theoretical and Computational Aspects of Simulated Annealing*, Centrum voor Wiskunde en Informatica, Amsterdam, Netherlands.

van Laarhoven, P.J.M. and Aarts, E.H.L. (1987) *Simulated Annealing: Theory and Applications*, D. Reidel, Kluwer Academic Publishers, Dordrecht, Boston, Norwell, Massachusetts.

Varanelli, J.M. and Cohoon, J.P. (1999) A fast method for generalized starting temperature determination in homogeneous two-stage simulated annealing systems. *Computers and Operations Research*, **26**, 481–503.

Yan, D. and Mukai, H. (1992) Stochastic discrete optimization. *SIAM Journal on Control and Optimization*, **30**, 594–612.

Yang, R.L. (2000) Convergence of the simulated annealing algorithm for continuous global optimization. *Journal of Optimization Theory and Applications*, **104**, 691–716.

Yao, X. (1995) A new simulated annealing algorithm. *International Journal of Computer Mathematics*, **56**, 161–168.

Yao, X. and Li, G. (1991) General simulated annealing. *Journal of Computer Science and Technology*, **6**, 329–338.

Zabinsky, Z.B., Smith, R.L., McDonald, J.F., Romeijn, H.E. and Kaufman, D.E. (1993) Improving hit-and-run for global optimization. *Journal of Global Optimization*, **3**, 171–192.

Chapter 11

ITERATED LOCAL SEARCH

Helena R. Lourenço
Universitat Pompeu Fabra, Barcelona, Spain
E-mail: helena.ramalhinho@econ.upf.es

Olivier C. Martin
Université Paris-Sud, Orsay, France
E-mail: martino@ipno.in2p3.fr

Thomas Stützle
Darmstadt University of Technology, Darmstadt, Germany
E-mail: stuetzle@informatik.tu-darmstadt.de

1 INTRODUCTION

The importance of high performance algorithms for tackling difficult optimization problems cannot be understated, and in many cases the only available methods are metaheuristics. When designing a metaheuristic, it is preferable that it be simple, both conceptually and in practice. Naturally, it also must be effective, and if possible, general purpose. If we think of a metaheuristic as simply a construction for guiding (problem-specific) heuristics, the ideal case is when the metaheuristic can be used without *any* problem-dependent knowledge.

As metaheuristics have become more and more sophisticated, this ideal case has been pushed aside in the quest for greater performance. As a consequence, problem-specific knowledge (in addition to that built into the heuristic being guided) must now be incorporated into metaheuristics in order to reach the state of the art level. Unfortunately, this makes the boundary between heuristics and *meta*heuristics fuzzy, and we run the risk of loosing both simplicity and generality. To counter this, we appeal to modularity and try to decompose a metaheuristic algorithm into a few parts, each with its own specificity. In particular, we would like to have a totally general purpose part, while any problem-specific knowledge built into the metaheuristic would be restricted to another part. Finally, to the extent possible, we prefer to leave untouched the embedded heuristic (which is to be "guided") because of its potential complexity. One can also consider the case where this heuristic is only available through an object module, the source code being proprietary; it is then necessary to be able to treat it as a "black-box" routine. Iterated local search provides a simple way to satisfy all these requirements.

The essence of the iterated local search metaheuristic can be given in a nut-shell: one *iteratively* builds a sequence of solutions generated by the embedded heuristic, leading to far better solutions than if one were to use repeated random trials of that heuristic. This simple idea [10] has a long history, and its rediscovery by many authors has lead to many different names for iterated local search like *iterated descent* [8,9], *large-step Markov chains* [49], *iterated Lin-Kernighan* [37], *chained local optimization* [48], or combinations of these [2]... Readers interested in these historical developments should consult the review [38]. For us, there are two main points that make an algorithm an iterated local search: (i) there must be a single chain that is being followed (this then excludes population-based algorithms); (ii) the search for better solutions occurs in a reduced space defined by the output of a black-box heuristic. In practice, local search has been the most frequently used embedded heuristic, but in fact any optimizer can be used, be-it deterministic or not.

The purpose of this review is to give a detailed description of iterated local search and to show where it stands in terms of performance. So far, in spite of its conceptual simplicity, it has lead to a number of state-of-the-art results without the use of too much problem-specific knowledge; perhaps this is because iterated local search is very malleable, many implementation choices being left to the developer. We have organized this chapter as follows. First we give a high-level presentation of iterated local search in Section 2. Then we discuss the importance of the different parts of the metaheuristic in Section 3, especially the subtleties associated with perturbing the solutions. In Section 4 we go over past work testing iterated local search in practice, while in Section 5 we discuss similarities and differences between iterated local search and other metaheuristics. The chapter closes with a summary of what has been achieved so far and an outlook on what the near future may look like.

2 ITERATING A LOCAL SEARCH

2.1 General Framework

We assume we have been given a problem-specific heuristic optimization algorithm that from now on we shall refer to as a local search (even if in fact it is not a true local search). This algorithm is implemented via a computer routine that we call LocalSearch. The question we ask is "Can such an algorithm be improved by the use of iteration?". Our answer is "YES", and in fact the improvements obtained in practice are usually significant. Only in rather pathological cases where the iteration method is "incompatible" with the local search will the improvement be minimal. In the same vein, in order to have the *most* improvement possible, it is necessary to have some understanding of the way the LocalSearch works. However, to keep this presentation as simple as possible, we shall ignore for the time being these complications; the additional subtleties associated with tuning the iteration to the local search procedure will be discussed in Section 3. Furthermore, all issues associated with the actual speed of the algorithm are omitted in this first section as we wish to focus solely on the high-level architecture of iterated local search.

Let C be the cost function of our combinatorial optimization problem; C is to be *minimized*. We label candidate solutions or simply "solutions" by s, and denote by $\mathcal{S}$ the set of all s (for simplicity $\mathcal{S}$ is taken to be finite, but it does not matter much). Finally, for the purposes of this high-level presentation, it is simplest to assume that the

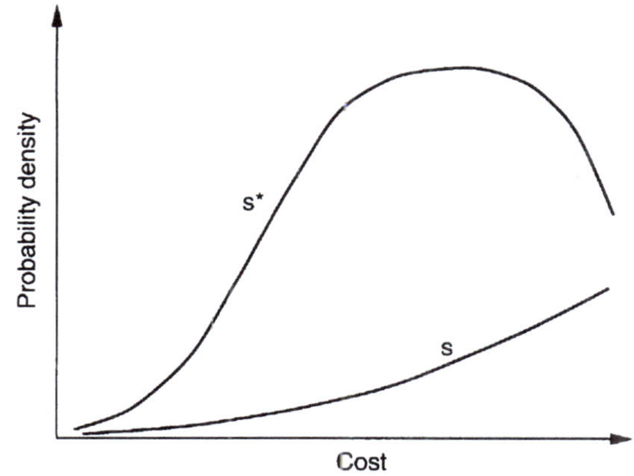

Figure 11.1. Probability densities of costs. The curve labeled s gives the cost density for all solutions, while the curve labeled s^* gives it for the solutions that are local optima.

local search procedure is deterministic and memoriless:[1] for a given input s, it always outputs the same solution s^* whose cost is less or equal to $C(s)$. LocalSearch then defines a many to one mapping from the set $\mathcal{S}$ to the smaller set $\mathcal{S}^*$ of locally optimal solutions s^*. To have a pictorial view of this, introduce the "basin of attraction" of a local minimum s^* as the set of s that are mapped to s^* under the local search routine. LocalSearch then takes one from a starting solution to a solution at the bottom of the corresponding basin of attraction.

Now take an s or an s^* at random. Typically, the distribution of costs found has a very rapidly rising part at the lowest values. In Figure 11.1 we show the kind of distributions found in practice for combinatorial optimization problems having a finite solution space. The distribution of costs is bell-shaped, with a mean and variance that is significantly smaller for solutions in $\mathcal{S}^*$ than for those in $\mathcal{S}$. As a consequence, it is much better to use local search than to sample randomly in $\mathcal{S}$ if one seeks low cost solutions. The essential ingredient necessary for local search is a neighborhood structure. This means that $\mathcal{S}$ is a "space" with some topological structure, not just a set. Having such a space allows one to move from one solution s to a better one in an intelligent way, something that would not be possible if $\mathcal{S}$ were just a set.

Now the question is how to go beyond this use of LocalSearch. More precisely, given the mapping from $\mathcal{S}$ to $\mathcal{S}^*$, how can one further reduce the costs found without opening up and modifying LocalSearch, leaving it as a "black box" routine?

2.2 Random Restart

The simplest possibility to improve upon a cost found by LocalSearch is to repeat the search from another starting point. Every s^* generated is then independent, and the use of multiple trials allows one to reach into the lower part of the distribution. Although such a "random restart" approach with independent samplings is sometimes a useful strategy (in particular when all other options fail), it breaks down as the instance

[1] The reader can check that very little of what we say really uses this property, and in practice, many successful implementations of iterated local search have non-deterministic local searches or include memory.

size grows because in that limit the tail of the distribution of costs collapses. Indeed, empirical studies [38] and general arguments [58] indicate that local search algorithms on large generic instances lead to costs that: (i) have a mean that is a fixed percentage excess above the optimum cost; (ii) have a *distribution* that becomes arbitrarily peaked about the mean when the instance size goes to infinity. This second property makes it impossible in practice to find an s^* whose cost is even a little bit lower percentage-wise than the typical cost. Note however that there do exist many solutions of significantly lower cost, it is just that *random* sampling has a lower and lower probability of finding them as the instance size increases. To reach those configurations, a biased sampling is necessary; this is precisely what is accomplished by a stochastic search.

2.3 Searching in $\mathcal{S}^*$

To overcome the problem just mentioned associated with large instance sizes, reconsider what local search does: it takes one from $\mathcal{S}$ where $\mathcal{C}$ has a large mean to $\mathcal{S}^*$ where $\mathcal{C}$ has a smaller mean. It is then most natural to invoke recursion: use local search to go from $\mathcal{S}^*$ to a smaller space $\mathcal{S}^{**}$ where the mean cost will be still lower! That would correspond to an algorithm with one local search nested inside another. Such a construction could be iterated to as many levels as desired, leading to a hierarchy of nested local searches. But upon closer scrutiny, we see that the problem is precisely how to formulate local search beyond the lowest level of the hierarchy: local search requires a neighborhood structure and this is not *à priori* given. The fundamental difficulty is to define neighbors in $\mathcal{S}^*$ so that they can be enumerated and accessed efficiently. Furthermore, it is desirable for nearest neighbors in $\mathcal{S}^*$ to be relatively close when using the distance in $\mathcal{S}$; if this were not the case, a stochastic search on $\mathcal{S}^*$ would have little chance of being effective.

Upon further thought, it transpires that one can introduce a good neighborhood structure on $\mathcal{S}^*$ as follows. First, one recalls that a neighborhood structure on a set $\mathcal{S}$ directly induces a neighborhood structure on *subsets* of $\mathcal{S}$: two subsets are nearest neighbors simply if they contain solutions that are nearest neighbors. Second, take these subsets to be the basins of attraction of the s^*; in effect, we are lead to identify any s^* with its basin of attraction. This then immediately gives the "canonical" notion of neighborhood on $\mathcal{S}^*$, notion which can be stated in a simple way as follows: s_1^* and s_2^* are neighbors in $\mathcal{S}^*$ if their basins of attraction "touch" (i.e., contain nearest-neighbor solutions in $\mathcal{S}$). Unfortunately this definition has the major drawback that one cannot in practice list the neighbors of an s^* because there is no computationally efficient method for finding all solutions s in the basin of attraction of s^*. Nevertheless, we can *stochastically* generate nearest neighbors as follows. Starting from s^*, create a randomized path in $\mathcal{S}$, $s_1, s_2, \ldots, s_i$, where s_{j+1} is a nearest neighbor of s_j. Determine the first s_j in this path that belongs to a different basin of attraction so that applying local search to s_j leads to an $s^{*\prime} \neq s^*$. Then $s^{*\prime}$ is a nearest-neighbor of s^*.

Given this procedure, we can in principle perform a local search[2] in $\mathcal{S}^*$. Extending the argument recursively, we see that it would be possible to have an algorithm implementing nested searches, performing local search on $\mathcal{S}, \mathcal{S}^*, \mathcal{S}^{**}$, etc in a hierarchical way. Unfortunately, the implementation of nearest neighbor search at the level of $\mathcal{S}^*$ is much too costly computationally because of the number of times one has to execute

[2]Note that the local search finds neighbors stochastically; generally there is no efficient way to ensure that one has tested *all* the neighbors of any given s^*.

LocalSearch. Thus we are led to abandon the (stochastic) search for *nearest neighbors* in $\mathcal{S}^*$; instead we use a weaker notion of closeness which then allows for a fast stochastic search in $\mathcal{S}^*$. Our construction leads to a (biased) sampling of $\mathcal{S}^*$; such a sampling will be better than a random one if it is possible to find appropriate computational ways to go from one s^* to another. Finally, one last advantage of this modified notion of closeness is that it does not require basins of attraction to be defined; the local search can then incorporate memory or be non-deterministic, making the method far more general.

2.4 Iterated Local Search

We want to explore $\mathcal{S}^*$ using a walk that steps from one s^* to a "nearby" one, without the constraint of using only nearest neighbors as defined above. Iterated local search (ILS) achieves this heuristically as follows. Given the current s^*, we first apply a change or perturbation that leads to an intermediate state s' (which belongs to $\mathcal{S}$). Then LocalSearch is applied to s' and we reach a solution $s^{*\prime}$ in $\mathcal{S}^*$. If $s^{*\prime}$ passes an acceptance test, it becomes the next element of the walk in $\mathcal{S}^*$; otherwise, one returns to s^*. The resulting walk is a case of a stochastic search in $\mathcal{S}^*$, but where neighborhoods are never explicitly introduced. This iterated local search procedure should lead to good biased sampling as long as the perturbations are neither too small nor too large. If they are too small, one will often fall back to s^* and few new solutions of $\mathcal{S}^*$ will be explored. If on the contrary the perturbations are too large, s' will be random, there will be no bias in the sampling, and we will recover a random restart type algorithm.

The overall ILS procedure is pictorially illustrated in Figure 11.2. To be complete, let us note that generally the iterated local search walk will not be reversible; in particular one may sometimes be able to step from s_1^* to s_2^* but not from s_2^* to s_1^*. However this "unfortunate" aspect of the procedure does not prevent ILS from being very effective in practice.

Since deterministic perturbations may lead to short cycles (for instance of length 2), one should randomize the perturbations or have them be adaptive so as to avoid this kind of cycling. If the perturbations depend on any of the previous s^*, one has a walk

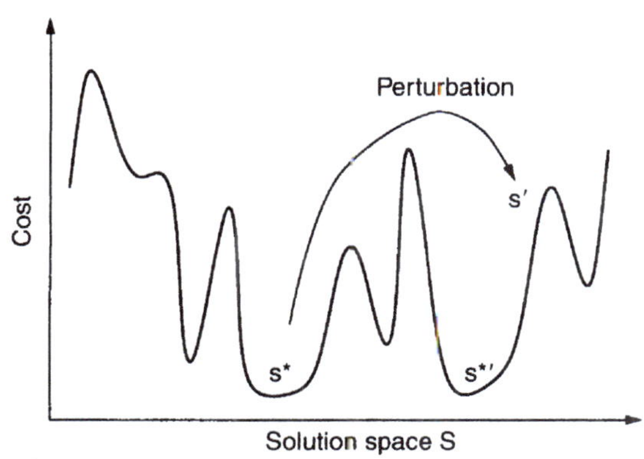

Figure 11.2. Pictorial representation of iterated local search. Starting with a local minimum s^*, we apply a perturbation leading to a solution s'. After applying LocalSearch, we find a new local minimum $s^{*\prime}$ that may be better than s^*.

in $\mathcal{S}^*$ with *memory*. Now the reader may have noticed that aside from the issue of perturbations (which use the structure on $\mathcal{S}$), our formalism reduces the problem to that of a stochastic search on $\mathcal{S}^*$. Then all of the bells and whistles (diversification, intensification, tabu, adaptive perturbations and acceptance criteria, etc.) that are commonly used in that context may be applied here. This leads us to define iterated local search algorithms as metaheuristics having the following high level architecture:

```
procedure Iterated Local Search
    s0 = GenerateInitialSolution
    s* = LocalSearch(s0)
    repeat
        s' = Perturbation(s*, history)
        s*' = LocalSearch(s')
        s* = AcceptanceCriterion(s*, s*', history)
    until termination condition met
end
```

In practice, much of the potential complexity of ILS is hidden in the history dependence. If there happens to be no such dependence, the walk has no memory:[3] the perturbation and acceptance criterion do not depend on any of the solutions visited previously during the walk, and one accepts or not $s^{*\prime}$ with a fixed rule. This leads to random walk dynamics on $\mathcal{S}^*$ that are "Markovian", the probability of making a particular step from s_1^* to s_2^* depending only on s_1^* and s_2^*. Most of the work using ILS has been of this type, though recent studies show unambiguously that incorporating memory enhances performance [61].

Staying within Markovian walks, the most basic acceptance criteria will use only the difference in the costs of s^* and $s^{*\prime}$; this type of dynamics for the walk is then very similar in spirit to what occurs in simulated annealing. A limiting case of this is to accept only improving moves, as happens in simulated annealing at zero temperature; the algorithm then does (stochastic) descent in $\mathcal{S}^*$. If we add to such a method a CPU time criterion to stop the search for improvements, the resulting algorithm pretty much has two nested local searches; to be precise, it has a local search operating on $\mathcal{S}$ embedded in a stochastic search operating on $\mathcal{S}^*$. More generally, one can extend this type of algorithm to more levels of nesting, having a different stochastic search algorithm for $\mathcal{S}^*, \mathcal{S}^{**}$ etc. Each level would be characterized by its own type of perturbation and stopping rule; to our knowledge, such a construction has never been attempted.

We can summarize this section by saying that the potential power of iterated local search lies in its *biased* sampling of the set of local optima. The efficiency of this sampling depends both on the kinds of perturbations and on the acceptance criteria. Interestingly, even with the most naïve implementations of these parts, iterated local search is much better than random restart. But still much better results can be obtained if the iterated local search modules are optimized. First, the acceptance criteria can be adjusted empirically as in simulated annealing without knowing anything about the problem being optimized. This kind of optimization will be familiar to any user of metaheuristics, though the questions of memory may become quite complex. Second, the Perturbation routine can incorporate as much problem-specific information as the

[3] Recall that to simplify this section's presentation, the local search is assumed to have no memory.

developer is willing to put into it. In practice, a rule of thumb can be used as a guide: "a good perturbation transforms one excellent solution into an excellent starting point for a local search". Together, these different aspects show that iterated local search algorithms can have a wide range of complexity, but complexity may be added progressively and in a modular way. (Recall in particular that all of the fine-tuning that resides in the embedded local search can be ignored if one wants, and it does not appear in the metaheuristic per-se.) This makes iterated local search an appealing metaheuristic for both academic and industrial applications. The cherry on the cake is speed: as we shall soon see, one can perform k local searches embedded within an iterated local search *much* faster than if the k local searches are run within random restart.

3 GETTING HIGH PERFORMANCE

Given all these advantages, we hope the reader is now motivated to go on and consider the more nitty-gritty details that arise when developing an ILS for a new application. In this section, we will illustrate the main issues that need to be tackled when optimizing an ILS in order to achieve high performance.

There are four components to consider: GenerateInitialSolution, LocalSearch, Perturbation, and AcceptanceCriterion. Before attempting to develop a state-of-the-art algorithm, it is relatively straight-forward to develop a more basic version of ILS. Indeed, (i) one can start with a random solution or one returned by some greedy construction heuristic; (ii) for most problems a local search algorithm is readily available; (iii) for the perturbation, a random move in a neighborhood of higher order than the one used by the local search algorithm can be surprisingly effective; and (iv) a reasonable first guess for the acceptance criterion is to force the cost to decrease, corresponding to a first-improvement descent in the set $\mathcal{S}^*$. Basic ILS implementations of this type usually lead to much better performance than random restart approaches. The developer can then run this basic ILS to build his intuition and try to improve the overall algorithm performance by improving each of the four modules. This should be particularly effective if it is possible to take into account the specificities of the combinatorial optimization problem under consideration. In practice, this tuning is easier for ILS than for memetic algorithms or tabu search to name but these metaheuristics. The reason may be that the complexity of ILS is reduced by its modularity, the function of each component being relatively easy to understand. Finally, the last task to consider is the overall optimization of the ILS algorithm; indeed, the different components affect one another and so it is necessary to understand their interactions. However, because these interactions are so problem dependent, we wait till the end of this section before discussing that kind of "global" optimization.

Perhaps the main message here is that the developer can choose the level of optimization he wants. In the absence of any optimizations, ILS is a simple, easy to implement, and quite effective metaheuristic. But with further work on its four components, ILS can often be turned into a very competitive or even state of the art algorithm.

3.1 Initial Solution

Local search applied to the initial solution s_0 gives the starting point s_0^* of the walk in the set $\mathcal{S}^*$. Starting with a good s_0^* can be important if high-quality solutions are to be reached *as fast as possible*.

Standard choices for s_0 are either a random initial solution or a solution returned by a greedy construction heuristic. A greedy initial solution s_0 has two main advantages over random starting solutions: (i) when combined with local search, greedy initial solutions often result in better quality solutions s_0^*; (ii) a local search from greedy solutions takes, on average, less improvement steps and therefore the local search requires less CPU time.[4]

The question of an appropriate initial solution for (random restart) local search carries over to ILS because of the dependence of the walk in $\mathcal{S}^*$ on the initial solution s_0^*. Indeed, when starting with a random s_0, ILS may take several iterations to catch up in quality with runs using an s_0^* obtained by a greedy initial solution. Hence, for short computation times the initial solution is certainly important to achieve the highest quality solutions possible. For larger computation times, the dependence on s_0 of the final solution returned by ILS reflects just how fast, if at all, the memory of the initial solution is lost when performing the walk in $\mathcal{S}^*$.

Let us illustrate the tradeoffs between random and greedy initial solutions when using an ILS algorithm for the permutation flow shop problem (FSP) [60]. That ILS algorithm uses a straight-forward local search implementation, random perturbations, and always applies Perturbation to the best solution found so far. In Figure 11.3 we show how the average solution cost evolves with the number of iterations for two instances. The averages are for 10 independent runs when starting from random initial solutions or from initial solutions returned by the NEH heuristic [57]. (NEH is one of the best performing constructive heuristics for the FSP.) For short runs, the curve for the instance on the right shows that the NEH initial solutions lead to better average solution quality than the random initial solutions. But at longer times, the picture is not so clear, sometimes random initial solutions lead to better results as we see on the instance on the left. This kind of test was also performed for ILS applied to the TSP [2]. Again it was observed that the initial solution had a significant influence on quality for short to medium sized runs.

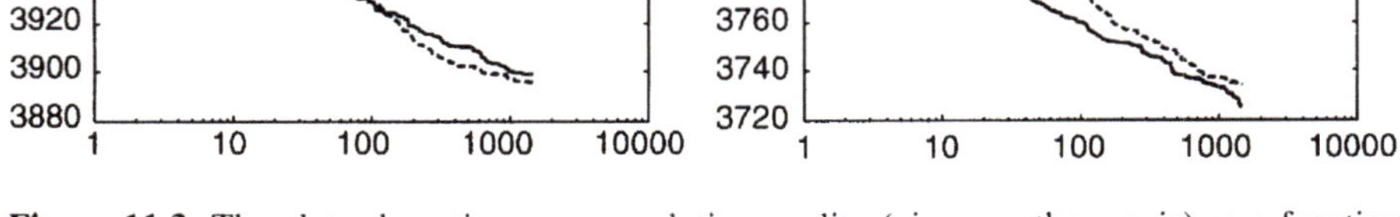

Figure 11.3. The plots show the average solution quality (given on the y-axis) as a function of the number of iterations (given on the x-axis) for an ILS algorithm applied to the FSP on instances `ta051` and `ta056`.

[4]Note that the best possible greedy initial solution need not be the best choice when combined with a local search. For example, in [38], it is shown that the combination of the Clarke-Wright starting tour (one of the best performing TSP construction heuristics) with local search resulted in worse local optima than starting from random initial solutions when using 3-opt. Additionally, greedy algorithms which generate very high quality initial solutions can be quite time-consuming.

In general, there will not always be a clear-cut answer regarding the best choice of an initial solution, but greedy initial solutions appear to be recommendable when one needs low-cost solutions quickly. For much longer runs, the initial solution seems to be less relevant, so the user can choose the initial solution that is the easiest to implement. If however one has an application where the influence of the initial solution does persist for long times, probably the ILS walk is having difficulty in exploring $\mathcal{S}^*$ and so other perturbations or acceptance criteria should be considered.

3.2 Perturbation

The main drawback of local descent is that it gets trapped in local optima that are significantly worse than the global optimum. Much like simulated annealing, ILS escapes from local optima by applying perturbations to the current local minimum. We will refer to the *strength* of a perturbation as the number of solution components which are modified. For instance for the TSP, it is the number of edges that are changed in the tour, while in the flow shop problem, it is the number of jobs which are moved in the perturbation. Generally, the local search should not be able to undo the perturbation, otherwise one will fall back into the local optimum just visited. Surprisingly often, a *random* move in a neighborhood of higher order than the one used by the local search algorithm can achieve this and will lead to a satisfactory algorithm. Still better results can be obtained if the perturbations take into account properties of the problem and are well matched to the local search algorithm.

By how much should the perturbation change the current solution? If the perturbation is too strong, ILS may behave like a random restart, so better solutions will only be found with a very low probability. On the other hand, if the perturbation is too small, the local search will often fall back into the local optimum just visited and the diversification of the search space will be very limited. An example of a simple but effective perturbation for the TSP is the *double-bridge move*. This perturbation cuts four edges (and is thus of "strength" 4) and introduces four new ones as shown in Figure 11.4. Notice that each bridge is a 2-change, but neither of the 2-changes individually keeps the tour connected. Nearly all ILS studies of the TSP have incorporated this kind of perturbation, and it has been found to be effective for all instance sizes. This is almost certainly because it changes the topology of the tour and can operate on quadruples

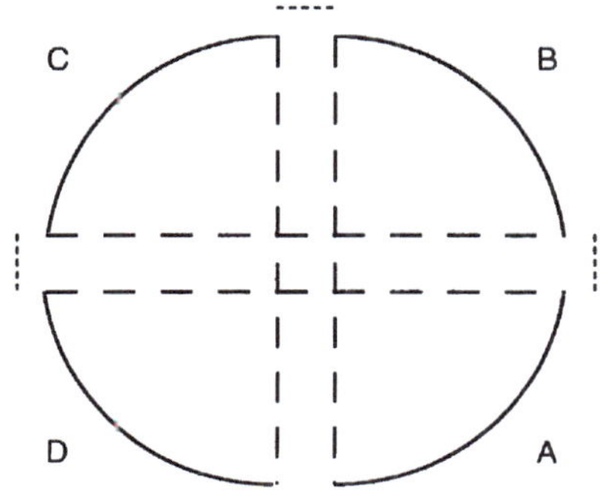

Figure 11.4. Schematic representation of the double-bridge move. The four dotted edges are removed and the remaining parts A, B, C, D are reconnected by the dashed edges.

of very distant cities, whereas local search always modifies the tour among nearby cities. (One could imagine more powerful local searches which would include such double-bridge changes, but the computational cost would be far greater than for the local search methods used today.) In effect, the double-bridge perturbation cannot be undone easily, neither by simple local search algorithms such as `2-opt` or `3-opt`, nor by Lin-Kernighan [43] which is currently the champion local search algorithm for the TSP. Furthermore, this perturbation does not increase much the tour length, so even if the current solution is very good, one is almost sure the next one will be good, too. These two properties of the perturbation—its small strength and its fundamentally different nature from the changes used in local search—make the TSP the perfect application for iterated local search. But for other problems, finding an effective perturbation may be more difficult.

We will now consider optimizing the perturbation assuming the other modules to be fixed. In problems like the TSP, one can hope to have a satisfactory ILS when using perturbations of fixed size (independent of the instance size). On the contrary, for more difficult problems, fixed-strength perturbations may lead to poor performance. Of course, the strength of the perturbations used is not the whole story; their nature is almost always very important and will also be discussed. Finally we will close by pointing out that the perturbation strength has an effect on the speed of the local search: weak perturbations usually lead to faster execution of LocalSearch. All these different aspects need to be considered when optimizing this module.

Perturbation strength For some problems, an appropriate perturbation strength is very small and seems to be rather independent of the instance size. This is the case for both the TSP and the FSP, and interestingly iterated local search for these problems is very competitive with today's best metaheuristic methods. We can also consider other problems where instead one is driven to large perturbation sizes. Consider the example of an ILS algorithm for the quadratic assignment problem (QAP). We use an embedded `2-opt` local search algorithm, the perturbation is a random exchange of the location of k items, where k is an adjustable parameter, and Perturbation always modifies the best solution found so far. We applied this ILS algorithm to QAPLIB instances[5] from four different classes of QAP instances [64]; computational results are given in Table 11.1. A first observation is that the best perturbation size is strongly

Table 11.1. The first column is the name of the QAP instance; the number gives its size n. The successive columns are for perturbation sizes $3, n/12, \ldots, n$. A perturbation of size n corresponds to random restart. The table shows the mean solution cost, averaged over 10 independent runs for each instance. The CPU-time for each trial is 30 s. for `kra30a`, 60 s. for `tai60a` and `sko64`, and 120 s. for `tai60b` on a Pentium III 500 MHz PC

Instance	3	$n/12$	$n/6$	$n/4$	$n/3$	$n/2$	$3n/4$	n
`kra30a`	2.51	2.51	2.04	1.06	0.83	0.42	0.0	0.77
`sko64`	0.65	1.04	0.50	0.37	0.29	0.29	0.82	0.93
`tai60a`	2.31	2.24	1.91	1.71	1.86	2.94	3.13	3.18
`tai60b`	2.44	0.97	0.67	0.96	0.82	0.50	0.14	0.43

[5]QAPLIB is accessible at http://serv1.imm.dtu.dk/~sk/qaplib/.

dependent on the particular instance. For two of the instances, the best performance was achieved when as many as 75% of the solution components were altered by the perturbation. Additionally, for a too small perturbation strength, the ILS performed worse than random restart (corresponding to the perturbation strength n). However, the fact that random restart for the QAP may perform—on average—better than a basic ILS algorithm is a bit misleading: in the next section we will show that by simply modifying a bit the acceptance criterion, ILS becomes far better than random restart. Thus one should keep in mind that the optimization of an iterated local search may require more than the optimization of the individual components.

Adaptive perturbations The behavior of ILS for the QAP and also for other combinatorial optimization problems [35,60] shows that there is no *à priori* single best size for the perturbation. This motivates the possibility of modifying the perturbation strength and adapting it *during* the run.

One possibility to do so is to exploit the search history. For the development of such schemes, inspiration can be taken from what is done in the context of tabu search [6,7]. In particular, Battiti and Protasi proposed [6] a reactive search algorithm for MAX-SAT which fits perfectly into the ILS framework. They perform a "directed" perturbation scheme which is implemented by a tabu search algorithm and after each perturbation they apply a standard local descent algorithm.

Another way of adapting the perturbation is to change deterministically its strength during the search. One particular example of such an approach is employed in the scheme called *basic variable neighborhood search* (basic VNS) [33,55]; we refer to Section 5 for some explanations on VNS. Other examples arise in the context of tabu search [31]. In particular, ideas such as strategic oscillations may be useful to derive more effective perturbations; that is also the spirit of the reactive search algorithm previously mentioned.

More complex perturbation schemes Perturbations can be more complex than changes in a higher order neighborhood. One rather general procedure to generate s' from the current s^* is as follows. (1) Gently modify the definition of the instance, e.g. via the parameters defining the various costs. (2) For this modified instance, run LocalSearch using s^* as input; the output is the perturbed solution s'. Interestingly, this is the method proposed in the oldest ILS work we are aware of: in [10], Baxter tested this approach with success on a location problem. This idea seems to have been rediscovered much later by Codenotti et al. in the context of the TSP. Those authors [18] first change slightly the city coordinates. Then they apply the local search to s^* using these perturbed city locations, obtaining the new tour s'. Finally, running LocalSearch on s' using the *unperturbed* city coordinates, they obtain the new candidate tour $s^{*\prime}$.

Other sophisticated ways to generate good perturbations consist in optimizing a sub-part of the problem. If this task is difficult for the embedded heuristic, good results can follow. Such an approach was proposed by Lourenço [44] in the context of the job shop scheduling problem (JSP). Her perturbation schemes are based on defining one- or two-machine sub-problems by fixing a number of variables in the current solution and solving these sub-problems, either heuristically [45] or to optimality using for instance Carlier's exact algorithm [15] or the early-late algorithm [45]. These schemes work well because: (i) local search is unable to undo the perturbations; (ii) after the perturbation, the solutions tend to be very good and also have "new" parts that are optimized.

Speed In the context of "easy" problems where ILS can work very well with weak (fixed size) perturbations, there is another reason why that metaheuristic can perform much better than random restart: *Speed*. Indeed, LocalSearch will usually execute much faster on a solution obtained by applying a small perturbation to a local optimum than on a random solution. As a consequence, iterated local search can run many more local searches than can random restart in the same CPU time. As a qualitative example, consider again Euclidean TSPs. $\mathcal{O}(n)$ local changes have to be applied by the local search to reach a local optimum from a random start, whereas empirically a nearly constant number is necessary in ILS when using the s' obtained with the double-bridge perturbation. Hence, in a given amount of CPU time, ILS can sample many more local optima than can random restart. This *speed factor* can give ILS a considerable advantage over other restart schemes.

Let us illustrate this speed factor quantitatively. We compare for the TSP the number of local searches performed in a given amount of CPU time by: (i) random restart; (ii) ILS using a double-bridge move; (iii) ILS using five simultaneous double-bridge moves. (For both ILS implementations, we used random starts and the routine AcceptanceCriterion accepted only shorter tours.) For our numerical tests we used a fast `3-opt` implementation with standard speed-up techniques. In particular, it used a fixed radius nearest neighbor search within candidate lists of the 40 nearest neighbors for each city and don't look bits [11,38,49]. Initially, all don't look bits are turned off (set to 0). If for a node no improving move can be found, its don't look bit is turned on (set to 1) and the node is not considered as a starting node for finding an improving move in the next iteration. When an arc incident to a node is changed by a move, the node's don't look bit is turned off again. In addition, when running ILS, after a perturbation we only turn off the don't look bits of the 25 cities around each of the four breakpoints in a current tour. All three algorithms were run for 120 s on a 266 MHz Pentium II processor on a set of TSPLIB[6] instances ranging from 100 up to 5915 cities. Results are given in Table 11.2. For small instances, we see that iterated local search ran between 2 and 10 times as many local searches as random restart. Furthermore, this advantage of ILS grows fast with increasing instance size: for the largest instance, the first ILS algorithm ran approximately 260 times as many local searches as random restart in our alloted time. Obviously, this speed advantage of ILS over random restart is strongly dependent on the strength of the perturbation applied. The larger the perturbation size, the more the solution is modified and generally the longer the subsequent local search takes. This fact is intuitively obvious and is confirmed in Table 11.2.

In summary, the optimization of the perturbations depends on many factors, and problem-specific characteristics play a central role. Finally, it is important to keep in mind that the perturbations also interact with the other components of ILS. We will discuss these interactions in Section 3.5.

3.3 Acceptance Criterion

ILS does a randomized walk in $\mathcal{S}^*$, the space of the local minima. The perturbation mechanism together with the local search defines the possible transitions between a current solution s^* in $\mathcal{S}^*$ to a "neighboring" solution $s^{*\prime}$ also in $\mathcal{S}^*$. The procedure AcceptanceCriterion then determines whether $s^{*\prime}$ is accepted or not as the new current

[6]TSPLIB is accessible at www.iwr.uni-heidelberg.de/iwr/comopt/software/TSPLIB95.

Table 11.2. The first column gives the name of the TSP instance which specifies its size. The next columns give the number of local searches performed when using: (i) random restart (#LS$_{RR}$); (ii) ILS with a single double-bridge perturbation (#LS$_{1-DB}$); (iii) ILS with a five double-bridge perturbation (#LS$_{5-DB}$). All algorithms were run 120 s. on a Pentium 266 MHz PC

Instance	#LS$_{RR}$	#LS$_{1-DB}$	#LS$_{5-DB}$
kroA100	17507	56186	34451
d198	7715	36849	16454
lin318	4271	25540	9430
pcb442	4394	40509	12880
rat783	1340	21937	4631
pr1002	910	17894	3345
pcb1173	712	18999	3229
d1291	835	23842	4312
fl1577	742	22438	3915
pr2392	216	15324	1777
pcb3038	121	13323	1232
fl3795	134	14478	1773
rl5915	34	8820	556

solution. AcceptanceCriterion has a strong influence on the nature and effectiveness of the walk in $\mathcal{S}^*$. Roughly, it can be used to control the balance between intensification and diversification of that search. A simple way to illustrate this is to consider a Markovian acceptance criterion. A very strong intensification is achieved if only better solutions are accepted. We call this acceptance criterion `Better` and it is defined for minimization problems as:

$$\texttt{Better}(s^*, s^{*\prime}, \mathit{history}) = \begin{cases} s^{*\prime} & \text{if } C(s^{*\prime}) < C(s^*) \\ s^* & \text{otherwise} \end{cases} \tag{1}$$

At the opposite extreme is the random walk acceptance criterion (denoted by `RW`) which always applies the perturbation to the most recently visited local optimum, irrespective of its cost:

$$\texttt{RW}(s^*, s^{*\prime}, \mathit{history}) = s^{*\prime} \tag{2}$$

This criterion clearly favors diversification over intensification.

Many intermediate choices between these two extreme cases are possible. In one of the first ILS algorithms, the large-step Markov chains algorithm proposed by Martin, Otto, and Felten [49,50], a simulated annealing type acceptance criterion was applied. We call it `LSMC`$(s^*, s^{*\prime}, \mathit{history})$. In particular, $s^{*\prime}$ is always accepted if it is better than s^*. Otherwise, if $s^{*\prime}$ is worse than $s^*, s^{*\prime}$ is accepted with probability $\exp\{(C(s^*) - C(s^{*\prime}))/T\}$ where T is a parameter called temperature and it is usually lowered during the run as in simulated annealing. Note that `LSMC` approaches the `RW` acceptance criterion if T is very high, while at very low temperatures `LSMC` is similar to the `Better` acceptance criterion. An interesting possibility for `LSMC` is to allow non-monotonic temperature schedules as proposed in [36] for simulated annealing or in tabu thresholding [28]. This can be most effective if it is done using memory: when further intensification no longer seems useful, increase the temperature to do diversification for a limited time, then resume intensification. Of course, just as in tabu search, it is desirable to do this in an automatic and self-regulating manner [31].

A limiting case of using memory in the acceptance criteria is to completely restart the ILS algorithm when the intensification seems to have become ineffective. (Of course this is a rather extreme way to switch from intensification to diversification.) For instance one can restart the ILS algorithm from a new initial solution if no improved solution has been found for a given number of iterations. The restart of the algorithm can easily be modeled by the acceptance criterion called `Restart`$(s^*, s^{*\prime}, history)$. Let i_{last} be the last iteration in which a better solution has been found and i be the iteration counter. Then `Restart`$(s^*, s^{*\prime}, history)$ is defined as

$$\texttt{Restart}(s^*, s^{*\prime}, history) = \begin{cases} s^{*\prime} & \text{if } C(s^{*\prime}) < C(s^*) \\ s & \text{if } C(s^{*\prime}) \geq C(s^*) \text{ and } i - i_{last} > i_r \\ s^* & \text{otherwise.} \end{cases} \tag{3}$$

where i_r is a parameter that indicates that the algorithm should be restarted if no improved solution was found for i_r iterations. Typically, s can be generated in different ways. The simplest strategy is to generate a new solution randomly or by a greedy randomized heuristic. Clearly many other ways to incorporate memory may and should be considered, the overall efficiency of ILS being quite sensitive to the acceptance criterion applied. We now illustrate this with two examples.

Example 11.1. *TSP: Let us consider the effect of the two acceptance criteria* `RW` *and* `Better`. *We performed our tests on the TSP as summarized in Table 11.3. We give the average percentage excess over the known optimal solutions when using 10 independent runs on our set of benchmark instances. In addition we also give this excess for the random restart* `3-opt` *algorithm. First, we observe that both ILS schemes lead to a significantly better average solution quality than random restart using the same local search. This is particularly true for the largest instances, confirming again the claims given in Section 2. Second, given that one expects the good solutions for the TSP to cluster (see Section 3.5), a good strategy should incorporate intensification. It is thus not surprising to see that the* `Better` *criterion leads to shorter tours than the* `RW` *criterion.*

The runs given in this example are rather short. For much longer runs, the `Better` *strategy comes to a point where it no longer finds improved tours and diversification should be considered again. Clearly it will be possible to improve significantly the results by alternating phases of intensification and diversification.*

Example 11.2. *QAP: Let us come back to ILS for the QAP discussed previously. For this problem we found that the acceptance criterion* `Better` *together with a (poor) choice of the perturbation strength could result in worse performance than random restart. In Table 11.4 we give results for the same ILS algorithm except that we now also consider the use of the* `RW` *and* `Restart` *acceptance criteria. We see that the ILS algorithm using these modified acceptance criteria are much better than random restart, the only exception being* `RW` *with a small perturbation strength on* `tai60b`.

Table 11.3. Influence of the acceptance criterion for various TSP instances. The first column gives the instance name and its size. The next columns give the excess percentage length of the tours obtained using: random restart (RR), iterated local search with `RW`, and iterated local search with `Better`. The data is averaged over 10 independent runs. All algorithms were run 120 s. on a Pentium 266 MHz PC

Instance	Δ_{avg}(RR)	Δ_{avg}(`RW`)	Δ_{avg}(`Better`)
`kroA100`	0.0	0.0	0.0
`d198`	0.003	0.0	0.0
`lin318`	0.66	0.30	0.12
`pcb442`	0.83	0.42	0.11
`rat783`	2.46	1.37	0.12
`pr1002`	2.72	1.55	0.14
`pcb1173`	3.12	1.63	0.40
`d1291`	2.21	0.59	0.28
`fl1577`	10.3	1.20	0.33
`pr2392`	4.38	2.29	0.54
`pcb3038`	4.21	2.62	0.47
`fl3795`	38.8	1.87	0.58
`rl5915`	6.90	2.13	0.66

This example shows that there are strong inter-dependences between the perturbation strength and the acceptance criterion. Rarely is this inter-dependence completely understood. But, as a general rule of thumb, when it is necessary to allow for diversification, we believe it is best to do so by accepting numerous small perturbations rather than by accepting one large perturbation.

Most of the acceptance criteria applied so far in ILS algorithms are either fully Markovian or make use of the search history in a very limited way. We expect that there will be many more ILS applications in the future making strong use of the search history; in particular, alternating between intensification and diversification is likely to be an essential feature in these applications.

3.4 Local Search

So far we have treated the local search algorithm as a black box which is called many times by ILS. Since the behavior and performance of the over-all ILS algorithm is quite sensitive to the choice of the embedded heuristic, one should optimize this choice whenever possible. In practice, there may be many quite different algorithms that can be used for the embedded heuristic. (As mentioned at the beginning of the chapter, the heuristic need not even be a local search.) One might think that the better the local search, the better the corresponding ILS. Often this is true. For instance in the context of the TSP, Lin-Kernighan [43] is a better local search than 3-opt which itself is better than 2-opt [38]. Using a fixed type of perturbation such as the double-bridge move, one finds that iterated Lin-Kernighan gives better solutions than iterated 3-opt which itself gives better solutions than iterated 2-opt [38,63]. But if we assume that the total computation time is fixed, it might be better to apply more frequently a faster

Table 11.4. Further tests on the QAP benchmark problems using the same perturbations and CPU times as before; given is the mean solution cost, averaged over 10 independent runs for each instance. Here we consider three different choices for the acceptance criterion. Clearly, the inclusion of diversification significantly lowers the mean cost found

Instance	Acceptance	3	$n/12$	$n/6$	$n/4$	$n/3$	$n/2$	$3n/4$	n
kra30a	Better	2.51	2.51	2.04	1.06	0.83	0.42	0.0	0.77
kra30a	RW	0.0	0.0	0.0	0.0	0.0	0.02	0.47	0.77
kra30a	Restart	0.0	0.0	0.0	0.0	0.0	0.0	0.0	0.77
sko64	Better	0.65	1.04	0.50	0.37	0.29	0.29	0.82	0.93
sko64	RW	0.11	0.14	0.17	0.24	0.44	0.62	0.88	0.93
sko64	Restart	0.37	0.31	0.14	0.14	0.15	0.41	0.79	0.93
tai60a	Better	2.31	2.24	1.91	1.71	1.86	2.94	3.13	3.18
tai60a	RW	1.36	1.44	2.08	2.63	2.81	3.02	3.14	3.18
tai60a	Restart	1.83	1.74	1.45	1.73	2.29	3.01	3.10	3.18
tai60b	Better	2.44	0.97	0.67	0.96	0.82	0.50	0.14	0.43
tai60b	RW	0.79	0.80	0.52	0.21	0.08	0.14	0.28	0.43
tai60b	Restart	0.08	0.08	0.005	0.02	0.03	0.07	0.17	0.43

but less effective local search algorithm than a slower and more powerful one. Clearly which choice is best depends on just how much more time is needed to run the better heuristic. If the speed difference is not large, for instance if it is independent of the instance size, then it usually worth using the better heuristic. This is the most frequent case; for instance in the TSP, 3-opt is a bit slower than 2-opt, but the improvement in quality of the tours are well worth the extra CPU time, be-it using random restart or iterated local search. The same comparison applies to using L-K rather than 3-opt. However, there are other cases where the increase in CPU time is so large compared to the improvement in solution quality that it is best not to use the "better" local search. For example, again in the context of the TSP, it is known that 4-opt gives slightly better solutions than 3-opt, but in standard implementations it is $O(n)$ times slower (n being the number of cities). It is then better not to use 4-opt as the local search embedded in ILS.[7]

There are also other aspects that should be considered when selecting a local search. Clearly, there is not much point in having an excellent local search if it will systematically undo the perturbation; however this issue is one of globally optimizing iterated local search, so it will be postponed till the next sub-section. Another important aspect is whether one can really get the speed-ups that were mentioned in Section 3.2. There we saw that a standard trick for LocalSearch was to introduce don't look bits. These give a large gain in speed if the bits can be reset also after the application of the perturbation. This requires that the developper be able to access the source code of LocalSearch. A state of the art ILS will take advantage of all possible speed-up tricks, and thus the LocalSearch most likely will not be a true black box.

Finally, there may be some advantages in allowing LocalSearch to sometimes generate worse solutions. For instance, if we replace the local search heuristic by tabu search or short simulated annealing runs, the corresponding ILS may perform better. This seems most promising when standard local search methods perform poorly. Such is indeed the case in the job-shop scheduling problem: the use of tabu search as the embedded heuristic gives rise to a very effective iterated local search [46].

3.5 Global Optimization of ILS

So far, we have considered representative issues arising when optimizing separately each of the four components of an iterated local search. In particular, when illustrating various important characteristics of one component, we kept the other components fixed. But clearly the optimization of one component depends on the choices made for the others; as an example, we made it clear that a good perturbation must have the property that it cannot be easily undone by the local search. Thus, at least in principle, one should tackle the *global* optimization of an ILS. Since at present there is no theory for analyzing a metaheuristic such as iterated local search, we content ourselves here with just giving a rough idea of how such a global optimization can be approached in practice.

If we reconsider the sub-section on the effect of the initial solution, we see that GenerateInitialSolution is to a large extent irrelevant when the ILS performs well and rapidly looses the memory of its starting point. Hereafter we assume that this is the case; then the optimization of GenerateInitialSolution can be ignored and we

[7] But see Ref. [29] for a way to implement 4-opt much faster.

are left with the joint optimization of the three other components. Clearly the best choice of Perturbation depends on the choice of LocalSearch while the best choice of AcceptanceCriterion depends on the choices of LocalSearch and Perturbation. In practice, we can approximate this global optimization problem by successively optimizing each component, assuming the others are fixed until no improvements are found for any of the components. Thus the only difference with what has been presented in the previous sub-sections is that the optimization has to be iterative. This does not guarantee global optimization of the ILS, but it should lead to an adequate optimization of the overall algorithm.

Given these approximations, we should make more precise what in fact we are to optimize. For most users, it will be the mean (over starting solutions) of the best cost found during a run of a given length. Then the "best" choice for the different components is a well posed problem, though it is intractable without further restrictions. Furthermore, in general, the detailed instance that will be considered by the user is not known ahead of time, so it is important that the resulting ILS algorithm be robust. Thus it is preferable not to optimize it to the point where it is sensitive to the details of the instance. This robustness seems to be achieved in practice: researchers implement versions of iterated local search with a reasonable level of global optimization, and then test with some degree of success the performance on standard benchmarks.

Search space characteristics At the risk of repeating ourselves, let us highlight the main dependencies of the components:

1. The perturbation should not be easily undone by the local search; if the local search has obvious short-comings, a good perturbation should compensate for them.
2. The combination Perturbation–AcceptanceCriterion determines the relative balance of intensification and diversification; large perturbations are only useful if they can be accepted, which occurs only if the acceptance criterion is not too biased towards better solutions.

As a general guideline, LocalSearch should be as powerful as possible as long as it is not too costly in CPU time. Given such a choice, then find a well adapted perturbation following the discussion in Section 3.2; to the extent possible, take advantage of the structure of the problem. Finally, set the AcceptanceCriterion routine so that S^* is sampled adequately. With this point of view, the overall optimization of the ILS is nearly a bottom-up process, but with iteration. Perhaps the core issue is what to put into Perturbation: can one restrict the perturbations to be weak? From a theoretical point of view, the answer to this question depends on whether the best solutions "cluster" in S^*. In some problems (and the TSP is one of them), there is a strong correlation between the cost of a solution and its "distance" to the optimum: in effect, the best solutions cluster together, i.e., have many similar components. This has been referred to in many different ways: "Massif Central" phenomenon [23], principle of proximate optimality [31], and replica symmetry [53]. If the problem under consideration has this property, it is not unreasonable to hope to find the true optimum using a biased sampling of S^*. In particular, it is clear that is useful to use intensification to improve the probability of hitting the global optimum.

There are, however, other types of problems where the clustering is incomplete, i.e., where very distant solutions can be nearly as good as the optimum. Examples of combinatorial optimization problems in this category are QAP, graph bi-section, and MAX-SAT. When the space of solutions has this property, new strategies have to be used. Clearly, it is still necessary to use intensification to get the best solution in one's current neighborhood, but generally this will not lead to the optimum. After an intensification phase, one must go explore other regions of $\mathcal{S}^*$. This can be attempted by using "large" perturbations whose strength grows with the instance. Other possibilities are to restart the algorithm from scratch and repeat another intensification phase or by oscillating the acceptance criterion between intensification and diversification phases. Additional ideas on the tradeoffs between intensification and diversification are well discussed in the context of tabu search (see, e.g., [31]). Clearly, the balance intensification—diversification is very important and is a challenging problem.

4 SELECTED APPLICATIONS OF ILS

ILS algorithms have been applied successfully to a variety of combinatorial optimization problems. In some cases, these algorithms achieve extremely high performance and even constitute the current state-of-the-art metaheuristics, while in other cases the ILS approach is merely competitive with other metaheuristics. In this section, we cover some of the most studied problems, with a stress on the traveling salesman problem and scheduling problems.

4.1 ILS for the TSP

The TSP is probably the best-known combinatorial optimization problem. *De facto*, it is a standard test-bed for the development of new algorithmic ideas: a good performance on the TSP is taken as evidence of the value of such ideas. Like for many metaheuristic algorithms, some of the first ILS algorithms were introduced and tested on the TSP, the oldest case of this being due to Baum [8,9]. He coined his method *iterated descent*; his tests used 2-opt as the embedded heuristic, random 3-changes as the perturbations, and imposed the tour length to decrease (thus the name of the method). His results were not impressive, in part because he considered the non-Euclidean TSP, which is substantially more difficult in practice than the Euclidean TSP. A major improvement in the performance of ILS algorithms came from the *large-step Markov chain* (LSMC) algorithm proposed by Martin et al. [49]. They used a simulated annealing like acceptance criterion (`LSMC`) from which the algorithm's name is derived and considered both the application of `3-opt` local search and the Lin-Kernighan heuristic (`LK`) which is the best performing local search algorithm for the TSP. But probably the key ingredient of their work is the introduction of the double-bridge move for the perturbation. This choice made the approach very powerful for the Euclidean TSP, and that encouraged much more work along these lines. In particular, Johnson [37,38] coined the term "iterated Lin-Kernighan" (ILK) for his implementation of ILS using the Lin-Kernighan as the local search. The main differences with the LSMC implementation are: (i) double-bridge moves are random rather than biased; (ii) the costs are improving (only better tours are accepted, corresponding to the choice `Better` in our notation). Since these initial studies, other ILS variants have been proposed, and Johnson and McGeoch [38] give a summary of the situation as of 1997.

Currently the highest performance ILS for the TSP is the chained `LK` code by Applegate, Bixby, Chvatal, and Cook which is available as a part of the Concorde software package at www.keck.caam.rice.edu/concorde.html. These authors have provided very detailed descriptions of their implementation, and so we refer the reader to their latest article [1] for details. Furthermore, Applegate et al. [2] performed thorough experimental tests of this code by considering the effect of the different modules: (i) initial tour; (ii) implementation choices of the `LK` heuristic; (iii) types of perturbations. Their tests were performed on very large instances with up to 25 million cities. For the double-bridge move, they considered the effect of forcing the edges involved to be "short", and investigated the random double-bridge moves as well. Their conclusion is that the best performance is obtained when the double-bridge moves are biased towards short edge lengths. However, the strength of the bias towards short edges should be adapted to the available computation time: the shorter the computation time, the shorter the edges should be. In their tests on the influence of the initial tour, they concluded that the worst performance is obtained with random initial tours or those returned by the nearest neighbor heuristic, while best results were obtained with the Christofides algorithm [17], the greedy heuristic [11] or the Quick-Boruvka heuristic proposed in that article. With long runs of their algorithm on TSPLIB instances with more than 10.000 cities they obtained an impressive performance, always obtaining solutions that have less than 0.3% excess length over the lower bounds for these instances. For the largest instance considered, a 25 million city instance, they reached a solution of only 0.3% over the estimated optimum.

Apart from these works, two new ILS algorithms for the TSP have been proposed since the review article of Johnson and McGeoch. The first algorithm is due to Stützle [61,63]; he examined the run-time behavior of ILS algorithms for the TSP and concluded that ILS algorithms with the `Better` acceptance criterion show a type of stagnation behavior for long run-times [61] as expected when performing a strong intensification search. To avoid such stagnation, restarts and a particular acceptance criterion to diversify the search were proposed. The goal of this latter strategy is to force the search to continue from a position that is beyond a certain minimal distance from the current position. This idea is implemented as follows. Let s_c be the solution from which to escape; s_c is typically chosen as s^*_{best}, the best solution found in the recent search. Let $d(s, s')$ be the distance between two tours s and s', that is the number of edges in which they differ. Then the following steps are repeated until a solution beyond a minimal distance $d_{\min}$ from s_c is obtained:

(1) Generate p copies of s_c.
(2) To each of the p solutions apply Perturbation followed by LocalSearch.
(3) Choose the best q solutions, $1 < q \leq p$, as candidate solutions.
(4) Let s^* be the candidate solution with maximal distance to s_c. If $d(s^*, s_c) \leq d_{\min}$ then repeat at (2); otherwise return s^*.

The purpose of step 3 is to choose good quality solutions, while step 4 guarantees that the point from which the search will be continued is sufficiently different (far) from s_c. The attempts are continued until a new solution is accepted, but one gives up after some maximum number of iterations. Computational results for this way of going back and forth between intensification and diversification show that the method is very effective, even when using only a `3-opt` local search [62,63].

The second ILS developed for the TSP since 1997 is that of Katayama and Narisha [39]. They introduce a new perturbation mechanism which they called a *genetic transformation*. The genetic transformation mechanism uses two tours, one of which is the best found so far s^*_{best}, while the second solution s' is a tour found earlier in the search. First a random 4-opt move is performed on s^*_{best}, resulting in $s^{*\prime}$. Then the subtours that are shared among $s^{*\prime}$ and s' are enumerated. The resulting parts are then reconnected with a greedy algorithm. Computational experiments with an iterated LK algorithm using the genetic transformation method instead of the standard double-bridge move have shown that the approach is very effective; further studies should be forthcoming.

4.2 ILS for Scheduling Problems

ILS has also been applied successfully to scheduling problems. Here we summarize the different uses of ILS for tackling these types of systems, ranging from single machine to complex multi-machine scheduling.

Single Machine Total Weighted Tardiness Problem (SMTWTP) Congram, Potts and van de Velde [19] have presented an ILS algorithm for the SMTWTP based on a dynasearch local search. Dynasearch uses dynamic programming to find a best move which is composed of a set of independent interchange moves; each such move exchanges the jobs at positions i and j, $j \neq i$. Two interchange moves are independent if they do not overlap, that is if for two moves involving positions i, j and k, l we have $\min\{i, j\} \geq \max\{k, l\}$ or vice versa. This neighborhood is of exponential size but dynasearch explores this neighborhood in polynomial time.

The perturbation consists of a series of random interchange moves. They also exploit a well-known property of the SMTWTP: there exists an optimal solution in which non-late jobs are sequenced in non-decreasing order of the due dates. This property is used in two ways: to diversify the search in the perturbation step and to reduce the computation time of the dynasearch. In the acceptance criterion, Congram et al. introduce a *backtrack step*: after β iterations in which every new local optimum is accepted, the algorithm restarts with the best solution found so far. In our notation, the backtrack step is a particular choice for the history dependence incorporated into AcceptanceCriterion.

Congram et al. used several different embedded LocalSearch, all associated with the interchange neighborhood. These heuristics were: (i) dynasearch; (ii) a local search based on first-improvement descent; (iii) a local search based on best-improvement descent. Then they performed tests to evaluate these algorithms using random restart and compared them to using iterated local search. While random restart dynasearch performed only slightly better than the two simpler descent methods, the ILS with dynasearch significantly outperformed the other two iterated descent algorithms, which in turn were far superior to the random restart versions. The authors also show that the iterated dynasearch algorithm significantly improves over the previously best known algorithm, a tabu search presented in [20].

Single and parallel machine scheduling Brucker et al. [12,13] apply the principles of ILS to a number of one-machine and parallel-machine scheduling problems. They introduce a local search method which is based on two types of neighborhoods. At each step one goes from one feasible solution to a neighboring one with respect to the secondary neighborhood. The main difference with standard local search methods is that this secondary neighborhood is defined on the set of locally optimal solutions with respect to the first neighborhood. Thus in fact this is an ILS with two nested neighborhoods; searching in their primary neighborhood corresponds to our local search phase; searching in their secondary neighborhood is like our perturbation phase. The authors also note that the second neighborhood is problem specific; this is what arises in ILS where the perturbation should be adapted to the problem. The search at a higher level reduces the search space and at the same time leads to better results.

Flow shop scheduling Stützle [60] applied ILS to the flow shop problem (FSP). The algorithm is based on a straightforward first-improvement local search using the insert neighborhood, where a job at position i is removed and inserted at position $j \neq i$. The initial schedule is constructed by the NEH heuristic [57] while the perturbation is generated by composing moves of two different kinds: swaps which exchange the positions of two adjacent jobs, and interchange moves which have no constraint on adjacency. Experimentally, it was found that perturbations with just a few swap and interchange moves were sufficient to obtain very good results. The article also compares different acceptance criteria; `ConstTemp`, which is the same as the `LSMC` acceptance criterion except that it uses a constant temperature T_c, was found to be superior to `Better`. The computational results show that despite the simplicity of the approach, the quality of the solutions obtained is comparable to that of the best performing local search algorithms for the FSP; we refer to [60] for a more detailed discussion.

ILS has also been used to solve a flow-shop problem with several stages in series. Yang et al. [67] presented such a method; at each stage, instead of a single machine, there is a group of identical parallel machines. Their metaheuristic has two phases that are repeated iteratively. In the first phase, the operations are assigned to the machines and an initial sequence is constructed. The second phase uses an ILS to find better schedules for each machine at each stage by modifying the sequence of each machine. (This part is very similar in spirit to the approach of Kreipl for the minimum total weighted tardiness job-shop problem [42] that is presented below.) Yang, Kreipl and Pinedo also proposed a "hybrid" metaheuristic: they first apply a decomposition procedure that solves a series of single stage sub-problems; then they follow this by their ILS. The process is repeated until a satisfactory solution is obtained.

Job shop scheduling Lourenço [44] and Lourenço and Zwijnenburg [46] used ILS to tackle the job shop scheduling problem (JSP). They performed extensive computational tests, comparing different ways to generate initial solutions, various local search algorithms, different perturbations, and three acceptance criteria. While they found that the initial solution had only a very limited influence, the other components turned out to be very important. Perhaps the heart of their work is the way they perform the perturbations. They consider relaxations of the problem at hand corresponding to the optimization of just some of the jobs. Then they use exact methods to solve these sub-problems, generating the perturbation move. This has the great advantage that much problem-specific knowledge is built into the perturbation. Such problem specific perturbations are difficult to generate from local moves only. Now, for the local search, three alternatives were considered: local descent, short simulated annealing runs, and short tabu search runs. Best results were obtained using the latter in the local search phase. Not surprisingly, ILS performed better than random restart given the same amount of time, for any choice of the embedded local search heuristic.

In more recent work on the job-shop scheduling problem, Balas and Vazacopoulos [4] presented a variable depth search heuristic which they called guided local search

(GLS). GLS is based on the concept of neighborhood trees, proposed by the authors, where each node corresponds to a solution and the child nodes are obtained by performing an interchange on some critical arc. In their work, the interchange move consists in reversing more than one arc and can be seen as a particular kind of variable depth interchange. They developed ILS algorithms by embedding GLS within the shifting bottleneck (SB) procedure by replacing the reoptimization cycle of the SB with a number of cycles of the GLS procedure. They call this procedure SB-GLS1. Later, they also proposed a variant of this method, SB-GLS2, which works as follows. After all machines have been sequenced, they iteratively remove one machine and apply GLS to a smaller instance defined by the remaining machines. Then again GLS is applied on the initial instance containing *all* machines. This procedure is an ILS where a perturbed solution is obtained by applying a (variable depth) local search to just part of an instance. The authors perform a computational comparison with other metaheuristics and conclude that SB-GLS (1 and 2) are robust and efficient, and provide schedules of high quality in a reasonable computing time. In some sense, both heuristics are similar to the one proposed by Lourenço [44], the main differences being: (i) Lourenço's heuristic applies perturbations to complete schedules whereas the SB-GLS heuristic starts by an empty (infeasible) schedule and iteratively optimizes it machine by machine until all machines have been scheduled, in a SB-style followed by a local search application; (ii) the local search algorithms used differ.

Recently, Kreipl applied ILS to the total weighted tardiness job shop scheduling problem (TWTJSP) [42]. The TWTJSP is closer to real life problems than the classical JSP with makespan objective because it takes into account release and due dates and also it introduces weights that indicate the importance of each job. Kreipl uses an ILS algorithm with the RW acceptance criterion. The algorithm starts with an initial solution obtained by the shortest processing time rule [34]. The local search consists in reversing critical arcs and arcs adjacent to these, where a critical arc has to be an element of at least one critical path (there may exist several critical paths). One original aspect of this ILS is the perturbation step: Kreipl applies a few steps of a simulated annealing type algorithm with the Metropolis acceptance criterion [52] but with a fixed temperature. For this perturbation phase a smaller neighborhood than the one used in the local search phase is taken: while in the local search phase any critical arc can be reversed, during the diversification phase only the critical arcs belonging to the critical path having the job with highest impact on the objective function are considered.[8] The number of iterations performed in the perturbation depends how good the incumbent solution is. In promising regions, only a few steps are applied to stay near good solutions, otherwise, a "large" perturbation is applied to permit the algorithm to escape from a poor region. Computational results with the ILS algorithm on a set of benchmark instances has shown a very promising performance compared to an earlier shifting bottleneck heuristic [59] proposed for the same problem.

4.3 ILS for Other Problems

Graph bipartitioning ILS algorithms have been proposed and tested on a number of other problems, though not as thoroughly as the ones we have discussed so far. We consider first the graph bipartitioning problem. Given a (weighted) graph and a

[8] It should be noted that the perturbation phase leads, in general, to an intermediate solution which is not locally optimal.

bisection or partition of its vertices into two sets A and B of equal size, call the cut of the partition the sum of the weights of the edges connecting the two parts. The graph partitioning problem is to find the partition with the minimum cut. Martin and Otto [47,48] introduced an ILS for this problem following their earlier work on the TSP. For the local search, they used the Kernighan-Lin variable depth local search algorithm (`KL`) [40] which is the analog for this problem of the `LK` algorithm. In effect, `KL` finds intelligently m vertices of one set to be exchanged with m of the other. Then, when considering possible perturbations, they noticed a particular weakness of the `KL` local search: `KL` frequently generates partitions with many "islands", i.e., the two sets A and B are typically highly fragmented (disconnected). Thus they introduced perturbations that exchanged vertices between these islands rather than between the whole sets A and B. This works as follows: choose at random one of the cut edges, i.e., an edge connecting A and B. This edge connects two "seed" vertices each belonging to their island. Around each seed, iteratively grow a connected cluster of vertices within each island. When a target cluster size or a whole island size is reached, stop the growth. The two clusters are then exchanged and this is the perturbation move. Finally, for the acceptance criterion, Martin and Otto used the `Better` acceptance criterion. The overall algorithm significantly improved over the embedded local search (random restart of `KL`); it also improved over simulated annealing if the acceptance criterion was optimized.

At the time of that work, simulated annealing was the state of the art method for the graph bisection problem. Since then, there have been many other metaheuristics [5,51] developed for this problem, so the performance that must be reached is much higher now. Furthermore, given that the graph bipartitioning problem has a low cost-distance correlation [51], ILS has difficulty in sampling all good low cost solutions. To overcome this, some form of history dependence most certainly would have to be built into the perturbation or the acceptance criterion.

MAX-SAT Battiti and Protasi present an application of *reactive search* to the MAX-SAT problem [6]. Their algorithm consists of two phases: a local search phase and a diversification (perturbation) phase. Because of this, their approach fits perfectly into the ILS framework. Their perturbation is obtained by running a tabu search on the current local minimum so as to guarantee that the modified solution s' is sufficiently different from the current solution s^*. Their measure of difference is just the Hamming distance; the minimum distance is set by the length of a tabu list that is adjusted during the run of the algorithm. For the LocalSearch, they use a standard greedy descent local search appropriate for the MAX-SAT problem. Depending on the distance between $s^{*\prime}$ and s^*, the tabu list length for the perturbation phase is dynamically adjusted. The next perturbation phase is then started based on solution $s^{*\prime}$—corresponding to the `RW` acceptance criterion. This work illustrates very nicely how one can adjust dynamically the perturbation strength in an ILS run. We conjecture that similar schemes will prove useful to optimize ILS algorithms in a nearly automatic way.

Prize-collecting Steiner tree problem The last combinatorial optimization problem we discuss is the prize-collecting Steiner tree problem on graphs. Canudo, Resende and Ribeiro [14] presented several local search strategies for this problem: iterative improvement, multi-start with perturbations, path-relinking, variable neighborhood search, and a algorithm based on the integration of all these. They showed that all these strategies are effective in improving solutions; in fact in many of their tests they found the optimal solution. One of their proposed heuristics, local search with perturbations, is in fact an ILS. In that approach, they first generated initial solutions by the primal-dual algorithm of Goemans and Williamson (GW) [32] but where the cost function is slightly modified. Canudo et al. proposed two perturbation schemes: perturbation by eliminations and perturbations by prize changes. In the first scheme, the perturbation is done by resetting to zero the prizes of some persistent node which appeared in the solution build by GW and remained at the end of local search in the previous iteration. In the second scheme, the perturbation consists in introducing noise into the node prize. This feature of always applying the perturbation to the last solution obtained by the local search phase is clearly in our notation the ILS-RW choice.

4.4 Summary

The examples we have chosen in this section stress several points that have already been mentioned before. First, the choice of the local search algorithm is usually quite critical if one is to obtain peak performance. In most of the applications, the best performing ILS algorithms apply much more sophisticated local search algorithms than simple best- or first-improvement descent methods. Second, the other components of an ILS also need to be optimized if the state of the art is to be achieved. This optimization should be global, and to succeed should involve the use of problem-specific properties. Examples of this last point were given for instance in the scheduling applications: there the good perturbations were not simply random moves, rather they involved re-optimizations of significant parts of the instance (c.f. the job shop case).

The final picture we reach is one where (i) ILS is a versatile metaheuristic which can easily be adapted to different combinatorial optimization problems; (ii) sophisticated perturbation schemes and search space diversification are the essential ingredients to achieve the best possible ILS performance.

5 RELATION TO OTHER METAHEURISTICS

In this section we highlight the similarities and differences between ILS and other well-known metaheuristics. We shall distinguish metaheuristics which are essentially variants of local search and those which generate solutions using a mechanism that is not necessarily based on an explicit neighborhood structure. Among the first class which we call *neighborhood based metaheuristics* are methods like simulated annealing (SA) [16,41], tabu search (TS) [26,27,31] or guided local search (GLS) [66]. The second class comprises metaheuristics like GRASP [22], ant colony optimization (ACO) [21], evolutionary algorithms (EA) [3,54], scatter search [30], variable neighborhood search (VNS) [33,55] and ILS. Some metaheuristics of this second class, like EAs and ACO, do not necessarily make use of local search algorithms; however a local search can be embedded in them, in which case the performance is usually enhanced [56,61]. The other metaheuristics in this class explicitly use embedded local search algorithms as an essential part of their structure. For simplicity, we will assume in what follows that all the metaheuristics of this second class do incorporate local search algorithms. In this case, such metaheuristics generate iteratively input solutions that are passed to a local search; they can thus be interpreted as multi-start algorithms, using the most general meaning of that term. This is why we call them here *multi-start based metaheuristics*.

5.1 Neighborhood Based Metaheuristics

Neighborhood based metaheuristics are extensions of iterative improvement algorithms and avoid getting stuck in locally optimal solutions by allowing moves to worse solutions in one's neighborhood. Different metaheuristics of this class differ mainly by their move strategies. In the case of simulated annealing, the neighborhood is sampled randomly and worse solutions are accepted with a probability which depends on a temperature parameter and the degree of deterioration incurred; better neighboring solutions are usually accepted while much worse neighboring solutions are accepted with a low probability. In the case of (simple) tabu search strategies, the neighborhood is explored in an aggressive way and cycles are avoided by declaring attributes of visited solutions as tabu. Finally, in the case of guided local search, the evaluation function is dynamically modified by penalizing certain solution components. This allows the search to escape from a solution that is a local optimum of the original objective function.

Obviously, any of these neighborhood based metaheuristics can be used as the LocalSearch procedure in ILS. In general, however, those metaheuristics do not halt, so it is necessary to limit their run time if they are to be embedded in ILS. One particular advantage of combining neighborhood based metaheuristics with ILS is that they often obtain much better solutions than iterative descent algorithms. But this advantage usually comes at the cost of larger computation times. Since these metaheuristics allow one to obtain better solutions at the expense of greater computation times, we are confronted with the following optimization problem when using them within an ILS:[9] "For how long should one run the embedded search in order to achieve the best tradeoff between computation time and solution quality?" This is very analogous to the question of whether it is best to have a fast but not so good local search or a slower but more powerful one. The answer depends of course on the total amount of computation time available, and on how the costs improve with time.

A different type of connection between ILS, SA and TS arises from certain similarities in the algorithms. For example, SA can be seen as an ILS without a local search phase (SA samples the original space S and not the reduced space S^*) and where the acceptance criteria is LSMC($s^*, s^{*\prime}$, *history*). While SA does not employ memory, the use of memory is the main feature of TS which makes a strong use of historical information at multiple levels. Given its effectiveness, we expect this kind of approach for incorporating memory to become widespread in future ILS applications.[10] Furthermore, TS, as one prototype of a memory intensive search procedure, can be a valuable source of inspiration for deriving ILS variants with a more direct usage of memory; this can lead to a better balance between intensification and diversification in the search.[11] Similarly, TS strategies may also be improved by features of ILS algorithms and by some insights gained from the research on ILS.

---

[9]This question is not specific to ILS; it arises for all multi-start type metaheuristics.

[10]In early TS publications, proposals similar to the use of perturbations were put forward under the name *random shakeup* [25]. These procedures where characterized as a "randomized series of moves that leads the heuristic (away) from its customary path" [25]. The relationship to perturbations in ILS is obvious.

[11]Indeed, in [26], Glover uses "strategic oscillation" strategies whereby one cycles over these procedures: the simplest moves are used till there is no more improvement, and then progressively more advanced moves are used.

5.2 Multi-start Based Metaheuristics

Multi-start based metaheuristics can be classified into *constructive* metaheuristics and *perturbation-based* metaheuristics.

Well-known examples of constructive metaheuristics are ant colony optimization and GRASP which both use a probabilistic solution construction phase. An important difference between ACO and GRASP is that ACO has an indirect memory of the search process which is used to bias the construction process, whereas GRASP does not have that kind of memory. An obvious difference between ILS and constructive metaheuristics is that ILS does not construct solutions. However, both generate a sequence of solutions, and if the constructive metaheuristic uses an embedded local search, both go from one local minimum to another. So it might be said that the perturbation phase of an ILS is replaced by a (memory-dependent) construction phase in these constructive metaheuristics. But another connection can be made: ILS can be used instead of the embedded "local search" in an algorithm like ant colony optimization or GRASP. This is one way to generalize ILS, but it is not specific to these kinds of metaheuristics: whenever one has an embedded local search, one can try to replace it by an iterated local search.

Perturbation-based metaheuristics differ in the techniques they use to actually perturb solutions. Before going into details, let us introduce one additional feature for classifying metaheuristics: we will distinguish between population-based algorithms and those that use a single current solution (the population is of size 1). For example, EA, scatter search, and ant colony optimization are population-based, while ILS uses a single solution at each step. Whether or not a metaheuristics is population-based is important for the type of perturbation that can be applied. If no population is used, new solutions are generated by applying perturbations to single solutions; this is what happens for ILS and VNS. If a population is present, one can also use the possibility of recombining several solutions into a new one. Such combinations of solutions are implemented by "crossover" operators in EAs or in the recombination of multiple solutions in scatter search.

In general, population-based metaheuristics are more complex to use than those following a single solution: they require mechanisms to manage a population of solutions and more importantly it is necessary to find effective operators for the combination of solutions. Most often, this last task is a real challenge. The complexity of these population-based local search hybrid methods can be justified if they lead to better performance than non-population based methods. Therefore, one question of interest is whether using a population of solutions is really useful. Unfortunately, there are very few systematic studies which address this issue [20,24,38,62,65]. Clearly for some problems such as the TSP with high cost-distance correlations, the use of a single element in the population leads to good results, so the advantage of population-based methods is small or nil. However, for other problems (with less cost-distance correlations), it is clear that the use of a population is an appropriate way to achieve search space diversification. Thus population based methods are desirable if their complexity is not overwhelming. Because of this, population-based extensions of ILS are promising approaches.

To date, several population-based extensions of ILS *have* been proposed [1,35, 61]. The approaches proposed in [35,61] keep the simplicity of ILS algorithms by maintaining unchanged the perturbations: one parent is perturbed to give one child.

But given that there is a population, the evolution depends on competition among its members and only the fittests survive. One can give up this simplicity as was done in the approach of Applegate et al. [1]. Given the solutions in a population that have been generated by an ILS, they define a smaller instance by freezing the components that are in common in all parents. (They do this in the context of the TSP; the subtours that are in common are then fixed in the sub-problem.) They then reoptimize this smaller problem using ILS. This idea is tested in [1], and they find very high quality solutions, even for large TSP instances.

Finally, let us discuss variable neighborhood search (VNS) which is the metaheuristic closest to ILS. VNS begins by observing that the concept of local optimality is conditional on the neighborhood structure used in a local search. Then VNS systemizes the idea of changing the neighborhood during the search to avoid getting stuck in poor quality solutions. Several VNS variants have been proposed. The most widely used one, *basic VNS*, can, in fact, be seen as an ILS algorithm which uses the `Better` acceptance criterion and a systematic way of varying the perturbation strength. To do so, basic VNS orders neighborhoods as $\mathcal{N}_1, \ldots, \mathcal{N}_m$ where the order is chosen according to the neighborhood size. Let k be a counter variable, $k = 1, 2, \ldots, m$, and initially set $k = 1$. If the perturbation and the subsequent local search lead to a new best solution, then k is reset to 1, otherwise k is increased by one. We refer to [33,55] for a description of other VNS variants.

A major difference between ILS and VNS is the philosophy underlying the two metaheuristics: ILS explicitly has the goal of building a walk in the set of locally optimal solutions, while VNS algorithms are derived from the idea of systematically changing neighborhoods during the search.

Clearly, there are major points in common between most of today's high performance metaheuristics. How can one summarize how iterated local search differs from the others? We shall proceed by enumeration as the diversity of today's metaheuristics seems to forbid any simpler approach. When comparing to ACO and GRASP, we see that ILS uses perturbations to create new solutions; this is quite different in principle and in practice from using construction. When comparing to EAs and scatter search, we see that ILS, as we defined it, has a population size of 1; therefore no recombination operators need be defined. We could continue like this, but we cannot expect the boundaries between all metaheuristics to be so clear-cut. Not only are hybrid methods very often the way to go, but most often one can smoothly go from one metaheuristic to another. In addition, as mentioned at the beginning of this chapter, the distinction between heuristic and metaheuristic is rarely unambiguous. So our point of view is not that iterated local search has essential features that are absent in other metaheuristics; rather, when considering the basic structure of iterated local search, some simple yet powerful ideas transpire, and these can be of use in most metaheuristics, being close or not in spirit to iterated local search.

6 CONCLUSIONS

ILS has many of the desirable features of a metaheuristic: it is simple, easy to implement, robust, and highly effective. The essential idea of ILS lies in focusing the search not on the full space of solutions but on a smaller subspace defined by the solutions that are locally optimal for a given optimization engine. The success of ILS lies in the *biased* sampling of this set of local optima. How effective this approach turns out to be depends mainly on the choice of the local search, the perturbations, and the acceptance criterion. Interestingly, even when using the most naïve implementations of these parts, ILS can do much better than random restart. But with further work so that the different modules are well adapted to the problem at hand, ILS can often become a competitive or even state of the art algorithm. This dichotomy is important because the optimization of the algorithm can be done progressively, and so ILS can be kept at any desired level of simplicity. This, plus the modular nature of iterated local search, leads to short development times and gives ILS an edge over more complex metaheuristics in the world of industrial applications. As an example of this, recall that ILS essentially treats the embedded heuristic as a black box; then upgrading an ILS to take advantage of a new and better local search algorithm is nearly immediate. Because of all these features, we believe that ILS is a promising and powerful algorithm to solve real complex problems in industry and services, in areas ranging from finance to production management and logistics. Finally, let us note that although all of the present review was given in the context of tackling combinatorial optimization problems, in reality much of what we covered can be extended in a straight-forward manner to continuous optimization problems.

Looking ahead towards future research directions, we expect ILS to be applied to new kinds of problems. Some challenging examples are: (i) problems where the constraints are very severe and so most metaheuristics fail; (ii) multi-objective problems, bringing one closer to real problems; (iii) dynamic or real-time problems where the problem data vary during the solution process.

The ideas and results presented in this chapter leave many questions unanswered. Clearly, more work needs to be done to better understand the interplay between the ILS modules GenerateInitialSolution, Perturbation, LocalSearch, and AcceptanceCriterion. In particular, we expect significant improvements to arise through the intelligent use of memory, explicit intensification and diversification strategies, and greater problem-specific tuning. The exploration of these issues has barely begun but should lead to higher performance iterated local search algorithms.

ACKNOWLEDGMENTS

O.M. acknowledges support from the Institut Universitaire de France.

This work was partially supported by the "Metaheuristics Network", a Research Training Network funded by the Improving Human Potential programme of the CEC, grant HPRN-CT-1999-00106. The information provided is the sole responsibility of the authors and does not reflect the Community's opinion. The Community is not responsible for any use that might be made of data appearing in this publication.

REFERENCES

[1] D. Applegate, R. Bixby, V. Chvátal and W. Cook (2000) Finding tours in the TSP. Preliminary version of a book chapter available via www.keck.caam.rice.edu/concorde.html.

[2] D. Applegate, W. Cook and A. Rohe (1999) Chained Lin-Kernighan for large traveling salesman problems. Technical Report No. 99887, Forschungsinstitut für Diskrete Mathematik, University of Bonn, Germany.

[3] T. Bäck (1996) *Evolutionary Algorithms in Theory and Practice*. Oxford University Press.

[4] E. Balas and A. Vazacopoulos (1998) Guided local search with shifting bottleneck for job shop scheduling. *Management Science*, **44**(2), 262–275.

[5] R. Battiti and A. Bertossi (1999) Greedy, prohibition, and reactive heuristics for graph-partitioning. *IEEE Transactions on Computers*, **48**(4), 361–385.

[6] R. Battiti and M. Protasi (1997) Reactive search, a history-based heuristic for MAX-SAT. *ACM Journal of Experimental Algorithmics*, **2**.

[7] R. Battiti and G. Tecchiolli (1994) The reactive tabu search. *ORSA Journal on Computing*, **6**(2), 126–140.

[8] E.B. Baum (1986) Iterated descent: A better algorithm for local search in combinatorial optimization problems. Technical report, Caltech, Pasadena, CA. manuscript.

[9] E.B. Baum (1986) Towards practical "neural" computation for combinatorial optimization problems. In: J. Denker (ed.), *Neural Networks for Computing*. AIP conference proceedings, pp. 53–64.

[10] J. Baxter (1981) Local optima avoidance in depot location. *Journal of the Operational Research Society*, **32**, 815–819.

[11] J.L. Bentley (1992) Fast algorithms for geometric traveling salesman problems. *ORSA Journal on Computing*, **4**(4), 387–411.

[12] P. Brucker, J. Hurink and F. Werner (1996) Improving local search heuristics for some scheduling problems—part I. *Discrete Applied Mathematics*, **65**(1–3), 97–122.

[13] P. Brucker, J. Hurink and F. Werner (1997) Improving local search heuristics for some scheduling problems—part II. *Discrete Applied Mathematics*, **72**(1–2), 47–69.

[14] S.A. Canuto, M.G.C. Resende and C.C. Ribeiro (2000) Local search with perturbations for the prize-collecting steiner tree problem in graphs. Networks (submitted).

[15] J. Carlier (1982) The one-machine sequencing problem *European Journal of Operational Research*, **11**, 42–47.

[16] V. Cerný (1985) A thermodynamical approach to the traveling salesman problem. *Journal of Optimization Theory and Applications*, **45**(1), 41–51.

[17] N. Christofides (1976) Worst-case analysis of a new heuristic for the travelling salesman problem. Technical Report 388, Graduate School of Industrial Administration, Carnegie-Mellon University, Pittsburgh, PA.

[18] B. Codenotti, G. Manzini, L. Margara and G. Resta (1996) Perturbation: An efficient technique for the solution of very large instances of the Euclidean TSP. *INFORMS Journal on Computing*, **8**, 125–133.

[19] R.K. Congram, C.N. Potts and S.L. Van de Velde (2000) An iterated dynasearch algorithm for the single–machine total weighted tardiness scheduling problem. *INFORMS Journal on Computing* (to appear).

[20] H.A.J. Crauwels, C.N. Potts and L.N. Van Wassenhove (1998) Local search heuristics for the single machine total weighted tardiness scheduling problem. *INFORMS Journal on Computing*, **10**(3), 341–350.

[21] M. Dorigo and G. Di Caro (1999) The ant colony optimization meta-heuristic. In: D. Corne, M. Dorigo and F. Glover (eds.), *New Ideas in Optimization*. McGraw Hill, pp. 11–32.

[22] T.A. Feo and M.G.C. Resende (1995) Greedy randomized adaptive search procedures. *Journal of Global Optimization*, **6**, 109–133.

[23] C. Fonlupt, D. Robilliard, P. Preux and E.-G. Talbi (1999) Fitness landscape and performance of meta-heuristics. In: S. Voss, S. Martello, I.H. Osman and C. Roucairol (eds.), *Meta-Heuristics: Advances and Trends in Local Search Paradigms for Optimization*. Kluwer Academic Publishers, Boston, MA, pp. 257–268.

[24] C. Glass and C. Potts (1996) A comparison of local search methods for flow shop scheduling. *Annals of Operations Research*, **63**, 489–509.

[25] F. Glover (1986) Future paths for integer programming and links to artificial intelligence. *Computers & Operations Research*, **13**(5), 533–549.

[26] F. Glover (1989) Tabu search—part I. *ORSA Journal on Computing*, **1**(3), 190–206.

[27] F. Glover (1990) Tabu search—part II. *ORSA Journal on Computing*, **2**(1), 4–32.

[28] F. Glover (1995) Tabu thresholding: Improved search by nonmonotonic trajectories. *ORSA Journal on Computing*, **7**(4), 426–442.

[29] F. Glover (1996) Finding a best traveling salesman 4-opt move in the same time as a best 2-opt move. *Journal of Heuristics*, **2**, 169–179.

[30] F. Glover (1999) Scatter search and path relinking. In: D. Corne, M. Dorigo and F. Glover (eds.), *New Ideas in Optimization*. McGraw Hill, pp. 297–316.

[31] F. Glover and M. Laguna (1997) *Tabu Search*. Kluwer Academic Publishers, Boston, MA.

[32] M.X. Goemans and D.P. Williamson (1996) The primal dual method for approximation algorithms and its application to network design problems. In: D. Hochbaum (ed.), *Approximation Algorithms for NP-hard Problems*. PWS Publishing, pp. 144–191.

[33] P. Hansen and N. Mladenović (1999) An introduction to variable neighborhood search. In: S. Voss, S. Martello, I.H. Osman and C. Roucairol (eds.), *Meta-Heuristics: Advances and Trends in Local Search Paradigms for Optimization*. Kluwer Academic Publishers, Boston, MA, pp. 433–458.

[34] R. Haupt (1989) A survey of priority rule-based scheduling. *OR Spektrum*, **11**, 3–6.

[35] I. Hong, A.B. Kahng and B.R. Moon (1997) Improved large-step Markov chain variants for the symmetric TSP. *Journal of Heuristics*, **3**(1), 63–81.

[36] T.C. Hu, A.B. Kahng and C.-W.A. Tsao (1995) Old bachelor acceptance: A new class of non-monotone threshold accepting methods. *ORSA Journal on Computing*, **7**(4), 417–425.

[37] D.S. Johnson (1990) Local optimization and the travelling salesman problem. In: *Proceedings of the 17th Colloquium on Automata, Languages, and Programming*, volume 443 of *LNCS*, Springer Verlag, Berlin, pp. 446–461.

[38] D.S. Johnson and L.A. McGeoch (1997) The travelling salesman problem: A case study in local optimization. In: E.H.L. Aarts and J.K. Lenstra (eds.), *Local Search in Combinatorial Optimization*. John Wiley & Sons, Chichester, England, pp. 215–310.

[39] K. Katayama and H. Narihisa (1999) Iterated local search approach using genetic transformation to the traveling salesman problem. In: *Proceedings of GECCO'99*, Vol. 1. Morgan Kaufmann, pp. 321–328.

[40] B.W. Kernighan and S. Lin (1970) An efficient heuristic procedure for partitioning graphs. *Bell Systems Technology Journal*, **49**, 213–219.

[41] S. Kirkpatrick, C.D. Gelatt Jr. and M.P. Vecchi (1983) Optimization by simulated annealing. *Science*, **220**, 671–680.

[42] S. Kreipl (2000) A large step random walk for minimizing total weighted tardiness in a job shop. *Journal of Scheduling*, **3**(3), 125–138.

[43] S. Lin and B.W. Kernighan (1973) An effective heuristic algorithm for the travelling salesman problem. *Operations Research*, **21**, 498–516.

[44] H.R. Lourenço (1995) Job-shop scheduling: Computational study of local search and large-step optimization methods. *European Journal of Operational Research*, **83**, 347–364.

[45] H.R. Lourenço (1998) A polynomial algorithm for a special case of the one-machine scheduling problem with time–lags. Technical Report Economic Working Papers Series, No. 339, Universitat Pompeu Fabra. *Journal of Scheduling* (submitted).

[46] H.R. Lourenço and M. Zwijnenburg (1996) Combining the large-step optimization with tabu-search: Application to the job-shop scheduling problem. In: I.H. Osman and J.P. Kelly (eds.), *Meta-Heuristics: Theory & Applications*. Kluwer Academic Publishers, pp. 219–236.

[47] O. Martin and S.W. Otto (1995) Partitoning of unstructured meshes for load balancing. *Concurrency: Practice and Experience*, **7**, 303–314.

[48] O. Martin and S.W. Otto (1996) Combining simulated annealing with local search heuristics. *Annals of Operations Research*, **63**, 57–75.

[49] O. Martin, S.W. Otto and E.W. Felten (1991) Large-step Markov chains for the traveling salesman problem. *Complex Systems*, **5**(3), 299–326.

[50] O. Martin, S.W. Otto and E.W. Felten (1992) Large-step Markov chains for the TSP incorporating local search heuristics. *Operations Research Letters*, **11**, 219–224.

[51] P. Merz and B. Freisleben (2000) Fitness landscapes, memetic algorithms and greedy operators for graph bi-partitioning. *Evolutionary Computation*, **8**(1), 61–91.

[52] N. Metropolis, A. Rosenbluth, M. Rosenbluth, A. Teller and M. Teller (1953) Equation of state calculations for fast computing machines. *Journal of Chemical Physics*, **21**, 1087–1092.

[53] M. Mézard, G. Parisi and M.A. Virasoro (1987) *Spin-Glass Theory and Beyond, volume 9* of *Lecture Notes in Physics*. World Scientific, Singapore.

[54] Z. Michalewicz and D.B. Fogel (2000) *How to Solve it: Modern Heuristics*. Springer-Verlag, Berlin.

[55] N. Mladenović and P. Hansen (1997) Variable neighborhood search. *Computers & Operations Research*, **24**, 1097–1100.

[56] H. Mühlenbein (1991) Evolution in time and space—the parallel genetic algorithm. In: *Foundations of Genetic Algorithms*. Morgan Kaufmann, San Mateo. pp. 316–337.

[57] M. Nawaz, E. Enscore Jr. and I. Ham (1983) A heuristic algorithm for the m-machine, n-job flow-shop sequencing problem. *OMEGA*, **11**(1), 91–95.

[58] G.R. Schreiber and O.C. Martin (1999) Cut size statistics of graph bisection heuristics. *SIAM Journal on Optimization*, **10**(1), 231–251.

[59] M. Singer and M. Pinedo (1997) A shifting bottleneck heuristic for minimizing the total weighted tardiness in a job shop. *IIE Scheduling and Logistics*, **30**, 109–118.

[60] T. Stützle (1998). Applying iterated local search to the permutation flow shop problem. Technical Report AIDA-98–04, FG Intellektik, TU Darmstadt, August.

[61] T. Stützle (1998) *Local Search Algorithms for Combinatorial Problems—Analysis, Improvements, and New Applications*. PhD thesis, Darmstadt University of Technology, Department of Computer Science.

[62] T. Stützle, A. Grün, S. Linke and M. Rüttger (2000) A comparison of nature inspired heuristics on the traveling salesman problem. In: Deb et al. (eds.), *Proceedings of PPSN-VI, volume 1917 of LNCS*. Springer Verlag, Berlin, pp. 661–670.

[63] T. Stützle and H.H. Hoos (2000) Analyzing the run-time behaviour of iterated local search for the TSP. Technical Report IRIDIA/2000-01, IRIDIA, Université Libre de Bruxelles. Available at http://www.intellektik.informatik.tu-darmstadt.de/~tom/pub.html.

[64] É.D. Taillard (1995) Comparison of iterative searches for the quadratic assignment problem. *Location Science*, **3**, 87–105.

[65] R.J.M. Vaessens, E.H.L. Aarts and J.K. Lenstra (1996) Job shop scheduling by local search. *INFORMS Journal on Computing*, **8**, 302–317.

[66] C. Voudouris and E. Tsang (1995) Guided Local Search. Technical Report Technical Report CSM-247, Department of Computer Science, University of Essex.

[67] Y. Yang, S. Kreipl and M. Pinedo (2000) Heuristics for minimizing total weighted tardiness in flexible flow shops. *Journal of Scheduling*, **3**(2), 89–108.

Chapter 12

MULTI-START METHODS

Rafael Martí
Dpto de Estadística e Investigación Operativa. Universitat de València
Dr. Moliner 50, 46100 Burjassot, Valencia, Spain
E-mail: Rafael.Marti@uv.es

Abstract Heuristic search procedures that aspire to find global optimal solutions to hard combinatorial optimization problems usually require some type of diversification to overcome local optimality. One way to achieve diversification is to re-start the procedure from a new solution once a region has been explored. In this chapter we describe the best known multi-start methods for solving optimization problems. We propose classifying these methods in terms of their use of randomization, memory and degree of rebuild. We also present a computational comparison of these methods on solving the linear ordering problem in terms of solution quality and diversification power.

Keywords: Optimization, Heuristic Search, Re-Starting

1 INTRODUCTION

Search methods based on local optimization that aspire to find global optima usually require some type of diversification to overcome local optimality. Without this diversification, such methods can become localized in a small area of the solution space, making it impossible to find a global optimum. In recent years many techniques have been suggested to avoid local optima. One way to achieve diversification is to re-start the search from a new solution once a region has been extensively explored. Multi-start strategies can then be used to guide the construction of new solutions in a long term horizon of the search process.

There are some problems in which we find it is more effective to construct solutions than to apply a local search procedure. For example, in constrained scheduling problems it is difficult to define neighborhoods to keep feasibility whereas solutions can be relatively easily constructed. Therefore, Multi-start methods provide an appropriate framework within which to develop algorithms to solve these problems.

The re-start mechanism can be super-imposed on many different search methods. Once a new solution has been generated, we can apply a simple greedy routine, slight perturbations or a complex metaheuristic to improve it. This chapter is focused on studying the different ways, strategies and methods of generating solutions to re-start a search for a global optimum.

2 AN OVERVIEW

Multi-start methods have two phases: the first one in which the solution is generated and the second one in which the solution is typically (but not necessarily) improved. Then, each global iteration produces a solution (usually a local optima) and the best overall is the algorithm's output.

Figure 12.1 shows a pseudo-code of the multi-start procedure. A solution x_i is constructed in Step 1 at iteration i. This is typically performed with a constructive algorithm. Step 2 is devoted to improving this solution, obtaining solution x_i'. A simple improvement method can be applied. However, this second phase has recently become more elaborate and, in some cases, is performed with a complex metaheuristic that may or may not improve the initial solution x_i (in this latter case we set $x_i' = x_i$).

In recent years, many heuristic algorithms have been proposed to solve some combinatorial optimization problems following the outline given in Figure 12.1. Some of them are problem-dependent and the ideas and strategies implemented are difficult to apply to different problems, while others are based on a framework that can be used directly to design solving methods for other problems. In this section we describe the most relevant procedures in terms of applying them to a wide variety of problems.

Tabu Search is by now a well-known metaheuristic for solving hard combinatorial optimization problems. One of the papers that contains a number of its foundation ideas (Glover, 1977), also focuses on applying these ideas within a framework of iterated re-starting. Adaptive memory designs can be used to retain and analyze features of selected solutions and thus, provide a basis for improving future executions of the constructive process. The authors propose different memory functions, like frequency and recency information, to design these restarting mechanisms.

Adaptive memory strategies introduced in this seminal paper have had widespread success in solving a variety of practical and difficult combinatorial optimization problems. They have been adapted to many different fields in the combinatorial optimization theory, since using such memory has proved to be very effective in most metaheuristic methods. Some of them are explicitly based on these memory structures like tabu search, while others, like simulated annealing or re-starting methods, have evolved incorporating these ideas. Some applications of probabilistic forms of re-starting based

Initialise $i = 1$
while(Stopping condition is not satisfied)
{
  Step 1. (Generation)
    Construct solution x_i
  Step 2. (Search)
    Apply a search method to improve x_i
    Let x_i' be the solution obtained
  if(x_i' improves the best)
    Update the best
  $i = i + 1$
}

Figure 12.1. Multi-start procedure.

on memory functions are given in Rochat and Taillard (1995) and Lokketangen and Glover (1996).

Early papers in multi-start methods are devoted to the Monte Carlo random re-start in the context of nonlinear unconstrained optimization, where the method simply evaluates the objective function at randomly generated points. The probability of success approaches one as the sample size tends to infinity under very mild assumptions about objective function. Many algorithms have been proposed that combine the Monte Carlo method with local search procedures (Rinnooy Kan and Timmer, 1989). Solis and Wets (1981) study convergence for random re-start methods in which the probability distribution used to choose the next starting point can depend on how the search evolves. Some extensions of these methods seek to reduce the number of complete local searches that are performed and increase the probability that they start from points close to the global optimum (Mayne and Meewella, 1988).

Ulder et al. (1990) combines genetic algorithms with local search strategies improving previous genetic approaches for the travelling salesman problem. We apply an iterative algorithm to improve each individual, either before or while being combined with other individuals to form a new solution "offspring". The combination of these three elements: *Generation, Combination* and *Local Search*, extends the paradigm of Re- Start and links with other areas of the metaheuristics such as Scatter Search (Glover et al., 2000) or Memetic Algorithms (Moscato, 1999).

From a theoretical point of view, Hu et al. (1994) study the combination of the "gradient" algorithm with random initializations to find a global optimum. Efficacy of parallel processing, choice of the restart probability distribution and number of restarts are studied for both discrete and continuous models. The authors show that the uniform probability is a good measure for restarting procedures.

Boese et al. (1994) analyze relationships among local minima "from the perspective of the best local minimum", finding convex structures in the cost surfaces. Based on the results of that study, they propose a multi-start method where starting points for greedy descent are adaptively derived from the best previously found local minima. In the first step, Adaptive Multi-start heuristics (AMS) generate r random starting solutions and run a greedy descent method from each one to determine a set of corresponding random local minima. In the second step, *adaptive starting solutions* are constructed based on the local minima obtained so far and improved with a greedy descent method. This improvement is applied several times from each adaptive starting solution to yield corresponding *adaptive local minima*. The authors test this method for the traveling salesman problem and obtain significant speedups over previous multi-start implementations. Hagen and Kahng (1997) apply this method for the iterative partitioning problem.

Moreno et al. (1995) proposed a stopping rule for the multi-start method based on a statistical study of the number of iterations needed to find the global optimum. The authors introduce two random variables that together provide a way of estimating the number of global iterations needed to find the global optima: the number of initial solutions generated and the number of objective function evaluations performed on finding the global optima. From these measures, the probability that the incumbent solution is the global optimum is evaluated via a normal approximation. Thus, at each global iteration, this value is computed and if it is greater than a prefixed threshold, the algorithm stops, otherwise a new solution is generated. The authors illustrate the method in the median p-hub problem.

Simple forms of multi-start methods are often used to compare other methods and measure their relative contribution. Baluja (1995) compares different genetic algorithms for six sets of benchmark problems commonly found in the GA literature: Traveling salesman problem, job-shop scheduling, knapsack, bin packing, neural network weight optimization, and numerical function optimization. The author uses the multi-start method (Multiple restart stochastic hill-climbing, MRSH) as a baseline in the computational testing. Since solutions are represented with strings, the improvement step consists of a local search based on random flip of bits. The results indicate that using genetic algorithms for the optimization of static functions does not yield a benefit, in terms of the final answer obtained, over simpler optimization heuristics. Other comparisons between MRSH and GAs can be found, for example, in Ackley (1987) or Wattenberg and Juels (1994).

One of the most well known Multi-start methods is the greedy adaptive search procedures (GRASP). The GRASP methodology was introduced by Feo and Resende (1995). It was first used to solve set covering problems (Feo and Resende, 1989). Each GRASP iteration consists of constructing a trial solution and then applying a local search procedure to find a local optimum (i.e., the final solution for that iteration). The construction step is an adaptive and iterative process guided by a greedy evaluation function. It is iterative because the initial solution is built considering one element at a time. It is greedy because the addition of each element is guided by a greedy function. It is adaptive because the element chosen at any iteration in a construction is a function of those previously chosen. (That is, the method is adaptive in the sense of updating relevant information from one construction step to the next.). At each stage, the next element to be added to the solution is randomly selected from a candidate list of high quality elements according to the evaluation function. Once a solution has been obtained, it is typically improved by a local search procedure. The improvement phase performs a sequence of moves towards a local optimum solution, which becomes the output of a complete GRASP iteration. Some examples of successful applications are given in Laguna et al. (1994), Resende (1998) and Laguna and Martí (1999).

Hickernell and Yuan (1997) present a multi-start algorithm for unconstrained global optimization based on *quasirandom samples*. Quasirandom samples are sets of deterministic points, as opposed to random, that are evenly distributed over a set. The algorithm applies an inexpensive local search (steepest descent) on a set of quasirandom points to concentrate the sample. The sample is reduced replacing worse points with new quasirandom points. Any point that is retained for a certain number of iterations is used to start an efficient complete local search. The algorithm terminates when no new local minimum is found after several iterations. An experimental comparison shows that the method performs favorably with respect to other global optimization procedures.

Hagen and Kang (1997) used an adaptive multi start method for the partitioning optimization VLSI problem where the objective is to minimize the number of signals which pass between components. The method consists of two phases: (1) To generate a set of random starting points and perform the iterative (local search) algorithm, thus determining a set of local minimum solutions; and (2) construct adaptive starting points that are central to the best local minimum solutions found so far. The authors add a preprocessing cluster module to reduce the size of the problem. The resulting Clustering Adaptive Multi Start method (CAMS) is fast and stable and improves upon previous partitioning results in the literature.

Fleurent and Glover (1999) propose some adaptive memory search principles to enhance multi-start approaches. The authors introduce a template of a constructive version of Tabu Search based on both, a set of elite solutions and the intensification strategies that rely on identifying of *strongly determined* and *consistent variables*. Strongly determined variables are those whose values cannot be changed without significantly eroding the objective function value or disrupting the values of other variables. A consistent variable is defined as one that receives a particular value in a significant portion of good solutions. The authors propose the inclusion of memory structures within the multi-start framework as those used in tabu search: *recency, frequency* and *attractiveness*. Computational experiments for the quadratic assignment problem disclose that these methods improve significantly over previous multi-start methods like GRASP and random restart that do not incorporate memory based strategies.

Multi-start procedures usually follow the global scheme given in Figure 1; but there are some applications in which Step 2 can be applied several times within a global iteration. In the *incomplete construction methods*, the improvement phase was periodically invoked during the construction process of the partial solution rather than the standard implementation after the complete construction. See Russell (1995) and Chiang and Russell (1995) for successful applications of this approach to vehicle routing.

Patterson et al. (1999) introduce a multi-start framework (Adaptive Reasoning Techniques, ART) based on memory structures. The authors implement the short term and long term memory functions, proposed in the tabu search framework, to solve the Capacitated Minimum Spanning Tree Problem. ART is an iterative, constructive solution procedure that implements learning methodologies on top of memory structures. ART derives its success from being able to learn about, and modify the behavior of a primary greedy heuristic. The greedy heuristic is executed repeatedly, and for each new execution we probabilistically introduce constraints that prohibit certain solution elements from being considered by the greedy heuristic. The active constraints are held in a short term memory. A long term memory holds information regarding which constraints were in the active memory for the best set of solutions.

Laguna and Martí (1999) introduced Path Relinking within GRASP as a way to improve Multi-start methods. Path Relinking has been suggested as an approach to integrate intensification and diversification strategies (Glover and Laguna, 1997) in the context of tabu search. This approach generates new solutions by exploring trajectories that connect high-quality solutions, by starting from one of these solutions and generating a path in the neighborhood space that leads toward the other solutions. This is accomplished by selecting moves that introduce attributes contained in the "guiding" solutions. The relinking in the context of GRASP consists of finding a path between a solution found after an improvement phase and the chosen elite solution. Therefore, the relinking concept has a different interpretation within GRASP, since the solutions found from one iteration to the next are not linked by a sequence of moves (as in the case of tabu search). The proposed strategy can be applied to any method that produces a sequence of solutions; specifically, it can be used in any multi-start procedure. Based on these ideas, Binato et al. (2001) proposed the Greedy Randomized Adaptive Path Relinking.

Glover (2000) proposes approaches for creating improved forms of constructive multi-start and strategic oscillation methods, based on new search principles: *persistent attractiveness* and *marginal conditional validity*. These concepts play a key role in deriving appropriate measures to capture information during prior search. Applied to constructive neighborhoods, strategic oscillation operates by alternating constructive and destructive phases, where each solution generated by a constructive phase is dismantled (to a variable degree) by the destructive phase, after which a new phase builds the solution anew. The conjunction of both phases and their associated memory structures provides the basis for an improved multi-start method.

Prais and Ribeiro (2000) propose an improved GRASP implementation, called reactive GRASP, for a matrix decomposition problem arising in the context of traffic assignment in communication satellites. The method incorporates a memory structure to record information about previously found solutions. In Reactive GRASP, the basic parameter which defines the restrictive-ness of the candidate list during the construction phase is self-adjusted, according to the quality of the previously found solutions. The proposed method matches most of the optimal solutions known.

An open question in order to design a good search procedure is whether it is better to implement a simple improving method that allows a great number of global iterations or, alternatively, to apply a complex routine that significantly improves a few generated solutions. A simple procedure depends heavily on the initial solution but a more elaborate method takes much more running time and therefore can only be applied a few times, thus reducing the sampling of the solution space. Some metaheuristics, such as GRASP, launch limited local searches from numerous constructions (i.e., starting points). In most of the tabu search implementations, the search starts from one initial point and if a restarting procedure is also part of the method, it is invoked only a limited number of times. However, the inclusion of re-starting strategies within the tabu search framework has been well documented in several papers (Glover, 1977; Glover and Laguna, 1997).

Martí et al. (2001) study the balance between restarting and search-depth (i.e., the time spent searching from a single starting point) in the context of the bandwidth matrix problem. They tested both alternatives and concluded that it was better to invest the time searching from a few starting points than restarting the search more often. Although we cannot draw a general conclusion from these experiments, the experience in the current context and in previous projects indicates that some metaheuristics, like tabu search, need to reach a critical search depth to be effective. If this search depth is not reached, the effectiveness of the method is severely compromised.

3 A CLASSIFICATION

We have found three key elements in multi-start methods that can be used for classification purposes: memory, randomization and degree of rebuild. The choices for each one of these elements are not restricted to the extreme case of "present" or "not present", but they represent the whole range between both extremes that can be labeled as *Memory/Memory-less*, *Systematic/Randomized* and *Rebuild/Build from scratch*, respectively.

The first element is the **Memory** and it is used to identify elements that are common to good previously generated solutions. As in the Tabu Search framework (Glover and Laguna, 1997), it provides a foundation for incentive-based learning, by means of incentives that reinforce actions leading to good solutions. Thus, instead of simply resorting to randomized re-starting processes, in which current decisions derive no benefit from knowledge accumulated during prior search, specific types of information are identified to exploit history. On the other hand, memory avoidance (*Memory-less*) is not as unreasonable as might be imagined since the construction of "unconnected" solutions may be interpreted as a means of strategically sampling the solution space. It should be noted that the meaning of good is not restricted to the objective function, but also includes the notion of diversity, as described later.

Starting solutions can be randomly generated or, on the contrary, they can be generated in a systematic way. **Randomization** is a very simple way of achieving diversification, but with no control over the diversity achieved since in some cases we can obtain very similar solutions. We can add some mechanisms to control the similarities in order to discard some solutions or generate the solutions in a deterministic way that guarantees a certain degree of difference. The extremes of this element can be described as *Randomized* where solutions are generated in a random way and *Systematic* where solutions are generated in a deterministic way. Between both extremes there are a great number of possibilities for combining random elements with deterministic rules. GRASP construction is an example of a combined method.

The **Degree of Rebuild** indicates the elements that remain fixed from one generation to another. Most applications *build* the solution at each generation *from scratch*, but recent implementations have fixed, for a certain number of iterations, some elements in the construction process that have appeared in previously generated solutions. Such an approach was proposed in the context of identifying and then iteratively exploiting 'strongly determined and consistent variables' in Glover (1977). This selective fixing of elements, by reference to their previous impact and frequency of occurrence in various solution classes, is a memory-based strategy of the type commonly used in tabu search. It can also be considered as an instance of Path Relinking (Glover and Laguna, 1993) which generates new solutions by exploring trajectories that connect high-quality solutions. This approach seeks to incorporate the attributes of elite solutions previously generated by creating inducements to favor these attributes in the solutions. In an extreme case all the elements in the new solution will be fixed by the information generated from the set of elite solutions considered. This is labeled as *Rebuild*.

The constructive algorithm depicted in Figure 12.2 has no memory structures, a combination between randomization and systematic construction rules and, at each iteration, the solution is built completely from scratch.

Given different re-starting methods for a problem, one way of comparing them is to generate a set of solutions with each and compare their quality and diversity. Since the quality is trivially measured by the objective function, we now propose two measures of diversity. We restrict our attention to solutions represented by permutations.

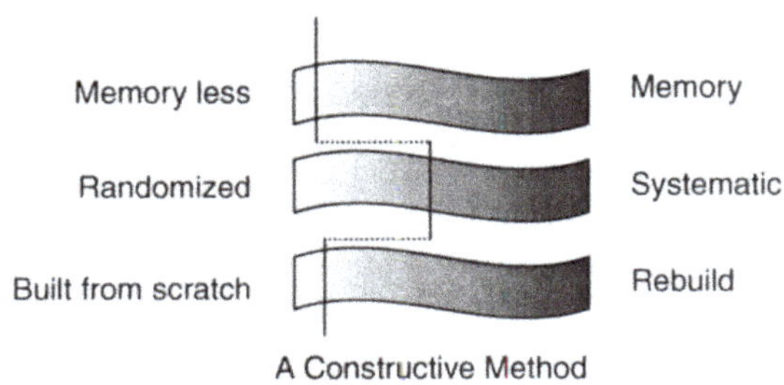

Figure 12.2. Multi-start classification.

3.1 Diversity Measures

The first measure consists of computing the distances between each solution and a "center" of the set of solutions P. The sum (or alternatively the average) of these $|P|$ distances provides a measure of the diversity of P. The second one is to compute the distance between each pair of solutions in P. The sum of these $|P \times P|$ distances provides another way of measuring the diversity of P.

The first diversity measure is calculated as follows:

1. Calculate the median position of each element i in the solutions in P.
2. Calculate the dissimilarity of each solution in the population with respect to the median solution. The dissimilarity is calculated as the sum of the absolute difference between the position of the elements in the solution under consideration and the median solution.
3. Calculate d as the sum of all the individual dissimilarities.

To illustrate, suppose that P consists of the following three orderings: (A, B, C, D), (B, D, C, A), (C, B, A, D). The median position of element A is therefore 3, since it occupies positions 1, 3 and 4 in the given orderings. In the same way, the median positions of B, C and D are 2, 3 and 4, respectively. Note that the median positions might not induce an ordering, as in the case of this example. The dissimilarity of the first solution is then calculated as follows:

$$d_1 = |1-3| + |2-2| + |3-3| + |4-4| = 2$$

In the same way, the dissimilarities of the other two solutions are $d_2 = 4$ and $d_3 = 2$. The diversity measure of P is then given by $d = 2 + 4 + 2 = 8$.

The second measure is calculated, for each pair of solutions in P, as the sum of the absolute differences between the positions of each element in both solutions. The sum of these $|P \times P|$ values provides the measure of the diversity of the set P. The value with solutions (A, B, C, D) and (B, D, C, A) in the previous example is computed as follows:

$$d'_{12} = |1-4| + |2-1| + |3-3| + |4-2| = 6$$

In the same way, the values of the other three pairs of solutions are $d'_{13} = 4$ and $d'_{23} = 6$. The diversity measure of P is then given by $d' = 6 + 4 + 6 = 16$.

We have computationally found that both measures are strongly correlated and provide the same information. Since the second measure is computationally more expensive than the first, we will use the first one (dissimilarity) in the following experiments.

It should be noted that a third measure could be added to evaluate a set of solutions. The notion of *influence* introduced by Glover (1990) in the context of Tabu Search, can be adapted to our study. The influence considers the potential and the structure of a solution in the search process. The authors propose memory functions that classify moves relative to their attractiveness within "distance classes" and other measures of their impact. Consider, for example, two solutions a and b with the same objective and diversity values, but a is close to a local optimum with a better objective function value than a and b, while b is itself a local optimum. Consequently, we probably obtain a better solution with a search from a rather than from b. Therefore it is more valuable to have a than b in the set of solutions since it has more influence in the search for the global optimum. Obviously we do not know *a priori* if a given solution is closer to a local optimum than another, but if we identify some properties of good solutions we will be able to define evaluators and measures to reflect the "importance" or influence of the solutions. Good starting points for this study are given by the solution structure, landscape and neighborhood induced by the local search method.

4 COMPUTATIONAL EXPERIMENTS

The linear ordering problem (LOP) has generated a considerable amount of research interest over the years, as documented in Grotschel et al. (1984) and Campos et al. (1999). Because of its practical and theoretical relevance, we use this problem as a test case for re-start mechanisms.

Given a matrix of weights $E = \{e_{ij}\}_{m \times m}$, the LOP consists of finding a permutation p of the columns (and rows) in order to maximize the sum of the weights in the upper triangle. In mathematical terms, we seek to maximize:

$$C_E(p) = \sum_{i=1}^{m-1} \sum_{j=i+1}^{m} e_{p(i)p(j)}$$

where $p(i)$ is the index of the column (and row) in position i in the permutation. Note that in the LOP, the permutation p provides the ordering of both the columns and the rows. The equivalent problem in graphs consists of finding, in a complete weighted graph, an acyclic tournament with a maximal sum of arc weights (Reinelt, 1985).

Instances of input–output tables from sectors in the European Economy can be found in the public-domain library LOLIB (1997). We employ these problem instances to compare different restarting methods.

We have tested 10 re-starting generation methods. Six of these methods are based on GRASP (Feo and Resende, 1995) constructions with a greedy function that selects sectors based on a measure of attractiveness.

G1 A GRASP construction where the attractiveness of a row is the sum of the elements in its corresponding row. The method randomly selects from a short list of the most attractive sectors and constructs the solution starting from the first position of the permutation.

G2 A GRASP construction where the attractiveness of a sector is the sum of the elements in its corresponding column. The method randomly selects from a short list of the most attractive sectors and constructs the solution starting from the first position of the permutation.

G3 A GRASP construction where the attractiveness of a sector is the sum of the elements in its corresponding row divided by the sum of the elements in its corresponding column. The method randomly selects from a short list of the most attractive sectors and constructs the solution starting from the first position of the permutation.

G4, G5 and G6 These methods are identical to the first three, except that sector selection is from a short list of the least attractive and the solution is constructed starting from the last position of the permutation.

MIX A mixed procedure derived from the previous six. The procedure generates an even number of solutions from each of the previous six methods and combines these solutions into a single set. That is, if n solutions are required, then each method Gi (for $i = 1, \ldots, 6$) contributes with $n/6$ solutions.

RND A random generator. This method simply generates random permutations.

DG A general purpose diversification generator suggested in Glover (1998) which generates diversified permutations in a systematic way without reference to the objective function.

FQ A method using frequency-based memory, as proposed in Tabu Search (Glover and Laguna, 1997). This method is based on modifying a measure of attractiveness with a frequency measure that discourages sectors from occupying positions that they have frequently occupied in previous solution generations. See Campos et al. (1999) for a description of the method.

In our first experiment we use the instance *stabu3* from the LOLIB. We have generated a set of N = 100 solutions with each of the 10 generation methods. Figures 12.3 and 12.4 show the box and whiskers plot of the objective function value and dissimilarity, respectively, of the solution set obtained with each method.

With both diagrams together (Figures 12.3 and 12.4) we can observe the performance of the 10 generators on the problem *stabu3*. We have repeated the same experiments on 10 other problems from the LOLIB, obtaining similar diagrams.

A good re-starting method must produce a set of solutions with high quality and high diversity. If we compare, for example, generators MIX and G3 we observe in Figure 12.3 that G3 produces slightly better solutions in terms of solution quality, but Figure 12.4 shows that MIX outperforms G3 in terms of diversity. Therefore, we will probably select MIX as a better method than G3.

In order to rank the methods we have computed the average of both measures across each set. Figure 12.5 shows in the x-axis the average of the dissimilarity and in the y-axis the average of the quality. A point is plotted for each method.

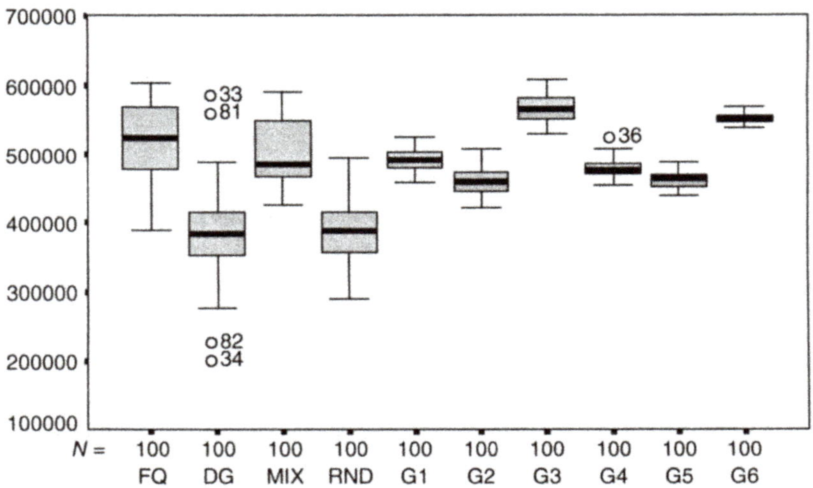

Figure 12.3. Objective function value box plot for each method.

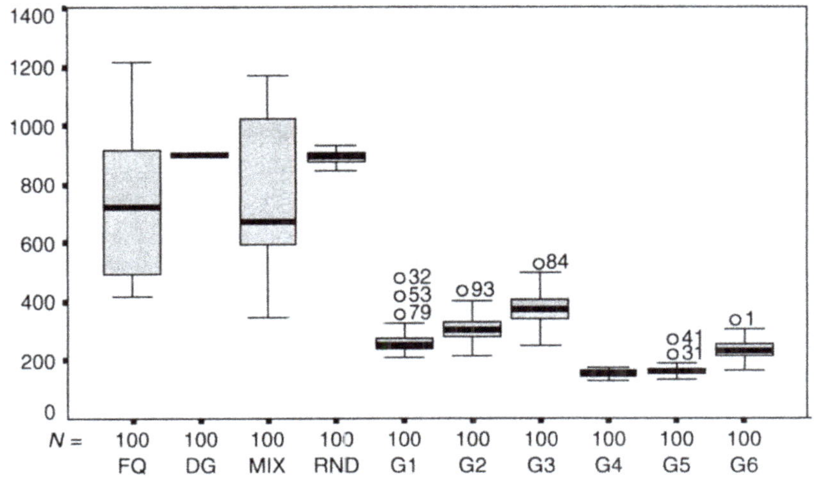

Figure 12.4. Dissimilarity box plot for each method.

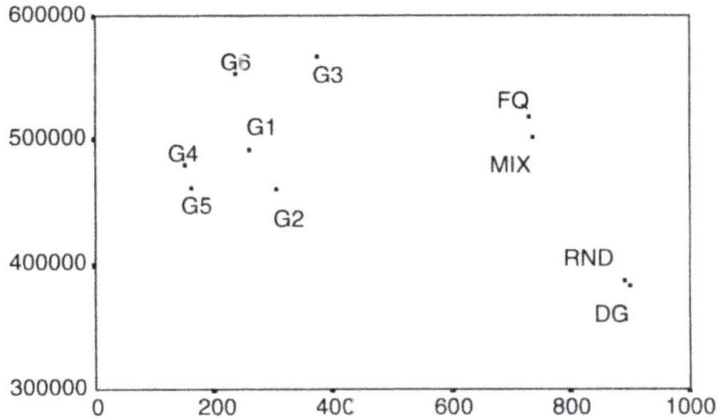

Figure 12.5. Quality and dissimilarity for each method.

As expected, the random generator (RND) produces the maximum diversity (as measured by the dissimilarity value). DG matches the diversity of RND using a systematic approach instead of randomness. The mixed method MIX provides a good balance between dissimilarity and quality, by the union of solutions generated with methods G1 to G6.

We have standardized both averages in order to directly compare them. We think that quality and diversity are equally important, so we have added both relative averages, obtaining the following ranking where the overall best is the FQ generator:

G5, G4, G2, G1, DG, RND, G6, G3, MIX and FQ.

These results are in line with previous works which show the inclusion of memory structures to be effective within the multi-start framework. However, one should note that this method ranking has been obtained considering both measures, quality and diversity, with equal weight. If we vary this criterion, the ranking would also change.

5 CONCLUSIONS

The objective of this study has been to extend and advance the knowledge associated to implementing multi-start procedures. Unlike other well-known methods, it has not yet become widely implemented and tested as a metaheuristic itself for solving complex optimization problems. We have shown new ideas that have recently emerged within the multi-start area that add a clear potential to this framework which has yet to be fully explored.

ACKNOWLEDGEMENTS

The author wishes to thank professor Vicente Campos for his valuable help in the computational testing.

REFERENCES

Ackley, D.H. (1987) An empirical study of bit vector function optimization. In: Davis (ed.), Genetic Algorithms and simulated annealing. Morgan Kaufmann Publishers.

Baluja, S. (1995) An Empirical Comparison of 7 iterative evolutionary function optimization heuristics. School of Computer Science, Carnegie Mellon University.

Binato, S., Faria Jr. Haroldo and Resende, M.G.C. (2001) Greedy randomized adaptive path relinking. MIC2001 conference proceedings.

Boese, K.D., Kahng, A.B. and Muddu, S. (1994) A new adaptive multi-start technique for combinatorial global optimisation. *Operations Research Letters*, **16**, 103–113.

Campos, V., Laguna, M. and Martí, R. (1999) Scatter search for the linear ordering problem. In: Corne, Dorigo and Glover (eds.), *New Ideas in Optimization*. Mc Graw Hill, pp. 331–341.

Chiang, W.C. and Russell, R.A. (1995) Simulated annealing metaheuristics for the vehicle routing problems with time windows. *Annals of Operations Research*, **60**.

Feo, T. and Resende, M.G.C. (1989) A probabilistic heuristic for a computationally difficult set covering problem. *Operations Research Letters*, **8**, 67–71.

Feo, T. and Resende, M.G.C. (1995) Greedy randomized adaptive search procedures. *Journal of Global Optimization*, **2**, 1–27.

Fleurent, C. and Glover, F. (1999) Improved constructive multi-start strategies for the Quadratic assignment problem using adaptive memory. *INFORMS Journal on Computing*.

Glover, F. (1977) Heuristics for integer programming using surrogate constraints. *Decision Sciences*, **8**, 156–166.

Glover, F. (1990) Tabu search: a tutorial. *Interfaces*, **20**, 74–94.

Glover, F. (2000) Multi-start and strategic oscillation methods—principles to exploit adaptive memory. In: Laguna and Gonzalez-Velarde (eds.), *Computing Tools for Modeling Optimization and Simulation*. Kluwer, pp 1–25.

Glover, F. and Laguna, M. (1993) Tabu search. In: C. Reeves (ed.), *Modern Heuristic Techniques for Combinatorial Optimization Problems*. Blackwell Scientific Publishing, Oxford, pp. 70–141.

Glover, F. and Laguna, M. (1997) *Tabu Search*. Kluwer Academic Publishers, Boston.

Glover, F., Laguna, M. and Martí, R. (2000) Scatter seach. In: A. Ghosh and S. Tsutsui (eds.), *Theory and Applications of Evolutionary Computation: Recent Trends*. Springer-Verlag (to appear).

Grotschel, M., Junger, M. and Reinelt, G. (1984) A cutting plane algorithm for the linear ordering problem. *Operations Research*, **32**(6), 1195–1220.

Hagen, L.W. and Kahng, A.B. (1997) Combining problem reduction and adaptive multi-start: a new technique for superior iterative partitioning. *IEEE Transactions on CAD*, **16**(7), 709–717.

Hickernell, F.J. and Yuan, Y. (1997) A simple multistart algorithm for global optimization. *OR Transactions*, **1**(2).

Hu, X., Shonkwiler, R. and Spruill, M.C. (1994) Random restarts in global optimization. Georgia Institute of technology, Atlanta.

Laguna, M., Feo, T. and Elrod, H. (1994) A greedy randomized adaptive search procedure for the 2-partition problem. *Operations Research*, **42**(4), 677–687.

Laguna, M. and Martí, R. (1999) GRASP and Path relinking for 2-layer straight line crossing minimization. *INFORMS Journal on Computing*, **11**(1), 44–52.

Lokketangen, A. and Glover, F. (1996) Probabilistic move selection in tabu search for 0/1 mixed integer programming problems. *Meta-Heuristics: Theory and Practice*. Kluwer, pp. 467–488.

LOLIB (1997) http://www.iwr.uni-heildelberg.de/iwr/comopt/soft/LOLIB/LOLIB.html

Martí, R., Laguna, M., Glover, F. and Campos, V. (2001) Reducing the bandwidth of a sparse matrix with tabu search. *European Journal of Operational Research*, **135**, 450–459.

Mayne, D.Q. and Meewella, C.C. (1988) A non-clustering multistart algorithm for global optimization. In: Bensoussan and Lions (eds.), *Analysis and Optimization of Systems, Lecture Notes in control and information sciences*, Vol. 111, Springer-Verlag.

Moreno, J.A., Mladenovic, N. and Moreno-Vega, J.M. (1995) An statistical analysis of strategies for multistart heuristic searches for p-facility location-allocation problems. Eighth Meeting of the EWG on Locational Analysis Lambrecht, Germany.

Moscato, P. (1999) Memetic algorithms. In: Corne, Dorigo and Glover (eds.), *New Ideas in Optimization*. McGraw Hill, pp. 219–235.

Osman, I.H. and Kelly, J.P. (1996) Meta-Heuristics: An Overview. *Meta-Heuristics: Theory and Practice*. Kluwer, pp. 1–23.

Prais, M. and Ribeiro C.C. (2000) Reactive GRASP: an application to a matrix decomposition problem in TDMA traffic assignment. **12**, 164–176.

Rochat and Taillard (1995) Probabilistic diversification and intensification in local search for vehicle routing. *Journal of Heuristics*, **1**(1), 147–167.

Resende, M.G.C. (1998) Computing approximate solutions for the maximum covering problem using GRASP. *Journal of Heuristics*, **4**, 161–171.

Patterson R., Pirkul, H. and Rolland, E. (1999) Adaptive reasoning technique for the capacitated minimum spanning tree problem. *Journal of Heuristics*, **5**, 159–180.

Rinnooy, Kan, A.H.G. and Timmer, G.T. (1989) Global optimization. In: Rinnoy Kan and Todds (eds.), Handbooks in Operations Research and Management Science, Vol. 1. North Holland, pp. 631–662.

Russell, R.A. (1995) Hybrid heuristics for the vehicle routing problem with time windows. *Transportation Science*.

Solis, F. and Wets, R. (1981) Minimization by random search techniques. *Mathematics of Operations Research*, **6**, 19–30.

Ulder, N.L.J., Aarts, E.H.L., Bandelt, H.J., Van Laarhoven, P.J.M. and Pesch, E. (1990) Genetic Local Search algorithms for the traveling salesman problem. In: Schwefel and Männer (eds.), *Parallel Problem Solving from Nature*, Springer-Verlag, pp. 109–116.

Wattenberg, M. and Juels, A. (1994) Stochastic Hillclimbing as a baseline method for evaluationg genetic algorithms. University of California, Berkeley, CSD94-834.

Chapter 13

LOCAL SEARCH AND CONSTRAINT PROGRAMMING

Filippo Focacci
ILOG S.A.
9, rue de Verdun
BP 85, 94253 Gentilly, France
E-mail: ffocacci@ilog.fr

François Laburthe
BOUYGUES – Direction des Technologies Nouvelles
1, avenue E. Freyssinet
78061 St-Quentin-en-Yvelines Cedex, France
E-mail: flaburthe@bouygues.com

Andrea Lodi
D.E.I.S., University of Bologna
Viale Risorgimento 2, 40136 Bologna, Italy
E-mail: alodi@deis.unibo.it

1 INTRODUCTION

Real-world combinatorial optimization problems have two main characteristics which makes them difficult: they are usually large (see, e.g., Caprara et al., 1997, which describes real-world crew scheduling applications), and they are not *pure*, i.e., they involve a heterogeneous set of side constraints (see, e.g., union contract regulations for crew scheduling and rostering again in Caprara et al., 1997).

Hence, in most cases, exact approaches cannot be applied to solve real-world problems, whereas incomplete methods, and among them *Local Search* (LS) ones, have been proved to obtain very good results in practice (see many examples throughout the book).

LS techniques are based on a simple and general idea. Let P be the combinatorial optimization problem we want to solve, and s a current solution which, for the moment, we assume to be feasible for P, and to have value $z(s)$. A *neighborhood* is defined for s with respect to a *move type* $\mathcal{N}$, i.e., a *function* mapping s (actually, any feasible solution of P) in a subset $\mathcal{N}(s)$ of the overall solution space. In other words, $\mathcal{N}(s)$ contains all the feasible solutions of P which can be reached from s by means of a move of type $\mathcal{N}$. Examples of moves are given throughout the paper, but, roughly speaking, the move is a manipulation of s whose effect is the transition to another solution $x \in \mathcal{N}(s)$. The LS framework explores the neighborhood by searching for the solution $x^* \in \mathcal{N}(s)$ such that $\delta z = z(s) - z(x^*)$ is maximized (for minimization problems). If $\delta z > 0$, then an improved solution has been found, and the process is iterated by considering x^* as new current solution. Otherwise, a local optimum has been reached, and several very effective techniques can be applied to escape from it (see almost all the chapters of this book).

If problem P presents relevant feasibility issues, i.e., it is not easy in practice to find a feasible initial solution, the same framework can be applied anyway by using as current solution an infeasible one. In order to drive the local search process towards feasible solutions, the cost function needs to be modified to measure the infeasibility of the candidate solution. In this sense, the evaluation of a move can be more complex (penalty functions are usually necessary), but the framework does not change substantially.

Three important issues must be taken into account when dealing with real-world problems.

1. Huge problems require large neighborhoods whose exploration can be computationally expensive.
2. When dealing with problems involving many heterogeneous side constraints it is often preferable to consider them as hard constraints rather than transforming them into penalty functions. In these cases, the neighborhood may contain few feasible solutions, and again large neighborhoods are required in order to avoid getting trapped into local minima too often.
3. Real-world applications typically lead to frequent update/addition of constraints (recall again union contract regulations), thus the algorithmic approach requires flexibility.

In the last decade Constraint Programming (CP) has shown its effectiveness in modeling and solving real-world combinatorial optimization problems. CP is a programming paradigm exploiting Constraint Satisfaction techniques (Mackworth, 1977), and in the following we restrict our attention to CP on *Finite Domains* (CP(FD)) which is the case of all constraint tools for discrete optimization such as CHIP (Aggoun and Beldiceanu, 1992), Ilog Solver (Solver, 2000), Eclipse (Schimpf et al., 1997), cc(FD) (van Hentenryck et al., 1993), CHOCO (Laburthe, 2000), etc.

The next paragraph briefly introduces the CP terminology that will be used in the remainder of the chapter. Complete definitions can be found in Marriott and Stuckey, 1998 (a complete textbook), and Focacci et al., 2000b (a basic introduction).

Combinatorial optimization problems are modeled through CP(FD) by means of a set of variables taking their value on a finite domain of integers, and are linked by a set of constraints. The constraints can be of mathematical or symbolic type, and in this second case, when they refer to a set of variables, are referred to as *global constraints*.

Global constraints typically model a well-defined part of the overall problem, where well-defined means that the same part has been recognized as a subproblem in several cases. A classic example of global constraint is the *all_different*$(x_1, \ldots, x_n)$ constraint which imposes that variables $x_1, \ldots, x_n$ must assume different values in a feasible solution.

To each constraint is associated a *propagation* algorithm aimed at deleting from variable domains the values that cannot lead to feasible solutions. Constraints interact through shared variables, i.e., as soon as a constraint has been propagated (no more values can be eliminated), and at least a value has been eliminated from the domain of a variable, say v, then the propagation algorithms of all the other constraints involving v are triggered (Mackworth, 1977).

Propagation algorithms are usually incomplete: once propagation is finished, there may remain some inconsistent values in the variable domains.[1] Therefore, unless the propagation phase ends with a fully instantiated solution or a failure (proving the problem inconsistent), a *search* phase is executed. One branching step is performed by partitioning the current problem (or the subproblem) into (easier) subproblems, e.g., by instantiating a variable to a feasible value in its domain. Propagation and search are interleaved in order to reach one or all feasible solutions.

As soon as a feasible solution improving the current best one is found, CP systems add a new constraint to the remaining search tree stating that further solutions should have a better value. This new constraint excludes leaf nodes from the remainder of the tree having a cost which is worse than the current one. Thus, CP solves a sequence of feasibility problems that improve the value of the objective function.

It is not difficult to see that the main advantage of using CP systems is flexibility: CP supports the design of declarative, compact and flexible models where the addition of new constraints is straightforward and does not affect the previous model. Indeed, the propagation of the previous constraints remain unchanged (since they locally model parts of the overall problem), and the previous constraints simply interact with the new ones through shared variables.

Thus many real-world combinatorial optimization problems may benefit from the efficiency of LS as well as from the flexibility of CP. Throughout this paper, we review hybrid methods that combine principles from both methods. A first set of such hybrids belongs to the family of local search methods (going from one solution to a neighbor one) and use CP as a way to efficiently explore large neighborhoods with side constraints. A second set belongs to the family of global search (tree search) methods and use LS as a way to improve some of the nodes of the search tree or to explore a set of paths close to the path selected by a greedy algorithm in the search tree. In short, LS may use ideas from CP in order to make large neighborhoods more tractable, while CP may use ideas from LS to explore a set of solutions close to the greedy path in a tree search and converge more quickly towards the optimum.

The chapter is organized as follows. In Section 2 we present the techniques to combine LS and CP within hybrid algorithms, and we briefly review the literature on the topic. In Section 3 we discuss in details a didactic case study by presenting examples of the techniques described in Section 2 with respect to this specific problem. Finally, some conclusions are drawn in Section 4.

2 BASIC TECHNIQUES

2.1 Constrained Local Search

As mentioned in Section 1, hybrid approaches combining LS and CP are particularly suitable for real-world combinatorial problems which are typically huge in size and involve side constraints. In these cases both the size of the problems and the presence of

[1] Note that in the general case forcing acr consistency even for a single constraint is NP-hard.

side constraints lead looking for optimality out of practice. However, even standard LS algorithm can get into trouble from a computational point of view with these problems since the size of the neighborhood grows very fast and/or testing solution feasibility is expensive.

2.1.1 Small Neighborhoods with Side Constraints

Fast LS algorithms typically uses neighborhoods of small size which can be explored with a relatively small computational effort. Classic examples are the neighborhoods defined by a move which simply exchange a pair of assignments of the current solution. This kind of move has a wide domain of application, it is referred as *2-opt*, and has been classically used by Lin and Kernighan, 1973 for the *Traveling Salesman Problem* (TSP). In the TSP case, given a Hamiltonian cycle (current solution), i.e., a sequence of edges connecting the cities in their order of visit, the 2-opt move simply deletes two of these edges by replacing them with two others in order to obtain a new feasible cycle.

It is well-known that 2-opt can be implemented so as to require an overall time complexity of $O(n^2)$ to find the move maximizing the improvement, i.e., to find the solution whose overall length is minimal among all neighbors of the current cycle.

Even if the neighborhood is in principle quite small, the addition of side constraints can considerably increase the computational effort required to explore it since it is often necessary to test feasibility.

A typical example is the time-constrained variant of TSP in which the salesman needs to visit the cities within specific *Time Windows* (TSPTW). In this case, in order to find the best 2-exchange move we need to test feasibility, which means testing, for each move, if the resulting solution violates some of the time window constraints. This is the standard way of addressing problems with side constraints in LS: the neighborhood of the pure problem is explored, and for each neighbor, the side constraints are iterated and, for each one of them, its satisfaction is tested. Thus, note that to each constraint must be associated an algorithm testing its satisfaction for a given solution. Note also that as soon as side constraints are added, the computational overhead of constraint checks for LS increases. However, checking feasibility only at the very end of the decision process is only a passive way of using constraints; constraints may be used in more active way by factoring out some of the constraint checks early on in the iteration. Therefore, single checks may discard (hopefully large) sets of neighbors, thus improving the overall efficiency of the neighborhood exploration.

In the example of the 2-opt neighborhood for the TSPTW, one check of time window constraints takes $O(n)$ time, therefore a straightforward implementation of 2-opt for the TSPTW takes $O(n^3)$ time. However, smart incremental computations can reduce the complexity of the TSPTW 2-opt to $O(n^2)$ by caching earliest arrival times and latest departure times (Kindervater and Savelsbergh, 1997). Such optimized neighborhood explorations and constraint checks require specialized code that must be substantially changed whenever new side constraints are considered.

The main point concerns, however, the effectiveness of small neighborhood for real-world applications. With the addition of side constraints the number of feasible solutions of the neighborhood becomes smaller, thus the local optimization process is more likely to get trapped into local optima. Therefore real-world problems require larger neighborhoods, and exploring them by simple enumeration becomes ineffective (the same holds since the size of the problems typically grows).

2.1.2 Exploring Large Neighborhood with CP

As neighborhoods grow larger, finding the best neighbor becomes an optimization problem on its own, thus the use of global search is preferable over blunt enumeration.

We review two possibilities for implementing a LS algorithm, using CP. In both cases, we suppose that we have at hand a current feasible solution s and a CP model of the problem.

The first method consists in keeping a fragment of the solution s (keeping the value assignment for a subset of the variables), erasing the value of all variables outside that fragment and solving the subproblem defined by the uninstantiated variables. This technique was introduced by Applegate and Cook (1991) for job-shop scheduling. The job sequence is kept on all machines but one, and the scheduling subproblem on that machine is solved to optimality by branch-and-bound. In this case, the fragment corresponds to the sequencing order on all machines but one. For scheduling, the approach was generalized to other fragments (time slice, ranks, sets of ordering decisions etc.) by Caseau and Laburthe (1996) and Nuijten and Le Pape (1998). The same approach was also applied to quadratic assignment problems by Mautor and Michelon (1997) and to vehicle routing by Shaw (1998). Such fragment-based LS methods are usually easy to implement once a first CP model has been built. A number of constraint based tools can be used to improve their efficiency:

- an optimization constraint can be set on the neighborhood imposing that only neighbors that strictly improve the objective value over the current solution are generated (see, e.g., Nuijten and Le Pape (1998));
- fragment based neighborhoods can be explored in a Variable Neighborhood Search (VNS, see, Mladenović and Hansen, 1997 for its definition, and see, Caseau and Laburthe (1996) for its application to fragment-based LS);
- in order to speed up the procedure, each neighborhood defined by a fragment can be explored by an incomplete search, such as Limited Discrepancy Search (see, Harvey and Ginsberg, 1995 and Section 2.2.5).

The second method, introduced in Pesant and Gendreau (1996) and Pesant and Gendreau (1999), consists in modeling the exploration of a neighborhood through CP variables and constraints. Roughly speaking, a CP model of the neighborhood is created such that every feasible solution of the CP problem represents a move that transforms the current solution into a neighbor solution. As an example one can consider the classical *swap* move that swaps the values of two variables x_i and x_j. The neighborhood associated to the move can obviously be explored by two nested loops over indices i and j. Alternatively, the neighborhood may be defined by a CP model with two domain variables I, J and one constraint $I < J$. Every feasible solution (i, j) of the problem defined by variables I, J and constraint $I < J$ uniquely identifies a swap move. With such a model, the exploration of the neighborhood by means of iterators (two nested loops) can be replaced by a tree search (such as branch-and-bound for finding the best move).

Formally, the neighborhood $\mathcal{N}(s)$ of a solution s is described by a CP model N_P such that there is a one-to-one mapping between the set of solutions of N_P and the set of neighbors $\mathcal{N}(s)$. We refer to N_P as the neighborhood model and to its decision variables as neighborhood variables. In the framework proposed by Pesant and Gendreau (1996),

Pesant and Gendreau (1999), local search is then described as a sequence of CP tree search on auxiliary problems N_P.

While searching for a neighbor, two CP models are active: the original model for P, and the neighborhood model for N_P. The two models communicate through *interface constraints* linking variables across P and N_P.

In the example given before, the interface constraints are:

$$x[I] = s[J] \wedge x[J] = s[I]$$

$$\forall k \quad I \neq k \wedge J \neq k \Rightarrow x[k] = s[k]$$

In addition, a cost function for the neighborhood model N_P can be defined, and branch-and-bound search can be used on N_P to find the best neighbor. These CP models support fast neighborhood explorations. Indeed, constraints are used not only for testing the feasibility of solutions (neighbors) once they have been generated, but also for removing during the search, through propagation, sets of infeasible neighbors. For instance, the values of the already instantiated neighborhood variables may cause the reduction of the domains of the problem variables through the interface constraints and the domain reductions for the problem variables may, in turn, back-propagate on other not yet instantiated neighborhood variables, removing the possibility to generate infeasible neighbors. Propagation can also reduce the search space when only improving neighbors or only the best neighbors are looked for: the bounding constraint on the cost of the move can propagate out non optimal neighbors. Propagation is thus able to discard infeasible or uninteresting portions of the neighborhood without actually iterating those sets of neighbors. The larger and the more constrained the problem, the more significant the reduction in neighborhood search provided by propagation.

Several other advantages can be identified in such a CP approach. First, a clear separation between problem modeling and problem solving is maintained. Modeling constraints for P are kept separate from the neighborhood model. This supports, e.g., the addition of side constraints to P without changing the neighborhood model nor the search methods. Second, any branching scheme may be used for building and exploring the neighborhood search tree. The simplest idea would only instantiate variables from N_P, but branching may also be performed on variables from P or on variables from both P and N_P. In addition, efficient exploration strategy like Limited Discrepancy Search may be used instead of Depth First Search. Few works have started taking advantage of this flexibility and the assessment of its interest is still an open research issue.

The main limitation of this approach lies in the overhead from the CP model and the propagation engine. CP models of the neighborhoods are of interest only when propagation produces a significant reduction of the search space; in such cases, the CP search of the neighborhood generates much fewer neighbors than the nested loop iteration. Moreover, the search tree exploration keeps instantiating and uninstantiating variables (upon backtracking). Searching the swap CP neighborhood with a simple branching scheme takes $O(n^3)$ time. Specific branching schemes have been proposed in Shaw et al. (2000) to reduce this complexity to $O(n^2 \log n)$.

2.2 Incomplete Global Search

Local search and constraint propagation can also be applied within a global search algorithm. Focusing on the family of constructive algorithms, this section shows that, on the scale from greedy algorithms to complete global search, incorporating ideas from local moves and neighborhoods within global search is useful for achieving interesting compromises between solution quality and search time.

2.2.1 *Constructive Algorithms*

A global search algorithm produces a solution by taking decisions and backtracking on failure. The decisions taken in a branch amount to adding a constraint to the problem. Some general branching scheme, such as the first-fail criterion (see, Haralick and Elliott, 1980) will select any variable from the model (that with the smallest number of values in its domain) and instantiate it: in such general cases, it is often difficult to interpret the state of the system before a solution has been reached. The situation is different for some branching schemes that are problem specific and where the decisions at each choice point build a small part of the final solution. For instance, in the case of vehicle routing, insertion algorithms consider customers one by one and decide the route that will visit them; for scheduling, ranking algorithms construct the schedule of a machine in a chronological manner by deciding which task should be sequenced first, which second, and so on; for time-tabling, assignment algorithms decide of the duty of a person (or a group of people) for one time-slot. Such global search algorithms are called *constructive search algorithms*: their states may indeed be interpreted as relevant partial solutions (routing plans for a subset of the customers, short-term schedules planning only a subset of the tasks or time-tables for a subset of the people) and it is easy to evaluate a bound of the objective function by adding the contribution from past decisions to an evaluation of the impact of the decisions to come.

2.2.2 *Greedy Constructive Algorithms*

The search in a constructive algorithm is guided by a heuristic:[2] at each choice point, a function h is evaluated for all possible choices and the choice are ranked by increasing values of h: the choice that minimizes h is considered the preferred decision. In a greedy constructive algorithm, the preferred branch is systematically followed and no backtracking takes place. For pure optimization problems where feasibility is not an issue, such greedy algorithms yield a solution in polynomial time. In case of feasibility issues, the algorithm may produce a partial solution (some customers are not assigned to a route, some tasks are not scheduled, some duties are not assigned in the time-table).

When the optimization system is granted more time than what is required by the greedy constructive algorithm, but not enough for performing a complete global search, local search may be an effective tool for improving the greedy solution. A first idea consists in using the greedy solution as starting point for a descent search or any random walk. However, interesting results can also be achieved by integrating notions from local search directly within the construction process.

The idea is that the construction process should explore a neighborhood of the greedy decision at each step of the construction process. Such a result can be reached in several ways: by performing a lookahead evaluation of the quality of branches (see Section 2.2.3), by considering a subset of the branches that are "close" to the best

[2]Note that, in the CP context, the word "heuristic" does not refer to an approximation algorithm, but to a function used to compare different branches at a choice point. In the remainder of the chapter, heuristic will always refer to that meaning.

branch selected by the heuristic (see Sections 2.2.4 and 2.2.5), or by trying to improve the current solution by a LS algorithm after each construction step (see Section 2.2.7).

2.2.3 Lookahead Algorithms

Simple heuristics for evaluating the interest of a branch are often "myopic" in the sense that they only assess a choice by some of its immediate consequences and not by long-term consequences on the planning process. For instance, in vehicle routing one may evaluate the insertion of a client in a route by the minimal distance between the client and any other client already in the route. A possibility for taking into account such far-reach effect consists in going down the branch, fully propagating the effects of the choice and evaluating a heuristic only thereafter. In the example of vehicle routing, this amounts to performing the insertion of the client at the best place in the route, propagating the consequences of the insertion, and returning the bound of the overall cost. One may also perform a deeper exploration of the subtree below each branch before evaluating and ranking them. Such lookahead heuristics are inspired from game theory where the players may select their moves by unrolling the game n moves ahead and choosing the branch from which the worst reachable situation is best. In the field of combinatorial optimization, lookahead evaluation is a common way of improving greedy algorithms in vehicle routing (see, e.g., Caseau and Laburthe, 1999) or scheduling (see, e.g., Dell'Amico and Trubian, 1993).

2.2.4 Restricted Candidate Lists

At each choice point, the heuristic provides a preferred branch, as well as an indication of the quality of the other branches. In case of binary branching schemes, the heuristic may indicate how close both possibilities are, with situations ranging from near ties to definite choices between one good option and a terrible one. In the case of wider (non binary) branching, the heuristic may consider that some branches are serious competitors to the favorite branch while others are not. In any case, one can explore a subset of all solutions by following only those paths in the tree that never consider a poor branch according to the heuristic. Let $b_1, \ldots, b_k$ be the possible choices (branches), b_1 the preferred one ($h(b_1) = \min_i (h(b_i))$) and b_k the worst one ($h(b_k) = \max_i (h(b_i))$). The idea of Restricted Candidate Lists (RCL, see Feo and Resende, 1995; Glover, 1995) is to retain only the good branches and to discard the bad ones. More precisely, given a parameter α such that $\alpha \in [0, 1]$, only those b_i such that $h(b_i) \leq h(b_1) + \alpha(h(b_k) - h(b_1))$ are kept in a RCL while the others are discarded.

The introduction of RCL thus defines a subtree of the overall search tree. For $\alpha = 0$, this subtree covers all solutions reachable by a greedy algorithm; on the opposite $\alpha = 1$ yields a complete tree covering all possible solutions. For intermediate values of α, the subtree contains only solutions whose construction paths are located "around" greedy paths.

Such a subtree can be explored either systematically or not. In both cases it is important to control the global amount by which a solution path will diverge from the heuristic. It is indeed favorable to generate first solutions that diverge little from the heuristic (following good branches from the RCL) over solutions that systematically diverge from the heuristic (following always the worst branches from the RCL).

The GRASP method (Feo and Resende, 1995) uses RCL within a randomized version of the greedy algorithm: a randomized version of the overall construction algorithm is run many times. For each construction, at each choice point, one branch is selected at random among the RCL and according to some probability distribution (with decreasing probabilities for $b_1, b_2, \ldots, b_k$). Thus, solutions that are globally closer to the heuristic are generated with an overall higher probability than solutions that are systematically far from the heuristic. Much room is left for tuning a GRASP algorithm, through the probability distributions or through the value of α. For instance, Prais and Ribeiro, 1998 showed that it is more efficient to consider varying values of α, starting with tight ones (around 0, in order to follow the heuristic), and progressively releasing their values to accept locally bad choices (higher values). Another possibility consists in allowing higher values of α early on in the construction process (at the first levels of the tree) and restricting the construction to quasi-greedy choices with low values of α in the end (deep in the tree); this is motivated by the fact that heuristics are often more reliable in the end of the construction process rather than at the beginning.

2.2.5 Discrepancy-based Search

The previous section introduced the notion of a subtree defined by RCL for a given heuristic. This subtree can either be explored by means of randomized construction procedures (GRASP) or by systematic search. This is, in essence, what discrepancy-based search procedures (xDS, Harvey, 1995) do: this subtree is explored with construction moves and backtrack moves.

xDS add two notions to RCL:

- it keeps track of all nodes where the algorithm has diverged from the heuristic. Such cases when a branch b_i, $i > 1$ is followed are called discrepancies;
- it keeps track of the paths already generated in order to avoid visiting them twice and relying on backtracking to avoid recomputing many times common intermediate nodes.

This global account of the amount of discrepancy from the greedy heuristic is used to drive the exploration towards solutions that diverge little from the heuristic (i.e., following most of the times the branch b_1) before solutions that diverge more from the heuristic (i.e., following many b_i branches with $i > 1$). Thus, xDS methods in a way ensure by explicit discrepancy bounding what GRASP ensures on average, through cumulated probabilities. The underlying principle is the same in both methods: it is assumed that good solutions are more likely to be constructed by following always but a few times the heuristic b_1, rather that by diverging often from it.

The global account of discrepancies can be performed by several manners:

- counting the number of times K the search did not follow the heuristic. Limited Discrepancy Search (LDS, Harvey and Ginsberg, 1995) explores the tree by generating all solutions for increasing values of K;
- counting the number of times the search did not follow the heuristic up to a certain depth. Depth-bounded Discrepancy Search (DDS, Walsh, 1997) explores the tree by generating all solutions that do not exceed a maximum number of discrepancies up to a certain depth, and then strictly follows the greedy algorithm;
- counting the total divergence in rank between the options taken and the preferred ones (with C denoting the sum of the rank discrepancy, i.e., $i-1$, for all branches b_i that are followed along the path). For two consecutive choices, this method

associates the same divergence to the path taking b_1 and then, b_3 and to a path taking twice the decision b_2. Credit Search, (Beldiceanu et al., 1999) generates all solutions for which C does not exceed a given limit.

2.2.6 Local search over priority list

In a similar spirit, in the case of priority-based heuristics, local moves can be applied directly to the priority list itself. Consider a problem P and a static heuristic for a constructive algorithm $L = (o_1, o_2, \ldots, o_n)$, where L is a list of decision objects o_i of P (e.g., a list of variables in a generic CSP, a list of activities in a scheduling problem). A constructive algorithm A sequentially makes decision on o_i by adding a constraint involving o_i. If the algorithm A stops at the first solution found, then it can be seen as a function mapping the constructive heuristic L into a solution S of the problem P. Given this interpretation, we can apply local search methods to the static heuristic instead of applying them directly to the solution S (e.g., by considering all lists generated by exchanging each pair of decision objects o_i, o_j in L). Note a similar technique is sometimes used in Genetic Algorithms (GA). A GA engine may work on an indirect representation of the problem P ("genotype"), and use a function to map each genotype to a "phenotype" representing a solution of the real problem. Since most of the times the mapping between genotype and phenotype needs to consider complex constraints, and it may be implemented as a constraint programming engine. In this case the static heuristic is generated by the GA engine; more generally, a static heuristic may be generated by a different algorithm A_1 working on a problem P_1 *related to* P. Two different ways can be used to explore the neighborhood of a static heuristic L: the first method applies local search directly to the list L. The second method interprets the rank i of a decision object o_i as a preference, and explore an incomplete tree search (for example using limited discrepancy search) within the preference-based framework as in (Junker, 2000). After having found a first solution by following the given order, preference based search is able to explore permutation of such an order by backtracking on a tree search while looking for new solutions.

2.2.7 Improving Solutions

A last possibility for enhancing an (incomplete) global search algorithm consists in applying some local search steps.

- Local search can be applied at a leaf of the global search tree for improving a solution. This is a straightforward generalization of local search methods which build a solution by a greedy algorithm. Global search is simply used as a way to generate several initial solutions on which a local search improvement phase is applied. It is interesting to generate starting solutions that are different enough for the overall exploration to be rather diversified. LDS is an interesting way of generating such a diversified initial set of solutions.
- Local search can be applied at internal nodes of the global search tree for repairing or improving a partial solution (see Prestwich, 2000). Designing such moves in the general case of any CP model may be hard. Indeed, such moves must handle partial assignments (producing a partial assignment similar to the current one). A simple idea of neighborhoods consists in selecting a set V_1 of variables that are instantiated in the current partial solution, produce a neighbor assignment

of V_1, and apply a propagation algorithm on the overall problem to reduce the domains of variables not in V_1. The global search process can then continue from this improved partial state. Such methods have proven successful on routing problems. Russell, 1995 introduced the idea of applying local moves every t steps of insertion: each local move tries to improve the partial plan (routes visiting a subset of the clients) by another partial plan, visiting the same clients, but in a different order, before continuing the insertion process. Caseau and Laburthe, 1999 compared the method that applies LS after each insertion step (Incremental Local Optimization, ILO, see also Gendreau et al., 1992) to the method that constructs a solution by greedy insertion and then improves it by LS. They showed that ILO was not only faster but also produced much better solutions.

A key element to be considered with LS is the evaluation of partial solutions. The quality of a move can easily be measured when LS is applied on completely instantiated solutions. On the other hand, when LS is applied on partial solutions, evaluating a move may be more difficult. It is usually interesting to consider bounds on the objective function, possibly with the addition of a term evaluating the difficulty of extending the partial solution into a complete one.

2.3 Other Related Methods

A variety of other methods have been proposed in the last decade to solve combinatorial problems with constraints using local search, focusing on solving over-constrained problems (problems where one wants to minimize a global account of penalties for violated constraints). For problems that are described as generic constraint satisfaction problems, the GSAT by Selman et al. (1992) and the Min-Conflict by Minton et al. (1992) methods start from a random infeasible assignment of values to variables and improve it by flipping the value of one variable that is involved in the biggest number of conflicts (violated constraints). Improvements have been proposed over this original framework, with the careful introduction of some randomization in the choice of the variable to be flipped, leading to the walkSat algorithm by Selman and Kautz, 1993. However, such methods are based on very simple models of the optimization problems, with constraints that are described by the list of their feasible assignments. There is thus no means of capturing the entire structure of the problem by such methods. The walkSAT algorithm has recently been generalized into the WSAT algorithm, capable of handling linear constraints over integer variables (see, Walser, 1999); however, as a generic method, it is unable to take advantage of the wealth of knowledge that has been developed for computing bounds over the cost of specific routing or assignment problems.

A completely different approach is followed by the localizer framework by Michel and van Hentenryck (1997). Problems are described by means of variables and formulas called invariants. Although it seems similar to a modeling language, the invariants do not specify feasible solutions, but are rather used to define the way in which data structures are updated when a move is performed. The approach is powered by powerful symbolic reasoning over the invariants and has proven efficient on some problems. However, it does not yet support global objects for modelling a route, a schedule or an assignment.

Thus, for optimization problems featuring a strong combinatorial structure as well as specific side constraints one is left with the need to craft a specific method from the model. For this reason, we will not discuss these methods any further.

3 PRACTICAL GUIDELINES THROUGH A CASE STUDY

The techniques for combining Local Search and Constraint Programming described in the previous section are applied here to a didactic case study: we address a variant of the classical *Vehicle Routing Problem* (VRP), (see, Toth and Vigo, 2002) in which several side constraints are considered modeling real-world requirements.

Informally, we are given a set of *clients* and a depot in which a fixed number of *trucks* are located. Each client produces a given amount of *goods* of a certain type and has to be visited within a *time window*. Early arrivals are allowed, in the sense that a truck can arrive before the time window lower bound, but, in this case, it has to wait until the client is ready for the beginning of the service. The service time, for each client, only depends on the type and quantity of goods to be collected (i.e., it does not depend on the truck), and each truck has two *capacitated bins* which each can contain a unique type of goods. The travel times and costs between each pair of clients, and between each client and the depot, are given.

This VRP variant, referred to as *didactic* Transportation Problem (dTP) in the following, calls for the determination of the set of trucks' routes minimizing the total travel cost, and satisfying the following constraints:

- each route is associated to a truck, and starts and ends at the depot;
- each client is visited exactly once, and within the time window;
- the bins' capacity constraints are respected.

3.1 A CP Model

The CP model of the transportation problem dTP can be formally stated by using the following notation:

- $i, j \in \{1, \ldots, N\}$ for the locations (0 denotes the depot) and, by extension, for the clients;
- $k \in \{1, \ldots, M\}$ for the trucks, and, by extension, for their routes;
- $h \in \{1, \ldots, 2M\}$ for the bins;
- $\ell \in \{1, \ldots, P\}$ for the types of goods.

For each location $i (i = 1, \ldots, N)$ the corresponding goods are denoted by $type_i$ and q_i, where $type_i \in \{1, \ldots, P\}$ indicates the type of goods to be collected by client i, while $q_i > 0$ is its quantity. Moreover, each client i has an associated time window $[a_i, b_i]$ representing the time frame during which the service must start. The fleet of vehicles located at the depot 0 is composed by M trucks: each truck k has two bins of identical capacity C, and each bin may collect a unique type of goods $\ell \in \{1, \ldots, P\}$. The duration of the service at location i is $d_i > 0$. Finally, for each pair of locations i and j, the travel time $tt_{ij} \geq 0$ and the travel cost c_{ij} is reported. A possible CP model for dTP is the following:

$$
\begin{array}{rll}
\min & totCost = \sum_{k=1}^{M} cost_k & \\
\text{on} & & \\
\forall k \in \{1, \dots, M\} & cost_k \geq 0, & \\
 & truck_k = UnaryResource(tt, c, cost_k) & \\
\forall h \in \{1, \dots, 2M\} & collects_h \in [1..P] & \\
\forall i \in \{1, \dots, N\} & start_i \in [a_i..b_i], & \\
 & service_i = Activity(start_i, d_i, i), & \\
 & visitedBy_i \in [1..M], & \\
 & collectedIn_i \in [1..2M] & \\
\text{subject to} & & \\
\forall i \in \{1, \dots, N\} & service_i \text{ requires } truck[visitedBy_i] & (1) \\
\forall h \in \{1, \dots, 2M\} & \sum_{i \mid collectedIn_i = h} q_i \leq C & (2) \\
\forall i \in \{1, \dots, N\} & collects[collectedIn_i] = type_i & (3) \\
\forall i \in \{1, \dots, N\} & visitedBy_i = \left\lceil \frac{collectedIn_i}{2} \right\rceil & (4)
\end{array}
$$

CP models for combinatorial optimization problems consist in the definition of three different sets of objects: an objective function, decision objects, and constraints. In the example, we explicitly separate the three sets with the keywords *min*, *on*, and *subject to*. Moreover, most Constraint Programming languages provide modeling objects such as *Activities*, *Resources*, etc. Using such objects rather than only variables yields more concise models. Even if the actual syntax of such objects may vary depending on the specific CP language at hand, we may safely assume that the model proposed can easily be coded using most CP languages.

3.1.1 Basic Model

We are given M trucks, $2M$ bins, and N clients that must be visited within a time window by exactly one truck. Each truck is a *UnaryResource* object containing the information on the travel time and cost among locations (clients), and a variable representing the total travel cost for the truck. The service at each client is an *Activity* object defined by a variable start-time, a constant duration, and a location. Constraint (1) '$service_i$ requires $truck_k$' enforces that from $start_i$ to $start_i + d_i$ the $truck_k$ *UnaryResource* is used by the *Activity* $service_i$ without interruption. A *UnaryResource* cannot be used simultaneously by more than one *Activity*, and, in addition, it is not used during the time needed to move from location to location. Moreover a given time and cost must be considered before the first and after the last activities are executed. In case tt satisfies the *triangle inequality* ($tt_{i_1,i_3} \leq tt_{i_1,i_2} + tt_{i_2,i_3}$), an equivalent model is that for every pair of *Activity* $service_i$, $service_j$ such that '$service_i$ requires $truck_k$', and '$service_j$ requires $truck_k$', $(start_i \geq start_j + d_j + tt_{ji}) \vee (start_j \geq start_i + d_i + tt_{ij})$. Note that the same objects used to model trucks and visits in dTP could also be used to model machines with maximal capacity equal to 1, and tasks with sequence dependent setup times and cost in scheduling problems. A client $service_i$ needs to be visited by exactly one of the M trucks, say the k-th, thus we model the alternative choice of trucks by defining, for each client, a variable $visitedBy_i$ referring to that index k. The requirement constraint is thus stated on the $truck_k$ array of alternative resources indexed by the variable $visitedBy_i$.

Each client i produces a quantity q_i of goods of type $type_i$. The variable $collectedIn_i$ identifies the bin used to serve client i. The bin capacity constraint (2) for each bin h simply states that the sum of all quantities q_i such that q_i is collected in h is less or equal to the maximal capacity C. The variable $collects_h$ identifies the type associated to bin h. The constraint requiring that each bin must contain goods of the same type can be stated using variables $collects_h$ and $collectedIn_i$ by imposing that for each client i the type associated to the bin used by the client i is equal to $type_i$ (3).

Bins with index $2k - 1$ and $2k$ are associated to truck k (bins 1 and 2 are placed on truck 1, bins 3 and 4 on truck 2, etc.). The link between bins and trucks is modeled with constraint (4).

The *basic model* correctly models dTP. Decision variables are start-time ($start_i$) and bin selection ($collectedIn_i$) variables for each client i. Once all variables $start_i$ and $collectedIn_i$ are instantiated to values satisfying all the constraints of the *basic model*, a solution is reached, since all the other variables of the model ($cost_k$, $visitedBy_i$, etc.) are instantiated by propagation.

Nevertheless, when dealing with routing problems, it may be convenient to explicitly manipulate the sequence of services performed by each truck. For this purpose an extension of the model is considered. This model, based on the abstraction of multiple path, is redundant with respect to the *basic model*, since it is not necessary for correctly modeling dTP.

3.2 Propagation

CP constraints embed domain propagation algorithms aimed at removing values from variable domains that are proven infeasible with respect to the constraint itself (see, Mackworth, 1977). Several propagation algorithms can be designed for the same CP constraint; we briefly sketch some algorithms that can be used to implement the propagation of the constraints used in the dTP model.

Concerning notation, given a variable v, in the following we will refer to the minimum (resp. maximum) value in its domain as `inf`(v) (resp. `sup`(v)). Moreover, the domain itself is referred to as `domain`(v), whereas once v has been instantiated its value is returned as `value`(v).

3.2.1 Disjunctive Relations

Consider the *UnaryResource* object used to model the trucks and the *require* constraints linking objects *Activity* and *UnaryResource*. As mentioned these constraints state that two activities must not be executed simultaneously on the same unary resource, and a transition time must be granted between two subsequent activities.

A simple propagation algorithm updates the start-time variable of activities by looking at all pairs of activities performed on the same resource and deducing precedence constraints. Consider two activities $service_i$, $service_j$, if $(\mathtt{inf}(start_j) + d_j + tt_{ji} > \mathtt{sup}(start_i)) \wedge (\mathtt{value}(visitedBy_i) = \mathtt{value}(visitedBy_j))$ then it can immediately be deduced that $service_i$ must precede $service_j$, i.e.,

$$\mathtt{inf}(start_j) := \max\{\mathtt{inf}(start_j), \mathtt{inf}(start_i) + d_i + tt_{ij}\}$$

and

$$\mathtt{sup}(start_i) := \min\{\mathtt{sup}(start_i) + d_i, \mathtt{sup}(start_j) - tt_{ij}\} - d_i.$$

This propagation rule finds new time bounds for the activities by considering the current time bounds, the capacity availability, and the resource assignments. Similarly, new possible assignments are deduced by considering current time bounds, and capacity available: if $(\mathtt{inf}(start_j) + d_j + tt_{ji} > \mathtt{sup}(start_i)) \wedge (\mathtt{inf}(start_i) + d_i + tt_{ij} > \mathtt{sup}(start_j)) \wedge (\mathtt{value}(visitedBy_i) = k)$ then $visitedBy_j \neq k$. More sophisticated propagation algorithms have been implemented in different CP languages, see e.g., Edge Finding techniques for unary and discrete resources, and constructive disjunctive techniques for alternative resources (see, Nuijten, 1994). These techniques have been successfully applied to Scheduling Problems and TSPTW (see, Pesant et al., 1998). In the proposed model, beside the modeling of capacity availability and transition times, the object *UnaryResource* is also responsible for maintaining a cost variable corresponding to the resource specific component of the total cost.

3.2.2 *Linking Trucks and Bins*

For its simplicity, it is worth describing in detail the constraint linking the variables used for the choice of trucks and bins, $visitedBy_i = \lceil collectedIn_i/2 \rceil$. The constraint is considered consistent iff the following condition hold:

$$\forall k \in \mathtt{domain}(visitedBy_i)\ \exists h \in \mathtt{domain}(collectedIn_i)\ | k = \lceil h/2 \rceil$$

and vice-versa

$$\forall h \in \mathtt{domain}(collectedIn_i)\ \exists k \in \mathtt{domain}(visitedBy_i)\ | k = \lceil h/2 \rceil.$$

A simple propagation algorithm iterates on all values k belonging to the domain of $visitedBy_i$ and checks for the existence of a value h in the domain of $collectedIn_i$ that satisfies the relation $k = \lceil h/2 \rceil$. If no value h is found, k is removed from the domain of variable $visitedBy_i$. The propagation algorithm also iterates on all values h belonging to the domain of $collectedIn_i$ and checks for the existence of a value k in the domain of $visitedBy_i$ that satisfy the relation $k = \lceil h/2 \rceil$. If no value k is found, h is removed from the domain of variable $collectedIn_i$.

3.2.3 *Propagating Costs*

In many cases the calculation of good lower bounds on the objective function variable is especially important. As shown in Focacci et al. (1999a) good lower bounds on the optimal cost can be used for discarding a priori uninteresting parts of the solution space. Moreover, lower bounds can also be used in greedy algorithms as described in Section 3.4.2 for guiding towards more promising parts of the solution space. In CP the link between the objective variable and the decision variables is maintained through a constraint.

In summary, three things should be stressed. First, whenever a simple or global constraint acts on several variables the modification of any of the variables triggers propagation algorithms that propagate the modification on the other variables involved. Second, for a given constraint several propagation algorithms can enforce different degrees of consistency. Finally, even without giving complexity results of the propagation algorithms, it is clear that some of them can be computationally more expensive than others. For example, checking a constraint for the feasibility on a set of instantiated variables is usually much easier than eliminating all values that can be proved to be infeasible on a set of yet uninstantiated ones. Therefore, the propagation algorithms to use may vary depending on the size of the problem, and on the type of solving procedure used.

3.3 Redundant Routing Model

Starting from the basic model of Section 3.1.1, a redundant model can be devised. The main advantage of using redundant models in CP consists in better filtering out inconsistent values. In this specific case, another important advantage is that many neighborhood structures can be easily defined by means of the additional variables ($next_i$, $succ_i$) introduced by the redundant models. Nevertheless, redundant models should be used with care. Depending on the problem size, the use of several linked models could eventually be computationally penalizing. In order to show a wide range of CP-based local search techniques, we used all redundant models simultaneously. Performance issues arising with very large instances could be addressed by using lighter models or other techniques such as decomposition.

The redundant *routing* model is the following:

$$\forall k \in \{1,\dots,M\} \quad \mathit{first}_k \in [1..N]$$

$$\forall i \in \{1,\dots,N\} \quad \mathit{next}_i \in [1..N+M],$$

$$\mathit{succ}_i \in [\{\}..\{1,\dots,N\}]$$

$$\mathit{multiPath}(\mathit{first}, \mathit{next}, \mathit{succ}, \mathit{visitedBy}) \tag{5}$$

$$\mathit{costPaths}(\mathit{first}, \mathit{next}, \mathit{succ}, c, \mathit{totCost}) \tag{6}$$

$$\forall i, j \in \{1,\dots,N\} \quad j \in \mathit{succ}_i \Leftrightarrow (\mathit{visitedBy}_i = \mathit{visitedBy}_j \wedge \mathit{start}_j > \mathit{start}_i) \tag{7}$$

Let $G = (V, A)$ be digraph, and a partition of V be defined by a set S of M *start* nodes, a set I of N *internal* nodes, and a set E of M *end* nodes. A multiple path constraint *multiPath* enforces that M paths will exist starting from a start node, ending in an end node and covering all internal nodes exactly once. All internal nodes have one direct predecessor (previous node) one direct successor (next node).

Internal nodes are labeled $1, 2, \dots, N$, end nodes are labeled $N+1, N+2, \dots, N+M$, while start nodes are labeled $N+M+1, N+M+2, \dots, N+2M$. The multiple path constraint makes use of four arrays of variables. Variable first_k is associated with the start node $N+M+k$, and identifies its next node in the path. Variables next_i, succ_i, and $\mathit{visitedBy}_i$ are associated with the internal node i. Variable next_i identifies the node next of i; variable $succ_i$ identifies the set of internal nodes occurring after i in the path containing i; finally, variable $visitedBy_i$ identifies the path containing node i.

The *multiPath* constraint can be used to model a subproblem of dTP: each internal node represents a client, the set of start and end nodes represents copies of the depot. The path k starting from the start node $N + M + k$, covering a set of internal nodes, and ending at an end node represents the route of $truck_k$. The *redundant routing model* and the *basic model* are linked by the variables $visitedBy_i$ used in both models, and by the constraint (7).

Note that the cost on each truck only depends on the sequence of services performed, therefore the multiple path structure can also be used to define the objective function. Since the variable $next_i$, and $succ_i$ will be extensively used as decision variables for dTP, it is indeed useful to explicitly link the multiple path model to the cost variables $cost_k$, as done with the constraint (6).

3.3.1 Propagation

Since the multi path model will be extensively used in the discussion of the local search methods, it is important to describe it in details. A model equivalent to the *multiPath* constraint based on only simple arithmetical and logical constraints is the following:

$$\forall i, j \in \{1, \dots, N\}, i > j \qquad next_i \neq next_j \tag{8}$$

$$\forall k_1, k_2 \in \{1, \dots, M\}, k_1 > k_2 \qquad first_{k_1} \neq first_{k_2} \tag{9}$$

$$\forall i \in \{1, \dots, N\}, \forall k \in \{1, \dots, M\} \qquad first_k \neq next_i \tag{10}$$

$$\forall i, j \in \{1, \dots, N\} \qquad next_i = j \Rightarrow succ_i = \{j\} \cup succ_j \tag{11}$$

$$\forall i, j \in \{1, \dots, N\} \qquad j \in succ_i \Rightarrow visitedBy_i = visitedBy_j \tag{12}$$

$$\forall k \in \{1, \dots, M\}, \forall i \in \{1, \dots, N\} \qquad first_k = i \Rightarrow visitedBy_i = k \tag{13}$$

$$\forall i \in \{1, \dots, N\} \qquad i \notin succ_i \tag{14}$$

As mentioned in the model description, the set variables $succ_i$ are used to enforce precedence relations among nodes. An integer set variable is a variable whose domain contains sets of integer values. In analogy with the definition of lower and upper bounds on integer variables, lower and upper bounds can also be defined for an integer set variable: the lower bound, called *required set*, is the intersection of all sets belonging to the domain of the variable; the upper bound, called *possible set*, is the union of all sets belonging to the domain of the variable. The variable is considered instantiated when the domain contains a single set of integer values, thus the required set is equal to the possible set. Required set, possible set, and value of the set variable v will be referred to as `req`(v), `poss`(v), and `value`(v).

Constraints (8)–(10) force all internal nodes $\{1, \dots, N\}$ to have exactly one incoming, and one outgoing arc, while all start nodes are forced to have exactly one outgoing arc belonging to the set of internal nodes. Constraints (11) and (12) link *succ* and *next*, and *succ* and *visitedBy* variables; constraint (11) links the *visitedBy* variable of the first internal node in a path starting from node $N + M + k$ to the value k. Finally the last constraint prevents cycles in the graph.

The global constraint *multiPath* could indeed be written using all the mentioned constraints. However, more effective and efficient propagation algorithms can be designed enforcing exactly the same semantic. Basic propagation on the *next* variables enforces that, whenever a variable $next_i$ is instantiated to a value t, the value t is removed from the domain of all variables $next_j, j \neq i$. A more powerful propagation can be obtained by using flow algorithms (see, Régin, 1994). Cycles can be avoided either by simple cycle removal algorithms, or by performing a more sophisticated propagation using strong connected components or isthmus detection. Some of the propagation that can be performed on the variables $succ_i$ exploit the transitivity of the *succ* array: $i_2 \in \mathtt{req}(succ_{i_1}) \wedge i_3 \in \mathtt{req}(succ_{i_2}) \Rightarrow \mathtt{req}(succ_{i_1}) := \mathtt{req}(succ_{i_1}) \cup \{i_3\}$. This propagation rule can be read as follows: if i_2 is necessarily after i_1, and i_3 is necessarily after i_2, then i_3 is necessarily after i_1.

Beside the simple link between *next* and *succ* variables in (11), more effective propagation can be performed. For example, if given two nodes i and j, the set of nodes that necessarily follows i has a non-empty intersection with the set of nodes that necessarily precedes j, then at least a node must exist in any path from i to j, therefore j cannot be the next of i. More formally, $\mathtt{req}(succ_i) \cap (\{1, \ldots, N\} \setminus \mathtt{poss}(succ_j)) \neq \emptyset \Rightarrow next_i \neq j$. Note that this last propagation rule is not part of the arithmetic model of the *multiPath* constraint (8)–(14); this is a simple example of the extra propagation that can be done by a global constraint.

3.4 Constructive Algorithms

The model presented above could be solved as such by complete global search. However, such a solution is impractical in terms of computing time for realistic instances (recall that the maximal size of plain VRPs that can currently be solved to optimality within minutes ranges from a few tens of clients for branch-and-bound methods to 80 for branch-and-cut methods, see Toth and Vigo, 2002). This section discusses constructive methods which are realistic for solving dTP instances with hundreds of clients.

3.4.1 Insertion Algorithms

In the dTP model, when all CP variables in the arrays *visitedBy* and *succ* are instantiated, then the solution is fully known. Indeed, variables arrays *first*, *next* and *cost* can be instantiated from *succ*. When all these variables are instantiated, the objective function is instantiated although the *collectedIn* and *start* variables may be left with an interval domain. A simple greedy algorithm can instantiate all these remaining variables. For example one could iteratively choose each client i and instantiate $collectedIn_i$ and $start_i$ to its lower bound.[3]

We propose a constructive algorithm based on this dominance property using ($visitedBy_i$) and ($succ_i$) as CP decision variables for dTP. Clients are considered one after the other: for each client i, the algorithm instantiates the variable $visitedBy_i$, and for all j such that $visitedBy_j$ has already been instantiated to the same value as $visitedBy_i$, the *succ* variables are reduced by enforcing one of the two decisions $j \in succ_i$ and $j \notin succ_i$.

[3] This is true with the current set of constraints while one has to be careful in the case of addition of other constraints.

Such an instantiation scheme implements a constraint-based version of the standard insertion heuristics for vehicle routing (see, Golden and Assad, 1988). Indeed, each time a client i is assigned to a truck, the ordering between i and all others clients assigned to that same route is settled, yielding a total order on all clients of the route. These ordering decisions implement the insertion of i between two consecutive clients in a route. Compared to standard insertion algorithms in Operations Research textbooks, this CP description adds two notions.

- The routing problem can be enriched with side constraints, such as time windows, bin assignments and bin capacity. When selecting the best possibility for insertion, feasibility is ensured. Rather than performing the insertion and checking for the satisfaction of all constraints, we rely on propagation to discard some of the infeasible assignments. Infeasible route assignments are removed from the domain of $visitedBy_i$, and infeasible orderings are discarded by reducing the domain of $succ_i$.
- The overall insertion process is seen as a monotonic process where variable domains are reduced. Indeed, the $next_i$ variables may remain uninstantiated until the very end of the construction process (until the routes are full). Until then, the routes are only partially sequenced with precedence decisions: the relative order among clients already assigned to a given route is settled ($\forall i, j,\ i \in succ_j \lor j \in succ_i$), but there is room to insert new clients between such i and j. This description of insertion within "open" routes supports the evaluation of lower bounds on the route costs: at any point in the algorithm, each route k has been assigned a subset of clients $i_1, \ldots, i_p$ ($visitedBy_{i_1} = visitedBy_{i_2} = \cdots = visitedBy_{i_p} = k$) which have been ordered: $i_2 \in succ_{i_1}, \cdots, i_p \in succ_{i_{p-1}}$. When the costs c_{ij} satisfy the triangle inequality, the cost of the route $cost_k$ can be bounded with $cost_k \geq c_{0,i_1} + c_{i_1,i_2} + \cdots + c_{i_{n-1},i_n} + c_{i_n,0}$.

The general framework for construction algorithms can thus be described formally as follows:

```
algorithm: INSERTION
 for all clients i
     for each possible value k in domain(visitedBy_i)
         branch on
             try the assignment visitedBy_i = k
             for all j such that domain(visitedBy_j) = {k}
                 branch on try j ∈ succ_i
                        or try j ∉ succ_i
     if some feasible insertions were found
        commit to one among them
     else return(failure)
 return(success)
```

In the end of the algorithm, either a feasible routing plan that services all clients (success) or no solution (failure) is obtained.[4]

[4] In case of failure, it is always possible to remove from the model those clients which could not be inserted by the solver and return a feasible routing plan covering only a subset of the clients.

3.4.2 Greedy Insertion

This general scheme can be refined into a greedy algorithm.

- In order to help the algorithm finding a routing plan covering all clients, it is wise to consider first the clients that are "difficult" to insert into a route (those for which the insertion into a partial route is most likely to be infeasible). We therefore sort all clients and use as priority criterion a function that returns higher values for clients far away from routes already built and with tight time windows over clients close to current routes and with a loose time window. The latter have indeed little propagation impact when assigned to a route while the former drastically augment the lower bound on the cost of the route and significantly reduce the possibilities of further assignments through time window constraints. An example of such a static criterion returning high values for nodes that should be considered first would be $priority(i) = tt_{0,i}/(b_i - a_i)$.
- Since all trucks are similar, each assignment of a client to a (still) empty route amounts to a route creation and is strictly equivalent. Therefore, when there exist several k such that for no client j, domain($visitedBy_j$) $= \{k\}$, then, the tentative assignment ($visitedBy_i = k$) is considered for only one k.
- When several possible insertions are found to be feasible for a client i, rather than selecting any of them, we can choose the assignment $visitedBy_i = k$ and the ordering (reductions on $succ_i$) that causes the lower bound of the overall objective z to increase the least: this turns the constructive method into a greedy algorithm.

The greedy algorithm is then as follows:

```
procedure: BEST_INS(i,k)
 oldInf := inf(cost_k)
 try visitedBy_i = k
 for all j such that domain(visitedBy_j)= {k}
     branch on try j ∈ succ_i
            or try j ∉ succ_i
 over all solutions, minimize Δ :=inf(cost_k) - oldInf

algorithm: GREEDY CONSTRUCTION
 sort all clients i by decreasing values of priority(i)
 for all clients i
 for k ∈ domain(visitedBy_i) | ∃j ≠ i, domain(visitedBy_j) = {k}
         try BEST_INS(i,k)
 if ∃k0 ∈ domain(visitedBy_i) | ∀j ≠ i, domain(visitedBy_j) ≠ {k0}
         try BEST_INS(i,k0)
     if some feasible insertion was found
        commit to the one that yielded the smallest Δ
     otherwise return(failure)
 return(success)
```

Adapting the greedy algorithm to a dynamic priority criterion is straightforward.

3.4.3 Discrepancy-based Search

As any greedy heuristic, the previous algorithm can be transformed into a search algorithm through the addition of *backtracking*. By adding a limited amount of backtracking, the algorithm explores a more substantial portion of the solution space at the price of an increased complexity. Discrepancy-based Search (see, Harvey, 1995) produces solutions by following the preferred branch of the heuristics always but a limited number of times (the "discrepancies"). Since we not only have a ranking of insertion choices, but also a valuation on each branch (the value of Δ for each possibility of insertion), we may precisely control the occurrences of the discrepancies. Pure LDS(K) (Harvey and Ginsberg, 1995) would allow for discrepancies at any choice point, granted that all solutions are generated with less that K discrepancies. For dTP, this can be refined in several manners:

- one of the interests of Discrepancy-based Search is to generate solutions that are spread across the overall solution space. It is therefore particularly interesting to branch on options that are radically different at each discrepancy. Insertions at different positions within the same route are considered similar; therefore the alternate choices for insertion that are considered involve different routes;
- it is not worthwhile considering the second best possibility of insertion when its impact on the cost is significantly worse than the preferred choice. Branching on discrepancies should be reserved for situations of near-ties. Thus, a discrepancy is generated only when $\Delta_{k_2} \leq \beta\Delta_{k_1}$ where Δ_{k_1} is the best (least) insertion cost and Δ_{k_2} the second best, for some ratio $\beta \geq 1.0$;
- the branching scheme compares insertions within routes to which some clients have already been assigned with an insertion into one empty route (route creation). Comparing such options by their impact on the cost lower bound is sometimes unfair to the second option. Indeed, let i be the current client, $\Delta_{k_0} = c_{0,i} + c_{i,0}$ whereas $\Delta_{k_1} = c_{j_1,i} + c_{i,j_2} - c_{j_1,j_2}$ if the best place of insertion for i in k_1 is between clients j_1 and j_2. Therefore, Δ_{k_0} accounts for two transitions between locations (back and forth between the depot, 0, and i), while Δ_{k_1} accounts for only one (replacing the direct transition from j_1 to j_2 by a transition from j_1 to i and one from i to j_2). In order to avoid such a bias against route creations, discrepancies where the alternate branch is a route creation are encouraged through a higher ratio threshold $\beta' > \beta$.

The overall Discrepancy-based Search *DISCREPANCY*(K), defined below, is a refined version of *LDS*(K). Its integer parameter, K, represents the maximal number of discrepancies allowed in a solution:

```
algorithm DISCREPANCY(K):
 sort all clients i by decreasing values of priority(i)
  DISC(1,K)

procedure: DISC(i,K):
 if i > N return(success)
 else
 for k ∈ domain(visitedBy_i) |∃j ≠ i, domain(visitedBy_j) = {k}
       try BEST_INS(i,k)
```

```
let k1 and k2 be the two branches with smallest Δk
if ∃k0 ∈ domain(visitedBy_i) | ∀j ≠ i, domain(visitedBy_j) ≠ {k0}
   try BEST_INS(i,k0)
if no feasible solution was found
   return(failure)
else
  branch on
     preferred choice:
       if Δk1 < Δk0 commit to BEST_INS(i,k1)
                        DISC(i + 1, K)
       else commit to BEST_INS(i,k0)
            DISC(i + 1, K)
     if K > 0 alternate branch:
       if Δk0 ≤ Δk1
          if Δk1 ≤ βΔk0 commit to BEST_INS(i,k1)
                           DISC(i + 1, K − 1)
       else if Δk0 ≤ β'Δk1 commit to BEST_INS(i,k0)
                              DISC(i + 1, K − 1)
       else if Δk2 ≤ βΔk1 commit to BEST_INS(i,k2)
                              DISC(i + 1, K − 1)
```

3.5 LS as Post-Optimization

This section illustrates various techniques for the exploration of constrained neighborhoods in the dTP case study. We assume to start with a feasible solution which can, for instance, be built through constructive methods presented in the previous section. We review four kinds of LS procedures on dTP: two simple adaptations of standard LS with constraint checks or inlined constraint checks and two descriptions of neighborhoods with CP, specifically, frozen fragments and CP models of the neighborhood.

Although we refer to the solution with the notations introduced in Section 3, it is never assumed that the variables of the model are CP domain variables. More precisely, in the next two sections (3.5.1 and 3.5.2), the easiest way to implement the LS methods consists in representing a solution with simple data structures (integer to represent variables modeling integer), whereas the approach in Sections 3.5.3 and 3.5.4 require that the variables of the model are implemented as CP variables with domains and active propagation.

3.5.1 LS + Constraint Checks

A first idea for improving a solution to dTP consists in using classical VRP neighborhoods, and checking, for each neighbor, the feasibility with respect to side constraints. This method is illustrated on two simple neighborhoods, specifically *3-opt* and *node-transfer*.

3-opt 3-opt is the straightforward extension of the 2-opt discussed in Section 2.1.1 in which three arcs, instead of two, are removed from the solution, and three others are added to obtain a new solution. Since this exchange is performed within the same route, the feasibility test for the new solution needs to be executed only on that route.

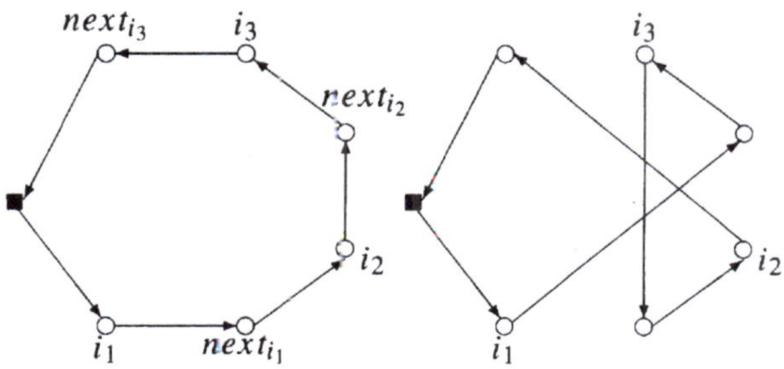

Figure 13.1. Example of 3-opt move.

The best move is the one with the largest decrease of z (if any). An example of 3-opt move is depicted in Figure 13.1.

More formally, let k be a route, and i_1, i_2, i_3 be three different clients such that $visitedBy_{i_1} = visitedBy_{i_2} = visitedBy_{i_3} = k$, and $i_2 \in succ_{i_1}$ and $i_3 \in succ_{i_2}$. A new solution can be obtained by re-assigning variables $next_{i_1}$, $next_{i_2}$, and $next_{i_3}$, and keeping unchanged all the other *next* variables. Since dTP incorporates side constraints, testing feasibility can be computationally expensive, but using constraints directly supports extensions to additional side constraints. For such 3-opt, since there is no exchange of clients among the routes, bin type and capacity constraints still hold, and only the time window constraints need to be checked.

The entire exploration of the neighborhood can be implemented as follows:

```
procedure: 3-opt_EXPLORE

 for each k ∈ {1,...,M}
     for each i1 | visitedBy_i1 = k
         for each i2 | visitedBy_i2 = k ∧ i2 ∈ succ_i1
             for each i3 | visitedBy_i3 = k ∧ i3 ∈ succ_i2
                 let j1 := next_i1, j2 := next_i2, j3 := next_i3
                 try next_i1 = j2, next_i2 = j3, next_i3 = j1
                 check feasibility for time windows
                 old := c_i1,j1 + c_i2,j2 + c_i3,j3
                 new := c_i1,j2 + c_i2,j3 + c_i3,j1
 over all neighbors, minimize Δ := old − new
```

Once the feasible solution with smallest Δ is found, the data structure *succ* must be updated, so as the objective function z. As mentioned at the beginning of Section 3.5, we do not consider here CP variables, but the constraints (in this case the time window ones) are used off-line to check feasibility and update the *start* variables.

Node-transfer Node-transfer moves a single client from a route to another by keeping feasibility of both routes, and decreasing the overall value of z. This move leads to more complicated feasibility issues than 3-opt. Specifically, three constraints need to be checked for the route in which the client is inserted: bin type compatibility,

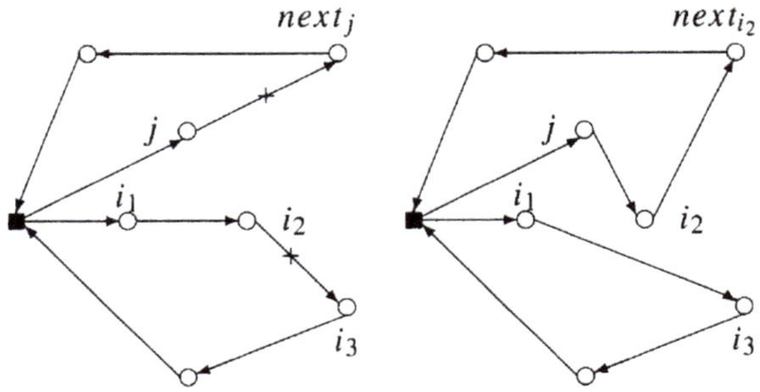

Figure 13.2. Example of node-transfer move.

bin capacity and the time window constraints. An example of node-transfer move is depicted in Figure 13.2.

More formally, let k_1 and k_2 be two routes, and let i_1, i_2 and i_3 be three clients such that $visitedBy_{i_1} = visitedBy_{i_2} = visitedBy_{i_3} = k_1$, and $next_{i_1} = i_2$, and $next_{i_2} = i_3$. Moreover, let j be one more client such that $visitedBy_j = k_2$.[5] A new solution is obtained by moving i_2 from k_1 to k_2 at position $next_j$.

A passive way of using constraints within a LS based on a single-client move is to have the following four nested loops:

```
procedure: node-transfer_EXPLORE
 for each k1 ∈ {1,...,M}
      for each i2 | visitedBy_i2 = k1
           for each k2 ∈ {1,...,M} | k2 ≠ k1
                for each j | visitedBy_j = k2
                     try visitedBy_i2 = k2, next_i2 = next_j,
                          next_j = i2, next_i1 = i3
                     check feasibility for all constraints
                     old := c_i1,i2 + c_i2,i3 + c_j,next_j
                     new := c_i1,i3 + c_j,i2 + c_i2,next_j
 over all neighbors, minimize Δ := old − new
```

Unlike the 3-opt case, more side constraints need to be checked, since the move changes the client-truck assignments, and again an updating of the data structures is needed.

3.5.2 Constraint Checks within the Neighborhood Iteration

In the pseudo-code presented for the *node-transfer* neighborhood exploration much of the time is wasted generating trivially infeasible neighbors. For instance, it is useless iterating j over a route k_2 covered by a truck whose bins are not compatible with $type_i$. Therefore, to speed up the exploration, we can anticipate some of the feasibility checks within the nested loop and before the final constraint checks. The infeasibility

[5]Since we do not have the variable $next_0$, if i_2 is the first client visited in route k_1, and/or if we want to insert it as first client in route k_2, then it is necessary to consider also variables $first_{k_1}$ and/or $first_{k_2}$.

(or ineffectiveness) of many of the solutions in the neighborhood can be easily detected by testing: (i) if the type of goods of client i_2 is compatible with the types collected by the bins of route k_2, (ii) if the residual capacity of the bins of route k_2 can accommodate the quantity q_{i_2}, and (iii) if the neighbor improves z, i.e., if $c_{i_1,i_3} + c_{j,i_2} + c_{i_2,next_j} - c_{i_1,i_2} - c_{i_2,i_3} - c_{j,next_j} < 0$.

Such an optimized neighborhood exploration can be stated as follows:

```
procedure: node-transfer_EXPLORE_updated
 for each h1 ∈ {1,...,2M}
     for each i2 | collectedIn_i2 = h1
         for each h2 ∈ {1,...,2M} |
                  h2 ≠ h1 ∧ collects_h1 = collects_h2 ∧ RC_h2 ≥ q_i2
             for each j | collectedIn_j = h2
                 old := c_i1,i3 + c_j,i2 + c_i2,next_j
                 new := c_i1,i2 + c_i2,i3 + c_j,next_j
                 if old - new < 0
                    try collectedIn_i2 = h2,
                        next_i2 = next_j, next_j = i2, next_i1 = i3,
                        visitedBy_i2 = visitedBy_j
                    check feasibility for time windows
 over all such neighbors, minimize Δ := old - new
```

where the RC_{h_2} represents the *residual capacity* of bin h_2, i.e.,

$$RC_{h_2} = C - \sum_{i \mid collectedIn_i = h_2} q_i$$

and must be updated (together with *succ*, etc.) once a feasible (improving) move is performed. Not only does this exploration generate less infeasible neighbors, it also needs to check less constraints (only time window ones) on the outcome of the loop.

3.5.3 Freezing Fragments

This section describes another kind of moves called *shuffle* (Applegate and Cook, 1991) or *Large Neighborhood Search* (LNS, Shaw, 1998). Such moves keep a fragment of the current solution (freezing the values of variables in that fragment), forget the value and restore the initial domain of variables outside the fragment. Finding the best completion of such a state into a solution is an optimization subproblem. The search space for this subproblem is smaller than for the original problem because some variables have been fixed. Enforcing part of the assignments from the previous solution is the opposite process as relaxation: the problem is strengthened with additional constraints. The frozen fragments must be selected carefully and should be:

- large enough so that the subproblem is more tractable than the original one;
- but not too large so that the subproblem contains enough flexibility to improve over the current solution;
- good enough so that the subproblem consists in good solutions in the overall search space.

In a way, the fragment plays the role of a strictly followed heuristic while building a solution. However, since the fragment that is kept corresponds to choices that were made at any point in the search tree, and not necessarily at the root of the tree, such a destruction-construction process differs from chronological backtracking.

Such LNS algorithms can be easily implemented within CP systems, since the neighborhood exploration consists in solving the original problem while some variables are instantiated and the domain of others are reduced. Moreover, insertion algorithms are natural candidates to LNS. Specifically, clients are partitioned into two sets: a set of clients for which the route assignment and relative sequencing is kept (the fragment from the reference solution that is frozen), and the remainder of the clients, for which all decisions (assignment and sequencing variables) are forgotten.

As described in Shaw (1998), the forgotten part of the solution should be of relative significant size and should be made of related clients (clients from the same routes or from a given geographical area).

In the case of dPT, several neighborhoods can be proposed for such LNS:

- keeping all decisions concerning a subset of the *routes* and re-optimizing all the remainder of the solution;
- keeping all decisions concerning a subset of the *types of goods* and re-optimizing all the remainder of the solution;
- keeping all decisions concerning a subset of the *clients* and re-optimizing all the remainder of the solution.

Note that such different neighborhoods may be combined through VNS (see, Mladenović and Hansen, 1997).

In the following, a simple example of LNS based algorithm is proposed. The algorithm iteratively freezes a part of the problem and tries to extend the partial solution. The iterative process stops when a time limit is reached.

A solution s is encoded through three arrays *sNext*, *sSucc*, and *sVisitedBy* containing the values of *next*, *succ*, and *visitedBy* in s. The algorithm selects randomly a client i^*, and forgets the decisions that were made on the set of clients that are *close to* i^*. The term *close to* is quantitatively defined by the parameter *radium*, and the travel time matrix tt: two clients i, j are considered close if $tt_{ij} \leq radium$. The subproblem obtained is solved by the GREEDY CONSTRUCTION algorithm described in Section 3.4.2. Whenever an improving solution is found, the arrays encoding the solution must be updated. The overall algorithm is as follows:

```
algorithm: LNS(radium)
 while time is still available
     select at random a client i*
     shuffleSet := {j | tt_{i*,j} ≤ radium}
     restore initial domains for variables not in shuffleSet
     for all clients i|i ∉ shuffleSet
         visitedBy_i = sVisitedBy_i
         if sNext_i ∉ shuffleSet
             next_i = sNext_i
         for all clients j|j ∈ sSucc_i
```

```
            if j ∉ shuffleSet
                j ∈ succ_i
    z < z*
    call algorithm GREEDY CONSTRUCTION
    if an improving solution is found
        for all clients i
            sNext_i := next_i
            sSucc_i := succ_i
            sVisitedBy := visitedBy_i
        z* := z
if any improving solution was found
    return(success)
otherwise return(failure)
```

Some interesting modifications of the LNS algorithm can be done.

- The *radium* parameter can start with small values and increase whenever the greedy algorithm was unable to find improving solutions for a given number of consecutive times. Leading to a Variable Neighborhood Search, this modification will help escaping from local optimal solutions;
- Any constructive algorithm could be used to extend the frozen fragment. The use of Discrepancy-based Search algorithms could indeed increase the likelyhood of finding solutions when the problem becomes harder due to the bounding constraint $z < z^*$;
- The pseudo code presented above forgot the assignments for all clients reachable within a given time from client i^*. This forgotten part of the solution could take into account a more sophisticated distance than mere travel time. For instance, tt_{ij} could be multiplied by a weight depending on the routes, the types of goods, and the time windows of clients i and j. Clients having equal type of goods or overlapping time windows should be considered relatively *closer* than clients with different type of goods or distant time windows.

3.5.4 CP Models for the Neighborhoods

In this section a neighborhood is modeled using CP as showed in Section 2.1.2; such a neighborhood is based on ejection chains: a set of clients belonging to different routes are chosen and re-located (see Figure 13.3).

A given solution s is encoded through four arrays *sNext*, *sSucc*, *sVisitedBy*, and *sPrev*; the first three arrays contain the values of *next*, *succ*, and *visitedBy* in solution s; array *sPrev* contain the inverse of *next* (if $next_i = j$ in s, then $sNext_i = j$, $sPrev_j = i$).

The following neighborhood structure can be defined: a move from s chooses $K \leq M$ clients $i_1, i_2, \ldots, i_K$ such that all the visits are performed in different trucks in solution s. Then it removes the edges $(sPrev_{i_j}, i_j)$, $(i_j, sNext_{i_j})$, with $j = 1, \ldots, K$, and re-connects the tours by rotating the nodes $i_1, i_2, \ldots, i_K$. For all $j = 1, \ldots, K-1$, i_{j+1} replaces i_j; the rotation is closed by replacing i_K with i_1. We may either look for any improving move or for the best move in the neighborhood.

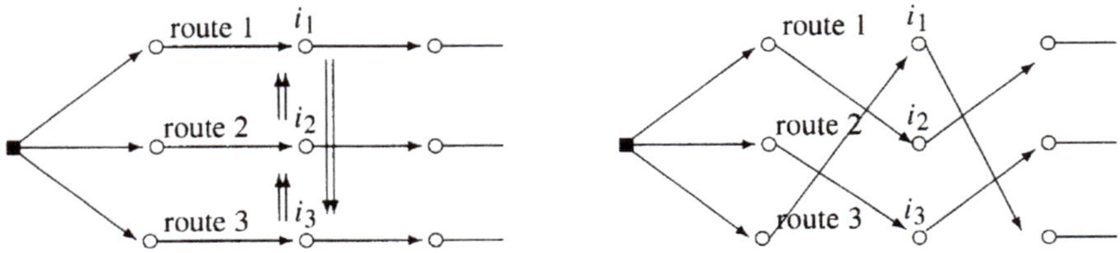

Figure 13.3. Example of ejection chain.

This neighborhood structure can be defined using CP by adding the following model to the existing dTP model:

$$\begin{aligned}
&\min && dtCost \\
&\text{on} && \\
& && dtCost < 0 \\
&\forall j \in \{1,\ldots,M\} && I_j \in [-1, 1 \ldots N] \\
& && size \in [2 \ldots M] \\
&\text{subject to} && \\
&\forall i,j \in \{1,\ldots,M\}, i<j && i \leq size \Leftrightarrow \\
& && sVisitedBy[I_i] \neq sVisitedBy[I_j] && (15)\\
&\forall j \in \{1,\ldots,M\} && j > size \Leftrightarrow I_j = -1 && (16)\\
&\forall j \in \{1,\ldots,M\}, i \in \{1,\ldots,N\} && I_j = i \Rightarrow \\
& && next[sPrev_\ell] = \ell,\ \forall \ell \mid i \in sSucc_\ell && (17)\\
&\forall j \in \{1,\ldots,M\}, i \in \{1,\ldots,N\} && I_j = i \Rightarrow \\
& && next[\ell] = sNext_\ell,\ \forall \ell \mid \ell \in sSucc_i && (18)\\
&\forall j \in \{1,\ldots,M-1\} && j < size \Rightarrow next[sPrev[I_j]] = I_{j+1} && (19)\\
&\forall j \in \{1,\ldots,M-1\} && j < size \Rightarrow next[I_{j+1}] = sNext[I_j] && (20)\\
& && next[sPrev[I_{size}]] = I_1 && (21)\\
& && next[I_1] = sNext[I_{size}] && (22)
\end{aligned}$$

$$\begin{aligned}
dtCost = {} & \sum_{i=1}^{size-1} (c[sPrev[I_i], I_{i+1}] + c[I_{i+1}, sNext[I_i]]) \\
& + c[sPrev[I_{size}], I_1] + c[I_1, sNext[I_{size}]] \\
& - \left(\sum_{i=1}^{size} (c[sPrev[I_i], I_i] + c[I_i, sNext[I_i]]) \right) \qquad (23)
\end{aligned}$$

A solution of the neighborhood model defines an ejection chain affecting *size* clients; since *size* can take any value between 2 and M, we introduce M variables representing the variable length ejection chain. The first *size* variables identify the clients changing route while the others are meaningless and are set to -1.

M domain variables $I_1, I_2, \ldots, I_M$, are used to identify the clients. The domain of variable I_j is $[-1, 1 \ldots N]$; the first *size* variables[6] will take values different from -1. Client I_2 goes in place of client I_1, client I_3 goes in place of client I_2, etc. finally client I_1 goes in place of client I_{size}. All variables I_j with $j > size$ take the value -1. The objective function of the neighborhood model is the difference between the cost of the current solution s and the cost of the tentative solution.

Constraints (15) and (16) define the neighborhood structure: *size* consecutive I_j variables identify the clients in the ejection chain, and the remaining $M - size$ are constrained to be equal to -1. The clients must be chosen from *size* different trucks (15).

Constraints (17)–(22) define the interface constraints between the dTP model and the neighborhood model. They are basically of two types: some of them are devoted to restoring of the unchanged part of the solution (17)–(18), some others are devoted to encoding the move (19)–(22). Finally, constraint (23) links the neighborhood variables to *dtCost*.

Defining the neighborhood structure via a CP model yields a stable neighborhood model that would remain valid if side constraints (such as precedence constraints, etc.) were added to dTP. All side constraints propagate from the problem variables to the neighborhood variables, removing some neighbors. For example, once variable I_1 is instantiated to a value i_1 visited by the first truck, all clients i_2 requiring a capacity exceeding the remaining capacity of the first truck are automatically removed from the domain of variable I_2 by propagation of the interface constraints.

3.6 LS during Construction

The last two sections showed how constructive algorithms from greedy to Discrepancy-based Search could yield one or several solutions, and how local moves from various neighborhoods could improve them. This section concerns the use of local moves at the very heart of the construction process. Following Caseau and Laburthe, 1999, we develop on the idea of Incremental Local Optimization (ILO, see Section 2.2.7) which applies improving local moves after each construction step.

3.6.1 Single Route Optimization

A standard move in routing problems consists in optimizing the sequence of visits for each route. Optimizing a route yields one small constrained TSP per route. This leads to a decomposition of the problem: the partition of clients by routes is kept, and the induced independent TSPs are re-solved. Each neighborhood for the TSP induces a neighborhood for dTP. A solution to dTP may, for instance, be transformed into a neighbor one by applying on any of the routes:

- a feasible 3-opt move (Reinelt, 1994);
- a feasible Lin and Kernighan move (Lin and Kernighan, 1973);
- a sequence of improving feasible 3-opt moves until no more such moves can be found;

[6] When I_j is used to index an *array*, and -1 is part of the domain of I_j, the value $(array_i)[-1]$ is conventionally considered equal to -1.

- a complete algorithm that sequences each route to optimality.

The last case is of particular interest. Indeed, many routing problems, although involving thousands of clients typically assign less than 15 clients per truck. It is thus easy to solve to optimality the constrained TSP associated to a route by branch and bound with constraint propagation. Such an approach was shown to be viable in Caseau and Laburthe (1996) and Focacci et al. (1999b).

After each insertion, the partial solution can be re-optimized in a state where each route is optimally sequenced. For the decomposition of dTP into a master partitioning sub-problem and an induced sequencing (routing) sub-problem, the sequencing sub-problem is solved to optimality. The complexity of the overall procedure remains tractable because the LS phase that is performed at each insertion step can be done much faster than a LS phase that would be performed in the end, after a complete solution has been built. Indeed, after the insertion of a node i in route k ($visitedBy_i = k$), TSP moves should be applied on route k only. Therefore, it is worthwhile resolving the TSP for route k only. This fast incremental exploration accounts for the overall efficiency of applying local optimization within the construction process.

3.6.2 Ejection Chains

The neighborhood introduced above is well suited for improving sequencing decisions among clients on the same route but it never questions the partition of clients into routes. Since the insertion algorithm inserts clients one at a time, it is likely to make assignments (of clients to routes) that seem good when only a subset of the clients is considered, but that later prove to be suboptimal when further clients are inserted. This section presents a neighborhood whose aim is to repair such partitioning decisions that turn out to be wrong along the insertion process.

Section 3.5.4 proposed a CP model for an ejection chain neighborhood. The idea of ejection chains, a standard technique for packing problems, is the following. Let $i_1, \ldots, i_r$ be a set of r clients currently assigned to different routes (say, $visitedBy_{i_1} = k_1, \ldots, visitedBy_{i_r} = k_r$) and i_0 be another client not assigned to k_1. The move changes the bin assignment into $visitedBy_{i_0} = k_1$, $visitedBy_{i_1} = k_2, \ldots,$ $visitedBy_{i_{r-1}} = k_r$ and resolves the induced sequencing problems on routes k_1 to k_r. Applied on the dTP, this move amounts to change the route assignment of a few nodes while respecting all constraints (satisfying capacity constraints with the new insertion $visitedBy_{i_j} = k_{j+1}$ is eased by the removal of i_{j+1} from k_{j+1}).

Such moves are particularly interesting in the case of promising infeasible insertions. For instance, one may foresee that the insertion $visitedBy_i = k$ would yield a very small insertion cost Δ_k, but that it is infeasible because of, say, a capacity constraint (there is no more room for client i in the route k). Instead of considering the insertion of i into routes that are geographically farther away (which may cause significant detours, thus higher insertion cost Δ), it is interesting to try to insert i into the route k by removing another client j from route k and transferring it to another route. This leads to the search for ejection chains initiated by the insertion of i in k. Such moves are particularly useful when capacity constraints on the routes are tight. Another situation where ejection chains prove useful is the case of travel optimization under a small number of trucks. In such a case, the routes must be tight in order to cover all clients. What usually happens in the insertion process is that the algorithm is not able to insert some of the clients towards the end of the construction (all tentative insertions turn out to be infeasible). Ejection chains can be used as a way to force these insertions: Caseau et al. (1999). An alternate chain of insertions and ejections is searched for such that i can be inserted into some route k_1 by transferring another client j from k_1 to k_2, and so on. Such neighborhoods can be explored through Breadth First Search in order to find the least length ejection chain (the one that involves the least number of clients), likely to be less disruptive of the current solution.

Finally, one should be careful about the fact that such ejection chain neighborhoods are very large neighborhoods and that their exploration may take significantly longer than the other neighborhoods presented so far (3-opt, single route direct optimization, etc.). It amounts to a form of LNS and should be used with care. Nevertheless, it complements very efficiently the short-sightness nature of constructive algorithms which build the solution one step at a time by repairing some of the early non-optimal route assignments.

4 CONCLUSIONS

This chapter has presented a variety of techniques for using local search methods with constraints. Hybrid combinations have been described for local search methods as well as for constructive global search methods.

Constraints can be blended with local search procedures in several ways:

- By expressing the problem as a standard problem with additional side constraints, iterating a neighborhood for the standard problem and, for each neighbor, checking feasibility for all side constraints; or by optimizing the neighborhood exploration procedure with inlined constraint checks. These methods can either be implemented with any programming language following the ideas described in this paper, or with a CP language using already defined specific data structures (such as constraints, neighborhood objects, and move operators) (De Backer et al., 2000).
- By using global search techniques for exploring the neighborhood defined by a fragment of the current solution; these methods are also described in Shaw (1998), Caseau et al. (1999) for routing problem, or in Nuijten and Le Pape (1998) for scheduling problems.
- By defining the search for the best neighbor as a constrained optimization problem on its own; originally introduced in Pesant and Gendreau (1996), Pesant and Gendreau (1999), it has been applied mainly to timetabling and routing problems in Pesant and Gendreau (1996), Pesant and Gendreau (1999), Pesant et al. (1997).

Local search techniques can be introduced within a constructive global search algorithm with the following techniques:

- Restricted Candidate Lists to filter the good branches of a choice point (see Cesta et al., 2000 for a recent application to scheduling problems);
- Discrepancy-based Search to generate near-greedy paths in a search tree;
- Incremental Local Optimization to improve the partial solutions at each step of a construction process such as a greedy insertion algorithm.

In all these cases, CP provides the user with clean formalism to combine technologies: use propagation to reduce the size of neighborhoods, use global search techniques to explore a neighborhood, control the divergence of a construction process from a greedy path, etc. In a sense, CP supports a clean engineering of many "good old tricks" from Operations Research. The first clean integration has been the expression of neighborhoods with CP models. CP models could be introduced for many other algorithm engineering purposes such as objective function combinations for multi-criteria optimization (see, e.g., Focacci et al., 2000a), or solution fragment combinations for population-based optimization methods. All these are open research topics. The definite impact of CP to Operations Research in general, and to LS in particular, is the introduction of structuring objects that provide a way to easily combine techniques.

Many languages embedding constraint propagation have been developed by research institutes, universities, and private companies; knowing that we are risking appearing unfair with respect to some of them, we believe it is important to give some pointers to few tools that demonstrated larger acceptance both in industries and academics. A practitioner desiring to experiment local search and constraint programming could take a closer look to CP tools such as CHIP (Aggoun and Beldiceanu, 1992), CHOCO (Laburthe, 2000), Eclipse (Schimpf et al., 1997), and ILOG Solver (Solver, 2000). All of them provide a constraint propagation engine, together with tree search exploration methods, and facilities to design local search methods.

Finally, it is worthwhile mentioning that recently the CP community has more and more showed interest in hybrid approaches for solving optimization problems. Some of the conferences and workshops that regularly present interesting papers on the subject are the *International Conference on Principles and Practice of Constraint Programming* (CP), the *International Workshop on Integration of Artificial Intelligence and Operations Research Techniques in CP for Combinatorial Optimization Problems* (CP-AI-OR), and the Constraint clusters at *INFORMS* conferences.

ACKNOWLEDGMENTS

We warmly thank Paul Shaw, Michela Milano, Fred Glover, Eric Bourreau, Etienne Gaudin, Cesar Rego, Gilles Pesant and Benoît Rottembourg for reading preliminary versions of this work and for many interesting discussions on the topic.

REFERENCES

Aggoun, A. and Beldiceanu, N. (1992) Extending CHIP in order to solve complex scheduling and placement problems. In: *Actes des Journees Francophones de Programmation et Logique*. Lille, France.

Applegate, D. and Cook, W. (1991) A computational study of the job-shop scheduling problem. *ORSA Journal on Computing*, **3**, 149–156.

Beldiceanu, N., Bourreau, E., Simonis, H. and Rivrau, D. (1999) Introducing metaheuristics in CHIP. In: *Proceedings of the 3rd Metaheuristics International Conference*. Angra do Reis, Brazil.

Caprara, A., Fischetti, M., Toth, P., Vigo, D. and Guida, P.-L. (1997) Algorithms for railway crew management. *Mathematical Programming*, **79**, 125–141.

Caseau, Y. and Laburthe, F. (1996) Improving branch and bound for job-shop scheduling with constraint propagation. In: M. Deza, R. Euler and Y. Manoussakis, (eds.), *Proceedings of Combinatorics and Computer Science, CCS'95, LNCS 1120*. Springer-Verlag, Berlin Heidelberg.

Caseau, Y. and Laburthe, F. (1999) Heuristics for large constrained routing problems. *Journal of Heuristics*, **5**, 281–303.

Caseau, Y., Laburthe, F. and Silverstein, G. (1999) A metaheuristic factory for vehicle routing problems. In: J. Jaffar (ed.), *Principle and Practice of Constraint Programming—CP'99, LNCS 1713*. Springer-Verlag, Berlin Heidelberg, pp. 144–158.

Cesta, A., Oddi, A. and Smith, S. (2000) A constraint-based method for project scheduling with time windows. *Journal of Heuristics* (to appear).

De Backer, B., Furnon, V., Shaw, P., Kilby, P. and Prosser, P. (2000) Solving vehicle routing problems using constraint programming and meta-heuristics. *Journal of Heuristics*, **6**, 481–500.

Dell'Amico, M. and Trubian, M. (1993) Applying tabu-search to the job-shop scheduling problem. *Annals of Operations Research*, **41**, 231–252.

Feo, T. and Resende, M. (1995) Greedy randomized adaptive search procedures. *Journal of Global Optimization*, **6**, 109–133.

Focacci, F., Laborie, P. and Nuijten, W. (2000a) Solving scheduling problems with setup times and alternative resources. In: *Proceedings of the Fifth International Conference on Artificial Intelligence Planning and Scheduling. AIPS'00*. AAAI Press.

Focacci, F., Lodi, A. and Milano, M. (1999a) Cost-based domain filtering. In: J. Jaffar (ed.), *Principle and Practice of Constraint Programming—CP'99, LNCS 1713*. Springer-Verlag, Berlin Heidelberg, pp. 189–203.

Focacci, F., Lodi, A. and Milano, M. (1999b) Solving TSP with time windows with constraints. In: D. De Schreye, (ed.), *Logic Programming—Proceedings of the 1999 International Conference on Logic Programming*. The MIT-press, Cambridge, Massachusetts, pp. 515–529.

Focacci, F., Lodi, A., Milano, M. and Vigo, D. (2000b) An introduction to constraint programming. *Ricerca Operativa*, **91**, 5–20.

Gendreau, M., Hertz, A. and Laporte, G. (1992) New insertion and postoptimization procedures for the traveling salesman problem. *Operations Research*, **40**, 1086–1094.

Glover, F. (1995) Tabu thresholding: Improved search by nonmonotonic trajectories. *ORSA Journal on Computing*, **7**, 426–442.

Golden, B. and Assad, A. (1988) *Vehicle Routing: Methods and Studies*. North-Holland, Amsterdam.

Haralick, R. and Elliott, G. (1980) Increasing tree search efficiency for constraint satisfaction problems. *Artificial Intelligence*, **14**, 263–313.

Harvey, W. (1995) *Nonsystematic Backtracking Search*. PhD thesis, Stanford University.

Harvey, W. and Ginsberg, M. (1995) Limited discrepancy search. In: *Proceedings of the 14th IJCAI*. Morgan Kaufmann, pp. 607–615.

Junker, U. (2000) Preference-based search for scheduling. In *Proceedings of the Seventeenth National Conference on Artificial Intelligence—AAAI-2000*, pp. 904–909.

Kindervater, G. and Savelsbergh, M. (1997) Vehicle routing: Handling edges exchanges. In: E. Aarts and J.K. Lenstra (eds.), *Local Search in Combinatorial Optimization*, J. Wiley & Sons, Chichester, pp. 337–360.

Laburthe, F. (2000) CHOCO: implementing a CP kernel. In: *CP'00 Post Conference Workshop on Techniques for Implementing Constraint programming Systems—TRICS*. Singapore.

Lin, S. and Kernighan, B. (1973) An effective heuristic for the traveling salesman problem. *Operations Research*, **21**, 498–516.

Mackworth, A. (1977) Consistency in networks of relations. *Artificial Intelligence*, **8**, 99–118.

Marriott, K. and Stuckey, P. (1998) *Programming with Constraints*. The MIT Press.

Mautor, T. and Michelon, P. (1997) MIMAUSA: A new hybrid method combining exact solution and local search. In: *Proceedings of the 2nd International Conference on Meta-Heuristics*. Sophia-Antipolis, France.

Michel, L. and van Hentenryck, P. (1997) Localizer: A modeling language for local search. In: G. Smolka (ed.), *Principle and Practice of Constraint Programming—CP'97, LNCS 1330*. Berlin Heidelberg, Springer-Verlag, pp. 237–251.

Minton, S., Johnston, M., Philips, A. and Laird, P. (1992) Minimizing conflicts: a heuristic repair method for constraint satisfaction and scheduling problems. *Artificial Intelligence*, **58**, 161–205.

Mladenović, N. and Hansen, P. (1997) Variable neighborhood search. *Computers & Operations Research*, **24**, 1097–1100.

Nuijten, W. (1994) *Time and Resource Constrainted Scheduling, a Constraint Satisfaction Approach*. PhD thesis, University of Eindhoven, The Netherlands.

Nuijten, W. and Le Pape, C. (1998) Constraint based job shop scheduling with ILOG scheduler. *Journal of Heuristics*, **3**, 271–286.

Pesant, G. and Gendreau, M. (1996) A view of local search in constraint programming. In: E. Freuder, (ed.), *Principle and Practice of Constraint Programming—CP'96, LNCS 1118*. Springer-Verlag, Berlin Heidelberg, pp. 353–366.

Pesant, G. and Gendreau, M. (1999) A constraint programming framework for local search methods. *Journal of Heuristics*, **5**, 255–279.

Pesant, G., Gendreau, M., Potvin, J. and Rousseau, J. (1998) An exact constraint logic programming algorithm for the travelling salesman problem with time windows. *Transportation Science*, **32**, 12–29.

Pesant, G., Gendreau, M. and Rousseau, J.-M. (1997) GENIUS-CP: A generic single-vehicle routing algorithm. In: G. Smolka (ed.), *Principle and Practice of Constraint Programming—CP'97, LNCS 1330*. Springer-Verlag, Berlin Heidelberg, pp. 420–433.

Prais, M. and Ribeiro, C. (1998) Reactive grasp: an application to a matrix decomposition problem in TDMA traffic assignment. Technical report, Catholic University of Rio de Janeiro, Department of Computer Science.

Prestwich, S. (2000) A hybrid search architecture applied to hard random 3-sat and low-autocorrelation binary sequences. In: R. Dechter (ed.), *Principle and Practice of Constraint Programming—CP2000, LNCS 1894*. Springer-Verlag, Berlin Heidelberg, pp. 337–352.

Régin, J. (1994) A filtering algorithm for constraints of difference in CSPs. In: *Proceedings of the Twelfth National Conference on Artificial Intelligence—AAAI'94*, pp. 362–367.

Reinelt, G. (1994) *The Traveling Salesman: Computational Solutions for TSP Applications*. Springer-Verlag.

Russell, R. (1995) Hybrid heuristics for the vehicle routing problem with time windows. *Transportation Science*, **29**, 156–166.

Schimpf, J., Novello, S. and Sakkout, H. (1997) *IC-Parc ECLiPSe Library Manual*.

Selman, B. and Kautz, H. (1993) Domain-independent extension to GSAT: Solving large structured satisfiability problems. In: *Proceedings of IJCAI-93, 13th International Joint Conference on Artificial Intelligence*. Sidney, AU, pp. 290–295.

Selman, B., Levesque, H. and Mitchell, D. (1992) A new method for solving hard satisfiability problems. In: P. Rosenbloom and P. Szolovits (eds.), *Proceedings of the Tenth National Conference on Artificial Intelligence*. AAAI Press, Menlo Park, California, pp. 440–446.

Shaw, P. (1998) Using constraint programming and local search methods to solve vehicle routing problems. In: M. Maher and J.-F. Puget (eds.), *Principle and Practice of Constraint Programming—CP'98, LNCS 1520*. Springer-Verlag, Berlin Heidelberg, pp. 417–431.

Shaw, P., Furnon, V. and De Backer, B. (2000) A lightweight addition to CP frameworks for improved local search. In: *Proceedings of CP-AI-OR'00*. Padderborn, Germany.

Solver (2000) *ILOG Solver 5.0 User's Manual and Reference Manual*. ILOG, S.A.

Toth, P. and Vigo, D. (2002) *The Vehicle Routing Problem*. Monographs on Discrete Mathematics and Applications. SIAM.

van Hentenryck, P., Saraswat, V. and Deville, Y. (1993) Evaluation of the constraint language cc(FD). Technical Report CS-93-02, Brown University.

Walser, J. (1999) *Integer Optimization by Local Search*, Volume 1637 of *Lecture Notes in Artificial Intelligence*. Springer Verlag.

Walsh, T. (1997). Depth-bounded discrepancy search. In: *Proceedings of the 15th International Joint Conference on Artificial Intelligence—IJCAI*. Morgan Kaufmann.

Chapter 14

CONSTRAINT SATISFACTION

Eugene C. Freuder and Mark Wallace
University College Cork and Imperial College

Abstract Many problems can be formulated in terms of satisfying a set of constraints. This chapter focuses on methods for modeling and solving such problems used in artificial intelligence and implemented in constraint programming languages.

Keywords: Constraint satisfaction, Constraint programming, Optimization, CSP

1 INTRODUCTION

Constraint satisfaction problems are ubiquitous. A simple example that we will use throughout the first half of this chapter is the following "scheduling problem": Choose "employees" A or B for each of three "tasks", X, Y, Z, subject to the "work rules" that the same employee cannot carry out both tasks X and Y, the same employee cannot carry out both tasks Y and Z, and only employee B is allowed to carry out task Z. (Many readers will recognize this as a simple "coloring problem".)

This is an example of a class of problems known as *Constraint Satisfaction Problems* or *CSPs*. CSPs consist of a set of *variables* (e.g., tasks), a *domain* of *values* (e.g., employees) for each variable, and *constraints* (e.g., work rules) among sets of variables. The constraints specify which combinations of value assignments are allowed (e.g., employee A for task X and employee B for task Y); these allowed combinations *satisfy* the constraints. A *solution* is an assignment of values to each variable such that all the constraints are satisfied [34].

CSPs can be represented as *constraint networks*, where the variables correspond to nodes and the constraints to arcs (Figure 14.1). The constraint network for our sample problem appears below. Constraints involving more than two variables can be modeled with hypergraphs, but most basic CSP concepts can be introduced with binary constraints involving two variables, and that is the route we will begin with in this chapter. We will say that a value for one variable is *consistent with* a value

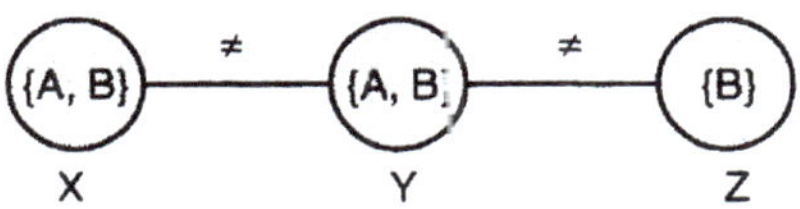

Figure 14.1. A constraint network representation of a sample constraint satisfaction problem.

for another if the pair of values satisfies the binary constraint between them. (This constraint could be the trivial constraint that allows all pairs of values; such constraints are not represented by arcs in the constraint network.) Note that specifying a domain of values for a variable can be viewed as providing a unary constraint on that single variable.

We stress that the basic CSP paradigm can be *extended* in many directions, e.g., variables can be added dynamically, domains of values can be continuous, constraints can have priorities, solutions can be optimal, not merely satisfactory.

Many application domains (e.g., design) naturally lend themselves to modeling as CSPs. Many forms of reasoning (e.g., temporal reasoning) can be viewed as constraint reasoning. Many disciplines (e.g., operations research) have been brought to bear on these problems. Many computational "architectures" (e.g., neural networks) have been utilized for these problems.

This chapter will focus on the methods developed in artificial intelligence and the approaches embodied in constraint programming languages. Of course, this brief chapter can only suggest some of the developments in these fields; it is not intended as a survey, only as an introduction. Rather than beginning with formal definitions, algorithms, and theorems, we will focus on introducing concepts through examples.

The constraint programming ideal is this: The programming is declarative; we simply state the problem as a CSP and powerful algorithms, provided by a constraint library or language, solve the problem. In practice, this ideal has, of course, been only partially realized, and expert constraint programmers are needed to refine modeling and solving methods for difficult problems.

2 INFERENCE

Inference methods make implicit constraint information explicit. Inference can reduce the effort involved in searching for solutions or even synthesize solutions without search. The most common form of inference is known as *arc consistency*. In our sample problem, we can infer that B is not a possible value for Y because there is *no* value for Z that, together with B, satisfies the constraint between Y and Z. This can be viewed as making explicit the fact that the unary constraint on the variable Y does not allow B.

This inference process can *propagate*: after deleting B from the domain of Y, there is no value remaining for Y that together with A for X will satisfy the constraint between X and Y, therefore we can delete A from the domain of X (see Figure 14.2). If we repeatedly eliminate inconsistent values in this fashion until any value for any variable is consistent with some value for all other variables, we have achieved arc consistency. Many algorithms have been developed to achieve arc consistency efficiently [4,24].

Eliminating inconsistent values by achieving arc consistency can greatly reduce the space we must search through for a solution. Arc consistency methods can also be

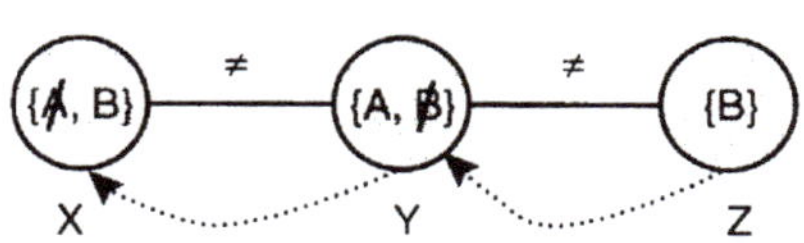

Figure 14.2. Arc consistency propagation.

interleaved with search to dynamically reduce the search space, as we shall see in the next section.

Beyond arc consistency lies a broad taxonomy of consistency methods. Many of these can be viewed as some form of *(i, j)-consistency*. A CSP is (i, j)-consistent if, given any consistent set of i values for i variables, we can find j values for any other j variables, such that the $i + j$ values together satisfy all the constraints on the $i + j$ variables. Arc consistency is (1,1)-consistency. $(k-1, 1)$-consistency, or *k-consistency*, for successive values of k constitutes an important constraint hierarchy [13].

More advanced forms of consistency processing often prove impractical either because of the processing time involved or because of the space requirements. For example, 3-consistency, otherwise known as *path consistency*, is elegant because it can be shown to ensure that given values for any two variables one can find values that satisfy all the constraints forming any given path between these variables in the constraint network. However, achieving path consistency means making implicit binary constraint information explicit, and storing this information can become too costly for large problems.

For this reason variations on *inverse consistency*, or $(1, j-1)$-consistency, which can be achieved simply by domain reductions, have attracted some interest [8]. Various forms of *learning* achieve *partial* k-consistency during search [9]. For example, if we modified our sample problem to allow only A for Z, and we tried assigning B to X and A to Y during a search for a solution to this problem, we would run into a "dead end": no value would be possible for Z. From that we could learn that the constraint between X and Y should be extended to rule out the pair (B,A), achieving partial path consistency.

Interchangeability provides another form of inference, which can also eliminate values from consideration. Suppose we modify our sample problem to add employees C and D who can carry out task X. Values C and D would be interchangeable for variable X because in any solution using one we can substitute the other. Thus we can eliminate one in our search for solutions (and if we want to, just substitute it back into any solutions we find). Just as with consistency processing there is a local form of interchangeability that can be efficiently computed. In a sense, inconsistency is an extreme form of interchangeability; all inconsistent values are interchangeable in the null set of solutions that utilize them [15].

3 MODELING

Modeling is a critical aspect of constraint satisfaction. Given a user's understanding of a problem, we must determine how to model the problem as a constraint satisfaction problem. Some models may be better suited for efficient solution than others [29].

Experienced constraint programmers may add constraints that are *redundant* in the sense that they do not change the set of solutions to the problem, in the hope that adding these constraints may still be cost effective in terms of reducing problem solving effort. Added constraints that do eliminate some, but not all, of the solutions, may also be useful, e.g., to break symmetries in the problem.

Specialized constraints can facilitate the process of modeling problems as CSPs, and associated specialized inference methods can again be cost-effective. For example, imagine that we have a problem with four tasks, two employees who can handle each, but three of these tasks must be undertaken simultaneously. This temporal constraint can be modeled by three separate binary inequality constraints between each pair of these tasks; arc consistency processing of these constraints will not eliminate any values from their domains. On the other hand an "all-different" constraint, that can apply to more than two variables at a time, not only simplifies the modeling of the problem, but an associated inference method can eliminate all the values from a variable domain, proving the problem unsolvable. Specialized constraints may be identified for specific problem domains, e.g., scheduling problems.

It has even proven useful to maintain multiple complete models for a problem "channeling" the results of constraint processing between the two [6]. As has been noted, a variety of approaches have been brought to bear on constraint satisfaction, and it may prove useful to model part of a problem as e.g., an integer programming problem. Insight is emerging into basic modeling issues, e.g., binary versus non-binary models [2].

In practice, modeling can be an iterative process. Users may discover that their original specification of the problem was incomplete or incorrect or simply impossible. The problems themselves may change over time.

4 SEARCH

In order to find solutions we generally need to conduct some form of search. One family of search algorithms attempts to build a solution by *extending* a set of consistent values for a subset of the problem variables, repeatedly adding a consistent value for one more variable, until a complete solution is reached. Another family of algorithms attempts to find a solution by *repairing* an inconsistent set of values for all the variables, repeatedly changing an inconsistent value for one variable, until a complete solution is reached. (Extension and repair techniques can also be combined.)

Often extension methods are systematic and *complete*, they will eventually try all possibilities, and thus find a solution or determine unsolvability, while often repair methods are stochastic and incomplete. The hope is that completeness can be traded off for efficiency.

4.1 Extension

The classic extension algorithm is *backtrack search*. Figure 14.3 shows a backtrack search tree representing a trace of a backtrack algorithm solving our sample problem.

A depth-first traversal of this tree corresponds to the order in which the algorithm tried to fit values into a solution. First the algorithm chose to try A for X, then A for Y. At this point it recognized that the choice of A for Y was inconsistent with the choice of A for X: it failed to satisfy the constraint between X and Y. Thus there was no need to try a choice for Z; instead the choice for Y was changed to B. But then B for Z was found to be inconsistent, and no other choice was available, so the algorithm "backed up" to look for another choice for Y. None was available so it backed up to try B for X. This could be extended to A for Y and finally to B for Z, completing the search.

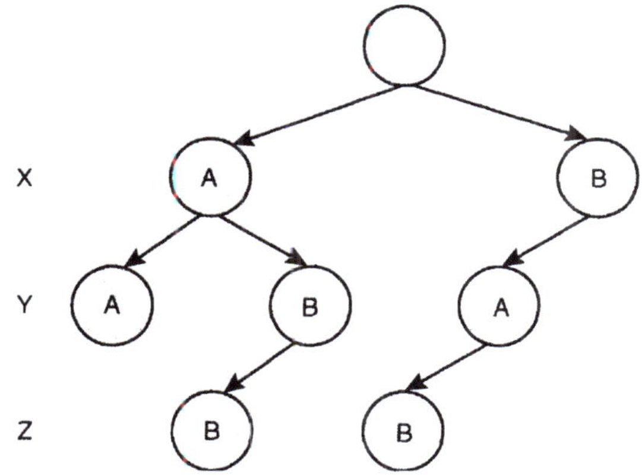

Figure 14.3. Backtrack search tree for example problem.

Backtrack search can prune away many potential combinations of values simply by recognizing when an assignment of values to a subset of the variables is already inconsistent and cannot be extended. However, backtrack search is still prone to "thrashing behaviour". A "wrong decision" early on can require an enormous amount of "backing and filling" before it is corrected. Imagine, for example, that there were 100 other variables in our example problem, and, after initially choosing A for X and B for Y, the search algorithm tried assigning consistent values to each of those 100 variables before looking at Z. When it proved impossible to find a consistent value for Z (assuming the search was able to get that far successfully) the algorithm would begin trying different combinations of values for all those 100 variables, all in vain.

A variety of modifications to backtrack search address this problem [22]. They all come with their own overhead, but the search effort savings can make the overhead worthwhile.

Heuristics can guide the search order. For example, the "minimal domain size" heuristic suggests that as we attempt to extend a partial solution we consider the variables in order of increasing domain size; the motivation there is that we are more likely to fail with fewer values to choose from, and it is better to encounter failure higher in the search tree than lower down when it can induce more thrashing behaviour. Using this heuristic in our example we would have first chosen B for Z, then proceeded to a solution without having to back up to a prior level in the search tree. While "fail first" makes sense for the order in which to consider the variables, "succeed first" makes sense for the order in which to try the values for the variables.

Various forms of inference can be used prospectively to prune the search space. For example, search choices can be interleaved with arc consistency maintenance. In our example, if we tried to restore arc consistency after choosing A for X, we would eliminate B from the domain of Z, leaving it empty. At this point we would know that A for X was doomed to failure and could immediately move on to B. Even when failure is not immediate, "look ahead" methods that infer implications of search choices can prune the remaining search space. Furthermore, "dynamic" search order heuristics can be informed by this pruning, e.g., the minimal domain size heuristic can be based on the size of the domains after look-ahead pruning. Maintaining arc consistency is an extremely effective and widely used technique [30].

Memory can direct various forms of "intelligent backtracking" [10]. For example, suppose in our example for some reason our search heuristics directed us to start the search by choosing B for Y followed by A for X. Of course B the only choice for Z would then fail. Basic backtrack search would back up "chronologically" to then try B for X. However, if the algorithm "remembers" that failure to find a value for Z was based solely on conflict with the choice for Y, it can "jump back" to try the alternative value A at the Y level in the search tree without unnecessarily trying B for X. The benefits of maintaining arc consistency overlap with those of intelligent backtracking, and the former may make the latter unnecessary.

Search can also be reorganized to try alternatives in a top-down as opposed to bottom-up manner. This responds to the observation that heuristic choices made early in the extension process, when the remaining search space is unconstrained by the implications of many previous choices, may be most prone to failure. For example, "limited discrepancy search" iteratively restarts the search process increasing the number of "discrepancies", or deviations from heuristic advice, that are allowed, until a solution is found [19]. (The search effort at the final discrepancy level dominates the upper bound complexity computation, so the redundant search effort is not as significant as it might seem.)

Extensional methods can be used in an incomplete manner. As a simple example, "random restart", starting the search over as soon as a dead end is reached, with a stochastic element to the search order, can be surprisingly successful [18].

4.2 Repair

Repair methods start with a complete assignment of values to variables, and work by changing the value assigned to a variable in order to improve the solution. Each such change is called a *move*, and the new assignment is termed a *neighbour* of the previous assignment. Genetic algorithms, which create a new assignment by combining two previous assignments, rather than by moving to a neighbour of a single assignment, can be viewed as a form of repair.

Repair methods utilize a variety of metaphors, physical (hill climbing, simulated annealing) and biological (neural networks, genetic algorithms). For example, we might start a search on our example problem by choosing value A for each variable. Then, seeking to "hill climb" in the search space to an assignment with fewer inconsistencies, we might choose to change the value of Y to B; and we would be done. Hill climbing, is a repair-based algorithm in which each move is required to yield a neighbour with a better cost than before. It cannot, in general, guarantee to produce an optimal solution at the point where the algorithm stops because no neighbour has a better cost than the current assignment.

Repair methods can also use heuristics to guide the search. For example, the *min-conflicts* heuristic suggests finding an inconsistent value and then changing it to the alternative value that minimizes the amount of inconsistency remaining [26].

The classic repair process risks getting "stuck" at a "local maximum", where complete consistency has not been achieved, but any single change will only increase inconsistency, or "cycling" through the same set of inconsistent assignments. There are many schemes to cope. A stochastic element can be helpful. When an algorithm has to choose between equally desirable alternatives it may do so randomly. When no good alternative exists it may start over, or "jump" to a new starting point. Simulated annealing allows moves to neighbours with a worse cost with a given probability. Memory can also be utilized to guide the search and avoid cycling (tabu search).

We do not emphasize repair-based approaches here because we expect this will be the greatest area of overlap with other chapters in this volume.

5 TRACTABILITY

CSPs are in general NP-hard. Analytical and experimental progress has been made in characterizing tractable and intractable problems. The results have been used to inform algorithmic and heuristic methods.

5.1 Theory

Tractable classes of CSPs have been identified based on the structure of the constraint network, e.g., tree structure, and on closure properties of sets of constraints [20], e.g., max-closed. Tractability has been associated with sets of constraints defining a specific class of problems, e.g., temporal reasoning problems defined by "simple temporal networks" [12].

If a constraint network is tree-structured, there will be a *width-one* ordering for the variables in which each variable is directly constrained by at most one variable earlier in the ordering. In our sample problem, which has a trivial tree structure, the ordering X, Y, Z is width-one: Y is constrained by X and Z by Y; the ordering X, Z, Y is not width-one: Y is constrained by both X and Z. If we achieve arc consistency and use a width-one ordering as the order in which we consider variables when trying to extend a partial solution, backtrack search will in fact be *backtrack-free*: for each variable we will be able to find a consistent value without backing up to reconsider a previously instantiated variable [14].

Max-closure means that if (a b) and (c d) both satisfy the constraint, then ((max (a c)), (max (b d))) will also satisfy the constraint. If all the constraints in a problem are max-closed, the problem will be tractable. The "less than" constraint is max-closed, e.g., $4 < 6, 2 < 9$ and $4 < 9$. Thus if we replaced the inequality constraints in our sample problem by less-than constraints, we would ensure tractability, even if we gave up the tree structure by adding a constraint between X and Z. In fact, there is a close relationship between the tractability of simple temporal networks and max closure.

5.2 Experiment

Intuitively it seems natural that many random CSPs would be relatively easy: loosely constrained problems would be easy to solve, highly constrained problems would be easy to prove unsolvable. What is more surprising is the experimental evidence that as we vary the constrainedness of the problems there is a sharp *phase transition* between solvable and unsolvable regions, which corresponds to a sharp spike of "really hard" problems [7]. ("Phase transition" is a metaphor for physical transitions, such as the one between water and ice.)

6 OPTIMIZATION

6.1 Modeling

6.1.1 Real-world Problems are Optimization Problems

Real-world combinatorial problems, as opposed to puzzles and benchmarks, must be solved—otherwise the organisations who face them would collapse! Accordingly the real-world requirement is rarely to find one of a small set of "ideal" solutions to a problem, but instead to find the best possible solution in the situation at hand. For example employee rostering, in the real world, involves sick leave and absenteeism as well as peaks in demand. Flexibility can be gained, at an extra cost, through overtime working. Thus the real-world rostering problem is an optimization problem, where the objective is to minimise extra costs such as overtime pay.

6.1.2 Constrained Optimization Problems (COPs)

Constrained optimization problems (COPs) are an extension of constraint satisfaction problems. The added feature is a cost variable that associates a numeric cost with each solution. The value of the cost variable is related to the values of the other decision variables by constraints.

An optimal solution to a COP is a feasible assignment to the decision and cost variables with optimal cost. The cost is optimal if no other feasible solution has a lower cost. In real-world practice, truly optimal solutions are rarely found. However whilst any solution to the underlying CSP is a solution to the COP, solutions with a lower cost are preferable.[1]

6.1.3 Mapping CSPs to COPs—Hard and Soft Constraints

It is often useful to weaken a CSP, if it is hard to solve, by modifying one or more constraints so that they each include a local cost variable. The modified constraint has local cost zero if, under the same assignment, the original constraint would have been satisfied; otherwise the local cost takes a value which represents the degree to which the original constraint would have been violated. As an example, the use of overtime in an employee rostering problem can be viewed as a weakening of the constraint that an employee can only be on duty during his/her normal working hours. In this way a CSP is transformed into a COP. The cost variable of the COP is constrained to be a function of the local cost variables associated with the individual constraints.

The constraints in a CSP are termed "hard" as they must be satisfied by any feasible solution to the problem. In the case of a COP, any constraint that has been weakened by adding a cost variable in the way described above, is termed "soft".

If the COP cost is the sum of the local costs, then finding the optimal solution of the COP also solves the CSP. Specifically, the optimal solution to the COP is zero if and only if the original CSP is satisfiable.

If the local cost of the weakened constraint is exactly one, whenever the original hard constraint is violated, then the COP cost is the number of constraints violated.

---

[1] Naturally there are maximisation COPs where higher cost (or "value") solutions are preferable, rather than lower. These can be mapped to a minimisation problem by simply negating the cost value in every constraint where it appears.

Solving a CSP by optimising this COP is a powerful technique introduced by Freuder [16] and Minton [26].

6.1.4 *Handling Projection in COPs*

Lastly we consider CSP's whose "solution variables" are a strict subset of the variables appearing in the CSP. Naturally this can be mapped to a COP with the same solution variables, plus the cost variable. However in this case a solution—which is an assignment to the solution variables—may not have a uniquely defined cost: the cost may also depend upon the assignment of the remaining variables. For such problems it is necessary to introduce new cost variables: one *global cost* associated with a complete assignment to all the variables in the problem, and another *projected cost* associated with the projection onto the solution variables.

The projected cost (of an assignment to the solution variables) must then be defined in terms of the global cost of each complete assignment which extends it. A standard definition is to make the projected cost the minimum of the global costs. However it is not possible to simply add a binary constraint which relates the projected cost to the global cost, because the constraint involves different values of the global cost under different assignments. Thus our mapping of CSPs to COPs cannot be extended to handle projection.

Instead we simply define two operations on costs: one which defines the global cost in terms of the local costs, and one which defines the projected cost in terms of the global cost. Assuming these operations have certain properties, they can be used to associate a unique cost with any assignment to any subset of the problem variables. These properties are satisfied if the operations on cost constitute a semi-ring [5].

6.2 Algorithms

6.2.1 *Depth-First Branch and Bound*

To find the optimal solution to a COP it suffices to find all solutions and pick (one of) the best. A more efficient way is to impose a new constraint, after each solution is found, that precludes finding any other solutions with a cost worse than that of the last solution found. This is termed depth-first "Branch and Bound".

After finding a solution, during depth-first branch and bound, the search can

1. either restart from the beginning, with the extra constraint [31], or
2. backtrack from the current leaf of the search tree, as if no solution had been found, and continue the search from there, with the new constraint imposed.

Surprisingly the first option often, though not always, yields a smaller search tree. It is also cleaner to implement.

Typically the rate at which new solutions are found, during both versions of depth-first branch and bound, is very unstable. A few solutions may be found quickly, and then there may be a long search before another, better, solution is found. Subsequently several more solutions may be found quite quickly again. It quite often happens that the "proof of optimality", when the search continues looking for a solution better than the optimal one and finally fails, is shorter than one of the intermediate searches between one solution and the next.

For this reason it is often very effective to use parallelism to speed up depth-first branch and bound. As soon as one of the processors has found a solution, all the other processors are notified and the new constraint is immediately applied to all the parallel searches. The speed up due to having n processors solving a branch and bound problem in parallel is often better than n. This kind of speedup is termed *super-linear* [38].

There are countless ways of implementing depth-first branch and bound search. For example the maximum time, or backtracks, between solutions may be limited to a fixed upper bound. After hitting this limit the current best solution may simply be returned, or another search may be started with a different cost bound.

One important complete search technique is to perform a "binary chop" on the cost bound. Initially upper and lower bounds on the cost are extracted by some independent method. The first constraint on the cost in the branch and bound search is that the cost should be better than the mid-point between the two bounds. If the search finds a solution (whose cost becomes the new upper bound *UB*), a new point is selected halfway between *UB* and the cost lower bound. On the other hand, if the search fails – proving there is no solution with cost better than the mid-point (which becomes the new lower bound *LB*), then a new point is selected halfway between the upper bound and *LB*. The search is now restarted with the cost constrained to be better than the new point. Clearly as search continues the upper and lower bounds on the cost are quickly narrowed. To solve the drawback that failed searches are less useful in practise than successful ones, the point chosen at each stage may be chosen to be nearer the previous upper bound, for example only one third of the way from the upper to the lower bound.

6.2.2 Best-First Branch and Bound

A limitation of depth-first branch and bound is that search is not necessarily directed towards high quality solutions, especially at early stages in the process. As a consequence a great deal of time and search effort may be spent in finding feasible solutions, which happen to have a very high cost.

If the cost variable is linked to the problem variables via constraints, constraint propagation can tighten the bounds on the cost variable during search. Indeed some global constraints have been expressly designed to extract as precise information as possible about the cost bounds [17]. With this information about cost it is possible, at an early stage, to select branches in the search tree, which, if they yield solutions, the solutions should be low cost.

In solving CSPs, in case there are no feasible solutions, the choice of which value to use first in labelling a variable has no effect on the size of the search tree. In solving COPs, however, almost all problems have solutions, and the choice of value can be critical in finding an optimal solution quickly. Finding a good solution first, rather than a bad one, may dramatically reduce the time and search effort needed to find an optimal solution and prove optimality.

One approach, which is used widely in the Operations Research community, is *best-first* branch and bound. Instead of making a choice at each search step which adds a branch below the current node of the search tree, best-first branch and bound examines *all* the nodes of the tree which have unopened branches. The node selected is one for which the cost lower bound is minimal. Naturally the first solution found by this technique is the one with minimal cost (since none of the other nodes could be extended to a solution with a smaller cost). This algorithm is known in Artificial Intelligence circles as A*.

There are, of course many variations on best-first search. Typically Operations Researchers start with depth-first search, until a solution is found, and only then revert to best-first search. Moreover the choice of node may use a cost estimate which is not a lower bound on cost. In this case it is no longer guaranteed that the first solution found by best-first search is indeed the best. But finding a non-optimal solution first, may still yield a faster proof of optimality overall

6.2.3 Search Control

Most real applications cannot be solved to optimality because they are simply too large and complex. In such cases it is crucial that the algorithm is directed towards areas of the search space where low-cost feasible solutions are most likely to be found.

Constraint propagation can yield very useful information, which can be used to guide the search for a solution. Not only can propagation exclude impossible choices "a priori", but it can also yield information about which choices would be optimal in the absence of certain (awkward) constraints.

Because constraint propagation occurs after each search step, the resulting information is used to support *dynamic* heuristics, where the next choice is contingent upon all the information about the problem gathered during the current search.

Constraint technology has proven that incomplete extension search techniques can produce high quality solutions to large complex industrial applications in areas such as transportation and logistics. The advantage, as against repair-based techniques, is that the solutions respect all the hard constraints and are therefore applicable in practice.

For example, in limited discrepancy search, while in theory the number of discrepancies can be allowed to grow until the complete search space has been covered, in practice the number of discrepancies is kept below a small limit, and thus it is typically used as an incomplete search technique. Specialised search control, such as this, is very important in obtaining high quality solutions within reasonable, or if necessary short, timescales.

6.2.4 Repair-Based Algorithms

Repair-based algorithms naturally extend to optimization problems. When *all* the constraints are soft, any complete assignment is a feasible solution to the problem. Basic hill climbing, for example, in this context would seek to move to a neighbour with a better cost than before. However, this cannot guarantee an optimal solution. We expect that repair-based optimization methods will be discussed at length elsewhere in this volume.

Constraint programmers use repair-based algorithms, in particular in conjunction with constraint reasoning. Some ways of combining constraint reasoning with repair-based algorithms are given in the Advanced Topics section below.

7 APPLICATIONS

7.1 Current Areas of Application

Constraint programming is based on logic. Consequently any formal specification of an industrial problem can be directly expressed in a constraint program. The drawbacks of earlier declarative programming paradigms have been

- that the programmer had to encode the problem in a way that was efficient to execute on a computer
- that the end user of the application could not understand the formal specification.

The first breakthrough of constraint programming has been to separate the logical representation of the problem form the efficient encoding in the underlying constraint solvers. This separation of logic from implementation has opened up a range of applications in the area of control, verification and validation.

The second breakthrough of constraint programming has been in the area of software engineering. The constraint paradigm has proven to accommodate a wide variety of problem solving techniques, and has enabled them to be combined into hybrid techniques and algorithms, suited to whatever problem is being tackled.

As important as the algorithms to the success of constraint technology, has been the facility to link models and solutions to a graphical user interface that makes sense to the end user. Having developers display the solutions in a form intelligible to the end users, forces the developers to put themselves into the shoes of the users.

Moreover not only are the final solutions displayed to the user: it is also possible to display intermediate solutions found during search, or even partial solutions. The ability to animate the search in a way that is intelligible to the end user means the users can put themselves into the shoes of the developers. In this way the crucial relationship and understanding between developers and end users is supported and users feel themselves involved in the development of the software that will support them in the future.

As a consequence, constraint technology has been applied very successfully in a range of combinatorial problem solving applications, extending those traditionally tackled using operations research.

The two main application areas of constraint programming are, therefore:

1. control, verification, and validation,
2. combinatorial problem solving.

7.2 Applications in Control, Verification and Validation

Engineering relies increasingly on software, not only at the design stage, but also during operation. Consider the humble photocopier. Photocopiers aren't so humble as they used to be—each system comprises a multitude of components, such as feeders, sorters, staplers and so on. The next generation of photocopiers will have orders of magnitude more components than now. The challenge of maintaining compatibility between the different components, and different versions of the components has become unmanageable.

Xerox has turned to constraint technology to specify the behaviour of the different components in terms of constraints. If a set of components are to be combined in a system, constraint technology is applied to determine whether the components will function correctly and coherently. The facility to specify behaviour in terms of constraints has enabled engineers at Xerox not only to simulate complex systems in software but also to revise their specifications before constructing anything and achieve compatibility first time.

Control software has traditionally been expressed in terms of finite state machines. Proofs of safety and reachability are necessary to ensure that the system only moves between safe states (e.g., the lift never moves while the door is open) and that required states are reached (the lift eventually answers every request). Siemens has applied constraint technology to validate control software, using techniques such as boolean unification to detect any errors. Similar techniques are also used by Siemens to verify integrated circuits.

Constraint technology is also used to prove properties of software. For example abstract interpretation benefits from constraint technology in achieving the performance necessary to extract precise information about concrete program behaviour.

Finally constraints are being used not only to verify software but to monitor and restrict its behaviour at runtime. *Guardian Agents* ensure that complex software, in medical applications for example, never behaves in a way that contravenes the certain safety and correctness requirements.

For applications in control, validation and verification, the role of constraints is to model properties of complex systems in terms of logic, and then to prove theorems about the systems. The main constraint reasoning used in this area is propositional theorem proving. For many applications, even temporal properties are represented in a form such that they can be proved using propositional satisfiability.

Nevertheless the direct application of abstract interpretation to concurrent constraint programs offers another way to prove properties of complex dynamic systems.

7.3 Combinatorial Problem Solving

Commercially constraint technology has made a huge impact in problem solving areas such as

- transportation
- logistics
- network optimization
- scheduling and timetabling
- production control
- design

and it is also showing tremendous potential in new application areas such as bio-informatics and virtual reality systems.

Starting with applications to transportation, constraint technology is used by airline, bus and railway companies, all over the world. Applications include timetabling, fleet scheduling, crew scheduling and rostering, stand, slot and platform allocation.

Constraints have been applied in the logistics area for parcel delivery, food, chilled goods, and even nuclear waste. As in other application areas, the major IT system suppliers (such as SAP and I2) are increasingly adopting constraint technology.

Constraints have been applied for Internet service planning and scheduling, for minimising traffic in banking networks, and for optimization and control of distribution and maintenance in water and gas pipe networks. Constraints are used for network planning (bandwidth, routing, peering points), optimising network flow and pumping energy (for gas and water), and assessing user requirements.

Constraint technology appears to have established itself as the technology of choice in the areas of short-term scheduling, timetabling and rostering. The flexibility and scalability of constraints was proven tested in the European market (e.g., at Dassault and Monsanto), but is now used worldwide.

It has been used for timetabling activities in schools and universities, for rostering staff at hospitals, call centres, banks and even radio stations. An interesting and successful application is the scheduling of satellite operations.

The chemical industry has an enormous range of complex production processes whose scheduling and control is a major challenge, currently being tackled with constraints. Oil refineries, and steel plants also use constraints in controlling their production processes. Indeed many applications of constraints to production scheduling, also include production monitoring and control.

The majority of commercial applications of constraint technology have, to date, used finite domain propagation. Finite domains are a very natural way to represent the set of machines that can carry out a task, the set of vehicles that can perform a delivery, or the set of rooms/stands/platforms where an activity can be carried out. Making a choice for one tasks, precludes the use of the same resource for any other task which overlaps with it, and propagation captures this easily and efficiently.

Naturally most applications involve many groups of tasks and resources with possibly complex constraints on their availability (e.g., personnel regulations may require that staff have two weekends off in three, that they must have a day off after each sequence of night-shifts, and that they must not work more than 40 hours per week). For complex constraints like this a number of special constraints have been introduced which not only enable these constraints to be expressed quite naturally, but also associate highly efficient specialised forms of finite domain propagation with each constraint.

7.4 Other Applications

- *Constraints and Graphics.* An early use of constraints was for building graphical user interfaces. Now these interfaces are highly efficient and scalable, allowing a diagram to be specified in terms of constraints so that it still carries the same impact and meaning whatever the size or shape of the display hardware. The importance of this functionality in the context of the Internet, and Mobile Computing, is very clear, and constrained-based graphics is likely to make a major impact in the near future. Constraints are also used in design, involving both spatial constraints and, in the case of real-time systems design, temporal constraints.
- *Constraint Databases.* Constraint databases have not yet made a commercial impact, but it is a good bet that future information systems will store constraints as well as data values. The first envisaged application of constraint databases is to geographical information systems. Environmental monitoring will follow, and

subsequently design databases supporting both the design and maintenance of complex artifacts such as airplanes.

8 CONSTRAINT LANGUAGES

8.1 Constraint Logic Programming

The earliest constraint programming languages, such as *Ref-Arf* and *Alice*, were specialised to a particular class of algorithms. The first general purpose constraint programming languages were constraint handling systems embedded in logic programming [21,36], called *Constraint Logic Programming* (CLP). Examples are CLP(fd), HAL, SICStus and ECLiPSe. Certainly logic programming is an ideal host programming paradigm for constraints, and constraint logic programming systems are widely used in industry and academia.

Logic programming is based on relations. In fact every procedure in a logic program can be read as a relation. However the definition of a constraint is exactly the same thing—a *relation*. Consequently the extension of logic programming to CLP is entirely natural. Logic programming also has backtrack search built-in, and this is easily modified to accommodate constraint propagation. CLP has been enhanced with some high-level control primitives, allowing active constraint behaviours to be expressed with simplicity and flexibility. The direct representation of the application in terms of constraints, together with the high-level control, results in short simple programs. Since it is easy to change the model and, separately, the behaviour of a program, the paradigm supports experimentation with problem solving methods. In the context of a rapid application methodology, it even supports experimentation with the problem (model) itself.

8.2 Languages for Search

One drawback of the logical basis is that repair-based search methods have not fitted naturally into the CLP paradigm. Recently a language has been introduced called Localizer [29] which is designed specifically to support the encoding of repair-based search algorithms such as Simulated Annealing and GSAT [33]. The fundamental contribution of Localizer is the concept of an "invariant", which is a constraint that retains information used during search. For GSAT, by way of example, an invariant is used to record, for each problem variable, the change in the number of satisfied propositions if the variable's value were to be changed. The invariant is specified as a constraint, but maintained by an efficient incremental algorithm. Another constraint-based language for specifying search is SALSA [23].

8.3 Modeling Languages

On the other hand, Operations Researchers have introduced a wide range of highly sophisticated specialised algorithms for different classes of problems. For many OR researchers CLP and Localizer are too powerful—they seek a modeling language rather than a computer programming language in which to encode their problems. Traditional mathematical modeling languages used by OR researchers have offered little control over the search and the constraint propagation. OPL [35] is an extension of such a modeling language to give more control to the algorithm developer. It represents a step towards a full constraint programming language.

By contrast a number of application development environments (e.g., Visual CHIP) have appeared recently that allow the developer to define and apply constraints graphically, rather than by writing a program. This represents a step in the other direction!

8.4 Global Constraints

One of the most important requirements of a programming system is support for reusability. Many complex models developed by Operations Researchers have made very little practical impact, because they are so hard to reuse. The concept of a global constraint is inherently quite simple. It is a constraint that captures a class of subproblems, with any number of variables. Global constraints have built-in algorithms, which are specialised for treating the problem class. Any new algorithm can be easily captured as a global constraint and reused. Global constraints have had a major impact, and are used widely and often as a tool in solving complex real-world problems. They are, arguably, the most important contribution that constraint programming can bring to Operations Research.

9 ADVANCED TOPICS

There are many topics that could be addressed in additional detail. This section briefly samples a few of these. Good starting points for further study include: the Constraint Programming conferences, the *Constraints* journal, and the Constraints Archive (http://www.4c.ucc.ie/archive).

9.1 Combining Constraint Reasoning with Repair-Based Algorithms

Constraint programmers use repair-based algorithms, in conjunction with constraint reasoning. One direct approach is to "compile" constraints into a local move operator, and then apply standard repair-based search [3]. A second approach is to apply repair-based search to a subset of the problem variables and to use extension search to complete each assignment and extract the cost. A third approach, which is now widely used, is to make large moves using extension search. For example in solving a vehicle routing problem a move may involve changing a delivery from one vehicle to another, and then replanning the routes of the affected vehicles [32]. Expanding the size of the neighbourhood is also used as a way of breaking out of local optima. Finding the best neighbour in a large neighbourhood is nicely handled using extension search [28].

9.2 Combining Constraint Propagation with Linear Programming

Many industrial applications involve several interdependent subproblems, such as resource allocation and temporal scheduling, or routing and rostering.

Traditionally the subproblems have been solved one at a time, by finding a good solution to one subproblem and using this solution to constrain the solution of the next. With the success of constraint technology comes the demand for a more global view of the problem.

The consequence for the technology is a major emphasis on techniques for combining different algorithms so as to elicit a global optimum for such many-faceted problems. An important example of such a hybrid is the combination of

- linear programming, and
- constraint propagation.

Linear programming is used to generate a solution to a relaxed problem (as it is in mixed integer programming), and this guides the next choice. After making a choice, constraint propagation, often involving global constraints with specialised constraint propagation algorithms, tightens the bounds on variables and detects inconsistencies.

9.3 Knowledge Acquisition and Explanation

As constraint programming becomes increasingly commercialized, increasing attention is drawn to "human factors". Issues faced by earlier "knowledge-engineering" technologies must be faced by constraint technology.

Acquiring domain-specific knowledge is obviously a key application issue. Provision needs to be made for interactive acquisition, e.g., in electronic commerce applications. Many problems, e.g., many configuration problems, change over time. While constraint programmers tout the advantages of their declarative paradigm for maintaining programs in the face of such change, acquiring and implementing new knowledge on a large scale still presents challenges.

Users may feel more comfortable when an "explanation" can accompany a solution. Explanation is particularly important when a problem is unsolvable. The user wants to know why, and can use advice on modifying the problem to permit a solution [1].

A related set of problems confronts the need constraint programmers have to better understand the solution process. Explanation and visualization of this process can assist in debugging constraint programs, computing solutions more quickly, and finding solutions closer to optimal [11].

9.4 Synthesizing Models and Algorithms

Ideally people with constraints to satisfy or optimize would simply state their problems, in a form congenial to the problem domain, and from this statement a representation suited to efficient processing and an appropriate algorithm to do the processing would be synthesized automatically. In practice, considerable human expertise is often needed to perform this synthesis. Less ambitiously, tools might be provided to assist the constraint programmer in this regard. Initial experiments with simple learning methods have proven surprisingly effective at producing efficient algorithms for specific problem classes [25].

9.5 Distributed Processing

Distributed constraint processing arises in many contexts. There are parallel algorithms for constraint satisfaction and concurrent constraint programming languages. There are applications where the problem itself is distributed in some manner. There are computing architectures that are naturally "distributed", e.g., neural networks.

There is considerable interest in the synergy between constraint processing and software agents. Agents have issues that are naturally viewed in constraint-based terms, e.g., negotiation. Agents can be used to solve constraint satisfaction problems [42].

9.6 Uncertainty

Real world problems may contain elements of uncertainty. Data may be problematic. The future may not be known. For example, decisions about fuel purchases may need to be made based on uncertain demand dependent on future weather patterns. We want to model and compute with constraints in the presence of such uncertainty [40].

10 PRACTICAL GUIDELINES

We consider a simple one-machine scheduling problem. The requirement is to schedule a set of tasks on a machine. Each task has a fixed duration, and each has an earliest start time (the "release date") and a latest end time (the "due date"). How should we schedule the tasks so as to finish soonest?

In constraint programming a problem is handled in three stages:

1. Initialise the problem variables;
2. Constrain the variables;
3. Search for values for the variables that satisfy the constraints.

10.1 Initialising Variables

For the one-machine scheduling problem, a variable is declared to represent the start time of each task. For each task t we introduce a variable St and we declare its lower bound (the release date) and its upper bound (the due date minus the duration).

For the first model of the problem we impose no further constraints on the start time, until search begins.

10.2 Search and Propagation

When the search begins one of the tasks, $t1$ with duration $d1$ is chosen to be first. Now the following constraints are posted:

The start time S_{t1} is constrained to take its lower bound value (in this case, its release date).

The start times of each of the remaining tasks are constrained to be greater than $S_{t1} + d1$.

As a result, the lower bounds of some or all of the remaining task start times may be increased.

After having "propagated" the new constraints, by computing the new lower bounds for all the start times, search resumes. Another task is chosen to be the first of the remaining tasks, and constraints 1 and 2 are posted as before.

10.3 Branch and Bound

When a solution is found the end time of the last task, *end* is recorded. The problem solving process is restarted, but now all tasks are constrained to end before *end*. This is captured by a constraint on each task ti that $S_{ti} + di$ is less than *end*.

If at any time the propagation on a start time makes its lower bound larger than its upper bound, then there is no solution which extends the current task ordering. The system therefore backtracks and chooses a different task to be first at the previous choice point.

10.4 Introducing Redundant Constraints

The first way to enhance this algorithm is by adding a global constraint, specialised for scheduling problems. The new constraint does not remove any solutions: it is logically redundant. However its powerful propagation behaviour enables parts of the search space, where no solutions lie, to be pruned. Consequently the number of search steps is reduced—dramatically for larger problems! This global constraint is called the *edge_finder* constraint. It constrains a whole set of tasks, with start times and durations, not to overlap. A specific algorithm is associated with the global constraint. Whenever the lower (or upper) bound of any start time is increased (respectively decreased), the algorithm propagates the effects of this change to the other start times, tightening their lower and upper bounds in turn. The algorithm was devised by operations researchers, but it has been encapsulated by constraint programmers as a single constraint.

10.5 Adding Search Heuristics

The next enhancement is to choose at each search step, first the task with the earliest due date. Whilst this does tend to yield feasible solutions, it does not necessarily produce good solutions, until the end time constraints become tight. Moreover, once an optimal end time has been found, the proof of optimality is no quicker than by choosing the first task at random!

10.6 Using Tentative Assignments for Better Search Heuristics

A more radical change is to focus on bottlenecks. After each search step, the task start times are tentatively set to their lower bounds, a propagation algorithm then elicits the biggest bottleneck, which is the start time at which the number of tasks which would be running at the same time is maximal. This tentative assignment is then dropped, but the information about bottlenecks is used as a heuristic for controlling the search.

Instead of choosing which task to place first at each search step, the algorithm simply orders two tasks, constraining one to end before the other starts. At each search step two tasks are chosen from the tightest bottleneck, and constrained not to overlap, so as to reduce the bottleneck. After making such a decision, propagation takes place as before. At an intermediate point during the search many tasks will have been ordered and propagation which narrows the bounds of one task may enable further propagation to take place narrowing the bounds of another task, sometimes yielding long "propagation sequences".

This search heuristic not only produces good solutions sooner than the previous one, but it also shortens the proof of optimality.

10.7 Using an Incomplete Search Technique

For very large problems, complete search may not be possible. In this case the backtrack algorithm may be controlled so as to limit the effort wasted in exploring unpromising parts of the search space. This can be done simply by limiting the number of times a non-preferred ordering of tasks is imposed during search and backtracking, using the LDS algorithm introduced earlier.

The above techniques combine very easily, and the combination is very powerful indeed. As a result constraint programming is currently the technology of choice for operational scheduling problems where task orderings are significant.

10.8 Code

A Constraint Logic Program for solving the one-machine scheduling problem is as follows. This code, written in ECLiPSe [39], implements the naïve algorithm:

```
% Define a data structure to hold info. about jobs
:- local struct(job(start,duration,end,release,due)).

% Load the finite domain solver
:- lib(fd).

% To solve a problem, first state the constraints and
% then encode the search procedure.
% Names starting with upper-case letters are variables.
job_schedule(Jobs,FinalEndTime) :-
        constrain(Jobs,FinalEndTime),
        minimize(label_starts(Jobs),FinalEndTime).

% Constrain the start and end time of each job
constrain(Jobs,FinalEndTime) :-
        ( foreach(Job,Jobs),
         param(FinalEndTime)
        do
% Each Job variable holds a data structure with the job
% details
            Job = job with [release:ReleaseTime,
                            due:DueTime,
                            duration:Duration,
                            start:StartTime,
                            end:EndTime
                           ],
% Constrain the start and end times
            EndTime #= StartTime + Duration,
            StartTime #>= ReleaseTime,
            EndTime #=< DueTime,
% Constrain the final end time to follow the end time of
% each job
            FinalEndTime #>= EndTime
        ).
% Stop when there are no more jobs to handle
label_starts([]).
% Select a job, make it start as early as possible
% and constrain the remaining jobs to start after it
```

```
label_starts(Jobs):-

        choose(Job,Jobs,Rest),
        Job = job with [start:StartTime,end:EndTime],
        fix_to_min(StartTime),
        propagate(EndTime,Rest),
        label_starts(Rest).

% Select any job from a non-empty list
choose(Job,[Job|Jobs],Jobs).
choose(Job,[NotThisJob|Jobs],[NotThisJob|Rest]):-
        choose(Job,Jobs,Rest).

% Constrain the remaining jobs to start after the given
% previous end time
propagate(PrevEndTime,Jobs):-
        ( foreach( job with start:StartTime, Jobs),
         param(PrevEndTime)
        do
           StartTime #>= PrevEndTime
        ).

% Make the variable Time take its smallest possible
% value
fix_to_min(Time):-
        mindomain(Time,Earliest),
        Time #= Earliest.
```

The Constraints Archive (http://www.4c.ucc.ie/archive) has pointers to constraint code libraries and constraint programming languages, both freely available software and commercial products.

ACKNOWLEDGEMENTS

Eugene Freuder is supported by a Principal Investigator Award from Science Foundation Ireland; some of his contribution to this chapter was prepared while he was at the University of New Hampshire.

REFERENCES

[1] J. Amilhastre, H. Fargier and P. Marquis (2002) Consistency restoration and explanations in dynamic CSPs—application to configuration. *Artificial Intelligence*, **135**, 199–234.

[2] F. Bacchus, X. Chen, P. van Beek and T. Walsh (2002) Binary vs. non-binary constraints. *Artificial Intelligence*.

[3] B. De Backer, V. Furnon, P. Prosser, P. Kilby and P. Shaw (1997) Local search in constraint programming: Application to the vehicle routing problem. Presented at the CP-97 Workshop on Industrial Constraint-based Scheduling.

[4] C. Bessiere, E. Freuder and J. Regin (1999) Using constraint metaknowledge to reduce arc consistency computation. *Artificial Intelligence*, **107**, 125–148.

[5] S. Bistarelli, H. Fargier, U. Montanari, F. Rossi, T. Schiex, and G. Verfaille (1996) Semiring-based CSPs and valued CSPs: Basic properties. In: M. Jampel, E.C. Freuder and M. Maher (eds.), *Over-Constrained Systems*, Volume 1106 of *Lecture Notes in Computer Science*. Springer, Berlin, pp. 111–150.

[6] B. Cheng, K. Choi, J. Lee and J. Wu (1999) Increasing constraint propagation by redundant modeling: an experience report. *Constraints*, **4**, 167–192.

[7] P. Cheeseman, B. Kanefsky and W. Taylor (1991) Where the really hard problems are. In: *Proceedings Twelth International Joint Conference in Artificial Intelligence*. Morgan Kaufmann, San Mateo, pp. 331–337.

[8] R. Debruyne and C. Bessière (2001) Domain filtering consistencies. *Journal of Artificial Intelligence Research*, **14**, 205–230.

[9] R. Dechter (1990). Enhancement schemes for constraint processing: backjumping, learning, and cutset decomposition. *Artificial Intelligence*, **41**, 273–312.

[10] R. Dechter and D. Frost (2002) Backjump-based backtracking for constraint satisfaction problems. *Artificial Intelligence*, **136**, 147–188.

[11] P. Deransart, M. Hermenegildo and J. Maluszynski (eds.) (2000) *Analysis and Visualization Tools for Constraint Programming. Lecture Notes in Computer Science No. 1870*. Springer, Berlin.

[12] R. Dechter, I. Meiri and J. Pearl (1991) Temporal constraint networks. *Artificial Intelligence*, **49**, 61–95.

[13] E. Freuder (1978) Synthesizing constraint expressions. *Communications of the ACM*, **11**, 958–966.

[14] E. Freuder (1982) A sufficient condition for backtrack-free search. *Journal of the Association for Computing Machinery*, **29**, 24–32.

[15] E. Freuder (1991) Eliminating interchangeable values in constraint satisfaction problems. In: *Proceedings Ninth National Conference on Artificial Intelligence*. AAAI Press/MIT, Menlo Park/Cambridge, pp. 227–233.

[16] E. Freuder and R. Wallace (1992). Partial constraint satisfaction. *Artificial Intelligence*, **58**, 1–70.

[17] F. Focacci, A. Lodi and M. Milano (1999) Cost-based domain filtering. In: J. Jaffar (ed.), *Principles and Practice of Constraint Programming* Volume 1713 of *Lecture Notes in Computer Science*. Springer.

[18] C. Gomes, B. Selman and N. Crato (1997) *Heavy-tailed Distributions in Combinatorial Search. Volume 1330 of Lecture Notes in Computer Science*. Springer.

[19] W. Harvey and M. Ginsberg (1995) Limited discrepancy search. In: *Proceedings of the Fourteenth International Joint Conference on Artificial Intelligence (IJCAI-95)*. Morgan Kaufmann, pp. 607–615.

[20] P. Jeavons, D. Cohen and M. Gyssens (1997) Closure properties of constraints. *Journal of the ACM*, **44**, 527–548.

[21] J. Jaffar and J.-L. Lassez (1987) Constraint logic programming. *Proceedings of the Annual ACM Symposium on Principles of Programming Languages (POPL)*. ACM, pp. 111–119.

[22] G. Kondrak and P. van Beek (1997) A theoretical evaluation of selected backtracking algorithms. *Artificial Intelligence*, **89**, 365–387.

[23] F. Laburthe and Y. Caseau (1998) SALSA: a language for search algorithms. *4th International Conference on the Principles and Practice of Constraint Programming (CP'98)*. Pisa, Italy, October.

[24] A. Mackworth (1977) Consistency in networks of relations. *Artificial Intelligence*, **8**, 99–118.

[25] S. Minton (1996) Automatically configuring constraint satisfaction programs. A case study. *Constraints*, **1**, 7–43.

[26] S. Minton, M.D. Johnston, A.B. Philips and P. Laird (1992) Minimizing conflicts: a heuristic repair method for constraint satisfaction and scheduling. *Artificial Intelligence*, **58**, 61–205.

[27] L. Michel and P. Van Hentenryck (1999) Localizer: a modeling language for local search. *INFORMS. Journal on Computing*.

[28] G. Pesant and M. Gendreau (1996) A view of local search in constraint programming. In: *Principles and Practice of Constraint Programming CP96: Proceedings of the Second International Conference*, Volume 1118 of *Lecture Notes in Computer Science*. Springer, Berlin, pp. 353–366.

[29] J.-C. Régin (2001) Minimization of the number of breaks in sports scheduling problems using constraint programming. In: E. Freuder and R. Wallace (eds.), *Constraint Programming and Large Scale Discrete Optimization, volume 57 of DIMACS Series in Discrete Mathematics and Theoretical Computer Science*. American Mathematical Society, Providence, RI, pp. 115–130.

[30] D. Sabin and E. Freuder (1997) Understanding and Improving the MAC Algorithm. In: *Principles and Practice of Constraint Programming—CP97: Proceedings of the Third International Conference, volume 1330 of Lecture Notes in Computer Science*. Springer, Berlin, pp. 167–181.

[31] H.M. Salkin (1970) On the merit of the generalized origin and restarts in implicit enumeration. *Operations Research*, **18**, 549–554.

[32] P. Shaw (1998) Using constraint programming and local search methods to solve vehicle routing problems. In: Michael Maher and Jean Francois Puget (eds.), *Principles and Practice of Constraint Programming CP98*, Volume 1520 of *Lecture Notes in Computer Science*. Springer. pp. 417–431.

[33] B. Selman, H. Levesque and D. Mitchell (1992) A new method for solving hard satisfiability problems. *Proceedings of the 10th National Conference on Artificial Intelligence (AAAI-92)*. San Jose, CA, USA, pp. 440–446.

[34] E. Tsang (1993) *Foundations of Constraint Satisfaction*. Academic Press, London.

[35] Pascal Van Hentenryck (1999) *The OPL Optimization Programming Language*. The MIT Press, Cambridge, MA.

[36] P. Van Hentenryck, H. Simonis and M. Dincbas (1992) Constraint satisfaction using constraint logic programming. *Artificial Intelligence*, **58**, 113–159.

[37] G. Verfaillie, M. Lemaitre and T. Schiex (1996) Russian doll search for solving constraint optimization problems. Proceedings of AAAI, pp. 181–187.

[38] A. Veron, K. Schuerman and M. Reeve (1993) Why and how in the ElipSys OR-parallel CLP system. Proceedings of PARLE.

[39] M. Wallace, S. Novello and J. Schimpf (1997) ECLiPSe—a platform for constraint programming. *ICL Systems Journal* **12**(1), 159–200.

[40] T. Walsh (2002) Stochastic constraint programming. *Proceedings of ECAI-2002*.

[41] M. Yokoo (1994) Weak-commitment search for solving constraint satisfaction problems. *Proceedings of AAAI*, pp. 313–318.

[42] M. Yokoo, E. Durfee, T. Ishida and K. Kuwabara (1998) The distributed CSP: Formalization and algorithms. *IEEE Transactions on Knowledge and Data Engineering*, **10**, 673–685.

[43] W. Zhang (1998) Complete anytime beam search. Proceedings of AAAI, pp. 425–430.

Chapter 15

ARTIFICIAL NEURAL NETWORKS FOR COMBINATORIAL OPTIMIZATION

Jean-Yves Potvin
Département d'informatique et de recherche opérationnelle
and Centre de recherche sur les transports
Université de Montréal
C.P. 6128, succursale Centre-ville
Montréal (Québec), Canada H3C 3J7
E-mail: potvin@iro.umontreal.ca

Kate A. Smith
School of Business Systems
Faculty of Information Technology
Monash University
P.O. Box 63B
Victoria 3800, Australia
E-mail: kate.smith@infotech.monash.edu.au

1 INTRODUCTION

Artificial neural networks (ANNs) are extremely simplified models of their biological counterpart, the network of natural neurons known as the human brain. Consequently, we are not interested here in the study of the brain, which is the subject of neuroscience. Rather, we are interested to know what these networks of artificial units are, what they can do and how. ANNs were originally developed to provide a fundamentally new and different approach to information processing, when an algorithmic procedure for solving a problem is not known. As opposed to programmed computing, ANNs are capable of internally developing information processing capabilities for solving a problem when fed with appropriate information about the problem. Thus, they are often referred to as learning or adaptive models.

The history of ANNs dates back to the paper of McCulloch and Pitts (1943), when simple types of neural networks were shown to be able to learn arithmetic or logical functions. Important successes were witnessed in the late 50's and early 60's, with the development of the perceptron model and the first neurocomputers (Rosenblatt, 1958). By the end of the 60's, however, the field collapsed: a book by Minsky and Papert (1969), demonstrating that even the simple exclusive-or logical function could not be implemented with a perceptron, was devastating and diverted away research funding. After a dark period, ANNs emerged again in the early 80's with the support of John Hopfield, a renowned scientist, and the publication of an important book by Rumelhart and McClelland (1986), which introduced the backpropagation neural network model to the scientific community. This model extended the capabilities of its ancestor, the perceptron, allowing it to learn a much larger class of functions (including the exclusive-or logical function). Since that time, the field has continually expanded.

The first successes with ANNs were reported for the most part in pattern recognition, classification and prediction tasks. Application of ANNs to combinatorial optimization problems (COPs) dates back to 1985 when Hopfield and Tank solved small instances of the Traveling Salesman Problem (TSP) with a Hopfield neural network (Hopfield and Tank, 1985). The TSP is a classical combinatorial optimization problem, which is simple to state but difficult to solve. Basically, the objective is to find the shortest possible tour (or Hamiltonian cycle) through a set of N vertices so that each vertex is visited exactly once. This problem is known to be NP-hard (Garey and Johnson, 1979; Nemhauser and Wolsey, 1988), and cannot be solved exactly in polynomial time. Because of the simplicity of its formulation, the TSP has always been a fertile ground for new solution ideas. Consequently, it is not surprising that many problem-solving approaches inspired by ANNs have been applied to the TSP (Potvin, 1993). The work of Hopfield and Tank was a first attempt in this direction and it generated much excitement in the neural network and operations research communities alike.

Many other types of COPs have since been tackled with ANNs in different application areas: routing and transportation, scheduling, cutting stock and packing, timetabling, telecommunications, and many others (Burke and Ignizio, 1992). In some cases, the results obtained are competitive with those reported with alternative techniques. In other cases, the results are not yet convincing. It is clear that the computational paradigm of ANNs, which is inherently parallel, distributed and adaptive cannot be fully exploited on current computer hardware. Their behavior must be simulated, thus leading to excessively large computation times. The development of suitable hardware for these models (often called neurocomputers) would thus be an important step toward their full recognition.

The classical backpropagation neural network model, although well suited for many learning tasks is not really indicated for combinatorial optimization. Consequently, ANNs applied to COPs are mostly based on three alternative models: Hopfield–Tank (H–T) and its variants, the elastic net (EN) and the self-organizing map (SOM). H–T provides a natural way to model many COPs and has been widely applied. EN and SOM provide an alternative, more efficient, approach for low-dimensional geometrical problems, like the TSP. These models are reviewed in the following.

2 HOPFIELD NEURAL NETWORKS

In his seminal paper of 1982, John Hopfield described a new way of modeling a system of neurons capable of performing computational tasks (Hopfield, 1982). Using a collection of binary-state neurons and a stochastic updating algorithm, these computational tasks were initially related to storage and retrieval of embedded memories. The computational capabilities of Hopfield's original model were expanded in Hopfield (1984) when he proposed a continuous version, and proved convergence of the model by demonstrating that the dynamics of the model minimized a constructed Liapunov function over time. From here, it became clear that Hopfield networks could be used to minimize any function provided the network parameters were set appropriately. The fact that the continuous version of the Hopfield network was designed to be implemented using electrical circuits also promised rapid computational ability.

This section first presents the two Hopfield neural network models: the discrete and stochastic model of 1982, and the continuous and deterministic model of 1984. The method of Hopfield and Tank (1985) for mapping a combinatorial optimization problem onto a Hopfield network is then described, using the TSP as an example. The section continues with a discussion of the criticisms of the approach. We then briefly review some of the many modifications and extensions that have been made to the original model and approach in an attempt to overcome these limitations. Finally, the reported performance of these Hopfield network models for combinatorial optimization across a range of benchmarked problems is discussed.

2.1 Discrete and Stochastic Hopfield Network

The original Hopfield network, as described in Hopfield (1982) comprises a fully interconnected system of n computational elements or *neurons*. In the following description, Hopfield's original notation has been altered where necessary for consistency. The strength of the connection, or weight, between neuron i and neuron j is determined by W_{ij}. This weight may be positive or negative depending on whether the neurons act in an excitatory or inhibitory manner (or zero if there is no interaction). Each neuron has an internal state u_i and an external state v_i. While the internal states are continuous valued, the external states are binary for this discrete model. The relationship between the internal and external states of the neurons can be shown as:

$$u_i(t+1) = \sum_{j=1}^{n} W_{ij} v_j(t) + I_i \tag{1}$$

$$v_i(t+1) = f(u_i) = \begin{cases} 1 & \text{if } u_i > 0 \\ 0 & \text{if } u_i \leq 0 \end{cases} \tag{2}$$

where I_i is a constant external input to neuron i and $f()$ is the transfer function between internal and external states. The connection weights $\mathbf{W}$ are also constant, and the only variable elements in the network are the internal and external states of the neurons that are updated over time. From equations (1) and (2) it is clear that the internal state of each neuron is calculated as the weighted sum of inputs from its connected neurons, with an additional constant input. The neuron will "fire" (as evidenced by an external state of 1), if it receives sufficient stimulation from its connecting neurons, otherwise the neuron's external state will be zero representing a dormant or "non-firing" state.

The neurons update themselves over time in a random sequence, thus the model is said to be discrete and stochastic. As the network updates, and provided the weight matrix is symmetric with non-negative diagonals, the following energy function is guaranteed to be minimized until the system converges to one of its stable states.

$$E_{\mathrm{d}} = -\frac{1}{2} \sum_{i=1}^{n} \sum_{j=1}^{n} W_{ij} v_i v_j - \sum_{i=1}^{n} I_i v_i \tag{3}$$

Using the terminology of operations research, the system of equations (1) and (2) perform a gradient descent on the energy function (3), with the neuron states $\boldsymbol{v}$ converging to one of its local minima. If the values of the weights $\boldsymbol{W}$ and external inputs $\boldsymbol{I}$ are fixed appropriately, this process can be used to minimize any quadratic function of binary variables.

2.2 Continuous and Deterministic Hopfield Network

Hopfield's subsequent modifications to the original 1982 model were driven by considerations of biological plausibility. In the biological system, u_i lags behind the instantaneous outputs v_j of the other neurons because of the input capacitance C_i of the cell membrane, the trans-membrane resistance R_i, and the finite impedance $R_{ij} = W_{ij}^{-1}$ between the output v_j and the cell body of neuron i. The external states of the neurons are now continuous valued between 0 and 1, rather than binary in the earlier model, and represent an average "firing rate". Hopfield (1984) modeled this more biologically based system using the following resistance-capacitance differential equation to determine the rate of change of u_i, and hence the time evolution of the continuous Hopfield network:

$$\frac{\mathrm{d}u_i}{\mathrm{d}t} = \sum_{j=1}^{n} W_{ij} v_j - \frac{u_i}{\tau} + I_i \tag{4}$$

$$v_i = f(u_i) \quad \text{and} \quad \tau = R_i C_i \tag{5}$$

where the transfer function $f()$ is now a continuous sigmoidal function such as:

$$v_i = f(u_i) = \frac{1}{2}\left(1 + \tanh\left(\frac{u_i}{T}\right)\right) \tag{6}$$

and T is a parameter used to control the slope of the transfer function. τ is the value of the time constant of the amplifiers, and without loss of generality can be assigned a value of unity, provided the time step of any discrete-time simulation of equation (4) is considerably smaller than unity. This same set of equations represents a resistively connected network of electronic amplifiers, and thus the system can be implemented with electrical circuits. We refer the interested reader to (Hopfield, 1984) for details of this implementation.

Similar to the original discrete model, the dynamics of this continuous model also minimize an energy function over time guaranteed to converge to stable states. This energy function is:

$$E_{\mathrm{c}} = -\frac{1}{2}\sum_{i=1}^{n}\sum_{j=1}^{n} W_{ij} v_i v_j - \sum_{i=1}^{n} I_i v_i + \int_0^{v_i} f_i^{-1}(\mathbf{v})\,\mathrm{d}\mathbf{v} \tag{7}$$

Hopfield (1984) showed that provided the weight matrix is symmetric, this function is a Liapunov function for the system of equations (4) and (5). Furthermore, if the slope of the transfer function (6) is particularly high (i.e., T is near zero), then the transfer function (6) approximates the behavior of the discrete version given by equation (2), and the integral term of equation (7) vanishes. Consequently, the local minima of E_{c} coincide with the local minima of E_{d}, and all these local minima lie at the vertices of the unit hypercube resulting in binary values for $\boldsymbol{v}$. Thus, for T near zero, the continuous Hopfield network converges to a 0–1 solution in $\boldsymbol{v}$ which minimizes the energy function E_d given by (3).

Thus, there are two Hopfield neural network models available: a discrete version and a continuous version. The continuous version can either be implemented using electrical circuits, or simulated on a digital computer using an approximation to the differential equation (4) such as the Euler approximation. The accuracy of this approximation depends on parameters like the time step of the discretization, and affects the degree to which the discretized dynamics converge on the Liapunov energy function. Clearly, a small time step will approximate the dynamics well, ensuring gradient descent on the energy function. This issue is discussed further in Section 4, along with a variety of other practical issues.

2.3 Adaptation to Solve Combinatorial Optimization Problems

In 1985, John Hopfield teamed together with David Tank to extend the applications of his model to include solving combinatorial optimization problems (Hopfield and Tank, 1985). Hopfield and Tank (H–T) realized that networks of neurons with this basic organization could be used to compute solutions to specific optimization problems by selecting weights and external inputs which appropriately represent the function to be minimized and the desired states of the problem. The updating of the neurons according to the differential equations given by (4) and (5) (or even the discrete versions (1) and (2)) ensures that both the energy function and the optimization problem are simultaneously minimized over time. The analog nature of the neurons and the hardware implementation of the updating procedure could be combined to create a rapid and powerful solution technique.

Using the method proposed by Hopfield and Tank, the network energy function is made equivalent to the objective function of the optimization problem needing to be minimized, while the constraints of the problem are included in the energy function as penalty terms.

Consider the quadratic formulation of the N-city TSP, given the binary decision variable

$$X_{ij} = \begin{cases} 1 & \text{if city } i \text{ is in position } j \\ 0 & \text{otherwise} \end{cases}$$

and the constant distance matrix d_{ik} representing the distance between cities i and k:

$$\text{minimise} \sum_{i=1}^{N} \sum_{\substack{k=1 \\ k \neq i}}^{N} \sum_{j=1}^{N} d_{ik} X_{ij} (X_{k,j+1} + X_{k,j-1}) \tag{8}$$

$$\text{subject to} \sum_{i-1}^{N} X_{ij} = 1 \quad \text{for all } j \tag{9}$$

$$\sum_{j=1}^{N} X_{ij} = 1 \quad \text{for all } i \tag{10}$$

$$X_{ij} \in \{0, 1\} \quad \text{for all } i, j \tag{11}$$

Apart from being a well benchmarked problem, the TSP is a useful problem to consider since its form is that of a quadratic assignment problem. Thus the methods used by Hopfield and Tank for mapping the optimization problem onto a Hopfield neural network can be generalized to a wide range of problems with similar constraint and objective types.

The first step is to construct an energy function representation of the complete optimization problem using a penalty parameter approach, so that all objective functions and constraints are integrated into a single function which needs to be minimized. This is achieved by observing that a constraint of the form (9) can be enforced by ensuring minimization of the quantity

$$\sum_{i=1}^{N} \sum_{\substack{k=1 \\ k \neq i}}^{N} X_{ij} X_{kj} \quad \text{for all } j.$$

That is, a constraint requiring a single "1" in each column can be enforced by minimizing the pairwise product of elements in each column. If there is no more than one "1" in the column, then this term will be at its minimum value of zero. If there is more than one "1" in the column, then this term will be greater than zero. A similar term can be constructed to enforce the row constraint (10). Naturally, these terms will also be zero if there are no "1"s in each row or column as well. Since we need exactly one "1" per column and row, we will also need an additional term to force N elements of the solution matrix $\boldsymbol{X}$ to be "1".

The complete set of constraints can therefore be enforced through minimization of penalty terms, and when we add the objective function to these terms, we arrive at the H–T energy function for the TSP:

$$E = \frac{A}{2} \sum_{j=1}^{N} \sum_{i=1}^{N} \sum_{\substack{k=1 \\ k \neq i}}^{N} X_{ij} X_{kj} + \frac{B}{2} \sum_{i=1}^{N} \sum_{j=1}^{N} \sum_{\substack{l=1 \\ l \neq j}}^{N} X_{ij} X_{il} + \frac{C}{2} \left(\sum_{i=1}^{N} \sum_{j=1}^{N} X_{ij} - N \right)^2 + \frac{D}{2} \sum_{i=1}^{N} \sum_{\substack{k=1 \\ k \neq i}}^{N} \sum_{j=1}^{N} d_{ik} X_{ij} (X_{k,i+1} + X_{k,i-1}) \qquad (12)$$

The first two terms enforce no more than one "1" per column and row respectively, the third term ensures that there are N elements "on" in the solution matrix, and the final term minimizes the tour length. The penalty parameters A, B, C and D need to be fixed at values that reflect the relative importance of these terms in the minimization process. If A, B and C are not large enough relative to D, then the resulting solution may be infeasible. Similarly, if D is not large enough, the solution may be feasible but the tour length may be larger than the optimal value. Hopfield and Tank (1985) used values of $A = B = D = 500$ and $C = 200$ to balance these terms.

Now that the energy function has been constructed, the next step is to derive the Hopfield network weights and external inputs so that the energy function is minimized by the network dynamics. For this we need to expand and rearrange the energy function (12) so that it is in the same form as the standard Hopfield energy function E_{d}

$$E_{\mathrm{d}} = -\frac{1}{2}\sum_{i=1}^{N}\sum_{j=1}^{N}\sum_{k=1}^{N}\sum_{l=1}^{N} W_{ijkl}X_{ij}X_{kl} - \sum_{i=1}^{N}\sum_{j=1}^{N} I_{ij}X_{ij} \tag{13}$$

which has been modified from (3) to reflect the fact that the neurons X_{ij} for our TSP problem are two dimensional, compared to the linear array of neurons v_i used in the standard Hopfield network. Once the forms of these two functions (12) and (13) are similar, the network weights $\boldsymbol{W}$ and external inputs $\boldsymbol{I}$ can be read as the coefficients of the quadratic and linear terms respectively. To ensure equivalence of the two functions, the summations of each term in (12) need to be extended across all relevant dimensions (i, j, k, l for quadratic terms and i, j for linear terms). Thus the Kronecker-Delta symbol is incorporated into each term of (12) where necessary:

$$\delta_{ab} = \begin{cases} 1 & \text{if } a = b \\ 0 & \text{if } a \neq b \end{cases}$$

Expanding (12) and rearranging the terms into quadratic, linear, and constant terms, thus yields:

$$\begin{aligned} E = -\frac{1}{2}\sum_{i=1}^{N}\sum_{j=1}^{N}\sum_{k=1}^{N}\sum_{l=1}^{N}[&-A\delta_{ik}(1-\delta_{jl}) - B\delta_{jl}(1-\delta_{ik}) - C \\ &- D\delta_{ik}(\delta_{l,j+1}+\delta_{l,j-1})]X_{ij}X_{kl} - \sum_{i=1}^{N}\sum_{j=1}^{N}[CN]X_{ij} + \frac{CN^2}{2} \end{aligned} \tag{14}$$

Comparing (14) to the standard Hopfield energy function (13) then, it is clear that the network parameters are:

$$\begin{aligned} W_{ijkl} &= -A\delta_{ik}(1-\delta_{jl}) - B\delta_{jl}(1-\delta_{ik}) - C - D\delta_{ik}(\delta_{l,j+1}+\delta_{l,j-1}) \\ I_{ij} &= CN \end{aligned} \tag{15}$$

The constant term in equation (14) can be ignored since it merely reflects a shift upwards and does not affect the location of the minima of the function.

Now that the network weights and external inputs are determined, the Hopfield network can be initialized (using random values for the initial states), and updated according to equations (4) and (5) (or equations (1) and (2) for a purely discrete version). This updating is guaranteed to minimize the Hopfield energy function (13), and since this function is equivalent to the TSP energy function (12) and (14), then the resulting solution matrix $\boldsymbol{X}$ will provide a local minima of the TSP energy function. The quality and feasibility of this solution depends on the choice of penalty parameters A, B, C and D, as well as the initialization of the neurons, and the accuracy with which the dynamics of the differential equation (4) can be simulated if the continuous model is chosen.

The complete procedure is summarized in pseudocode form below:

```
Step 0: Preliminary Tasks
    0.1 Construct an energy function for the optimization
        problem using a penalty parameter approach.
    0.2 Expand energy function and infer network weights and
        external inputs.
Step 1: Initialization Tasks
    1.1 Initialize neuron states to random values.
    1.2 Select A,B,C,D.
    1.3 Select T, the parameter of the continuous transfer
        function, and the value of the discrete time step if
        simulating the continuous model.
Step 2: If energy function has converged to local minimum
        proceed to Step 5, otherwise proceed to step 3
Step 3: Repeat n times:
    3.1 Randomly choose a neuron i to update (if using
        discrete time dynamics).
    3.2 Update u_i and v_i using equations (1)-(2) or (3)-(4).
Step 4: Go back to Step 2.
Step 5: Examine final solution matrix and determine
        feasibility and optimality.
Step 6: Adjust parameters A,B,C,D if necessary to obtain a
        satisfactory solution, re-initialize neuron states,
        and repeat from Step 2.
```

Clearly, one of the main limitations of the H–T approach to solving combinatorial optimization is the difficulty in choosing appropriate penalty parameters. In addition to this difficulty, the dynamics of the Hopfield network perform a gradient descent on the energy function, and thus converge to the first local minimum they encounter. Coupling these two issues, it seems likely that the H–T approach may yield solutions of poor quality. Wilson and Pawley (1988) first published these findings nearly three years after Hopfield and Tank's original paper was published. In doing so, they raised serious doubts as to the validity of the H–T approach to solving optimization problems, which seemingly served to dampen the enthusiasm surrounding the field.

2.4 Extensions

Since Wilson and Pawley's results were published, it has been widely recognized that the H–T formulation is not ideal, even for problems other than the TSP. The problem of optimally selecting the penalty parameters is not trivial and much work has been done to try to facilitate this process (Hedge et al., 1988; Kamgar-Parsi, 1992; Lai and Coghill, 1992). Many other researchers believed that the H–T energy function needed to be modified before any progress would be made, and considerable effort has also been spent in this area (Brandt et al., 1988; Van den Bout and Miller, 1988). One obvious improvement to the H–T approach to solving the TSP is to reduce the number of terms needed to represent the constraints by using the form $(\sum_{i=1}^{N} X_{ij} - 1)^2$ to represent the column constraints, for example. This eliminates the need for the third term in equation (12), thus the penalty parameter C is also eliminated.

Perhaps the most important breakthrough in the field, however, came from the valid subspace approach of Aiyer et al. (1990), and the subsequent work of Gee (1993). Their idea is to represent the constraint set as a hyperplane, and encourage the solution to lie upon it. This is achieved by including a single term in the energy function for the constraints which attempts to minimize the deviation between the solution matrix and the constraint plane, or valid subspace. A single penalty parameter needs to be selected, which if large enough, will guarantee the feasibility of the final solution.

Some researchers have also attempted to address the limitations of the H–T approach by considering alternative representations of constraints, suitable values for penalty parameters, and other modeling issues. The majority of other researchers in the field, however, have focused on the limitation of the Hopfield network dynamics. By extending the network dynamics to include stochasticity and hill-climbing capabilities, various methods have emerged that attempt to avoid the many local minima of the energy function.

The variations of the Hopfield network that have been proposed can be broadly categorized as either deterministic or stochastic. The deterministic approaches include problem specific enhancements such as the "divide and conquer" method of Foo and Szu (1989) for solving the TSP, deterministic hill-climbing such as the "rock and roll" perturbation method of Lo (1992), and the use of alternative neuron models within the Hopfield network such as the winner-take-all neurons used by Amartur et al. (1992) to improve the feasibility of the solutions. Stochastic approaches address the problem of poor solution quality by attempting to escape from local minima. There are basically four main methods found in the literature to embed stochasticity into the Hopfield network:

1. replace sigmoidal transfer function with a stochastic decision-type function;
2. add noise to the weights of the network;
3. add noise to the external inputs of the network;
4. any combination of the above methods.

The *Boltzmann machine* (Aarts and Korst, 1989; Hinton et al., 1984) utilizes the first method based on a discrete Hopfield model. The inputs are fixed, but the discrete transfer function is modified to become probabilistic. Much like simulated annealing (Kirkpatrick et al., 1983), the consequence of modifying the binary transfer level of each neuron is evaluated according to the criteria of the Boltzmann probability factor. This model is able to escape from local minima, but suffers from extremely large computation times. In order to improve the efficiency and speed of the Boltzmann machine, Akiyama et al. (1989) proposed Gaussian machines which combine features of continuous Hopfield networks and the Boltzmann machine. Gaussian machines have continuous outputs with a deterministic transfer function like the Hopfield network, but random noise is added to the external input of each neuron. This noise is normally distributed (or Gaussian) with a mean of zero and a variance controlled by a temperature parameter T. However, based upon Szu's fast simulated annealing (Szu and Hartley, 1987) which uses Cauchy noise to generate new search states and requires only a $T/\log(T)$ cooling schedule, the Cauchy machine (Szu, 1988; Takefuji and Szu, 1989) was proposed as an improvement to solution quality. The Cauchy distribution is thought to yield a better chance of convergence to the global minimum than the Gaussian distribution. Furthermore, Cauchy noise produces both local random walks and larger random leaps in solution space, whereas Gaussian noise produces only local random walks (Takefuji and Szu, 1989). The noise is incorporated into the transfer function, while the outputs of the Cauchy machine are binary. In the high-gain limit of the stochastic transfer function (T near zero), the Cauchy machine approaches the behavior of the discrete and deterministic Hopfield network. Another stochastic approach which has been very successful is mean-field annealing (Peterson and Soderberg, 1989; Van den Bout and Miller, 1989, 1990), so named because the model computes the mean activation levels of the stochastic binary Boltzmann machine.

2.5 Performance

Hopfield and Tank successfully applied their approach to several optimization problems including an analog-to-digital converter, a signal decision circuit, and a linear programming model (Tank and Hopfield, 1986). It was, however, their results for the combinatorial TSP that attracted the most attention. Hopfield and Tank (1985) simulated a network of 10 cities (100 neurons), chosen at random on the interior of a 2-dimensional unit square. Their results for small-sized problems were quite encouraging. For a 10 city problem, and for 20 random starts, 16 converged to valid tours. About 50% of the trials produced one of the two known shortest tours. Hopfield and Tank then studied a 30 city (900 neuron) problem. Since the time required to simulate the differential equations on a computer scales worse than $O(n^3)$, their results were fragmentary. They were unable to find appropriate penalty parameters to generate valid tours, and commented that "parameter choice seems to be a more delicate issue with 900 neurons than with 100". In fact, their best solution was around 40% away from the best known solution of Lin and Kernighan (1973) on the same 30 city problem.

Since then, the many modifications to the original H–T approach have seen considerable improvement in these results. A recent fuzzy modification of Aiyer's subspace approach yielded nearest-city quality tours for up to 100 randomly generated cities (Wolfe, 1999). Peterson and Soderberg reported solutions for 200 cities using a mean field annealing neural network that were only slightly worse than simulated annealing results (Peterson and Soderberg, 1993). These results are still a long way from those that can be obtained by well known heuristics. For example, the iterated Lin-Kernighan heuristic can routinely find solutions within 1% of optimal for problems with thousands of cities (Johnson, 1990). Even other neural network approaches such as the deformable template methods discussed in the next section yield considerably better results than the Hopfield variations seem capable of.

The advantage of the H–T approach to combinatorial optimization however lies in its generalization abilities. The H–T approach can be applied to any combinatorial optimization problem that can be formulated within quadratic terms. It does not rely on the geometry of the problem like many of the TSP heuristics or the deformable template methods. The many variations of the Hopfield network that have emerged over the last decade or so have been applied to a wide range of classical combinatorial optimization problems including assignment problems, constraint satisfaction problems, graph problems, integer programming, and scheduling problems to name a few. We refer the interested reader to Smith (1999) for a survey of these and other applications. Many of the results are competitive with other metaheuristic approaches. One of the deficiencies of the literature in this area, however, is the fact that few studies are established as comparative analyses, aimed to determine the competitiveness of the proposed neural network approach with the best known heuristic or metaheuristic approaches to the same problem. This makes a true evaluation of the performance of Hopfield-type models difficult. As Looi (1992) noted, "although there is a large collection of operations research based and other methods for solving all of these problems, comparisons between neural network methods with existing methods have been lacking". Solutions to this problem in the form of guidelines for experiments have now been published (Barr et al., 1995; Hooker, 1995) and we hope that researchers will soon provide enough studies of this nature that an accurate evaluation of the performance and potential of Hopfield-type neural networks on a wide variety of problems can be established.

3 DEFORMABLE TEMPLATES

Elastic nets (EN) and Self-organizing maps (SOM), often referred to as deformable templates, provide alternatives for solving low-dimensional problems with a geometric interpretation, like the Euclidean TSP. These models are fundamentally different from H–T, as they evolve in a low-dimensional continuous search space. In the following, we describe both models for solving the Euclidean TSP. We then establish some relationships between the two models and present a few extensions.

3.1 Elastic Net

The elastic net (EN) of Durbin and Willshaw (1987), originated from a previous work by Willshaw and von der Malsburg (1979). It is an iterative procedure where M points, with M typically larger than the number of vertices (or cities) N, are lying on a circular ring or "rubber band" originally located at the center of gravity of the vertices. The rubber band is gradually elongated until it is sufficiently close to each vertex to define a tour. During that process two forces apply: one for minimizing the length of the ring, and the other for minimizing the distance between the vertices and the ring. These forces are gradually adjusted as the procedure evolves.

Figure 15.1(a)–(c) show how the elastic net typically evolves over time. In the figure, the small black circles are the points located on the ring which are migrating towards the vertices in the Euclidean plane. When there is a point on the ring sufficiently close to each vertex, a solution is obtained, as shown in Figure 15.1(d). This model will now be presented more formally, using a pseudocode notation. Let X_i be the coordinates of vertex $i, i = 1, \ldots, N$, Y_j the coordinates of ring point $j, j = 1, \ldots, M$, and $d_{X_iY_j}$ the Euclidean distance between i and j. We have:

Step 0: $K \leftarrow K_0; Y_j \leftarrow Y_j^0, j = 1, \ldots, M$;
Step 1: Repeat *rep* times
  1.1 Update the coordinates Y_j of ring point $j, j = 1, \ldots, M$;
  1.2 If $\min_{j=1,\ldots,M} d_{X_iY_j} \leq \varepsilon, i = 1, \ldots, N$, then STOP;
Step 2: $K \leftarrow \alpha K (0 < \alpha < 1)$;
Step 3: Go back to *Step 1*.

Step 0 initializes the scale parameter K (see below) and selects an initial location for the points on the ring. In Step 1, the points migrate towards the vertices through

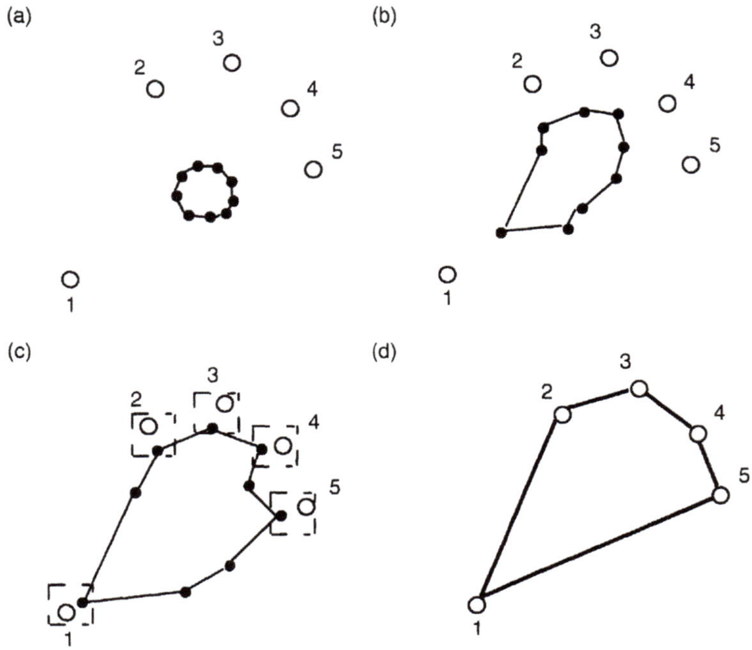

Figure 15.1. Evolution of the elastic net over time (a)–(c) and the final tour 1-2-3-4-5 (d).

an iterative procedure governed by parameter K. After a fixed number of iterations, related to the size of the problem, the value of parameter K is slightly reduced and the migration process is pursued with this new value. This is repeated until there is a point on the ring sufficiently close to each vertex, as specified by the tolerance ε. An alternative or additional stopping criterion for EN is to iterate until some preset $K_{\min}$ value is reached and then, to associate each vertex with the closest ring point. Parameter K is reminiscent of the temperature parameter in the simulated annealing algorithm, as its value must be progressively reduced according to a prespecified "cooling schedule" to obtain a good solution to the problem.

In Step 1.1, the coordinates of each ring point j are updated as follows:

$$Y_j \leftarrow Y_j + \Delta Y_j$$

where

$$\Delta Y_j = \alpha \sum_{i=1,\ldots,N} w_{ij}\left(X_i - Y_j\right) + \beta K\left(Y_{j+1} + Y_{j-1} - 2Y_j\right) \tag{16}$$

$$w_{ij} = \frac{\phi\left(d_{X_i Y_j}, K\right)}{\sum_{k=1,\ldots,M} \phi\left(d_{X_i Y_k, K}\right)}, \quad i = 1, \ldots, N \tag{17}$$

$$\phi(d, K) = e^{-d^2/2K^2} \tag{18}$$

where α, β are constant parameters and w_{ij} is a normalized measure of the "attraction" of vertex i on ring point j. In equation (16), the α term drives the points on the ring

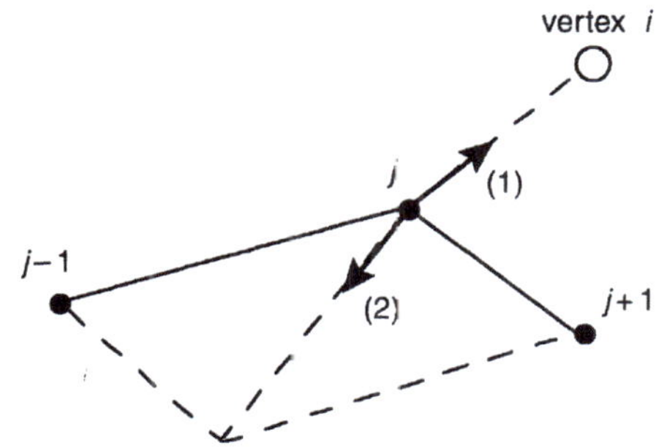

Figure 15.2. Forces that apply on ring point j.

towards the vertices, and the β term keeps neighboring points on the ring together during the migration process to produce a short tour (i.e., neighboring points are associated with vertices that are close in distance). These two forces are illustrated in Figure 15.2.

In this figure, force (1) is derived from the α term, and drives point j towards vertex i. Force (2), which is derived from the β term, is more easily understood by considering the equivalence

$$Y_{j+1} + Y_{j-1} - 2Y_j = (Y_{j+1} - Y_j) + (Y_{j-1} - Y_j). \tag{19}$$

It thus defines a tension on the ring that keeps neighboring points together. Through parameters α and β, the relative strength of the two forces can be regulated.

It is work noting that the update equations (16) can be expressed as the derivative of an appropriate energy function E_K, namely

$$\Delta Y_j = -K \frac{\partial E_K}{dY_j} \tag{20}$$

where

$$E_K = -\alpha K \sum_{i=1,\dots,N} \ln \sum_{j=1,\dots,M} \phi\left(d_{X_i Y_j}, K\right) + \frac{\beta}{2} \sum_{j=1,\dots,M} d^2_{Y_j Y_{j+1}} \tag{21}$$

This algorithm thus finds a local minimum of this energy function by performing a gradient descent in a continuous two-dimensional Euclidean space. When K approaches 0 and the ratio of M to N approaches infinity, minimizing the energy is equivalent to minimizing the total length of the ring and, thus, the solution value. Since the shape of the energy function and its local minima change with K, the function is gradually modified through a slow reduction of the value of parameter K until the minima correspond to good TSP tours.

3.2 Self-organizing Map

A self-organizing map is an instance of the so-called competitive neural networks (Kohonen, 1982, 1988). It is composed of a layer of input units fully connected to a layer of output units, the latter being organized according to a particular topology, such as a ring structure. Self-organizing maps basically produce topological mappings

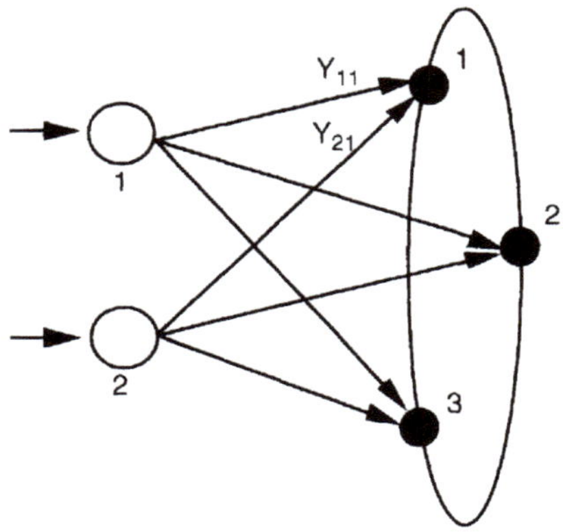

Figure 15.3. A SOM.

from high-dimensional input spaces to low-dimensional output spaces. In the case of a ring structure, the p-dimensional input vectors are associated with output units or 1-dimensional positions on the ring. The mapping is such that two input vectors that are close in the input space will be associated with units that are close on the ring.

In Figure 15.3, a SOM with $P = 2$ white input units and $M = 3$ black output units (indexed from 1 to 3) is illustrated.

In Figure 15.3, Y_{11} and Y_{21} denote the weights on the connections from the two input units to output unit 1, and $Y_1 = (Y_{11}, Y_{21})$ is the weight vector associated with output unit 1. In a TSP context, each input vector corresponds to the coordinates of a vertex. We then want vertices that are close in distance to be associated with units that are close on the ring to produce a short tour.

This is obtained through the following iterative adjustment of the connection weights. Let assume that we have a SOM with two input units and M output units on a ring, each with weight vector $Y_j = (Y_{1j}, Y_{2j}), j = 1, \ldots, M, M \geq N$. Let $X_i = (x_i, y_i)$ be the coordinates of vertex $i, i = 1, \ldots, N$ and $d_{X_i Y_j}$, the Euclidean distance between vertex i and output unit j. Then, we have:

Step 0: Initialization. $i \leftarrow 0; \quad Y_j \leftarrow Y_j^0; \quad j = 1, \ldots, M;$
Step 1: Competition.
  1.1 $i \leftarrow (i + 1) \bmod (N + 1)$;
  1.2 $o_j \leftarrow d_{X_i Y_j}, j = 1, \ldots, M$;
  1.3 $o_j \leftarrow \min_{j=1,\ldots,M}\{o_j\}$;
Step 2: Weight adjustment.
  $Y_j \leftarrow Y_j + \mu f(j, j^*)(X_i - Y_j), j = 1, \ldots, M, 0 < \mu < 1;$
Step 3: If $\min_{j=1,\ldots,M} d_{X_i Y_j} \leq \varepsilon, i = 1, \ldots, N$, then STOP;
Step 4: Go back to Step 1.

In Step 1, the winning output unit j^* is the one with the closest weight vector (in Euclidean distance) to the current vertex. In Step 2, function f is typically a decreasing function of the lateral distance between output units j and j^* on the ring (i.e., if there are k units on the ring between the two, the lateral distance is $k+1$) and its range is the interval [0,1]. Thus, the weight vector of the winning unit j^* and the weight vectors of units that are close to j^* on the ring all move towards the current vertex, but with decreasing intensity as the lateral distance to the winning unit increases. Typically, function f is modified as the algorithm unfolds to gradually reduce the magnitude of the weight adjustment. At the start, all units that are close to the winning unit on the ring "follow" that unit in order to move in the same area. At the end, only the weight vector of the winning unit significantly moves towards the current vertex.

This iterative adjustment procedure is repeated, through multiple passes over the set of vertices (c.f., the modulo operator in Step 1.1) until there is a weight vector sufficiently close to each vertex. Other or additional stopping criteria may be considered like a fixed number of passes through the vertices or a stable competition, when it is observed that the winning unit for each vertex does not change anymore from one pass to another. The association of each vertex with a weight vector (i.e., an output unit with an index or position on the ring) produces a tour. If we consider the two-dimensional weight vector of each output unit on the ring as the location of that unit in the Euclidean space, Figure 15.1 is still a good way of visualizing the evolution of the SOM with the output units on the ring migrating towards the vertices.

3.3 Extensions

Since both EN and SOM explcit the geometry of the problem, they have mostly been applied to the TSP. A few extensions are reported in the literature for the mutiple TSP (Goldstein, 1990) and vehicle routing problems (Ghaziri, 1991, 1996; Matsuyama, 1991; Vakhutinsky and Golden, 1994; Potvin and Robillard, 1995). In these applications, mutiple tours are formed through the migration in parallel of multiple rings. In the case of the VRP, the capacity or time constraints typically break the geometrical nature of the problem and the algorithm must be modified accordingly. The SOM described in (Ghaziri, 1991), for example involves competition for the current vertex at two different levels: one within each ring based on geometric properties and the other among the rings to take into account the capacity constraints. Some recent papers have also generalized the SOM for applications in non geometric contexts, like the multidimensional knapsack problem and the generalized quadratic assignment problem (Glover, 1994; Smith, 1995). These approaches depart from the traditional ring structure and exploit more complex mappings and topologies.

3.4 Performance

Both SOM and EN are closely related. They both involve migration of a ring towards vertices, although the mechanism for updating the location of the ring points is different. In the case of EN, the update is defined through the minimization of an appropriate energy function. The SOM does not require such a function.

It is difficult to get an accurate picture of the performance of the two models from the current literature. The results are scattered in different journals from different research area. Computation times are often missing and comparisons with alternative methods are rare. Furthermore, the problems are not taken from standard benchmarks. Overall, the literature indicates that both EN and SOM outperform the Hopfield model on the TSP. For problems with up to a few hundred cities, EN often ends up with slightly better solutions than SOM (Angeniol et al., 1988). However, it is also more computationally expensive, as it requires more iterations to converge (c.f., the slow cooling schedule of scale parameter K). SOM scales up better and has been applied on much larger problems. In (Favata and Walker, 1991), for example, the authors report results on problems with up to 10,000 vertices. The solutions obtained were about 5% worse than those produced by a simulated annealing heuristic.

4 PRACTICAL ISSUES

While the previous sections presented a review of Hopfield neural networks, elastic nets and self-organizing maps, this section aims to discuss some of the more practical issues affecting their performance. These have been separated into issues affecting the efficiency of the models, and further issues that have proven to be problematic for researchers and need careful consideration.

4.1 Hopfield Model

4.1.1 Efficiency Issues

The efficiency of the Hopfield network for solving combinatorial optimization depends significantly on the way in which the problem is mapped onto the network. The encoding of the variables, constraints, and objective function into the Hopfield energy function and the values of the penalty parameters combine to determine the complexity of the energy surface. This, in turn, affects the degree to which the Hopfield network dynamics are able to search for local minima.

Problem Representation It is interesting to note that all of the Hopfield network applications to the TSP have been based on the formulation used by Hopfield and Tank which uses a decision variable denoting if a city belongs to a given position in the tour sequence. The operations research community however has based many of its TSP methods on another formulation: here the decision variable X_{ij} denotes if city i follows city j in the tour sequence. This alternative formulation results in a linear objective function, but some complex constraints are needed to avoid sub-tours. Hopfield and Tank's method cannot readily be applied to the linear formulation due to the difficulty of encoding this constraint (Smith et al., 1996). Nevertheless, this raises the issue of the existence of alternative formulations for a given problem, and the degree to which this impacts the efficiency of any resulting Hopfield networks. A recent study of a Hopfield network application to school timetabling has shown that, where two formulations exist for a problem, the chosen formulation greatly impacts the dimensionality of the neurons, the number of constraints needed to represent the problem, and the complexity of the energy surface (Smith et al., 1999).

Energy Function There are frequently several different ways of enforcing constraints in an energy function. As was shown in Section 2.4, Hopfield and Tank's TSP row and column constraints can be rewritten to explicitly state that each row and column must sum to one, thus eliminating the need for their third term in the energy function (Brandt et al., 1988). For other constraint types also, there are often several different approaches to incorporating them into the energy function (Peterson and Soderberg, 1993). Other researchers avoid the need for representing constraints in the energy function by modifying the network dynamics. This is the approach taken by mean field annealing (Van den Bout and Miller, 1989), where a row constraint stating that the neurons must sum to one is handled by normalizing the neuron states in each row. The aim in exploring different representations of constraints is to reduce the number of terms and parameters needed in the energy function, as well as perhaps to reduce the number of local minima generated by these terms. Other terms often added to the energy function include one of the form $\sum_{i=1}^{N} v_i(1-v_i)$ which is designed to encourage convergence to binary valued neuron states. This term essentially adds a concave term to the original energy surface, thus altering the convexity and driving the solution towards the vertices of the unit hypercube.

Penalty Factors The choice of penalty factor values affect the contour of the energy function surface, and thus greatly affect the ability of the Hopfield network to find local minima of the optimization problem. For many types of constraints, the penalty factors can be treated as equivalent (e.g., penalty factors for row and column constraints in the TSP should be identical, since these constraints are equally important and equally difficult to satisfy). This observation can often reduce the search space for the optimal penalty factor combination. Many researchers have attempted to eliminate the need for trial and error parameter selection by examining the theoretical balancing of terms in the energy function. For the TSP, Hedge et al. (1988) showed that, while some regions of parameter space can be identified that yield better quality results, the size of these regions diminishes rapidly as the problem size increases. Kamgar-Parsi and Kamgar-Parsi (1992) developed a systematic method for selecting the penalty factors based on analyzing the dynamical stability of feasible solutions. Trial and error searching however does not necessarily preclude a systematic method. The efficiency of searching for optimal penalty parameter values can be improved by adopting the following systematic approach: first find values for the penalty factors that provide a feasible solution, holding the objective function penalty factor constant at unity. Once a combination of penalty factors has been found that consistently yields feasible solutions, slowly start to increase the objective function factor in an attempt to produce less expensive feasible solutions. As soon as feasibility is lost, the bounds on this parameter can be established. This much reduced search space can then be explored in more detail to obtain the combination of penalty factors that yields consistently feasible and optimal solutions.

4.1.2 Problematic Issues

There are a number of important issues that researchers should be aware of when developing Hopfield network models for solving combinatorial optimization problems. Some of these issues are rarely addressed in the available literature, yet they dramatically affect the degree to which the Hopfield network follows its theoretical behavior.

Diagonal Weights A common misconception in the literature is that zero diagonals of the weight matrix ($W_{ii} = 0$) are necessary for the stable states of the continuous model to coincide with those of the original discrete model. This belief has no doubt evolved due to Hopfield's original simplifying assumption that $W_{ii} = 0$. In fact, there are no restrictive conditions on the diagonals of $\mathbf{W}$ for the continuous model to converge to a minimum of E_c. If $W_{ii} < 0$ however, that minimum may lie in the interior of the hypercube, due to the convexity of the energy function. In this case, annealing techniques are usually employed to drive the solution trace towards the vertices.

Unlike the continuous model however, non-zero diagonal elements of the discrete Hopfield network do not necessarily allow Liapunov descent to local minima of E_d.

This is because the change in energy E_d due to a change in output level v_i is

$$\Delta E_d = -(u_i \Delta v_i) - \tfrac{1}{2} W_{ii} (\Delta v_i)^2 \tag{22}$$

Since $u_i \geq 0$ results in $\Delta v_i \geq 0$, and $u_i < 0$ results in $\Delta v_i \leq 0$ under the discrete model, the first term on the right-hand side of (22) is always negative. The second term is positive however for $W_{ii} < 0$. Consequently, ΔE_d is only negative provided

$$||\Delta v_i|| \leq 2 \frac{||u_i||}{||W_{ii}||} \tag{23}$$

Thus, for non-negative diagonal elements, convergence of the discrete model is guaranteed. For negative diagonals, however, convergence may not be possible since Δv_i is large for discrete dynamics. The continuous model does not suffer from these problems if implemented using the differential equation system (4) and (5). When these equations are simulated on a computer, however, the time discretization means that this model too may become unstable if the changes to neuron states are too large.

Consequently, it is important to be aware that, for problems with negative diagonals of the weight matrix (this includes many practical applications, as well as Hopfield and Tank's original TSP formulation), strict Liapunov descent for even the continuous model will become harder to simulate and cannot be guaranteed during simulation in the extreme high-gain limit (i.e., T near zero) of the continuous transfer function (which approximates the discrete case). If $W_{ii} < 0$, the best we can do for simulation purposes is to make sure the gain of the transfer function is not so high that we are approximating the discrete model too effectively. We can then use an annealing technique to drive the network towards a vertex solution. Of course, these problems are not encountered in a hardware implementation where the differential equation (4) does not need to be simulated.

There is also a relationship worth noting between the diagonal elements of the weight matrix and the shape of the energy surface. As mentioned above, adding a term of the form $\lambda \sum_{i=1}^{N} v_i(1 - v_i)$ to the energy function serves to encourage convergence to the vertices of the unit hypercube. The process of hysteretic annealing proposes starting with a negative value for λ and slowly increasing to positive so that the energy surface is converted from convex to concave, driving the solutions out to the vertices. The idea of slowly decaying away the diagonal weight elements has also been shown to lead to chaotic dynamics, which have been useful for escaping local minima of the energy function (Chen and Aihara, 1995).

Simulation While the continuous Hopfield network is designed to be implemented in hardware, most researchers simulate its behavior on a computer. This involves using a discrete-time approximation to the differential equation (4), such as the Euler approximation:

$$u_i(t+1) = u_i(t) + \Delta t \frac{du_i}{dt} = u_i(t)\left[1 - \frac{\Delta t}{\tau}\right] + \sum_{j=1}^{n} W_{ij} v_j(t) + I_i \tag{24}$$

If the time-step Δt is too large, the dynamics of the simulated network will not closely approximate the continuous model, and Liapunov descent cannot be guaranteed. This can be seen also by equations (22) and (23) since a large time step produces a large change in neuron states $\boldsymbol{v}$. If the time-step is too small, convergence can be guaranteed but the time required to simulate the network will be increased. For a TSP with N cities, it has been shown (Kamgar-Parsi and Kamgar-Parsi, 1987) that a time-step for the simulation of (4) as small as $\Delta t < O(1/N^2)$ is required, slowing down simulations immensely for large problem sizes, and making the discretization of the continuous model quite ineffective. The value chosen for the transfer function parameter T also needs to be selected with care, since it too can produce large changes in neuron states $\mathbf{v}$ and result in oscillations. Kamgar-Parsi and Kamgar-Parsi (1987) have proposed a discrete-time Hopfield network model with continuous dynamics that satisfies the condition in (23) and produces quick convergence.

Initial States Since the Hopfield network is a gradient descent technique, the initial states of the neurons play a large role in determining the final solution quality. Certainly, the discrete-time versions of the networks incorporate some stochasticity since neurons are selected for updating at random (only the continuous model implemented in hardware is truly deterministic). Nevertheless, this degree of stochasticity is rarely enough to avoid the high dependence of the solution on the initial states. Most researchers report their results based on multiple random and unbiased initializations to avoid this dependence. Some work has been conducted to identify heuristic methods for more effective initializations (Lai and Coghill, 1994; Naphade and Tuzun, 1995; Schwartz et al., 1991).

Updating Mode For any discrete-time version of the Hopfield network (discrete model or simulated continuous model), there are two ways of updating the neurons: asynchronously (sequentially) or synchronously (in parallel). Asynchronous updating is the method described in Section 2, whereby a neuron is chosen at random and its internal and external states are updated before the next random neuron is chosen. Liapunov descent is guaranteed under these conditions (provided equation (23) holds true). For synchronous updating however, all neurons update their internal states, and then the updated values are used to simultaneously update the external states. Under this updating mode, convergence cannot be guaranteed and the network may become trapped in a 2-cycle (Bruch, 1990). Most researchers use the asynchronous mode of updating to avoid such oscillations.

Solution Integrality For combinatorial optimization problems the final solution needs to be binary, even if the method used approaches this solution from within the unit hypercube. Using the discrete Hopfield model is appropriate for ensuring binary solutions, but the problems with convergence for many practical problems where $W_{ii} < 0$ have resulted in most researchers turning to the continuous model. There are several approaches to encouraging binary solutions within the continuous model. Firstly, we can use a gradient of the transfer function with T near zero (which approximates the discrete transfer function (2)), but this risks producing convergence problems like the discrete model, as discussed above. Secondly, we can add an additional term to the energy function to drive towards the vertices. While this introduces another parameter to select, some additional advantages can be found due to the chaotic effect on the dynamics and the hill-climbing this creates (Chen and Aihara, 1995). Other methods have been proposed including a variety of rounding approaches and heuristics for interpretation of final non-integer solutions (Wolfe, 1999).

Termination Criteria One of the factors that has made it difficult for researchers to reproduce Hopfield and Tank's original results is that they omitted certain critical details in their paper about the method used to simulate the differential equation, and the termination criteria. Wilson and Pawley (1988) experimented with three different termination criteria in an effort to reproduce the results: the network simulation was terminated if (i) a valid tour was found, (ii) the network had frozen as measured by no neuron $\boldsymbol{v}$ values changing by more than 10^{-35} since the last update, and (iii) more than 1000 updating iterations had elapsed (a "time-out" test, useful for catching cyclic and oscillatory convergence). Wilson and Pawley found that 1000 iterations were sufficient for their experiments, and increasing the "time-out" range to 10,000 iterations did not produce any improvement in results. It is important to be aware, however, that the quality of the reported results is usually affected greatly by the termination criteria selected, and researchers need to be sure to report these accurately.

4.2 Elastic Net

4.2.1 Efficiency Issues

In the EN algorithm, updating the position of the ring points is computationally expensive, as the update of a single point depends on the position of every vertex, through the attraction coefficients w_{ij}. Furthermore, these coefficients must be recomputed at each iteration. For $M \approx N$, the complexity of each iteration is thus $O(N^2)$. Different approaches have been proposed to reduce this burden.

Filtering A natural way of addressing the problem without degrading too much solution quality is to consider only vertices that have a significant impact on the ring points (Boeres et al., 1992; Vakhutinsky and Golden, 1995). In other words, attraction coefficients that are not of sufficient magnitude, because their vertices are too far from the corresponding ring points, are filtered out. A large number of coefficients may be eliminated in this way because the function $\phi(d, K)$ decreases quickly as the square of the distance d^2 grows.

Hierarchical EN The idea of the hierarchical EN (Vakhutinsky and Golden, 1995) is to divide the area containing the vertices into smaller subareas and to replace the vertices in each subarea by a single vertex located at their center of gravity. As the algorithm unfolds, the subareas are progressively reduced until each subarea contains a single vertex. It is reported that working on smaller aggregated problems at the start allows the algorithm to find the general shape of the solution more quickly.

4.2.2 Problematic Issues

When implementing an elastic net algorithm, the following issues should be taken care of:

Normalization of Coordinate Vectors The behavior of EN may be quite different from one problem instance to another. Some authors have noted that a more robust algorithmic behavior is obtained by normalizing the coordinate vectors of the vertices (Favata and Walker, 1991).

Initial Ring The number of ring points M to be used is clearly related to the number of vertices N. However, using too many points leads to a loss of efficiency. In the literature, M is usually set around $2.5N$. The initial position of the ring is typically around the center of gravity of the vertices. Good results are also reported when the points on the ring correspond to the convex hull of the vertex set (Burke, 1994).

Ring Migration When the value of parameter K is large, the energy function is rather smooth, but as this value is reduced a multimodal energy landscape emerges,

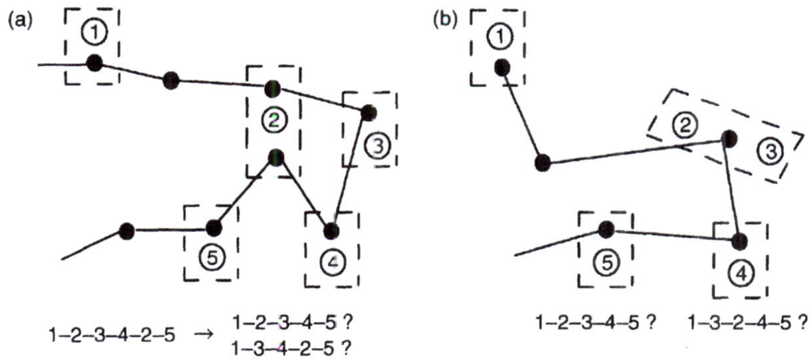

Figure 15.4. Solution ambiguity.

where good local minima should correspond to good tours. In order to obtain this result, parameter K must be slowly reduced to avoid some form of twisting or crossover of the ring (which typically leads to long tours). For example, a problem with 100 vertices was solved in (Durbin and Willshaw, 1987) by setting K to an initial value of 0.2 and by reducing it by 1% every 25 iterations until a value in the range 0.01–0.02 was obtained.

Interpretation of the Final Configuration Two or more different ring points may be associated with the same vertex if they are all within the tolerance of that vertex (this is sometimes referred to as a spike). Conversely, a ring point may be associated with two or more vertices. In such cases, the solution is not well defined. In Figure 15.4(a), two ring points are within the tolerance of vertex 2, and two different sequences 1-2-3-4-5 and 1-3-4-2-5 are obtained depending of the ring point chosen. In Figure 15.4(b), a single ring point is associated with vertices 2 and 3. Hence, it is impossible to know if vertex 2 is visited before or after vertex 3. One possible way of solving these problems is through postprocessing. For example, all valid solutions obtainable with the current configuration can be considered and the best overall solution is taken. Trying to avoid this phenomenon during the course of the algorithm is rather difficult. Some theoretical studies indicate that an appropriate setting of the β/α ratio (so that it is about one-half of the average inter-point distance on the ring) is likely to lead to fully-specified solutions (Simmen, 1991).

4.3 Self-organizing Map

4.3.1 Efficiency Issues

As noted by different authors (Burke and Damany, 1992; Favata and Walker, 1991), significant gains in efficiency are obtained by reducing the number of ring points that move towards the current vertex (i.e., those that are close the the winning point) and also by reducing the magnitude of the move. In Step 2, the update mechanism is governed by function $f(j, j^*)$ which has a form like:

$$\begin{aligned} f(j, j^*) &= \left(1 - \frac{d''_{jj^*}}{L}\right)^{\beta}, \quad if\ d''_{jj^*} < L \\ &= 0, \qquad\qquad\qquad otherwise \end{aligned} \tag{25}$$

where $d''_{jj^*} = \min(|j - j^*|, m - |j - j^*|)$.

In this definition, d'' is the lateral distance on the ring between point j and winning point j^* (assuming that the points on the ring are indexed from 1 to M), and $|x|$ is the absolute value of x. This function is such that it always returns a value of 1 for the winning point, and this value decreases as the lateral distance from the winning point increases; when the lateral distance goes beyond parameter L, the points do not move at all. The value of parameter β is increased from one pass to another to progressively reduce the magnitude of the move of neighboring units. At the end, when β is sufficiently large, only the winning unit moves towards the current vertex and separates from its neighbors to fix the solution.

4.3.2 Problematic Issues

Since the SOM is closely related to EN, many issues mentioned above for EN still hold here. We thus focus on a specific issue, which is related to the mechanics of the SOM for updating the ring location.

Ring Point Freezing Solution ambiguity can occur, either because many ring points fall within the tolerance of a given vertex or a single ring point falls within the tolerance of two or more vertices. However, due to the specific methodology adopted for stretching the ring, which is based on a competition between ring points, it may also happen that a number of points will freeze at their initial location, because they never win any competition. Consequently, partially-defined tours are obtained, where a number of vertices are ⟨⟨orphans⟩⟩ (i.e., do not have any close ring points). Different techniques have been proposed to alleviate this problem:

— In Angeniol et al. (1988), the implementation is based on the distinctive feature that ring points are dynamically created and deleted. A point is duplicated if it wins for two different vertices after a complete pass through the set of vertices. It is deleted, if it does not win after three complete passes. Through this mechanism, vertices are less likely to end up alone. Starting with a single point on the ring, the authors report that up to twice as many points as vertices may be created during the procedure.

— In Burke and Damany (1992), a conscience mechanism proposed by Desieno (1988) replaces the dynamic creation and deletion of ring points. A penalty is added to the distance between a ring point and a vertex, based on the number of times that point has won the competition in the past. Consequently, frequent winners are heavily penalized in favor of other units. Basically, Step 1.2 (competition) in the SOM algorithm of Section 3.2 is modified as follows for a given vertex i and ring point j:

$$o_j \leftarrow d_{X_i Y_j} + \gamma b_j,$$

where b_j is the penalty or bias associated with ring point j. This penalty is typically the fraction of competitions won by ring point j in the past. Good results are reported with a number of ring points now equal to the number of vertices, leading to substantial savings in computation time.

— Parameter γ that weighs the penalty with respect to the true distance in the conscience mechanism, is reportedly difficult to tune. In (Burke, 1994), a vigilant net is proposed where ring points are turned off if they win too often to let others win. Basically, the number of wins is recorded for each unit and that unit is turned off for the remaining of the pass through the set of vertices, if this number exceeds some threshold value (known as the vigilant parameter). At the start of the next pass, the winning score of all units is reset to zero. The vigilance parameter is large initially, to allow the vertices to win freely at the start, and is progressively reduced until it reaches a value of one, to let the ring points separate and converge towards distinct vertices

5 CONCLUDING REMARKS

This chapter has reviewed the two main types of neural network models that can be used for combinatorial optimization: Hopfield networks and the deformable template models of elastic nets and self-organizing maps. The review has covered both the theoretical aspects of their application to combinatorial optimization problems, as well as discussing a variety of practical considerations that affect the performance of the models.

From a metaheuristics viewpoint, neural networks can be seen as an alternative technique with the current potential to match the performance of better known algorithms such as tabu search and simulated annealing. This potential relies on due consideration of the aforementioned range of issues affecting the success and efficiency of the methods. The deformable template methods are well suited to solving low dimensional problems with geometric interpretation like the TSP. The Hopfield network method generalizes to a broad range of combinatorial problems, but the cost of this generalization is a reduction in efficiency and scalability. Certainly, current developments in hardware implementation of neural architectures should see some of these limitations relaxed in future years. The advantage of neural networks over other metaheuristic techniques could then be more fully demonstrated.

ACKNOWLEDGMENTS

This work was partly supported by the Canadian Natural Sciences and Engineering Research Council (NSERC) and the Quebec Fonds pour la Formation de Chercheurs et l'Aide a la Recherche (FCAR). This support is gratefully acknowledged.

REFERENCES

Aarts, E.H.L. and Korst, J. (1989) *Simulated Annealing and Boltzmann Machines.* John Wiley & Sons., Essex.

Aiyer, S.V.B., Niranjan, M. and Fallside, F. (1990) A theoretical investigation into the performance of the Hopfield model. *IEEE Transactions on Neural Networks*, **1**, 204–215.

Akiyama, Y., Yamashita, A., Kajiura, M. and Aiso, H. (1989) Combinatorial optimization with gaussian machines. *Proceedings IEEE International Joint Conference on Neural Networks*, **1**, 533–540.

Amartur, S.C., Piraino, D. and Takefuji, Y. (1992) Optimization neural networks for the segmentation of magnetic resonance Images. *IEEE Transactions on Medical Imaging*, **11**, 215–220.

Angeniol, B., Vaubois, G. and Le Texier, J.Y. (1988) Self-organizing feature maps and the travelling salesman problem. *Neural Networks*, **1**, 289–293.

Barr, R.S., Golden, B.L., Kelly, J.P., Resende, M.G.C. and Stewart, W.R. (1995) Designing and reporting on computational experiments with heuristic methods. *Journal of Heuristics*, **1**, 9–32.

Boeres, M.S.C., de Carvalho, L.A.V. and Barbosa, V.C. (1992) A faster elastic net algorithm for the traveling salesman problem. In: *Proceedings of the International Joint Conference on Neural Networks*. Baltimore, U.S.A., II-215-220.

Brandt, R.D., Wang, Y., Laub, A.J. and Mitra, S.K. (1988) Alternative networks for solving the travelling salesman problem and the list-matching problem. *Proceedings International Conference on Neural Networks* Vol. 2, 333–340.

Bruch, J. (1990) On the convergence properties of the Hopfield model. *Proceedings of the IEEE*, **78**(10), 1579–1585.

Burke, L.I. (1994) Adaptive neural networks for the traveling salesman problem: insights from operations research. *Neural Networks*, **7**, 681–690.

Burke, L.I. and Damany, P. (1992) The guilty net for the traveling salesman problem. *Computers & Operations Research*, **19**, 255–265.

Burke, L.I. and Ignizio, J.P. (1992) Neural networks and operations research: an overview. *Computers & Operations Research*, **19**, 179–189.

Chen, L. and Aihara, K. (1995) Chaotic simulated annealing by a neural network model with transient chaos. *Neural Networks*, **8**(6), 915–930.

Desieno, D. (1988) Adding a conscience mechanism to competitive learning. In: *Proceedings of the IEEE International Conference on Neural Networks*, San Diego, U.S.A., I-117-124.

Durbin, R. and Willshaw, D.J. (1987) An analogue approach to the traveling salesman problem using an elastic net method. *Nature*, **326**, 689–691.

Favata, F. and Walker, R. (1991) A study of the application of Kohonen-type neural networks to the traveling salesman problem. *Biological Cybernetics*, **64**, 463–468.

Foo, Y.P.S. and Szu, H. (1989) Solving large-scale optimization problems by divide-and-conquer neural networks. *Proceedings IEEE International Joint Conference on Neural Networks*, **1**, 507–511.

Garey, M.R. and Johnson, D.S. (1979) *Computers and Intractability*. W.H. Freeman, New York.

Gee, A.H. (1993) *Problem Solving with Optimization Networks*, Ph.D. Dissertation, Queen's College, Cambridge University, U.K.

Ghaziri, H. (1991) Solving routing problems by a self-organizing map. In: T. Kohonen, K. Makisara, O. Simula and J. Kangas (eds.), *Artificial Neural Networks*. North-Holland, Amsterdam, pp. 829–834.

Ghaziri, H. (1996) Supervision in the self-organizing feature map: application to the vehicle routing problem. In: I.H. Osman and J.P. Kelly (eds.), *Meta-Heuristics: Theory & Applications*. Kluwer, Boston, pp. 651–660.

Glover, F. (1994) Optimization by ghost image processes in neural networks. *Computers & Operations Research*, **21**, 801–822.

Goldstein, M. (1990) Self-organizing feature maps for the multiple traveling salesmen problem. In: *Proceedings of the International Neural Network Conference*. Paris, France, pp. 258–261.

Hegde, S., Sweet, J. and Levy, W. (1988) Determination of parameters in a hopfield/tank computational network. *Proceedings IEEE International Conference on Neural Networks*, **2**, 291–298.

Hinton, G.E., Sejnowski, T.J. and Ackley, D.H. (1984) Boltzmann machines: constraint satisfaction networks that learn. Carnegie Mellon University Technical Report CMU-CS-84-119.

Hopfield, J.J. (1982) Neural networks and physical systems with emergent collective computational abilities. *Proceedings of the National Academy of Sciences*, **79**, 2554–2558.

Hopfield, J.J. (1984) Neurons with Graded response have collective computational properties like those of two-state neurons. *Proceedings of the National Academy of Sciences*, **81**, 3088–3092.

Hopfield, J.J. and Tank, D.W. (1985) Neural computation of decisions in optimization problems. *Biological Cybernetics*, **52**, 141–152.

Hooker, J.N. (1995) Testing heuristics: we have it all wrong. *Journal of Heuristics*, **1**, 33–42.

Johnson, D.S. (1990) Local optimization and the traveling salesman problem. In: G. Goos and J. Hartmanis (eds.), *Automata, Languages and Programming, Lecture Notes in Computer Science 443*. Springer-Verlag, Berlin, pp. 446–461.

Kamgar-Parsi, B. and Kamgar-Parsi, B. (1987) An efficient model of neural networks for optimization. *Proceedings IEEE International Conference on Neural Networks*, **3**. 785–790.

Kamgar-Parsi, B. and Kamgar-Parsi, B. (1992) Dynamical stability and parameter selection in neural optimization. *Proceedings IEEE International Conference on Neural Networks*, **4**, 566–571.

Kirkpatrick, S., Gelatt, C.D. and Vecchi, M.P. (1983) Optimization by simulated annealing. *Science*, **220**, 671–680.

Kohonen, T. (1982) Self-organized formation of topologically correct feature maps. *Biological Cybernetics*, **43**, 59–69.

Kohonen, T. (1988) *Self-Organization and Associative Memory*. Springer, Berlin.

Lai, W.K. and Coghill, G.G. (1992) Genetic breeding of control parameters for the Hopfield/Tank neural net. *Proceedings International Joint Conference on Neural Networks*, **4**, 618–623.

Lai, W.K. and Coghill, G.G. (1994) Initialising the continuous Hopfield net. *Proceedings IEEE International Conference on Neural Networks*, **7**, 4640–4644.

Lin, S. and Kernighan, B.W. (1973) An effective heuristic algorithm for the travelling salesman problem. *Operations Research*, **21**, 498–516.

Lo, J.T.-H. (1992) A new approach to global optimization and its applications to neural networks. *Proceedings IEEE International Joint Conference on Neural Networks*, **4**, 600–605.

Looi, C.K. (1992) Neural network methods in combinatorial optimization. *Computers & Operations Research*, **19**, 191–208.

Matsuyama, Y. (1991) Self-organization via competition, cooperation and categorization applied to extended vehicle routing problems. In: *Proceedings of the International Joint Conference on Neural Networks*. Seattle, U.S.A., I-385-390.

McCulloch, W.S. and Pitts, W. (1943) A logical calculus of ideas immanent in nervous activity. *Bulletin of Mathematical Biophysics*, **5**, 115–133.

Minsky, M. and Papert, S. (1969) *Perceptrons*. MIT Press, Cambridge, MA.

Naphade, K. and Tuzun, D. (1995) Initializing the Hopfield–Tank network for the TSP using a convex hull: a computational study. *Intelligent Engineering Systems Through Artificial Neural Networks*. vol. 5. ASME Press, New York, pp. 399–404.

Nemhauser, G.L. and Wolsey, L.A. (1988) *Integer and Combinatorial Optimization*. John Wiley & Sons, Canada.

Peterson, C. and Soderberg, B. (1989) A new method for mapping optimization problems onto neural networks. *International Journal of Neural Systems*, **1**, 3–22.

Peterson, C. and Soderberg, B. (1993) Artificial neural networks. In: C.R. Reeves (ed.), *Modern Heuristic Techniques for Combinatorial Optimisation*, Chapter 5. Blackwell Scientific Publishers, Oxford, UK.

Potvin, J.Y. (1993) The traveling salesman problem: a neural network perspective. *ORSA Journal on Computing*, **5**, 328–348.

Potvin, J.Y. and Robillard, C. (1995) Clustering for vehicle routing with a competitive neural network model. *Neurocomputing*, **8**, 125–139.

Rosenblatt, F. (1958) The perceptron: a probabilistic model for information storage and organization in the brain. *Psychological Review*, **65**, 386–408.

Rumelhart, D.E. and McClelland, J.L. (1986) *Parallel distributed processing: explorations in the microstructure of cognition*, I & II. MIT Press, Cambridge, MA.

Schwartz, B.L., Das, P. and Koch, J.B. (1991) Initialization in Hopfield networks. *Neurocomputing*, **3**(3), 135–145.

Simmen, M.W. (1991) Parameter sensitivity of the elastic net approach to the traveling salesman problem. *Neural Computation*, **3**, 363–374.

Smith, K.A. (1995) Solving the generalized quadratic assignment problem using a self-organizing process. In: *Proceedings IEEE International Conference on Neural Networks 4*, Perth, Australia, pp. 1876–1879.

Smith, K.A. (1999) Neural networks for combinatorial optimization: a review of more than a decade of research. *INFORMS Journal on Computing*, **11**, 15–34.

Smith, K.A., Abramson, D. and Duke, D. (1999) Efficient timetabling formulations for Hopfield neural networks. In: C. Dagli et al. (eds.), *Smart Engineering System Design: Neural Networks, Fuzzy Logic, Evolutionary Programming, Data Mining, and Complex Systems*, vol. 9. ASME Press, pp. 1027–1032.

Smith, K.A., Palaniswami, M., Krishnamoorthy, M. (1996) A hybrid neural approach to combinatorial optimization. *Computers & Operations Research*, **23**, 597–610.

Szu, H. and Hartley, R. (1987) Fast simulated annealing. *Physics Letters A*, **122**, 157–162.

Szu, H. (1988) Fast TSP algorithm based on binary neuron output and analog input using zero-diagonal interconnect matrix and necessary and sufficient conditions of the permutation matrix. *Proceedings IEEE International Conference on Neural Networks*, **2**, 259–266.

Takefuji, Y. and Szu, H. (1989) Design of parallel distributed Cauchy machines *Proceedings IEEE International Joint Conference on Neural Networks*, **1**, 529–532.

Tank, D.W. and Hopfield, J.J. (1986) Simple neural optimization networks: an A/D converter, signal decision circuit and a linear programming circuit. *IEEE Transactions on Circuit Systems*, **33**, 533–541.

Vakhutinsky, A.I. and Golden, B.L. (1994) Solving vehicle routing problems using elastic nets. *Proceedings of the IEEE International Conference on Neural Networks*, **7**, 4535–4540.

Vakhutinsky, A.I. and Golden, B.L. (1995) A hierarchical strategy for solving traveling salesman problems using elastic nets. *Journal of Heuristics*, **1**, 67–76.

Van Den Bout, D.E. and Miller, T.K. (1988) A travelling salesman objective function that works. *Proceedings IEEE International Conference on Neural Networks*, **2**, 299–303.

Van Den Bout, D.E. and Miller, T.K. (1989) Improving the performance of the Hopfield–Tank neural network through normalization and annealing. *Biological Cybernetics*, **62**, 129–139.

Van Den Bout, D.E. and Miller, T.K. (1990) Graph partitioning using annealed neural networks. *IEEE Transactions on Neural Networks*, **1**, 192–203.

Willshaw, D.J. and von der Malsburg, C. (1979) A marker induction mechanism for the establishment of ordered neural mappings: its application to the retinotectal problem. *Philosophical Transactions of the Royal Society, Series B*, **287**, 203–243.

Wilson, G.V. and Pawley, G.S. (1988) On the stability of the TSP algorithm of Hopfield and Tank. *Biological Cybernetics*, **58**, 63–70.

Wolfe, W.J. (1999) A fuzzy Hopfield–Tank traveling salesman problem model. *INFORMS Journal on Computing*, **11**(4), 329–344.

Chapter 16

HYPER-HEURISTICS: AN EMERGING DIRECTION IN MODERN SEARCH TECHNOLOGY

Edmund Burke, Graham Kendall and Jim Newall
The University of Nottingham, UK

Emma Hart, Peter Ross and Sonia Schulenburg
Napier University, UK

Abstract This chapter introduces and overviews an emerging methodology in search and optimisation. One of the key aims of these new approaches, which have been termed *hyper-heuristics*, is to raise the level of generality at which optimisation systems can operate. An objective is that hyper-heuristics will lead to more general systems that are able to handle a wide range of problem domains rather than current meta-heuristic technology which tends to be customised to a particular problem or a narrow class of problems. Hyper-heuristics are broadly concerned with intelligently choosing the right heuristic or algorithm in a given situation. Of course, a hyper-heuristic can be (often is) a (meta-)heuristic and it can operate on (meta-)heuristics. In a certain sense, a hyper-heuristic works at a *higher* level when compared with the typical application of meta-heuristics to optimisation problems, i.e., a hyper-heuristic could be thought of as a (meta)-heuristic which operates on *lower* level (meta-)heuristics. In this chapter we will introduce the idea and give a brief history of this emerging area. In addition, we will review some of the latest work to be published in the field.

Keywords: Hyper-heuristic, Meta-heuristic, Heuristic, Optimisation, Search

1 INTRODUCTION

Meta-heuristics have played a key role at the interface of Artificial Intelligence and Operational Research over the last 10–15 years or so. The investigation of meta-heuristics for a wide and diverse range of application areas has strongly influenced the development of modern search technology. Indeed, applications of meta-heuristic development can be found in such diverse areas as scheduling, data mining, stock cutting, medical imaging and bio-informatics, and many others. However, while such developments have deepened our scientific understanding of the search process and the automatic solution of large complex problems, it is true to say that the practical impact in commercial and industrial organisations has not been as great as might have been expected some years ago. Many state-of-the-art meta-heuristic developments are too *problem-specific* or too *knowledge-intensive* to be implemented in cheap, easy-to-use computer systems. Of course, there are technology provider companies that have brought such developments to market but such products tend to be expensive and their development tends to be very resource intensive. Often, users employ simple heuristics which are easy to implement but whose performance is often rather poor. There is a spectrum which ranges from cheap but fragile heuristics at one extreme and knowledge-intensive methods that can perform very well but are hard to implement and maintain at the other extreme. Many small companies are not interested in solving their optimisation problems to optimality or even close to optimality. They are more often interested in "*good enough—soon enough—cheap enough*" solutions to their problems. There is a current school of thought in meta-heuristic and search technology that contends that one of the main goals of the discipline over the next few years is to raise the level of generality at which meta-heuristic and optimisation systems can operate. This would facilitate the development of easy to implement (and cheap to implement) systems that can operate on a range of related problems rather than on one narrow class of problems. Of course, many papers in the literature discuss meta-heuristic and heuristic development on just one problem instance. Such papers provide a valuable insight into meta-heuristic development but they offer little assistance to a small company that simply cannot afford the significant amount of resources that are required to *tailor make* special purpose meta-heuristics for the company's own version of whatever optimisation problem it has.

This chapter is concerned with ***hyper-heuristics*** which is an emerging search technology that is motivated, to a large extent, by the goal of raising the level of generality at which optimisation systems can operate. The term has been defined to broadly describe the process of using (meta-)heuristics to *choose* (meta-)heuristics to solve the problem in hand. The majority of papers in the (meta-)heuristics area investigate the use of such approaches to operate directly on the problem. For example, most of the papers on Evolutionary Computation in timetabling consider populations of timetables and the basic idea is that the population will evolve over a number of generations with the aim of generating a strong population. However, a hyper-heuristic Evolutionary approach to timetabling [1,18] would deal with a population of (meta-)heuristics for the timetabling problem and, over a number of generations, it is these (meta-)heuristics that would evolve. Another example of the use of hyper-heuristics is presented in [2] where a genetic algorithm evolves the choice of heuristic (in open shop scheduling) whenever a task is to be added to the schedule under construction. Of course, in 1994, the term hyper-heuristic did not exist and the process was called "evolving heuristic choice." These examples are discussed in more detail later in the chapter. There are many more examples of such approaches and indeed one of the main purposes of this chapter is to give an overview of them.

One of the main motivations for studying hyper-heuristic approaches is that they should be cheaper to implement and easier to use than problem specific *special purpose* methods and the goal is to produce good quality solutions in this more general framework. Of course, the overall aim of the hyper-heuristic goal goes beyond meta-heuristic technology. Indeed, there is current work which is investigating the development of machine learning approaches (such as case-based reasoning) to intelligently select heuristics according to the situation in hand [3,4]. Such an approach is very much concerned with the same goal that motivates hyper-heuristic development. Actually, there is considerable scope for hybridising meta-heuristics with such machine learning approaches to intelligent heuristic selection [4].

This chapter is not intended to be an intensive survey of all the scientific papers which are related to the basic hyper-heuristic definition and that seek to satisfy the same objective. Rather, it is intended to give a brief overview of the idea and to set it within the context of the current state-of-the-art in meta-heuristic technology.

2 THE EMERGENCE OF HYPER-HEURISTICS

2.1 The Concept and its Origins

This section builds on the concept of hyper-heuristics and describes some early examples.

For many real-world problems, an exhaustive search for solutions is not a practical proposition. The search space may be far too big, or there may not even be a convenient way to enumerate the search space. For example, there may be elaborate constraints that give the space of feasible solutions a very complex shape. It is common then to resort to some kind of heuristic approach, sacrificing a guarantee of finding an optimal solution for the sake of speed and perhaps also a guarantee of obtaining at least a certain level of solution quality. Enormous numbers of heuristics have been developed over the years, each typically justified either by experimental results or by an argument based on the specific problem class for which the heuristic in question had been tailored.

The term 'heuristic' is sometimes used to refer to a whole search algorithm and is sometimes used to refer to a particular decision process sitting within some repetitive control structure. Viewing heuristics as search algorithms, some authors have occasionally tried to argue for the absolute superiority of one heuristic over another. This practice started to die out when in 1995 Wolpert and MacReady [5] published their "No Free Lunch Theorem" which showed that, when averaged over all problems defined on a given finite search space, all search algorithms had the same average performance. This is an intuitively natural result since the vast majority of possible problems have no exploitable structure whatsoever, such as some form of global or local continuity, differentiability or regularity. They can only be defined by a complete lookup table. The "No Free Lunch Theorem" helped to focus attention on the question of what sorts of problems any given algorithm might be particularly useful for.

Long before the theorem was published it was clear that individual heuristics, however valuable, could have interesting quirks and limitations. Consider, for example, the topic of one-dimensional bin-packing. In its simplest incarnation there is an unlimited supply of identical bins and there is a set of objects to be packed into as few bins as possible. Each object has an associated scalar (think of it as the object's weight) and a bin cannot carry more than a certain total weight. The general task of finding an optimal assignment of objects to bins is NP-hard. A commonly used heuristic is 'largest first, first fit': sort the objects into decreasing order of weight, then, taking them in this order, put each object into the first bin into which it will fit (the bins in use are ordered too, according to when they first came into use). This heuristic has the benefits of simplicity and cheapness; it may not produce a solution that uses the minimal number M of bins, but it is known that it will not use more than $11M/9+4$ bins [6]. See [7] for a good survey of such results. A worst-case performance guarantee of this sort can be very reassuring

Table 16.1. A bin-packing problem

442	252	127	106	37	10	10
252	252	127	106	37	10	9
252	252	127	85	12	10	9
252	127	106	84	12	10	
252	127	106	46	12	10	

if, for example, money is at stake. However, the following example due to Ron Graham of AT&T Labs shows that even this simple heuristic contains surprises. In this problem, the bins have capacity 524 and the 33 objects have the weights shown in Table 16.1.

You can easily verify that the 'largest first, first fit' heuristic will produce a solution that uses seven bins, exactly filling all of them. However, if the object of weight 46 is removed from the problem, leaving only 32 objects to pack, the algorithm now uses eight bins. It is counter-intuitive that the heuristic should produce a worse result on a sub-problem than it does on the strictly larger problem.

As an example of heuristics that are tailored to specific types of problem, consider another one-dimensional bin-packing heuristic due to Djang and Finch [8], called by them 'Exact Fit'. The heuristic fills one bin at a time, as follows. Objects are taken largest first, and placed in the bin until the bin is at least one-third full. It then sets an allowable wastage w, initially $w = 0$. The heuristic searches for one object that fills the bin to within w of its capacity. If this fails it searches for any two objects that fill the bin to within w. If this fails it searches for any three objects that fill the bin to within w. If this also fails it sets $w \Leftarrow w+1$ and repeats. As reported by the authors, this outperforms a variety of earlier heuristics on a large set of benchmark problems. However, these benchmark problems are of a particular sort: all the objects have a weight that is a significantly large fraction of a bin's capacity. Benchmark problems tend not to involve many objects with very small weights, because those objects can be treated as a form of 'sand' that can be used to fill up otherwise wasted space. It should be clear that the 'Exact Fit' heuristic can perform very badly on problems in which the objects are all very small. For example, consider a problem in which the bins have capacity 100 and there are 1000 objects each of weight 1. In the optimal solution ten bins are needed; the 'Exact Fit' heuristic will put at most 37 objects in any one bin and so will use 28 bins.

It would be plausible to argue that any self-respecting bin-packing heuristic should not start a new bin if there were existing partially-filled bins still capable of holding further items. Such heuristics would never use more than $2M$ bins because, if they did, there would be at least two bins whose combined contents fitted into one bin and so the heuristic should have at least been able to combine their contents or otherwise ensure that at least one of them was better filled. So there is a large class of heuristics whose worst-case performance on a large class of problems is better than that of 'Exact Fit', even though 'Exact Fit' is very successful on benchmark problems that are generally acknowledged to be hard (but see [9] for a dissenting view).

2.2 The Concept of Hyper-heuristics

Since different heuristics have different strengths and weaknesses, it makes sense to see whether they can be combined in some way so that each makes up for the weaknesses of another. A simplistic way of doing this would be as shown in Figure 16.1.

```
If (problemType(P) == p1)
    apply(heuristic1, P);
else if (problemType(P) == p2)
    apply(heuristic2, P);
else ...
```

Figure 16.1. A naive way to combine heuristics.

One logical extreme of such an approach would be an algorithm containing an infinite switch statement enumerating all finite problems and applying the best known heuristic for each. There are many more practical problem-solving frameworks than this, such as the greedy randomised adaptive search procedure GRASP [10] which repeatedly fabricates a candidate solution C from parts on a so-called 'restricted candidate list', conducts a local search starting from C to find a locally optimal answer, and uses that to update information about desirability of parts and thus revise the restricted candidate list.

The key idea in hyper-heuristics is to use members of a set of known and reasonably understood heuristics to transform the state of a problem. The key observation is a simple one: the strength of a heuristic often lies in its ability to make some good decisions on the route to fabricating an excellent solution. Why not, therefore, try to associate each heuristic with the problem conditions under which it flourishes and hence apply different heuristics to different parts or phases of the solution process? For example, it should be clear from the preceding discussion that in bin-packing, some combination of the 'Exact Fit' procedure and the 'largest first, first fit' procedure should be capable of outperforming either of them alone.

The alert reader will immediately notice an objection to this whole idea. Good decisions are not necessarily easily recognizable in isolation. It is a sequence of decisions that builds a solution, and so there can be considerable epistatis involved—that is, a non-linear interdependence between the parts. However, many general search procedures such as evolutionary algorithms can cope with a considerable degree of epistasis, so the objection is not necessarily fatal. And, on the positive side, there are some real potential benefits as far as real-world use is concerned. For example, in attempting to find a way to combine heuristic ingredients it can be possible to start from a situation in which a single, pure and unalloyed heuristic is used throughout the solution process. If it fails to survive the search process that is attempting to combine ingredients from different heuristics, it is because it wasn't good enough; the process has discovered something better.

Here, therefore, is one possible framework for a hyper-heuristic algorithm:

1. start with a set H of heuristic ingredients, each of which is applicable to a problem state and transforms it to a new problem state. Examples of such ingredients in bin-packing are a single top-level iteration of 'Exact Fit' or a single top-level iteration of 'largest first, first fit';
2. let the initial problem state be S_0
3. if the problem state is S_i then find the ingredient that is in some sense most suitable for transforming that state. Apply it, to get a new state of the problem S_{i+1};
4. if the problem is solved, stop. Otherwise go to 3.

There could be many variants of this, for example in which the set H varies as the algorithm runs or in which suitability estimates are updated across the iterations or in which the size of a single state transformation varies because the heuristic ingredients are dynamically parameterised. There is very considerable scope for research here.

2.3 Some Historical Notes

Intellectually, the concept of hyper-heuristics owes a debt to work within the field of Artificial Intelligence on automated planning systems. The earliest of such systems tried to devise a series of actions to achieve a given goal, usually a goal composed of a conjunct of required state features, by finding actions which would reduce the difference between the current state of the world and the desired state. This hill-climbing approach suffered all the familiar problems of such a method. Later systems such as the DART logistical planning system [11] were much more sophisticated; DART was used in the Gulf War and was later judged by the US Chamber of Commerce to have saved more money than the US Government had spent on funding all forms of AI research over the previous 30 years. The focus eventually turned towards the problem of learning control knowledge; perhaps the best-known example is Minton's PRODIGY system [12] which used explanation-based learning to learn what action would be best to apply at each decision point.

This thread of AI research on planning and scheduling led to one of the earliest examples of a hyper-heuristic approach, the LR-26 scheduler within the COMPOSER system [13] was used for planning satellite communication schedules involving a number of earth-orbiting satellites and three ground stations. The problems involved are far from trivial. For example, certain satellites must communicate at some length with a ground station several times per day, with a predetermined maximum interval between communications, and yet the communication windows and choice of ground stations are constrained by the satellites' orbits. LR-26 treats the problem as a 0–1 integer programming problem involving hundreds of variables and thousands of linear constraints. The system handles many of the constraints by Langrangian relaxation, i.e., by converting them to weighted components of the objective function that is to be maximised, so that if a constraint is violated the consequence is that the objective function value will be reduced by some amount that depends on the associated weight. The scheduler works by finding a partial schedule that may not satisfy all constraints, finding uncommitted variables within unsatisfied constraints and proposing values for them, and searching through a stack of such proposals to find good extensions to the partial schedule.

There are various possible heuristics used for each of several decision steps in this process. In particular, there are four different weight-adjustment heuristics, 9 primary and 9 secondary heuristic methods of ordering the set of unsatisfied constraints, 2 heuristic methods for proposing possible solutions to unsatisfied constraints and 4 heuristic methods for stacking these proposals for consideration. There are thus $4 \times 9 \times 9 \times 2 \times 4 = 2592$ possible strategies. To evaluate any one strategy properly meant testing it on 50 different problems which, the authors calculate, would have meant spending around 450 CPU days to evaluate all the strategies. Instead, COMPOSER applied a simple hill-climbing strategy thus restricting the search to $4 + 2 + (9 \times 4) + 9 = 51$ strategies, at a tolerable cost of 8.85 CPU days. The outcome was "a significant improvement in performance", in terms of solution speed and quality and in the number of problems that could be solved at all, compared to the originally-used strategy. A potential disadvantage of the approach lies in the assumption that the training set is in some way representative of future problems. In the area of satellite communication scheduling this is unlikely to be true—new satellites can have new orbits and different and potentially less demanding communication window requirements, for example.

A second example of the use of a hyper-heuristic approach concerns open-shop scheduling problems. In such problems there are, say, j jobs each consisting of a certain number of tasks. A task consists of visiting a certain machine for a certain task-specific length of time. The tasks associated with a job can be done in any order—if it was a fixed job-specific order then it would be a job-shop problem instead. Fang et al. [2] used a genetic algorithm which built solutions as follows. A chromosome was a series of pairs of integers $[t_0, h_0, t_1, h_1, \ldots]$ interpreted from left to right and meaning, for each i, 'consider the t_i-th uncompleted job (regarding the list of uncompleted jobs as circular, so that this is always meaningful) and use heuristic h_i to select a task to insert into the schedule in the earliest place where it will fit'. Examples of heuristics used included:

- choose the task with largest processing time;
- choose the task with shortest processing time;
- of those tasks which can be started as early as possible, choose the one with largest processing time;
- among those operations which can be inserted into any gap in the schedule, pick the one that best fills a gap (that is, leaves as little idle time as possible);
- and so on.

This approach, termed *evolving heuristic choice*, provided some excellent results on benchmark problems including some new best results at that time. Nevertheless, there are some caveats. First, what is the nature of the space being searched? It is likely that in many cases, several heuristics might lead to the same choice of task, so it may be that the search space is not nearly as large as it might first seem. Second, is the genetic algorithm necessary or might some simpler, non-population-based search algorithm do as well? Third, if there are n tasks then there are effectively $n - 1$ pairs of genes (there is no choice when it comes to inserting the very last task), so if n is large the chromosome will be very long and there is a real risk that genetic drift will have a major impact on the result. Fourth, the process evolves solutions to individual problems rather than creating a more generally applicable algorithm.

Schaffer [14] carried out an early investigation into the use of a genetic algorithms which select heuristics. The paper describes the Philips FCM SMD robot and the heuristic selection genetic algorithm.

In Hart and Ross [15] considered job-shop scheduling problems. The approach there relies on the fact that there is an optimal schedule which is *active*—an active schedule is one in which, to get any task completed sooner you would be forced to alter the sequence in which tasks get processed on some machine, and to do that you would force some other task to be delayed. The optimal schedule might even be *non-delay*, that is, not only active but also such that no machine is ever idle when there is some task that could be started on it. There is a widely-used heuristic algorithm due to Giffler and Thompson [16] that generates active schedules:

1. let C = the set of all tasks that can be scheduled next
2. let t = the minimum completion time of tasks in C, and let m = machine on which it would be achieved
3. let G = the set of tasks in C that are to run on m whose *start* time is $< t$
4. choose a member of G, insert it in the schedule
5. go to 1.

Note that step 4 involves making a choice. This algorithm can be simplified so as to generate non-delay schedules by only looking at the earliest-starting tasks:

1. let C = the set of all tasks that can be scheduled next
2. let G = the subset of C that can start at the earliest possible time
3. choose a member of G, insert it in the schedule
4. go to 1.

Note that step 3 also involves making a choice. The idea in [15] is to use a chromosome of the form $[a_1, h_1, a_2, h_2, \ldots]$, where the chromosome is again read from left to right and a_i is 0 or 1 and indicates whether to use an iteration of the Giffler and Thompson algorithm or an iteration of the non-delay algorithm to place one more task into the growing schedule, and h_i indicates which of twelve heuristics to use to make the choice involved in either algorithm. Again, this produced very good results on benchmark problems when tested on a variety of criteria. The authors looked at the effective choice at each stage and found that fairly often, there were only one or two tasks to choose between, and never more than four. The space being searched is thus much smaller than the chosen representation would suggest; many different chromosomes represent the same final schedule. The authors also observed that, in constructing a schedule, it is the early choices that really matter. For example, if the first 50% of choices are made according to what the evolved sequence of heuristic choices suggests, but the other 50% are made at random, then the result is still a very satisfactory schedule. This suggests at least a partial answer to the worry raised above about using very long chromosomes: quite simply, do not do it. Instead, use a much shortened chromosome just to evolve the early choices and then resort to using a fixed heuristic to complete the construction.

A real-world example of using a hyper-heuristic approach is described in [17], where the problem is to schedule the collection and delivery of live chickens from farms all over Scotland and Northern England to one of two processing factories, in order to satisfy the set of orders from supermarkets and other retailers. The customer orders would change week by week and sometimes day by day. The task was to schedule the work done by a set of 'catching squads' who moved around the country in mini-buses, and a set of lorries who would ferry live chickens from each farm back to one of the factories. The principal aim was to keep the factories supplied with work without requiring live chickens to wait too long in the factory yard, for veterinary and legal reasons. This was complicated by many unusual constraints. For example, there were several types of catching squad, differentiated by which days they worked, when they started work, what guaranteed minimum level of work they had been offered and what maximal amount they could do. There were constraints on the order in which farms could be visited, to minimise potential risks of spreading diseases of chickens. There were constraints on the lorry drivers and on how many chickens could be put into a single 'module' (a tray-like container) and variations in the number of such modules different lorries could carry, and so on. Overall, the target was not to produce optimal schedules in cost terms, because the work requirements could anyway change at very short notice but it was not generally practicable to make very large-scale changes to staff schedules at very short notice. Instead, the target was to create good schedules satisfying the many constraints, that were also generally similar to the kinds of work pattern that the staff were already familiar with, and to do so quickly and reliably. The eventual solution used two genetic algorithms. One used a heuristic selection approach to decompose the total set of customer orders into individual tasks and assign those tasks to catching squads. The other started with these assignments and evolved the schedule of lorry arrivals at each factory; from such schedules it was possible to reason backwards to determine which squad and lorry had to arrive at each farm at which times, and thus construct a full schedule for all participants. In the first genetic algorithm, the chromosome specified a permutation of the customer orders and then two sequences of heuristic choices. The first sequence of choices worked through the permutation and split each order into loads using simple heuristics about how to partition the total customer order into convenient workload chunks; the second sequence of choices suggested how to assign those chunks to catching squads. The end result did meet the project's requirements but, like many other practically-inspired problems, it was not feasible to do a long-term study to determine just how crucial each of the ingredients was to success. However, the authors report that a more conventional permutation-based genetic algorithm approach to this scheduling task had not been successful.

In the above examples (apart from the hill-climbing approach used in the LR-26 satellite communication scheduler), a genetic algorithm was used to evolve a fine-grained sequence of choices of heuristics, with one choice per step in the process of constructing a complete solution. Although the end results were generally good, this is still somewhat unsatisfactory because the method evolves solutions only to specific instances of problems and may not handle very large problems well. A different kind of hyper-heuristic approach that begins to tackle such objections is described in [18]. That paper is concerned with solving large-scale university exam timetabling problems. There are a number of fixed time-slots at which exams can happen, and there are rooms of various sizes. Two exams cannot happen at the same time if there is any student who takes both; more than one exam can happen in a room at the same time if no student takes both and the room is large enough. Certain exams may be constrained to avoid certain time-slots when, for example, an appropriate invigilator is unavailable. There are also some 'soft' constraints which it would be good to respect but which can be violated if necessary. For example, it is desirable that no student should have to sit exams in consecutive time-slots on the same day and it is often desirable that very large exams should happen early so as to permit more time for marking them before the end of the whole exam period. Such problems can involve thousands of exams and tens of thousands of students.

The approach taken in [18] is to suppose that there is an underlying timetable construction algorithm of the general sort shown in Figure 16.2.

The idea is to use a genetic algorithm to evolve the choices of H1, H2, H3 and H4 and the condition X which determines when to switch from the first phase to the

```
/* Do first phase of construction */
while(condition(X) == FALSE)
{
    event = apply_event_heuristic(H1);
    timeslot = apply_slot_heuristic(H2,event);
    add_to_timetable(event, timeslot);
}
/* Do second phase of construction */
while(not(completed()))
{
    event = apply_event_heuristic(H3);
    timeslot = apply_slot_heuristic(H4,event);
    add_to_timetable(event, timeslot);
}
```

Figure 16.2. A two-phase timetable building algorithm.

second. Some of the heuristics and possible conditions involved further parameters, such as whether and how much to permit backtracking in making a choice, or switching to phase 2 after placing N events. The rationale for using such an algorithm is that many timetabling problems necessitate solving a certain sort of problem initially (for example, a bin-packing problem to get the large exams well packed together if room space is in short supply) but a different sort of problem in the later stage of construction. Soft constraints are handled within the heuristics, for example, in order to cut down the chances of a student having exams in adjacent time-slots an heuristic might consider time-slots in some order that gave adjacent slots a very low priority.

A chromosome was evaluated by constructing the timetable and assessing the quality of the result. Interestingly, this method solved even very large timetabling problems very satisfactorily in under 650 evaluations. The authors also conducted a brute-force search of the space of chromosomes in order to check whether the genetic algorithm was delivering very good-quality results (at least as far as the chosen representation would permit) whilst visiting only a tiny proportion of the search space, and were able to confirm the truth of this. However, they did not also examine whether the discovered instances of the framework described in Figure 16.2 could be applied successfully to other problems originating from the same university; this is a topic for further research.

3 HYPER-HEURISTIC FRAMEWORK

This section describes a particular hyper-heuristic framework that has been presented in [19,20,25–28]. As has been discussed earlier, a metaheuristic typically works on the problem directly, often with domain knowledge incorporated into it. However, this hyper-heuristic framework operates at a higher level of abstraction and often has no knowledge of the domain. It only has access to a set of low level heuristics that it can call upon, but with no knowledge as to the purpose or function of a given low level heuristic. The motivation behind this suggested approach is that once a hyper-heuristic algorithm has been developed then new problem domains can be tackled by only having

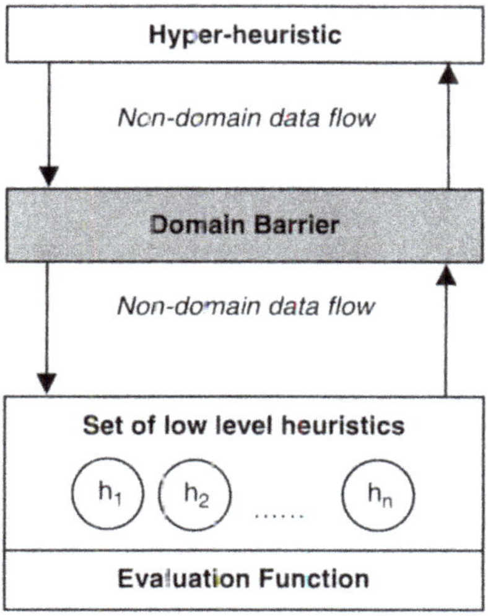

Figure 16.3. Hyper-heuristic framework.

to replace the set of low level heuristics and the evaluation function, which indicates the quality of a given solution.

A diagram of a general hyper-heuristic framework is shown in Figure 16.3.

The figure shows that there is a barrier between the low level heuristics and the hyper-heuristic. Domain knowledge is not allowed to cross this barrier. Therefore, the hyper-heuristic has no knowledge of the domain under which it is operating. It only knows it has n low level heuristics on which to call and it knows it will be passed the results of a given solution once it has been evaluated by the evaluation function.

There is, of course, a well defined interface between the hyper-heuristic and the low level heuristics. The reasons for this are two-fold

1. It allows the hyper-heuristic to communicate with the low level heuristics using a standard interface, otherwise the hyper-heuristic would need a separate interface for each low level heuristic which is obviously nonsensical. In addition, it facilitates the passing of non-domain data between the low level heuristics and the hyper-heuristic (and vice versa). For example, the interface developed in [25] includes components such as the result of the evaluation function and the CPU time taken by the low level heuristic (which, equally, could be calculated by the hyper-heuristic). In addition, we have also included a component that allows us to "tell" a low level heuristic how long it has to run. The motivation behind this idea is that we call each heuristic in turn giving it a specified amount of time and the heuristic that performs the best, in the time allowed, is the one that is applied to the current solution. In conjunction with this component, the interface also defines if the low level heuristic should apply its changes to the current solution or if it should just report what effect it would have *if* it did apply those changes. The idea is that the hyper-heuristic can ask each low level heuristic how well it will do against a given solution. The hyper-heuristic can then decide which heuristic (or set of heuristics) should be allowed to update the solution. We have

not fully explored these ideas yet but the interface will allow us to investigate them at an appropriate time.

2. It allows rapid development for other domains. When implementing a new problem, the user has to supply a set of low level heuristics and a suitable evaluation function. If the low level heuristics follow a standard interface the hyper-heuristic does not have to be altered in any way. It is able to simply start solving the new problem as soon as the user has supplied the low level heuristics and the evaluation function. That is, as stated above, we aim for a hyper-heuristic to operate at a higher level of abstraction than a meta-heuristic approach.

An example of a hyper-heuristic operating at a higher level of abstraction can be seen in the work of Cowling, Kendall and Soubeiga [19,25–27]. In [25] a hyper-heuristic approach was developed and applied to a sales summit problem (a problem of matching suppliers to potential customers at a sales seminar). In [25] the hyper-heuristic had access to 10 low level heuristics which included *removing a delegate from a meeting with a particular supplier, adding a delegate to supplier to allow them to meet* and *remove meetings from a supplier who has more than their allocation.* In [26] the hyper-heuristic was modified so that it automatically adapted some of its parameters but, again, the sales summit problem was used as a test bench. In [27], the hyper-heuristic remained the same but a new problem was introduced. This time, the problem was scheduling third year undergraduate project presentations for a UK university. Eight low level heuristics were developed, which included *replace a staff member in a given session, move one presentation from one session to another* and *swap the 2nd marker for a given session*. The changes were such that only the low level heuristics and the evaluation function was changed. The hyper-heuristic remained the same in the way that it chose which low level heuristic to call next. Good quality solutions were produced for this different problem domain.

This idea was further developed in [19] when the same hyper-heuristic approach was applied to nurse rostering.

The only information that the hyper-heuristic has access to in this framework is data which is common to all problem types and which it decides to record as part of its internal state. For example, the hyper-heuristic might store the following:

- How much CPU time a given heuristic used when it was last called?
- The change in the evaluation function when the given heuristic was called?
- How long (possibly in CPU time) has elapsed since a given heuristic has not been called?

The important point is that the hyper-heuristic has no knowledge as to the function of each heuristic. For example, it will not know that one of the heuristics performs 2-opt for the Traveling Salesman Problem. Indeed, it does not even know that the problem being optimised is the Traveling Salesman Problem.

Of course, the hyper-heuristic designer is allowed to be as imaginative as he/she wishes within the constraints outlined above. For example, the internal state of the hyper-heuristic could store how well pairs of heuristics operate together. If the hyper-heuristic calls one heuristic followed by another, does this lead to a worsening of the evaluation function after the first call but a dramatic (or even slight) improvement after the second heuristic has been called? Therefore, one hyper-heuristic idea might be to store data about pairs (triples etc.) of heuristics that appear to work well when called consecutively but, if called in isolation, each heuristic, in general, performs badly.

Using its internal state, the hyper-heuristic has to decide which low level heuristic(s) it should call next. Should it call the heuristic that has led to the largest improvement in the evaluation function? Should it call the heuristic that runs the fastest? Should it call the heuristic that has not been called for the longest amount of time? Or, and more likely, should it attempt to balance all these factors (and more) to make an informed decision as to which low-level heuristic (or pairs, triples etc.) it should call next.

The internal state of the hyper-heuristic is a matter for the designer, as is the process to decide on which heuristic to call next, but once a suitable hyper-heuristic is in place then the hope is that it will perform reasonably well across a whole range of problems and not just the one for which it was originally implemented.

The hyper-heuristic framework described above is the framework adopted in [25,26,27], where the hyper-heuristic maintains an internal state which is accessed by a *choice* function to decide which low level heuristic to call next. The choice function is a combination of terms which considers recent performance of each low level heuristic (denoted by f_1), recent performance of pairs of heuristics (denoted by f_2) and the amount of time since a given heuristic has been called (denoted by f_3). Thus we have

$$f(H_k) = \alpha f_1(H_k) + \beta f_2(H_j, H_k) + \delta f_3(H_k)$$

where H_k is the kth heuristic, α, β and δ are weights which reflect the importance of each term. It is these weights that are adaptively changed in [26], $f_1(H_k)$ is the recent performance of heuristic H_k, $f_2(H_j, H_k)$ is the recent performance of heuristic pair H_j, H_k and $f_3(H_k)$ is a measure of the amount of time since heuristic H_k was called. f_1 and f_2 are aiming to intensify the search while f_3 attempts to add a degree of diversification. The maximal value, $f(H_k)$, dictates the heuristic which is called next.

A different type of hyper-heuristic is described in [28] where a genetic algorithm is used to evolve a sequence of calls to the low level heuristics. In effect, the genetic algorithm replaces the choice function described above and each member of the population (that is, each chromosome) is evaluated by the solution it returns when applying the heuristics in the order denoted by the chromosome. In later work by Han, Kendall and Cowling [20] the chromosome length is allowed to adapt and longer chromosomes are penalised on the basis that they take longer to evaluate.

However, in [20] and [28] the concept of the domain barrier remains and the genetic algorithm has no knowledge of the problem. It simply tries to evolve a good sequence of heuristic calls.

4 MODERN HYPER-HEURISTIC APPROACHES

There has been much recent research directed to the various aspects of Hyperheuristics. An example that follows the framework described in Section 2.2 can be found in [21]. The focus here is on learning solution processes applicable to many problem instances rather than learning individual solutions. Such process would be able to choose one of various simple, well-understood heuristics to apply to each state of a problem, gradually transforming the problem from its initial state to a solved state. The first use of such model has been applied to the one-dimensional bin-packing problem described in Section 2.1. In this work, an accuracy-based Learning Classifier System (XCS) [22] has been used to learn a set of rules that associates characteristics of the current state of a problem with, in this case, eight different heuristics, two of which have been explained in Section 2.1 (largest-first-first-fit and exact-fit). The set of rules is used as follows: given the initial problem characteristics P, a heuristic H is chosen to pack a bin, gradually altering the characteristics of the problem that remains to be solved. At each step, a rule appropriate to the current problem state P' is chosen, and the process repeats until all items have been packed.

The approach is tested using 890 benchmark bin-packing problems, of which 667 were used to train the XCS and 223 for testing. The combined set provides a good test of whether the system can learn from a very varied collection of problems. The method (HH) achieved optimal results on 78.1% of the training problems, and 74.6% of the remaining test solutions. This compares well with the best single heuristic (the author's improved version of exact-fit) which achieved optimality 73% of the time. Largest-first-first-fit, for instance, achieves optimality in 62.2%, while another one of the heuristics used named 'next-fit' reaches optimality in 0% of the problems. Even though the results of the best heuristic might seem close to HH, it is also noteworthy that when HH is trained purely on some of the harder problems (which none of the component heuristics could solve to optimality alone), it manages to solve seven out of ten of these problems to optimality.

Improvements over this initial approach are reported in [23], where a new heuristic (R) that uses a random choice of heuristics has been introduced to compare results of HH. R solves only 56.3% of the problems optimally, while HH reaches 80%. This work also looks more closely at the individual processes evolved arising from two different reward schemes presented during the learning process of HH. The behaviour of HH is further analysed and compared to the performance of the single heuristics. For instance, in HARD9, one of the very difficult problems used in the set, HH learned that the combination of only two of the eight single heuristics (here called $h1$ to $h8$) reaches optimality when none of the heuristics used individually achieved it. The solution found by HH used only 56 bins, while the best of the individual heuristics used alone needs 58 bins. The best reported result, using a method not included in any of the heuristics used by HH, used 56 bins (see [24], Data Set 3). Improvements of this type were found in a large number of other problems, including all the HARD problems. The approach looks promising and further research is being carried out to evaluate it in other problem domains.

Cowling, Han, Kendall and Soubeiga (previous section) propose a hyper-heuristic framework existing at a higher level of abstraction than meta-heuristic local search methods, where different neighbourhoods are selected according to some "choice function". This choice function is charged with the task of determining which of these different neighbourhoods is most appropriate for the current problem. Traditionally the correct neighbourhoods are selected by a combination of expert knowledge and/or time consuming trial and error experimentation, and as such the automation of this task offers substantial potential benefits to the rapid development of algorithmic solutions. It naturally follows that an effective choice function is critical to the success of this method, and this is where most research has been directed [19,25–27]. See the previous section for a more detailed discussion.

The approach of Cowling, Kendall and Han to a trainer scheduling problem [28] also fits into this class of controlling the application of low level heuristics to construct solutions to problems. In this approach a "Hyper-genetic algorithm" is used to evolve an ordering of 25 low-level heuristics which are then applied to construct a solution. The fitness of each member of the genetic algorithm population is determined by the quality of the solution it constructs. Experimental results show that the method outperformed the tested conventional genetic and memetic algorithm methods. It also greatly outperformed any of the component heuristic methods, albeit in a greatly multiplied amount of CPU time.

Another approach proposed by Burke and Newall to examination timetabling problems [29] uses an adaptive heuristic to try and improve on an initial heuristic ordering. The adaptive heuristic functions by first trying to construct a solution by initially scheduling exams in an order dictated by the original heuristic. If using this ordering means that an exam cannot be acceptably scheduled, it is promoted up the order in a subsequent construction. This process continues either until the ordering remains static (all exams can be scheduled acceptably), or until a pre-defined time limit expires. The experiments showed that the method can substantially improve quality over that of the original heuristic. The authors also show that even when given a poor initial heuristic acceptable results can still be found relatively quickly.

Other approaches that attempt to harness run-time experience include the concept of Squeaky wheel optimisation proposed by Joslin and Clements [30]. Here a greedy constructor is applied to a problem, followed by an analysis phase that identifies problematic elements in the produced solution. A prioritiser then ensures that the greedy constructor concentrates more on these problematic elements next time, or as the authors phrase it: "The squeaky wheel gets the grease". This cycle is iterated until some stopping criteria are met. Selman and Kautz propose a similar modification to their GSAT procedure [31]. The GSAT procedure is a randomised local search procedure for solving propositional satisfiability problems. It functions by iteratively generating truth assignments and then successively "flips" the variable that leads the greatest increase in clauses satisfied, in a steepest descent style. The proposed modification increases the "weight" of clauses if they are still unsatisfied at the end of the local search procedure. This has the effect that on subsequent attempts the local search element will concentrate more on satisfying these clauses and hopefully in time lead to full satisfaction of all clauses.

Burke et al. [3] investigate the use of the case based reasoning paradigm in a hyper-heuristic setting for selecting course timetabling heuristics. In this paper, the system maintains a case base of information about which heuristics worked well on previous course timetabling instances. The training of the system employs knowledge discovery techniques. This work is further enhanced by Petrovic and Qu [4] who integrate the use of Tabu Search and Hill Climbing into the Case Based Reasoning system.

5 CONCLUSIONS

It is clear that hyper-heuristic development is going to play a major role in search technology over the next few years. The potential for scientific progress in the development of more general optimisation systems, for a wide variety of application areas, is significant. For example, Burke and Petrovic [32] discuss the scope for hyper-heuristics for timetabling problems and this is currently an application area that is seeing significant research effort [1,3,4,18,25,29,32]. Indeed, in 1997, Ross et al. [1] said, "However, all this naturally suggests a possibly worthwhile direction for timetabling research involving Genetic Algorithms. We suggest that a Genetic Algorithm might be better employed in searching for a good algorithm rather than searching for a specific solution to a specific problem." Ross, Hart and Corne's suggestion has led to some important research directions in timetabling. We think that this suggestion can be generalised further and we contend that a potentially significant direction for metaheuristic research is to investigate the use of hyper-heuristics for a wide range of problems. As this chapter clearly shows, this work is already well underway.

ACKNOWLEDGEMENTS

The authors are grateful for support from the UK Engineering and Physical Sciences Research Council (EPSRC) under grants GR/N/36660, GR/N/36837 and GR/M/95516.

REFERENCES

[1] P. Ross, E. Hart and D. Corne (1997) Some observations about GA-based exam timetabling. In: E.K. Burke and M. Carter (eds.), LNCS 1408, *Practice and Theory of Automated Timetabling II: Second International Conference, PATAT 1997*, Toronto, Canada, selected papers. Springer-Verlag, pp. 115–129.

[2] H.-L. Fang, P.M. Ross and D. Corne (1994) A promising hybrid GA/heuristic approach for open-shop scheduling problems. In: A. Cohn (ed.), *Proceedings of ECAI 94: 11th European Conference on Artificial Intelligence*. John Wiley and Sons Ltd, pp. 590–594.

[3] E.K. Burke, B.L. MacCarthy, S. Petrovic and R. Qu (2002) Knowledge discovery in a hyper-heuristic for course timetabling using case based reasoning. In: Proceedings of the *Fourth International Conference on the Practice and Theory of Automated Timetabling (PATAT'02)*, Ghent, Belgium (to appear).

[4] S. Petrovic and R. Qu (2002) Case-Based Reasoning as a Heuristic Selector in a Hyper-Heuristic for Course Timetabling. In: Proceedings of the *Sixth International Conference on Knowledge-Based Intelligent Information & Engineering Systems (KES'2002)*, Crema, Italy (to appear).

[5] D. Wolpert and W.G. MacReady (1997) No free lunch theorems for optimization. *IEEE Transactions on Evolutionary Computation*, **1**(1), 67–82.

[6] D.S. Johnson (1973) *Near-optimal Bin-packing Algorithms*. Ph.D. thesis. MIT Department of Mathematics, Cambridge, MA.

[7] E.G. Coffman, M.R. Garey and D.S. Johnson (1996) Approximation algorithms for bin packing: a survey. In: D. Hochbaum (ed.), *Approximation Algorithms for NP-Hard Problems*. PWS Publishing, Boston, pp. 46–93.

[8] P.A. Djang and P.R. Finch. Solving one dimensional bin packing problems. Available as http: //www.zianet.com/pdjang/binpack/paper.zip.

[9] I.P. Gent (1998) Heuristic solution of open bin packing problems. *Journal of Heuristics*, **3**(4), 299–304.

[10] L.S. Pitsoulis and M.G.C. Resende (2001) Greedy randomized adaptive search procedures. In: P.M. Pardalos and M.G.C. Resende (eds.), *Handbook of Applied Optimization*. OUP, pp. 168–181.

[11] S.E. Cross and E. Walker (1994) Dart: applying knowledge-based planning and scheduling to crisis action planning. In: M. Zweben and M.S. Fox (eds.), *Intelligent Scheduling*. Morgan Kaufmann.

[12] S. Minton (1998) *Learning Search Control Knowledge: An Explanation-based Approach*. Kluwer.

[13] J. Gratch, S. Chein and G. de Jong (1993) Learning search control knowledge for deep space network scheduling. In: *Proceedings of the Tenth International Conference on Machine Learning*. pp. 135–142.

[14] J.D. Schaffer (1996) Combinatorial optimization by genetic algorithms: the value of the phenotype/genotype distinction. In: E.D. Goodman, V.L. Uskov, W.F. Punch III (eds.), *First International Conference on Evolutionary Computing and its Applications (EvCA'96)*, Russian Academy of Sciences, Moscow, Russia, June 24–27, Institute for High Performance Computer Systems of the Russian Academy of Sciences, Moscow, Russia, pp. 110–120.

[15] E. Hart and P.M. Ross (1998) A heuristic combination method for solving job-shop scheduling problems. In: A.E. Eiben, T. Back, M. Schoenauer and H.-P. Schwefel (eds.), *Parallel Problem Solving from Nature V*, LNCS 1498, Springer-Verlag, pp. 845–854.

[16] B. Giffler and G.L. Thompson (1960) Algorithms for solving production scheduling problems. *Operations Research*, **8**(4), 487–503.

[17] E. Hart, P.M. Ross and J. Nelson (1998) Solving a real-world problem using an evolving heuristically driven schedule builder. *Evolutionary Computation*, **6**(1), 61–80.

[18] H. Terashima-Marín, P.M. Ross and M. Valenzuela-Rendón (1999) Evolution of constraint satisfaction strategies in examination timetabling. In: W. Banzhaf et al. (eds.), *Proceedings of the GECCO-99 Genetic and Evolutionary Computation Conference*. Morgan Kaufmann, pp. 635–642.

[19] P. Cowling, G. Kendall and E. Soubeiga (2002) Hyperheuristics: a robust optimisation method applied to nurse scheduling. Technical Report NOTTCS-TR-2002-6 (submitted to PPSN 2002 Conference), University of Nottingham, UK, School of Computer Science & IT.

[20] L. Han, G. Kendall and P. Cowling (2002) An adaptive length chromosome hyperheuristic genetic algorithm for a trainer scheduling problem. Technical Report NOTTCS-TR-2002-5 (submitted to SEAL 2002 Conference), University of Nottingham, UK, School of Computer Science & IT.

[21] P. Ross, S. Schulenburg, J.G. Marín-Blázquez and E. Hart (2002) Hyper-heuristics: learning to combine simple heuristics in bin-packing problems. Accepted for *Genetic and Evolutionary Computation Conference (GECCO 20020) 2002*, July 9–13, *New York*.

[22] S. Wilson (1998) Generalisation in the XCS classifier system. In: J. Koza (ed.), *Proceedings of the Third Genetic Programming Conference*. Morgan Kaufmann, pp. 665–674.

[23] S. Schulenburg, P. Ross, J.G. Marín-Blázquez and E. Hart. A hyper-heuristic approach to single and multiple step environments in bin-packing problems. *Proceedings of the Fifth International Workshop on Learning Classifier Systems 2002 (IWLCS-02)* (to appear).

[24] http://bwl.tu-darmstadt.de/bwl3/forsch/projekte/binpp.

[25] P. Cowling, G. Kendall, E. Soubeiga (2000) A hyperheuristic approach to scheduling a sales summit. In: E.K. Burke and W. Erben (eds.), *LNCS 2079, Practice and Theory of Automated Timetabling III: Third International Conference, PATAT 2000*, Konstanz, Germany, August, selected papers, Springer-Verlag, pp. 176–190.

[26] P. Cowling, G. Kendall and E. Soubeiga (2001) A parameter-free hyperheuristic for scheduling a sales summit. In: *Proceedings of 4th Metahuristics International Conference (MIC 2001), Porto Portugal*, 16–20 July, pp. 127–131.

[27] P. Cowling, G. Kendall and E. Soubeiga (2002) Hyperheuristics: a tool for rapid prototyping in scheduling and optimisation. In: S. Cagoni, J. Gottlieb, E. Hart, M. Middendorf and R. Günther (eds.), *LNCS 2279, Applications of Evolutionary Computing: Proceedings of Evo Workshops 2002, Kinsale, Ireland*, April 3–4, ISSN 0302-9743, ISBN 3-540-43432-1, Springer-Verlag, pp. 1–10.

[28] P. Cowling, G. Kendal and L. Han (2002) An investigation of a hyperheuristic genetic algorithm applied to a trainer scheduling problem. In: *Proceedings of Congress on Evolutionary Computation (CEC2002)*, Hilton Hawaiian Village Hotel, Honolulu, Hawaii, May 12–17, pp. 1185–1190, ISBN 0-7803-7282-4.

[29] E.K. Burke and J.P. Newall (2002) A new adaptive heuristic framework for examination timetabling problems. Technical Report NOTTCS-TR-2001-5 (submitted to Annals of Operations Research), University of Nottingham, UK, School of Computer Science & IT.

[30] D.E. Joslin and D.P. Clements (1999) Squeaky wheel optimization. *Journal of Artificial Intelligence Research*, **10**, 353–373.

[31] B. Selman and H. Kautz (1993) Domain-independent extensions to GSAT: Solving large structured satisfiability problems. In: *Proceedings of the 13th International Joint Conference on Artificial Intelligence*, pp. 290–295.

[32] E.K. Burke and S. Petrovic (2002) Recent Research Directions in Automated Timetabling. *European Journal of Operational Research* (to appear).

Chapter 17

PARALLEL STRATEGIES FOR META-HEURISTICS

Teodor Gabriel Crainic
Département de management et technologie
Université du Québec à Montréal and Centre de recherche sur les transports
Université de Montréal
C.P. 6128, Succursale Centre-ville
Montréal (QC) Canada H3C 3J7
E-mail: theo@crt.umontreal.ca

Michel Toulouse
Department of computer science
University of Manitoba
Winnipeg (MB) Canada R3T 2N2
E-mail: toulouse@cs.umanitoba.ca

Abstract We present a state-of-the-art survey of parallel meta-heuristic developments and results, discuss general design and implementation principles that apply to most meta-heuristic classes, instantiate these principles for the three meta-heuristic classes currently most extensively used—genetic methods, simulated annealing, and tabu search, and identify a number of trends and promising research directions.

Keywords: Parallel computation, Parallelization strategies, Meta-heuristics, Genetic methods, Simulated annealing, Tabu search, Co-operative search

1 INTRODUCTION

Meta-heuristics are widely acknowledged as essential tools to address difficult problems in numerous and diverse fields, as this volume eloquently demonstrates. In fact, meta-heuristics often offer the only practical approach to solving complex problems of realistic scale.

Even using meta-heuristics, the limits of what may be solved in "reasonable" computing times are still reached rapidly, however, at least much too rapidly for the growing needs of research and industry alike. Heuristics do not, in general, guaranty optimality. Moreover, the performance often depends on the particular problem setting and data. Consequently, a major issue in meta-heuristic design and calibration is not only how to build them for maximum performance, but also how to make them *robust*, in the sense of offering a consistently high level of performance over a wide variety of problem settings and characteristics.

Parallel meta-heuristics aim to address both issues. Of course, the first goal is to solve larger problem instances in reasonable computing times. In appropriate settings, such as co-operative multi-thread strategies, parallel meta-heuristics also prove to be much more robust than sequential versions in dealing with differences in problem types and characteristics. They also require less extensive, and expensive, parameter calibration efforts.

The objective of this paper is to paint a general picture of the parallel meta-heuristic field. Specifically, the goals are to (1) present a state-of-the-art survey of parallel meta-heuristic developments and results, (2) discuss general design and implementation principles that apply to most meta-heuristic classes, (3) instantiate these principles for the three meta-heuristic classes currently most extensively used: genetic methods, simulated annealing, and tabu search, and (4) identify a number of trends and promising research directions.

The parallel meta-heuristic field is a very broad one, while the space available for this paper imposes hard choices and limits the presentation. In addition to the references provided in the following sections, a number of surveys, taxonomies, and syntheses have been proposed and may prove of interest: Greening (1990), Azencott (1992), Mühlenbein (1992), Shonkwiler (1993), Voß (1993), Lin et al. (1994), Pardalos et al. (1995), Ram et al. (1995), Verhoeven and Aarts (1995), Laursen (1996), Crainic et al. (1997), Glover and Laguna (1997), Holmqvist et al. (1997), Cantù-Paz (1998), Crainic and Toulouse (1998), Crainic (2002), Cung et al. (2002).

The paper is organized as follows. Section 2 introduces the notation, describes a generic meta-heuristic framework, and sets genetic, simulated annealing, and tabu search methods within this framework. Section 3 is dedicated to a brief introduction to parallel computing and the presentation of three main strategies used to build parallel meta-heuristics. Sections 4, 5, and 6 are dedicated to the survey and discussion of issues related to the parallelization of genetic approaches, simulated annealing, and tabu search, respectively. Section 7 briefly treats a number of other meta-heuristic approaches, draws a number of general conclusions, and points to research directions and challenges.

2 HEURISTICS AND META-HEURISTICS

Sequential and parallel meta-heuristics are used in many disciplines—mathematics, operations research, artificial intelligence—and numerous applications: design, planning, and operation of complex systems and networks (e.g., production, transportation, telecommunication, etc.); management, allocation, scheduling, and utilization of scarce resources; speech and image recognition and enhancement; VLSI design; and so on. To simplify the presentation, and with no loss of generality, in the following we adopt the notation and vocabulary of *combinatorial optimization* formulations.

Given a set of objects, the value associated to each, and the rules specifying how objects may be joined together, the combinatorial optimization formulation aims to select a subset of objects such that the sum of their contributions is the highest/lowest among all possible combinations. Many problems of interest may be cast as combinatorial optimization formulations, including design, location, routing, and scheduling. In most cases, such formulations are extremely difficult to solve for realistically-sized problem instances, the main issue being the number of feasible solutions—combinations of objects—that grows exponentially with the number of objects in the initial set.

Combinatorial optimization problems are usually formulated as (mixed) integer optimization programs. To define notation, assume that one desires to minimize (or maximize) a function $f(x)$ subject to $x \in \mathcal{X} \subseteq \mathbb{R}^n$. The objective function $f(x)$ may be linear or not. The set $\mathcal{X}$ summarizes constraints on the *decision variables* x and defines the *feasible domain*. Decision variables are generally non-negative and all or part of the elements of x may be compelled to take discrete values. One seeks a globally *optimal solution* $x^* \in \mathcal{X}$ such that $f(x^*) \leq f(x)$ for all $x \in \mathcal{X}$.

Once various methods have been applied to re-formulate the problem and to bound the region where the optimal solution is to be found, most solution methods are based on some form of exploration of the set of feasible solutions. Explicit enumeration is normally out of the question and the search for the optimal solution proceeds by implicit enumeration. Branch-and-bound (and price, and cut, ...) methods are both typical of such approaches and one of the methods of choice used in the search for optimal solutions to combinatorial problems. Unfortunately, these methods fail for many instances, even when parallel implementations are used. Thus, heuristics have been, and continue to be, an essential component of the methodology used to address combinatorial optimization formulations.

A *heuristic* is any procedure that identifies a feasible solution $\tilde{x} \in \mathcal{X}$. Of course, one would like $\tilde{x}$ to be identical to x^* (if the latter is unique) or $f(\tilde{x})$ to be equal to $f(x^*)$. For most heuristics, however, one can only hope (and for some, prove) that $f(\tilde{x})$ is "close" to $f(x^*)$. Heuristics have a long and distinguished track record in combinatorial optimization. Often, heuristics are the only practical alternative when dealing with problem instances of realistic dimensions and characteristics.

Many heuristics are *improving* iterative procedures that *move* from a given solution to a solution in its *neighbourhood* that is better in terms of the objective function value (or some other measure based on the solution characteristics). Thus, at each iteration, such a *local search* procedure identifies and evaluates solutions in the neighbourhood of the current solution, selects the best one relative to given criteria, and implements the transformations required to establish the selected solution as the current one. The procedure iterates until no further improvement is possible.

Formally, let $\mathcal{N} \subseteq \mathcal{X}$ represent the set of neighbours of a given solution x that may be reached by a simple transformation (e.g., complement the value of an integer-valued variable) or a given sequence of operations (e.g., λ-opt modifications of routes in a vehicle routing problems). Let $m(x)$ denote the application that corresponds to these moves and that yields a solution $y \in \mathcal{N}(s)$. Then, Figure 17.1 illustrates a simple steepest descent heuristic where the objective function value is the only neighbour evaluation criterion.

A major drawback of classical heuristic schemes is their inability to continue past the first encountered local optimum. Moreover, such procedures are unable to react and adapt to particular problem instances. Re-starting and randomization strategies, as well as combinations of simple heuristics offer only partial and largely unsatisfactory answers to these issues. The class of modern heuristics known as *meta-heuristics* aims to address these challenges.

Meta-heuristics have been defined as master strategies (heuristics) to guide and modify other heuristics to produce solutions beyond those normally identified by local

1. Identify an initial solution $x^0 \in \mathcal{X}$; $k = 0$;
2. $k = k + 1$;
3. Find $\tilde{x} = argmin\{f(x)|x \in \mathcal{N}(x^k)\}$;
4. If $f(\tilde{x}) \geq f(x^k)$ Stop.
5. Otherwise, $x^{k+1} = m(\tilde{x})$; Goto 2.

Figure 17.1. Simple local search/steepest descent heuristic.

search heuristics (Glover, 1986; see also Glover and Laguna, 1993). Compared to exact search methods, such as branch-and-bound, meta-heuristics cannot generally ensure a systematic exploration of the entire solution space. Instead, they attempt to examine only parts thereof where, according to certain criteria, one believes good solutions may be found. Well-designed meta-heuristics avoid getting trapped in local optima or sequences of visited solutions (*cycling*) and provide reasonable assurance that the search has not overlooked promising regions.

Meta-heuristics for optimization problems may be described summarily as a "walk through neighbourhoods", a search trajectory through the solution domain of the problem at hand. Similar to classical heuristics, these are iterative procedures that *move* from a given solution to another solution in its *neighbourhood*. Thus, at each iteration, one evaluates moves towards solutions in the neighbourhood of the current solution, or in a suitably selected subset. According to various criteria (objective value, feasibility, statistical measures, etc.), a number of good moves are selected and implemented. Unlike classical heuristics, the solutions implemented by meta-heuristics are not necessarily improving, however. Tabu search and simulated annealing methods usually implement one move at each iteration, while genetic methods may generate several new moves (individuals) at each iteration (generation). Moves may belong to only one type (e.g., add an element to the solution) or to several quite different categories (e.g., evaluate both add and drop moves). Moves may marginally modify the solution or drastically inflect the search trajectory. The first case is often referred to as *local search*. The diversification phase of tabu search or the application of mutation operators in an evolutionary process are typical examples of the second alternative. This last case may also be described as a change in the "active" neighbourhood.

Each meta-heuristic has its own behaviour and characteristics. All, however, share a number of fundamental components and perform operations that fall within a limited number of categories. To facilitate the comparison of parallelization strategies for various meta-heuristic classes, it is convenient to define these common elements:

1. *Initialization.* A method to create an initial solution or set of problem configurations that may be feasible or not.
2. *Neighbourhoods.* To each solution x corresponds a set of neighbourhoods and associated moves: $\{\mathcal{N}_1, \mathcal{N}_2, \ldots, \mathcal{N}_q\}$, where $\mathcal{N}_i(x) = \{y = m_i(x), y \in \mathcal{X}\}$, $i = 1, \ldots, q$.
3. A *neighbourhood selection criterion* is defined when more than one neighbourhood is included. This criterion must specify not only what neighbourhood to choose but also when to select it. Alternatives range from "each iteration" (e.g.,

genetic methods) to "under given conditions" (e.g., the diversification moves of tabu search).

4. *Candidate selection.* Neighbourhoods may be very large. Then, often, only a subset of moves are examined at each iteration. The corresponding *candidate list* $\mathcal{C}(x) \subseteq \mathcal{N}(x)$ may be permanent and updated from iteration to iteration (e.g., tabu search) or it may be constructed at each new iteration (e.g., genetic methods). In all cases, a selection criterion specifies how solutions are picked for inclusion in the candidate list.
5. *Acceptance criterion.* Moves are evaluated by applying a function $g(x, y)$ based on one or several attributes of the two solutions: objective function value, distance from certain constraints, penalties for violating some others, etc. External factors, such as random terms or biases from aggregated characteristics of past solutions may also be included in the evaluation. The best solution with respect to this criterion

$$\tilde{x} = argopt\{g(x, y);\ y \in \mathcal{C}(x)\}$$

   is selected and implemented (unless forbidden by cycling-prevention mechanisms).
6. *Stopping criteria.* Meta-heuristics may be stopped on a variety of criteria: computing time, number of iterations, rate of improvement, etc. More than one criterion may be defined to control various phases of the search (usually corresponding to various neighbourhoods).

With these definitions, we introduce a generic meta-heuristic procedure illustrated in Figure 17.2 and use it to describe the three main classes of meta-heuristics: *genetic methods*, *simulated annealing*, and *tabu search*. These methodologies have been, and continue to be, most oftenly used and parallelized. They are therefore treated in more detail in the following sections. Other methods, such as *scatter search*, GRASP, *ant colony systems*, and *variable neighbourhood search* have also been proposed and we briefly discuss related parallelization issues in Section 7.

Genetic algorithms belong to the larger class of evolutionary methods and were inspired by the evolution processes of biological organisms. In biology, when natural populations are studied over many generations, they appear to *evolve* according to the principles of *natural selection* and *survival of the fittest* to produce "well adapted" individuals. Genetic algorithms mimic this process, attempting to *evolve* solutions to

1. Initialization: x^0
2. Neighbourhood selection $\mathcal{N} \in \{\mathcal{N}_1, \ldots, \mathcal{N}_q\}$
3. Candidate selection $\mathcal{C}(x) \subseteq \mathcal{N}(x)$
4. Move evaluation/neighbourhood exploration $g(x, y),\ y \in \mathcal{C}(x)$
5. Move implementation $\tilde{x} = argopt\{g(x, y)\}$
6. Solution evaluation, update search parameters
7. Test stopping criteria: Stop or Goto 3 (continue local search phase) or Goto 2 (initiate new search phase)

Figure 17.2. Generic meta-heuristic.

optimization problems (Holland, 1975; Goldberg, 1989; Whitley, 1994; Fogel, 1994; Michalewicz, 1992; Michalewicz and Fogel, 2000). In recent years, the genetic algorithm paradigm was considerably enriched, as it evolved to include hybridization with local improvement heuristics and other meta-heuristics. Although the specialized literature is frequently replacing the term "genetic algorithms" with "evolutionary methods", we use the former in this paper to distinguish these methods from other strategies where a population of solutions evolves through an iterative process (e.g., scatter search and most co-operative parallel methods described later in this paper).

Genetic methods work on a population of solutions that evolves by generating new individuals out of combinations of existing individuals. At each iteration a *selection* operator applied to the current population identifies the parents to be used for the generation of new individuals. Thus, in a genetic context, the candidate selection and move evaluation is based solely on the value $g(x)$, the *fitness*, of the parents (this may be contrasted to the evaluations used in, for example, simulated annealing, tabu search, and scatter search). *Crossover* operators are used to generate the new individuals. *Mutation* and *hill climbing* (local search) operators modify the definition or characteristics of the new individuals to improve their fitness and the diversity of the population. In several variants, especially so for parallel genetic methods, the implementation of the move is completed by a *survival* operation that determines which of the parents and offspring advances to the next generation. Figure 17.3 displays the functions of the classical genetic operators, while Figure 17.4 summarizes the main steps of a generic genetic algorithm.

Simulated annealing methods are inspired by the *annealing* process of cooling materials in a heat bath. Here, solid materials are first heated past melting point. They are then gradually cooled back to a solid state, the *rate of cooling* directly influencing on the

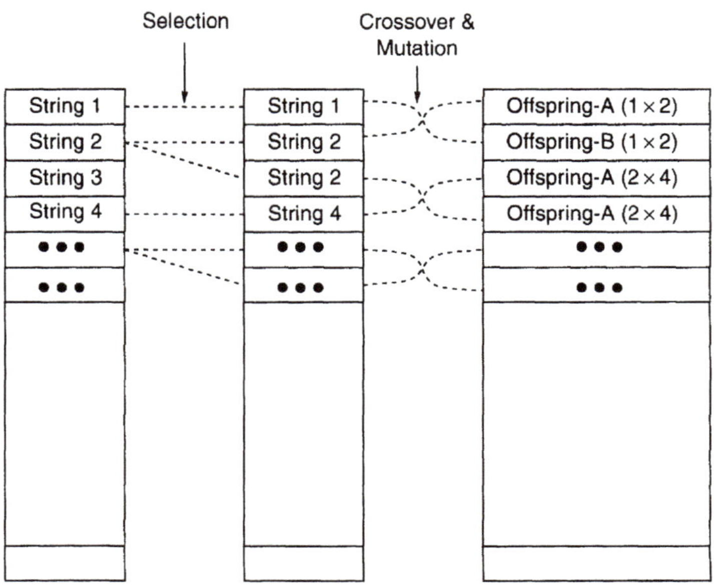

Figure 17.3. Genetic evolutionary algorithm operators.

1. Initialization: generation of the initial population.
2. Neighbourhood selection: selection of crossover and mutation operators.
3. Candidate—*parent*—selection: application of a selection operator to the current population
4. Move evaluation/neighbourhood exploration: none
5. Move implementation: application of crossover, mutation, hill climbing, and offspring and parent selection operators to obtain a new population
6. If stopping criteria not met, Goto 3 (continue the evolution) or Goto 2 (modify the evolution criteria)

Figure 17.4. A generic genetic algorithm.

structural properties of the final product. The materials may be represented as systems of particles and the whole process of heating and cooling may be simulated to evaluate various cooling rates and their impact on the properties of the finished product. The *simulated annealing* metaphor attempts to use similar statistical processes to guide the search through feasible space (Metropolis et al., 1953; Kirkpatrick et al., 1983; Laarhoven and Aarts, 1989; Aarts and Korst, 1989, 2002; etc.). A randomized scheme, the *temperature* control, determines the probability of accepting non-improving solutions. This mechanism aims to allow escaping from local optima. The *cooling schedule* determines how this probability evolves: many non-improving solutions are accepted initially (*high temperature*) but the temperature is gradually reduced such that few (none) inferior solutions are accepted towards the end (*low temperature*). A generic simulated annealing procedure is displayed in Figure 17.5.

One of the most appropriate tabu search metaphors is the capability of the human brain to store, recall, and process information to guide and enhance the efficiency of repetitive processes. *Memory* and *memory hierarchy* are major concepts in tabu search, as are the memory-based strategies used to guide the procedure into various search phases. Here, as elsewhere, a heuristic locally explores the domain by moving from one solution to the best available solution in its neighbourhood. Inferior quality solutions are accepted as a strategy to move away from local optima. Short-term *tabu status* memories record recent visited solutions or their attributes to avoid repeating or inverting recent actions. The tabu status of a move may be lifted if testing it against an *aspiration criterion* signals the discovery of a high quality solution (typically, the best one encountered so far). Medium to long-term memory structures record various informations and statistics relative to the solutions already encountered (e.g., frequency of certain attributes in the best solutions) to "learn" about the solution space and guide the search. *Intensification* of the search around a good solution and its *diversification* towards regions of the solution space not yet explored are two main ingredients in tabu search. These two types of moves are based on medium and long-term memories and are implemented using specific neighbourhoods. More details on the basic and advanced features of tabu search may be found in Glover (1986, 1989, 1990, 1996), Glover and Laguna (1993, 1997), Gendreau (2002). Figure 17.6 displays a generic tabu search procedure.

1. Initialization: Select
   - Initial state (solution) $x = x_0$;
   - Initial temperature $\tau = \tau_0$;
   - Temperature reduction function α;
2. Neighbourhood and candidate selections: none (generally);
   Replaced by
3. Selection of number of iterations L to approximate equilibrium at temperature τ
4. Move evaluation/neighbourhood exploration:
   Randomly select $y \in \mathcal{X}$.
5. Move implementation

   $\Delta f := f(y) - f(x)$;
   if $\Delta f \leq 0$ **then** $x := y$;
   else if $g(x, y) = exp\,(-\Delta f/\tau) > random(0, 1)$ **then** $x := y$;
6. Solution evaluation
7. Test stopping criteria
   - If less than L iterations, Goto 4
   - If convergence not verified, $\tau = \alpha(\tau)$; Goto 3

Figure 17.5. Generic simulated annealing procedure.

1. Initialization: x_0
2. Neighbourhood selection: local search, intensification, diversification, ...
3. Candidate selection $\mathcal{C}(x) \subseteq \mathcal{N}(x)$
4. Move evaluation/neighbourhood exploration:
   tabu criteria, aspiration criterion
5. Move implementation
6. Update of memories and tabu status
7. Test stopping criteria; If it fails Goto 3 (continue local search) or Goto 2 (change search phase)

Figure 17.6. A generic tabu search.

This very brief summary of three major meta-heuristics emphasizes the similarities of the main activities used by the various methodologies to explore the solution space of given problems. This similarity translates into "similar" requirements when strategies for parallelization are contemplated. For example, all meta-heuristic procedures encompass a rather computationally heavy stage where the neighbourhood (or the population) is explored. Fortunately, the computational burden may be reduced by performing the exploration in parallel and most implementations of the first parallelization strategy discussed in the following section address this issue. This observation explains our choice of discussing parallelization strategies not according to particular meta-heuristic characteristics but rather following a few general principles.

3 PARALLEL COMPUTATION

The central goal of parallel computing is to speed up computation by dividing the work load among several processors. From the view point of algorithm design, "pure" parallel computing strategies exploit the partial order of algorithms (i.e., the sets of operations that may be executed concurrently in time without modifying the solution method and the final solution obtained) and thus correspond to the "natural" parallelism present in the algorithm. The partial order of algorithms provides two main sources of parallelism: *data* and *functional* parallelism.

To illustrate, consider the multiplication of two matrices. To perform this operation, one must perform several identical operations executing sums of products of numbers. It is possible to overlap the execution of these identical operations on different input data. Among computer architectures with several arithmetic and logic units (ALUs), *Single Instruction stream, Multiple Data stream* (*SIMD*) computers are particularly suited to this type of parallelism as they can load the same operation on all ALUs (single flow of instructions) and execute it on different input data (multiple flows of data). The total number of computer operations required to compute the matrix product is not reduced, but given the concurrency of several operations, the total wall-clock computation time is reduced proportionally to the average number of overlapping sets of operations during the computation. This is data parallelism.

Computations may also be overlapped even when operations are different. It is usually inefficient to exploit this parallelism at the fine-grain level of a single instruction. Rather, the concurrent execution of different operations typically occurs at the coarse-grain level of procedures or functions. This is functional parallelism. For example, one process can compute the first derivative vector of a function while another computes the second derivative matrix. The two processes can overlap at least partially in time. When computations are complex and dimensions are large, this partial overlap may yield interesting speedups. Parallel computers that are well adapted to perform functional parallelism usually follow a *MIMD* (*Multiple Instructions stream, Multiple Data stream*) architecture where both data and instructions flow concurrently in the system. MIMD computers are often made up of somewhat loosely connected processors, each containing an ALU and a memory module.

Parallel computation based on data or functional parallelism is particularly efficient when algorithms manipulate data structures that are strongly regular, such as matrices in matrix multiplications. Algorithms operating on irregular data structures, such as graphs, or on data with strong dependencies among the different operations remain difficult to parallelize efficiently using only data and functional parallelism. Meta-heuristics generally belong to this category of algorithms that are difficult to parallelize. Yet, as we will see, parallelizing meta-heuristics offers opportunities to find new ways to use parallel computers and to design parallel algorithms.

3.1 Parallelizing Meta-heuristics

From a computational point of view, meta-heuristics are just algorithms from which we can extract functional or data parallelism. Unfortunately, data and functional parallelism are in short supply for many meta-heuristics. For example, the local search loop (Steps 3–7) of the generic tabu search in Figure 17.6 displays strong data dependencies between successive iterations, particularly in the application of the tabu criterion and the update of memories and tabu status. Similarly, the passage from one generation to another in standard genetic methods is essentially a sequential process, while the replacement of the current solution of the generic simulated annealing procedure (Step 5 in Figure 17.5) cannot be done in parallel, forcing the sequential execution of the inner loop (Steps 4–6). As in other types of algorithms, however, operations inside one step may offer some functional or data parallelism. Moreover, the exploration of the solution space based on random restarts can be functionally parallelized since there are no dependencies between successive runs. The set of visited solutions, as well as the outcome of the search made up of random restarts are identical to those obtained by the sequential procedure provided the set of initial solutions is the same for both the sequential and parallel runs.

Meta-heuristics as algorithms may have limited data or functional parallelism but, as problem solving methods, they offer other opportunities for parallel computing. To illustrate, consider the well-known Branch-and-Bound technique. The branching heuristic is one of the main factors affecting the way Branch-and-Bound algorithms explore the search tree. Two Branch-and-Bound algorithms, each using a different branching heuristic, will most likely perform different explorations of the search tree of one problem instance. Yet, both will find the optimum solution. Thus, the utilization of different Branch-and-Bound search patterns does not prevent the technique from finding the optimal solution. This critical observation may be used to construct parallel Branch-and-bound methods. For example, the parallel exploration of the search tree based on distributing sub-trees will modify the data available to the branching heuristic and thus, for the same problem instance, the parallel and sequential search patterns will differ. Yet, the different search strategies will all find the optimal solution. Consequently, the exploration of sub-trees may be used as a source of parallelism for Branch-and-Bound algorithms. This source of parallelism is not related to data parallelism, since the data (the variables of the optimization problem) is not partitioned. It is not functional parallelism either, because the two computations, sequential and parallel, are different. Although this difference makes comparative performance analyzes more difficult to perform (since the parallel implementation does not do the same work as the sequential one), sub-tree distribution remains a valuable and widely used parallelization strategy for Branch-and-Bound algorithms.

Similar observations can be made relative to new sources of parallelism in meta-heuristics. A meta-heuristic algorithm started from different initial solutions will almost certainly explore different regions of the solution space and return different solutions. The different regions of the solution space explored can then become a source of parallelism for meta-heuristic methods. However, the analysis of parallel implementation of meta-heuristic methods becomes more complex because often the parallel implementation does not return the same solution as the sequential implementation. Evaluation criteria based on the notion of *solution quality* (i.e., does the method find a better solution?) have then to be used to qualify the more classical acceleration (speedup) measures.

We have classified the parallelization strategies applied to meta-heuristics according to the source of parallelism used:

***Type 1*:** This source of parallelism is usually found within an iteration of the heuristic method. The limited functional or data parallelism of a move evaluation is exploited or moves are evaluated in parallel. This strategy, also called *low-level* parallelism, is rather straightforward and aims solely to speed up computations, without any attempt at achieving a better exploration (except when the same total wall-clock time required by the sequential method is allowed to the parallel process) or higher quality solutions.

***Type 2*:** This approach obtains parallelism by partitioning the set of decision variables. The partitioning reduces the size of the solution space, but it needs to be repeated to allow the exploration of the complete solution space. Obviously, the set of visited solutions using this parallel implementation is different from that of the sequential implementation of the same heuristic method.

***Type 3*:** Parallelism is obtained from multiple concurrent explorations of the solution space.

Type 1 parallelism Type 1 parallelizations may be obtained by the concurrent execution of the operations or the concurrent evaluation of several moves making up an iteration of a search method. Type 1 parallelization strategies aim directly to reduce the execution time of a given solution method. When the same number of iterations are allowed for both sequential and parallel versions of the method and the same operations are performed at each iteration (e.g., the same set of candidate moves is evaluated and the same selection criterion is used), the parallel implementation follows the same exploration path through the problem domain as the sequential implementation and yields the same solution. As a result, standard parallel performance measures apply straightforwardly. To illustrate, consider the computation of the average fitness of a population for genetic methods. Because the sequence used to compute the fitness of the individuals is irrelevant to the final average fitness of the population, it can be spartitioned and the partial sums of each subpopulation can be computed in parallel. Both the parallel and sequential computations yield the same average fitness, the parallel implementation just runs faster.

Some implementations modify the sequential method to take advantage of the extra computing power available, but without altering the basic search method. For example, one may evaluate in parallel several moves in the neighbourhood of the current solution instead of only one. In Figure 17.5, one can choose several y variables in Step 4 and then perform Step 5 in parallel for each selected y. In tabu search, one may *probe* for a few moves beyond each immediate neighbour to increase the available knowledge when selecting the best move (Step 4 of Figure 17.6). The resulting search patterns of the serial and parallel implementations are different in most cases. Yet, under certain conditions, the fundamental algorithmic design is not altered, therefore these approaches still qualify as Type 1 parallelism.

Type 2 parallelism In Type 2 strategies, parallelism comes from the decomposition of the decision variables into disjoint subsets. The particular heuristic is applied to each subset and the variables outside the subset are considered fixed. Type 2 strategies are generally implemented in some sort of master-slave framework:

- A *master* process partitions the decision variables. During the search, the master may modify the partition. Modifications may be performed at intervals that are either fixed before or determined during the execution, or, quite often, are adjusted when restarting the method.
- *Slaves* concurrently and independently explore their assigned partitions. Moves may proceed exclusively within the partition, the other variables being considered fixed and unaffected by the moves which are performed, or the slaves may have access to the entire set of variables.
- When slaves have access to the entire neighbourhood, the master must perform a more complex operation of combining the partial solutions obtained from each subset to form a complete solution to the problem.

Note that a decomposition based on partitioning the decision variables may leave large portions of the solution space unexplored. Therefore, in most applications, the partitioning is repeated to create different segments of the decision variable vector and the search is restarted.

Type 3 parallelism The first two parallelization strategies yield a single search path. Parallelization approaches that consist of several concurrent searches in the solution space are classified as Type 3 strategies. Each concurrent thread may or may not execute the same heuristic method. They may start from the same or different initial solutions and may communicate during the search or only at the end to identify the best overall solution. The latter are known as *independent search* methods, while the former are often called *co-operative multi-thread* strategies. Communications may be performed synchronously or asynchronously and may be event-driven or executed at predetermined or dynamically decided moments. These strategies belong to the *p-control* class according to the taxonomy proposed by Crainic, Toulouse, and Gendreau (1997), and are identified as *multiple-walks* by Verhoeven and Aarts (1995).

To speed up computation by using a multi-thread strategy, one generally tries to make each thread perform a shorter search than the sequential procedure. This technique is implemented differently for each class of meta-heuristic. Let p be the number of processors. For tabu search, each thread performs T/p iterations, where T is the number of iterations of the corresponding sequential procedure. For simulated annealing, the total number of iterations of the inner loop (Steps 4 to 6 in Figure 17.5) is reduced proportionally from L to L/p. For genetic algorithms, it is not the number of generations which is generally reduced. Rather, the size N of the sequential population is reduced to N/p for each genetic thread.

Type 3 parallelization strategies are often used to perform a more thorough exploration of the solution space. Several studies have shown that multi-thread procedures yield better solutions than the corresponding sequential meta-heuristics, even when the exploration time permitted to each thread is significantly lower than that of the sequential computation. Studies have also shown that the combination of several threads that implement different parameter settings increases the robustness of the global search relative to variations in problem instance characteristics. We review some of these results in Sections 4–6.

It is noteworthy that the application of classical performance measures (e.g., Barr and Hickman, 1993) to multi-thread, parallel meta-heuristics is somewhat problematic. For example, it is generally difficult to eliminate or control the overlap between the search paths (to adequately control the search overlap would involve such high levels of search synchronization and information exchanges that all benefits of parallelization would be lost). Thus, one cannot measure correctly the search efficiency in terms of the work performed. Moreover, many Type 3 parallelizations are based on asynchronous interactions among threads. As asynchronous computations are time dependent, such computations can produce different outputs for the same input. Classical speedup measures are ill-defined to compare the performances of asynchronous parallel meta-heuristics with sequential ones. In fact, several asynchronous Type 3 parallel meta-heuristics are so different from the original sequential procedure that one can hardly consider the two implementations to belong to the same meta-heuristic class.

3.2 Other Taxonomies

The classification described above is sufficiently general to apply to almost any meta-heuristic and parallelization strategy. Moreover, the lessons learned by comparing within the same class the implementations and performances of particular meta-heuristics are of general value and may be extended to search methods not covered in depth in this paper. Most taxonomies proposed in the literature are, however, related to a specific type of meta-heuristic.

Greening (1990) divides simulated annealing parallelization strategies according to the degree of accuracy in the evaluation of the cost function associated with a move. Parallel algorithms that provide an error-free evaluation are identified as *synchronous* while the others are *asynchronous*. The synchronous category is divided further between parallel simulated annealing algorithms that maintain the convergence properties of the sequential method (*serial-like*) and those that have an accurate cost evaluation but differ from the sequential computation in the search path generation (*altered generation*). Type 1 parallelism completely covers the serial-like category. The altered generation category overlaps with the Type 1 and Type 2 strategies. Asynchronous algorithms are those that tolerate some error in the cost function in order to get better speedups and correspond to a subset of the Type 2 category. No Type 3 algorithms are considered in this work.

Cantù-Paz (1995) provides a classification of parallel genetic algorithms. The first category, called *global* parallelization is identical to the Type 1 parallelization. Two other categories classify genetic algorithms according to the size of the populations that evolve in parallel, the so-called *coarse-grained* and *fine-grained* parallelization strategies. There is also a class for *hybrid* genetic algorithm parallelizations. For example, global parallelization applied to subpopulations of a coarse-grained parallel algorithm is one instance of an hybrid algorithm. The union of these three groups forms the Type 3 category described in this paper. No Type 2 strategies are considered in Cantù-Paz's taxonomy.

Verhoeven and Aarts (1995) define local search as the class of approximation methods based on the exploration of neighbourhoods of solutions, including tabu search, simulated annealing, and genetic algorithms. Their taxonomy divides parallel methods between *single-walk* and *multiple-walk* strategies. The former corresponds to Type 1 and Type 2 parallelism. The latter includes *multiple independent walks* and *multiple interacting walks* and thus corresponds to the Type 3 parallelism of this paper. Single-walk methods are further classified as *single-step* or *multiple-step* strategies. The former corresponds to the simple parallel neighbourhood evaluation of Type 1. The latter includes probing and Type 2 strategies. The taxonomy explicitly distinguishes between synchronous and asynchronous approaches. Cung et al. (2002) present a classification of parallel meta-heuristic strategies based on that of Verhoeven and Aarts (1995).

Currently, the most comprehensive taxonomy of parallel tabu search methods is offered by Crainic et al. (1997) and Crainic (2002). The classification has three dimensions. The first dimension, *control cardinality*, explicitly examines how the global search is controlled, by a single process (as in master-slave implementations) or collegially by several processes. The four classes of the second dimension indicate the *search differentiation*: do search threads start from the same or different solutions and do they make use of the same or different search strategies? Finally, the *control type* dimension addresses the issue of information exchange and divides methods into four classes: *rigid synchronization* (e.g., simple neighbourhood evaluation in Type 1 strategies), *knowledge synchronization* (e.g., probing in Type 1 strategies and synchronous information exchanges in Type 3 methods), and, finally, *collegial* and *knowledge collegial*. The last two categories correspond to Type 3 strategies but attempt to differentiate methods according to the quantity and quality of information exchanged, created, and shared. Although introduced for tabu search, this classification applies to many other meta-heuristics and may form the basis for a comprehensive taxonomy of meta-heuristics. It refines, in particular, the classification used in this paper, which is based on the impact of parallelization on the search trajectory.

4 GENETIC ALGORITHMS

At each iteration k, genetic algorithms compute the average fitness $\bar{g}_k = \sum_{i=1}^{i=N} g(x_i^k)/N$ of the N strings in the current population (Step 3 in Figure 17.4). The time-consuming part of this operation is performing the summation $\sum_{i=1}^{i=N} g(x_i^k)$. Obviously, this computation can be distributed over several processors and, since there are no dependencies among operations, it is a good candidate for efficient data, Type 1 parallelization. This summation has indeed been the first component of genetic algorithms to be parallelized (Grefenstette, 1981).

Intuitively, one expects almost linear speedups from this parallelization of the average fitness evaluation. Surprisingly, however, most experiments report significant sub-linear speedups due to the latencies of low-speed communication networks (Fogarty and Huang, 1990; Hauser and Männer, 1994; Chen et al., 1996; Abramson and Abela, 1992; Abramson et al., 1993). The implementations of the selection and crossover operators are also based on simple iterative loops, but each involves relatively few computations. Therefore, considering the impact of the communication overhead on a time-consuming operation like the fitness evaluation, Type 1 parallelization of the genetic operators has received little attention.

Genetic algorithms are acknowledged to be inherently parallel. This inherent parallelism is limited to Type 1, however. We are not aware of any Type 2 parallelization strategy for genetic algorithms and, although many known parallel genetic algorithms are of Type 3, most of them are not strictly derived from the standard genetic paradigm. Standard genetic methods are based on single panmictic population, and computation is usually initiated on a new generation only after the old one has died out, thus preventing the occurrence of parallelism across generations. Parallel genetic models, on the other hand, find more opportunities for parallelism such as, concurrent computations across different generations or among different subpopulations. The literature on parallel genetic methods often identifies two categories of Type 3 approaches: *coarse-grained* and *fine-grained*. However, some Type 3 parallel genetic algorithms do not display such clear cut characterization (e.g., Moscato and Norman, 1992).

Coarse-grained parallelizations usually refer to methods where the same sequential genetic algorithm is run on p subpopulations (each of size N/p), although some researchers (e.g., Schlierkamp-Voosen and Mühlenbein, 1994; Herdy, 1992) have pondered the possibility of using different strategies for each subpopulation. In such models, each subpopulation is relatively small in comparison with the initial population. This has an adverse impact on the diversity of the genetic material, leading to premature convergence of the genetic process associated to each subpopulation. To favor a more diversified genetic material in each subpopulation, a new genetic operator, the *migration* operator, is provided.

The migration operator defines strategies to exchange individuals among subpopulations. This operator has several parameters: the selection process that determines which individuals will migrate (e.g., best-fit, randomly, randomly among better than average individuals), the migration rate that specifies the number of strings migrated, the migration interval that determines when migration may take place (usually defined in terms of a number of generations), and the immigration policy that indicates how individuals are replaced in the receiving subpopulation. Information exchanges are further determined by the neighbourhood structure. In the *island* model, individuals may migrate towards any other subpopulation, while in the *stepping-stone* model only direct neighbours are reachable. Often the connection structure of the parallel computer determines how subpopulations are logically linked. Finally, migration may be performed either synchronously or asynchronously.

Since a genetic algorithm is associated with each subpopulation, coarse-grained parallel strategies can exploit parallelism across generations, provided each generation is related to a different subpopulation. And, it is always possible to combine Type 1 and Type 3 parallelism by computing the average fitness within each subpopulation using the Type 1 strategy as described above. MIMD computers are well adapted to coarse-grained parallel genetic methods. Migration rate and frequency are such that, in general, the quantity of data exchanged is small and can be handled efficiently even by low-speed interconnection networks. Furthermore, since the workload of each processor is significant (running a sequential genetic algorithm), latencies due to communications (if any) can be hidden by computations in asynchronous implementations. Therefore, linear speedups could be expected. Yet, few reports detail the speedups of coarse-grained parallel genetic algorithms. To some extent, this is explained by the fact that speedups do not tell the whole story regarding the performance of coarse-grained parallelizations, since one also needs to consider the associated convergence behavior. Therefore, to this day, most efforts regarding coarse-grained parallelizations have been focused on studying the best migration parameter settings. Most studies conclude that migration is better than no migration, but that the degree of migration needs to be controlled.

Schnecke and Vornberger (1996) analyze the convergence behaviour of coarse-grained parallel genetic algorithms using a Type 3 strategy where a different genetic algorithm is assigned to each subpopulation and search strategies, rather than solutions, are migrated between subpopulations. At fixed intervals, the different genetic methods are ranked (using the *response to selection* method of Mühlenbein and Schlierkamp-Voosen, 1994) and the search strategies are adjusted according to the "best" one by importing some of the "best" one's characteristics (mutation rate, crossover rate, etc). The paper contains references to several other works where self-adapting parallel genetic evolution strategies are analyzed. Lis (1996), in particular, applies self-adaptation to the mutation rate. The author implements a farming model where a master processor manages the overall population and sends the same set of best individuals to slave processors, each of which has a different mutation probability. Periodically, according to the mutation rate of the process that obtained the best results, the mutation rates of all slave processors are shifted one level up or down and populations are recreated by the master processor using the best individuals of the slave processors. Starkweather et al. (1991; see also Whitley and Starkweather, 1990a,b) also suggest that an adaptive mutation rate might help achieve better convergence for coarse-grained parallel genetic algorithms.

A more conceptually focused approach to improve the convergence of coarse-grained parallel genetic strategies may be derived from co-evolutionary genetic algorithm ideas. Schlierkamp-Voosen and Mühlenbein (1994), e.g., use competing subpopulations as a means to adapt the parameters controlling the genetic algorithm associated with each subpopulation. In the general setting of co-evolutionary genetic methods, each subpopulation may have a different optimization function that either competes with other subpopulations (as in a prey-predator or host-parasite relationship, Hillis, 1992) or may co-operate with the other subpopulations by specializing on sub-problems that are later combined to yield the full solution (Potter and De Jong, 1994). In the competitive scheme proposed by Hillis, a population of solutions competes with a population of evolving problems (test cases). Fitness and selection pressure favors individuals that make life difficult in the other population. For example, fit individuals in the population of solutions are those that can solve many test cases, while fit test cases are those that only few individuals in the solution population can solve correctly. In the co-operative scheme, the selection pressure favors individuals that co-operate well to solve the global problem. The co-evolutionary setting provides the concept of complementary sub-problems as a way to improve the convergence of coarse-grained parallel genetic algorithms. To this day, however, this avenue has not been widely explored.

Fine-grained strategies for parallel genetic algorithms divide the population into a large number of small subsets. Ideally, subsets are of cardinality one, each individual being assigned to a processor. Each subset is connected to several others in its neighbourhood. Together, a subset and its neighbouring subsets form a subpopulation (or *deme*). Genetic operators are applied using asynchronous exchanges between individuals in the same deme only. Demes may be defined according to a fixed topology (consisting of individuals residing in particular processors), obtained by a random walk (applying a given Hamming distance among individuals), etc. Neighbourhoods overlap to allow propagation of individuals or individual characteristics and mutations across subpopulations. This overlapping plays a similar role to that of the migration operator for coarse-grained parallel genetic algorithms. Fine-grained parallel genetic algorithms are sometimes identified as *cellular algorithms*, because a fine-grained method with fixed topology deme and relative fitness policy may be shown to be equivalent to finite cellular automata with probabilistic rewrite rules and an alphabet equal to the set of strings in the search space (see Whitley, 1993).

Fine-grained parallel genetic algorithms evolve a single population that spawns over several generations. This enables parallelism across generations. A "generation gap" (signaled by different iteration counts) tends to emerge in the population because the selection and crossover operators in one deme are not synchronized with the other demes. In effect, it is still a single population, due to the overlap among demes. Yet, the global dynamics of fine-grained parallel genetic algorithms are quite different from those of general genetic methods. In a single panmictic population, individuals are selected based on a global average fitness value and the selected individuals have the same probability of interacting with each other through the crossover operator. In fine-grained parallel strategies, average fitness values (or whatever stand for them) are local to demes. Consequently, individuals in the population do not have the same probability to mate and the genetic information can only propagate by diffusion through overlapping demes.

Diffusion is channeled by the overlapping structure of the demes, which is often modeled on the interconnection network of the parallel computer. Consequently, network topology is an important issue for fine-grained parallelization strategies, because the diameter of the network (the maximum shortest path between two nodes) determines how long it takes for good solutions to propagate over all of the demes. Long diameters isolate individuals, giving them little chance of combining with other good individuals. Short diameters prevent genotypes (solution vectors) from evolving, since good solutions rapidly dominate, which leads to premature convergence. Individual fitness values are relative to their deme and thus individuals on processing units not directly connected may have no chance to be involved together in the same crossover operator. Schwehm (1992) implemented a fine-grained parallel genetic algorithm on a massively parallel computer to investigate which network topology is best-suited to fine-grained parallel genetic algorithms. Compared with a ring and three cubes of various dimensions, a torus yielded the best results. Baluja (1993) conducted studies regarding the capability of different topologies to prevent demes of fine-grained parallel genetic algorithms to be dominated by the genotype of strong individuals. Three different topologies were studied and numerical results suggest that 2D arrays are best suited to fine-grained parallel genetic algorithms. See also the recent work of Kohlmorgen et al. (1999).

Fine-grained parallel genetic methods have been hybridized with hill-climbing strategies. Mühlenbein et al. (1987, 1988), among others, have designed hybrid strategies for several optimization problems and obtained good performance. Memetic algorithms (e.g., Moscato, 1989; Moscato and Norman, 1992) belong to the same category. Hybrid schemes construct selection and crossover operators in a similar manner to regular fine-grained parallelizations but a hill-climbing heuristic is applied to each individual. When the computational cost of the hill-climbing heuristic (or any other heuristic) is substantial (in Memetic algorithms, for example), the population size has to be small and the computing units powerful. Such hybrids appear to be closer to coarse-grained parallelism, except that the selection and crossover operators are those usually associated with fine-grained parallelization mechanisms. Interesting comments about fine-grained parallel genetic strategies, their design and application as well as the role of hill-climbing heuristics, can be found in Mühlenbein (1991, 1992, 1992a).

Research on parallel genetic algorithms is still active and prolific. Unlike other meta-heuristics, parallelizations of Type 1, that exploit the inherent parallelism of standard genetic methods, are still quite competitive in terms of performance, degree of parallelism, adaptation to current parallel computer architectures, and ease of implementation. In terms of research directions, innovation in algorithmic design, and capacity for hybridization with other search methods, Type 3 parallelism is currently the most active area. However, good models to compare the performance of different Type 3 parallel strategies for genetic algorithms are still missing.

5 SIMULATED ANNEALING

A simulated annealing iteration consists of four main steps (Steps 4 to 6 in Figure 17.5): select a move, evaluate the cost function, accept or reject the move, update (replace) the current solution if the move is accepted. Two main approaches are used to obtain Type 1 parallel simulated annealing algorithms: *single-trial* parallelism where only one move is computed in parallel, and *multiple-trial* strategies where several moves are evaluated simultaneously.

The evaluation of the cost function for certain applications may be quite computationally intensive, thus suggesting the possible exploitation of functional parallelism. Single-trial strategies exploit functional parallelism by decomposing the evaluation of the cost function into smaller problems that are assigned to different processors. Single-trial strategies do not alter the algorithmic design nor the convergence properties of the sequential simulated annealing method. The resulting degree of parallelism is very limited, however, and thus single-trial parallelization strategies do not speedup computation significantly.

Multiple-trial parallelizations distribute the iterations making up the search among different processors. Each processor fully performs the four steps of each iteration mentioned above. This distribution does not raise particular issues relative to the first three steps since these tasks are essentially independent with respect to different potential moves. Solution replacement is, however, a fundamentally sequential operation. Consequently, the concurrent execution of several replacement steps may yield erroneous evaluations of the cost function because these evaluations could be based on outdated data.

Type 1 multiple-trial strategies for simulated annealing enforce the condition that parallel trials always result in an *error-free* cost function evaluation. This may be achieved when solution updating is restricted to a single accepted move or to moves that do not interact with each other. The latter approach, referred to as the *serializable subset* method, accepts only a subset of moves that always produces the same result when applied to the current state of the system, independent of the order of implementation (a trivial serializable subset contains only rejected moves). To implement the former approach, one processor is designated to be the holder of the current

1. Initialization (Master: Processor 0):
   (a) $x^0; \mathcal{N}_0^0, \mathcal{N}_0^1, \ldots, \mathcal{N}_0^{p-1}; L_0 = L_0/p; \tau_0; k = 0;$
   (b) Send $\mathcal{N}_0^0, \mathcal{N}_0^1, \ldots, \mathcal{N}_0^{p-1}$, τ_0, L_0, k to the slave processors
2. Slave processor i (concurrent processing):
   (a) Perform search: Steps 4 to 6 of Figure 17.5 for L_k iterations
   (b) Send partial configuration x_i^k to processor 0
3. Master processor 0:
   (a) Solution update: Combine partial solutions
   (b) $L_k = L_k/p; \tau_k; \mathcal{N}_k^0, \mathcal{N}_k^1, \ldots, \mathcal{N}_k^{p-1}$ such that $\mathcal{N}_k^l \neq \mathcal{N}_{k-1}^l$, $l = 0, 1, \ldots, p-1;$
   (c) If stopping criterion not true, send $\mathcal{N}_k^0, \mathcal{N}_k^1, \ldots, \mathcal{N}_k^{p-1}$, τ_k, and L_k, k to the slave processors

Figure 17.7. Type 2 parallel simulated annealing.

solution. Once a processor accepts a move, it sends the new solution to the holder, which then broadcasts it to all processors. Any new move accepted during the update of the current solution is rejected. Performance varies with the temperature parameter. At high temperatures, when many potential moves are accepted, communications, synchronization, and rejection of moves generate substantial overheads. At low temperatures, fewer moves are accepted and speedups improve. Yet, performance is not very satisfactory in most cases.

Most multiple-trial parallelization strategies for simulated annealing follow a Type 2 approach and partition the variables into subsets. A master-slave approach is generally used, as illustrated in Figure 17.7. To initiate the search, the master processor partitions the decision variables into p initial sets. The appropriate set is sent to each processor together with the initial values for the temperature τ_0 and the number of iterations L_0 to be executed. In Step 2, each slave processor i executes the simulated annealing search at temperature τ_k on its set of variables $\mathcal{N}_i^k$ and sends its partial configuration of the entire solution to the master processor. Once the information from all slaves has been received, the master processor merges the partial solutions into a complete solution and verifies the stopping criterion. If the search continues, it generates a new partition of the variables such that $\mathcal{N}_l^k \neq \mathcal{N}_l^{k-1}$ and sends it to the slave processors together with new values for L_k and τ_k.

Felten et al. (1985) applied this strategy to a 64-city *travelling salesman problem (TSP)* using up to 64 processors of a hypercube computer. An initial tour was randomly generated and partitioned into p subsets of adjacent cities, which were assigned to p processors. Each processor performed local swaps on adjacent cities for a given number of iterations, followed by a synchronization phase where cities were rotated among processors. Parallel moves did not interact due to the spatial decomposition of the decision variables. Moreover, each synchronization ensured the integrity of the global state. Hence, there was no error and almost linear speedups were observed.

In most cases, however, error-free strategies cannot be efficiently implemented. It is often difficult, for example, to partition variables such that parallel moves do not interact. Therefore, the two main issues in Type 2 parallel simulated annealing are "how important is the error?" and "how can errors be controlled?".

Algorithms executed on shared-memory systems can regularly and quite efficiently update the global state of the current solution so that errors do not accumulate during the computation. However, the issue is significantly more difficult for distributed memory systems because each processor has its own copy of the data, including the "current" solution, and global updates are costly in communication time. Trade-offs, therefore, must be made between the frequency of the global state updates and the level of error one is ready to tolerate during the parallel simulated annealing computation, while acknowledging the possibility that significant errors might accumulate locally. It has been observed, however, that errors tend to decrease as temperatures decrease, because solution updates occur less frequently at low temperatures. Jayaraman and Darema (1988) and Durand (1989) specifically address the issue of error tolerance for parallel simulated annealing. As expected, they conclude that the error increases as the frequency of synchronizations decreases and the number of processors increases. In their studies, the combined error due to synchronization and parallelism had a significant impact on the convergence of the simulated annealing algorithm. Of the two factors, parallelism emerged as the most important.

One of the reasons for partitioning variables among processors is to prevent the same variable from being simultaneously involved in more than one move. This goal can also be achieved by *locking* the variables involved in a move. A locking mechanism permits only the processor that owns the lock to update a given variable. Any other processor that attempts to execute a move involving a locked variable must either wait for the variable to become available or attempt a different move. However, the use of locks results in a communication overhead which increases with the number of processors (e.g., Darema et al., 1987).

Rather than having several processors execute moves from the same current solution or subset of decision variables, processors could work independently using different initial solutions and cooling schedules, or simply using the probabilistic nature of simulated annealing to obtain different search paths. This is Type 3 parallelism. Numerous efforts to develop Type 3 parallel simulated annealing strategies using independent or co-operative search threads are reported in the literature. An interesting recent development involves the systematic inclusion of principles and operators from genetic algorithms in simulated annealing multi-thread procedures. This hybrid approach tends to perform very well.

The first Type 3 parallelization strategy to emerge was the *division strategy* proposed by Aarts et al. (1986). Let L be the number of iterations executed by a sequential simulated annealing program before reaching equilibrium at temperature τ. The division strategy executes L/p iterations on p processors at temperature τ. Here, a single initial solution and cooling strategy is used and it is assumed that the search paths will not be the same due to the different probabilistic choices made by each processor. At the L/p-th iteration, processors can either synchronize and choose one of the solutions as the initial configuration for the next temperature, or continue from their last configurations at the preceding temperature level. When synchronization is used, the procedure corresponds to a synchronous co-operative scheme with global exchange of information (otherwise, it is equivalent to an independent search approach). Unfortunately, the length of the chain can not be reduced arbitrarily without significantly affecting the convergence properties of the method. This is particularly true at low temperatures, where many steps are required to reach equilibrium. To address this problem, the authors clustered the processors at low temperatures and applied multi-trial parallelism of Type 1 within each cluster. Kliewer and Tschöke (2000) have addressed practical issues such as the proper length of parallel chains and the best time to cluster processors.

An alternative to the division strategy is to run each processor with its own cooling schedule in an independent search framework. The Multiple Independent Runs (MIR, Lee 1995) and the Multiple Markov Chains (MMC, Lee and Lee 1996) schemes are Type 3 parallelizations based on this approach. When there are no interactions among processors, performance is negatively affected by idle processors which are waiting for the longest search path to terminate. The MIR strategy addresses this problem by calculating estimates of the total run length, and then using these estimates to end computation on all processing units. The MMC scheme addresses the same issue by allowing processes to interact synchronously and asynchronously at fixed or dynamic intervals. The authors of this co-operating multi-thread strategy observe that communication overheads from co-operation are largely compensated for by the reduction of processor idle time.

A different Type 3 initiative to increase the degree of parallelism of simulated annealing algorithms consists of moving the methodology closer to genetic algorithms by considering a population of simulating annealing threads. Laursen (1994) proposed such a population scheme based on the selection and migration operators of parallel genetic algorithms. Each processor concurrently runs k simulated annealing procedures for a given number of iterations. Processors are then paired and each processor migrates (copies) its solutions to its paired processor. Thus, after the migration phase, each processor has $2k$ initial solutions and this number is reduced to k by selection. These new k solutions become the initial configurations of the k concurrent simulated annealing threads, and the search restarts on each processor. Pairing is dynamic and depends on the topology of the parallel machine. For example, in a grid topology, processors can pair with any of their corner neighbours. Because processors are dynamically paired and neighbourhoods overlap, information propagates in the network of processors similar to the stepping-stone coarse-grained model for parallel genetic methods. Mahfoud and Goldberg (1995) also propose to evaluate concurrently a population of n Markov chains. The general idea proceeds as follows: after $n/2$ iterations, two parents are selected from the population of the n current solutions. Two children are generated using a genetic crossover operator, followed by a mutation operator. Probabilistic trial competitions are held between children and parents and the replacement step is performed according to the outcome of the competition. The temperature is lowered when the population reaches equilibrium. There are different ways to parallelize this algorithm. The asynchronous parallelization described in Mahfoud and Goldberg (1995) follows the Type 3 coarse-grained parallel genetic algorithm approach. The population of n Markov chains is divided into p subpopulations of n/p Markov chains. Crossover, mutation, and probability trials are applied to individuals of each local subpopulation. Asynchronous migration enables sharing of individuals among subpopulations.

The literature on parallel simulated annealing methods has continued to flourish in recent years. Recent research focuses on applying parallel simulated annealing to new problems and developing software packages, rather than on discovering new parallelization strategies. The relatively poor performance of Type 1 and Type 2 approaches have been noticed and, consequently, there have been few applications of these strategies. The most actively applied parallelization strategies are of Type 3: hybridation with hill-climbing (e.g., Du et al., 1999) or with genetic methods (e.g., Kurbel et al., 1995); co-operative multi-threads (e.g., Chu et al., 1999); and massive parallelism (e.g., Mahfoud and Goldberg, 1995; Bhandarkar et al., 1996). We believe methods of Type 3 will continue to offer the best performance for parallel simulated annealing.

6 TABU SEARCH

Tabu search has proved a fertile ground for innovation and experimentation in the area of parallel meta-heuristics. Most parallel strategies introduced in Section 3 have been applied to tabu search for a variety of applications, and a number of interesting parallelization concepts have been introduced while developing parallel tabu search methods.

Similar to most other meta-heuristics, low-level, Type 1 parallelism has been the first strategy to be applied to tabu search methods. The usual target in this case is the acceleration of the neighbourhood exploration (Step 4 of Figure 17.6). Following the ideas summarized in Section 3, most Type 1 implementations correspond to a master process that executes a sequential tabu procedure and dispatches, at each iteration, the possible moves in the neighbourhood of the current solution to be evaluated in parallel by slave processes. Slaves may either evaluate only the moves in the set they receive from the master process, or may probe beyond each move in the set. The master receives and processes the information resulting from the slave operations and then selects and implements the next move. The master also gathers all the information generated during the tabu exploration, updates the memories, and decides whether to activate different search strategies or stop the search.

The success of Type 1 strategies for tabu search appears more significant than for genetic or simulated annealing methods. Indeed, very interesting results have been obtained when neighbourhoods are very large and the time to evaluate and perform a given move is relatively small, such as in *quadratic assignment* (*QAP*: Chakrapani and Skorin-Kapov, 1992, 1993, 1995; Taillard (1991, 1993a), *travelling salesman* (Chakrapani and Skorin-Kapov, 1993a) and *vehicle routing* (*VRP*: Garcia et al., 1994) applications. For the same quality of solution, near-linear speedups are reported using a relatively small number of processors. Moreover, historically (the first half of the 90's), Type 1 parallel tabu search strategies permitted improvements to the best-known solutions to several problem instances proposed in the literature.

Similarly to the other meta-heuristics, Type 1 tabu search implementations depend heavily upon the problem characteristics. Thus, performance results are less interesting when the time required by one serial iteration is relatively important compared to the total solution time, resulting in executions with only a few hundred moves compared to the tens of thousands required by a typical VRP tabu search procedure. This was illustrated by the comparative study of several synchronous tabu search parallelization strategies performed by Crainic, Toulouse, and Gendreau (1995a) for the location-allocation problem with balancing requirements. With respect to Type 1 parallelization approaches, two variants were implemented: (1) slaves evaluate candidate moves only; (2) *probing*: slaves also perform a few local search iterations. The second variant performed marginally better. However, both variants were outperformed by co-operative multi-thread (Type 3) implementations, which attempt a more thorough exploration of the solution space.

Typical tabu search implementations of Type 2 parallelization strategies partition the vector of decision variables and perform a search on each subset. This approach was part of the preliminary experimentation in the study by Crainic, Toulouse, and Gendreau (1995a). It performed poorly, mainly because of the nature of the class of problems considered; multicommodity location with balancing requirements requires a significant computation effort to evaluate and implement moves, resulting in a limited number of moves that may be performed during the search.

As with Type 1 implementations, Type 2 parallel methods were more successful for problems for which numerous iterations may be performed in a relatively short time and restarting the method with several different partitions does not require unreasonable computational efforts. TSP and VRP formulations belong to this class of applications. Fiechter (1994) proposed a method for the TSP that includes an intensification phase during which each process optimizes a specific slice of the tour. At the end of the intensification phase, processes synchronize to recombine the tour and modify (shift part of the tour to a predetermined neighbouring process) the partition. To diversify, each process determines from among its subset of cities a candidate list of most promising moves. The processes then synchronize to exchange these lists, so that all build the same final candidate list and apply the same moves. Fiechter reports near-optimal solutions to large problems (500, 3000 and 10000 vertices) and almost linear speedups (less so for the 10000 vertex problems). Porto and Ribeiro (1995, 1996) studied the task scheduling problem for heterogeneous systems and proposed several synchronous parallel tabu search procedures where a master process determines and modifies partitions, synchronizes slaves, and communicates best solutions. Interesting results were reported, even for strategies involving a high level of communications. Almost linear speedups were observed, better performances being observed for larger problem instances.

Taillard (1993) studied parallel tabu search methods for vehicle routing problems. In Taillard's approach, the domain is decomposed into polar regions, to which vehicles are allocated, and each subproblem is solved by an independent tabu search. All processors synchronize after a certain number of iterations (according to the total number of iterations already performed) and the partition is modified: tours, undelivered cities, and empty vehicles are exchanged between adjacent processors. Taillard reports very good results for the epoch. However, enjoying the benefit of hindsight, the main contribution of this paper is to mark the evolution towards one of the most successful sequential meta-heuristics for the VRP: a tabu search method called *adaptive memory* (Rochat and Taillard, 1995; Glover, 1996).

According to an adaptive memory approach, cities are initially separated into several subsets, and routes are built using a construction heuristic. Initial routes are then stored in a structure called an *adaptive memory*. Then, a combination procedure builds a complete solution using the routes in the memory, and the solution is further improved using a tabu search method. The routes of "good" solutions are then deposited into the same memory, which thus adapts to reflect the current state of knowledge of the search. The process then re-starts with a new solution built from the routes stored in the adaptive memory. The method stops when a pre-specified number of calls to the adaptive memory have been performed. This approach clearly implements the principles of Type 2 decomposition using a serial procedure; See also the interesting developments in vocabulary building strategies for tabu search proposed by Glover (1996). Adaptive memory principles have now been successfully applied to other problem classes and are opening interesting research avenues (Glover, 1996). However, interestingly, most parallel applications of this approach are now found in co-operative multi-thread strategies (Type 3).

Type 3 parallelizations for tabu search methods follow the same basic pattern described in Section 3: p threads search through the same solution space, starting from possibly different initial solutions and using possibly different tabu (or other) search strategies. Historically, independent and synchronous co-operative multi-thread methods were proposed first. However, currently, asynchronous procedures are being increasingly developed. Consequently, one observes an increased awareness of the issues related to the definition and modelling of co-operation.

Battiti and Tecchiolli (1992, for the QAP) and Taillard (the main study is found in his 1994 paper on parallel tabu methods for job shop scheduling problems) studied independent multi-thread parallelization schemes, where the independent search processes start the exploration from different, randomly generated, initial configurations. Both studies empirically established the efficiency of independent multi-thread procedures when compared to the best heuristics proposed at the time for their respective problems. Both studies also attempted to establish some theoretical justifications for the efficiency of independent search. Battiti and Tecchiolli derived probability formulas that tended to show that the probability of "success" increases, while the corresponding average time to "success" decreases, with the number of processors (provided the tabu procedure does not cycle). On the other hand, Taillard showed that the conditions required for the parallel method to be "better" than the sequential one are rather strong, where "better" was defined as "the probability the parallel algorithm achieves success with respect to some condition (in terms of optimality or near-optimality) by time t is higher than the corresponding probability of the sequential algorithm by time pt". However, the author also mentions that, in many cases, the empirical probability function of iterative algorithms is not very different from an exponential one, implying that independent multi-thread parallelization is an efficient strategy. The results for the job shop problem seemed to justify this claim. Similar results may also be found in Eikelder et al. (1999).

Malek et al. (1989, for the TSP), De Falco et al. (1994, for the QAP), and De Falco et al. (1995, for the mapping problem) proposed co-operative parallel strategies where the individual search threads are rather tightly synchronized. The implementation proposed by Malek et al. (1989) proceeds with one main process that controls the co-operation, and four child processes that run serial tabu search algorithms with different tabu conditions and parameters. The child processes are stopped after a specified time interval, solutions are compared, bad areas of solution space are eliminated, and the searches are restarted with a good solution and an empty tabu list. This implementation was part of a comparative study of serial and parallel simulated annealing and tabu search algorithms for the TSP. The authors report that the parallel tabu search implementation outperformed the serial one and consistently produced comparable or better results than sequential or parallel simulated annealing. De Falco and colleagues implemented a multi-thread strategy, where each process performs a local search from its best solution. Then, processes synchronize and best solutions are exchanged between processes that run on neighbouring processors. Local best solutions are replaced with imported ones only if the latter are better. The authors indicate that better solutions were obtained when co-operation was included compared to an independent thread strategy. Super-linear speedups are reported.

Rego and Roucairol (1996) proposed a tabu search method for the VRP based on ejection chains and implemented an independent multi-thread parallel version, each thread using a different set of parameter settings but starting from the same solution. The method is implemented in a master-slave setting, where each slave executes a complete sequential tabu search. The master gathers the solutions found by the threads, selects the overall best, and reinitializes the threads for a new search. Low-level (Type 1) parallelism accelerates the move evaluations of the individual searches, as well as the post-optimization phase. Experiments show the method to be competitive on the standard VRP problem set (Christofides et al., 1979).

Asynchronous co-operative multi-thread search methods are being proposed in continuously increasing numbers. All such developments we have identified use some form of *central memory* for inter-thread communications. Each individual search thread starts from a different initial solution and generally follows a different search strategy. Exchanges are performed asynchronously and are done through the central memory. One may classify co-operative multi-thread search methods according to the type of information stored in the central memory: complete or partial solutions. In the latter case, one often refers to *adaptive memory* strategies, while *central memory*, *pool of solutions*, or *solution warehouse* methods are used for the former.

Very successful Type 3 co-operative multi-thread parallel tabu search methods are based on *adaptive memory* concepts. This strategy has been particularly used for real-time routing and vehicle dispatching problems (Gendreau et al., 1999), as well as for VRP with time window restrictions (Taillard et al., 1997; Badeau et al., 1997). A general implementation framework of adaptive memory strategies begins with each thread constructing an initial solution and improving it through a tabu search or any other procedure. Each thread deposits the routes of its improved solution into the adaptive memory. Each thread then constructs a new initial solution out of the routes in the adaptive memory, improves it, communicates its routes to the adaptive memory, and repeats the process. A "central" process manages the adaptive memory and oversees communication among the independent threads. It also stops the procedure based on the number of calls to the adaptive memory, the number of successive calls which show no improvement in the best solution, or a time limit. In an interesting development, Gendreau et al. (1999) also exploited parallelism within each search thread by decomposing the set of routes along the same principles proposed in Taillard's work (1993). Good results have been obtained by using this approach on a network of workstations, especially when the number of processors is increased. Another interesting variant on the adaptive memory idea may be found in the work of Schulze and Fahle (1999). Here, the pool of partial solutions is distributed among processes to eliminate the need for a "master". The elements of the best solutions found by each thread are broadcast to ensure that each search has still access to all the information when building new solutions. Implemented on a limited number of processors, the method performed well (it is doubtful, however, that it would perform equally well for a larger number of processors).

As far as we can tell, Crainic, Toulouse, and Gendreau (1997) proposed the first central memory strategy for tabu search as part of their taxonomy. The authors also presented a thorough comparison of various parallelization strategies based on this taxonomy (Crainic et al., 1995a,b). The authors implemented several Type 1 and 2 strategies, one independent multi-thread approach, and a number of synchronous and asynchronous co-operative multi-thread methods. They used the multicommodity location problem with balancing requirements for experimentation. The authors report that the parallel versions achieved better quality solutions than the sequential ones and that, in general, asynchronous methods outperformed synchronous strategies. The independent threads and the asynchronous co-operative approaches offered the best performance.

Crainic and Gendreau (2001) proposed a co-operative multi-thread parallel tabu search for the fixed cost, capacitated, multicommodity network design problem. In their study, the individual tabu search threads differed in their initial solution and parameter settings. Communications were performed asynchronously through a central memory device. The authors compared five strategies of retrieving a solution from the pool when requested by an individual thread. The strategy that always returns the overall best solution displayed the best performance when few (4) processors were used. When the number of processors was increased, a probabilistic procedure, based on the rank of the solution in the pool, appears to offer the best performance. The parallel procedure improves the quality of the solution and also requires less (wall clock) computing time compared to the sequential version, particularly for large problems with many commodities (results for problems with up to 700 design arcs and 400 commodities are reported). The experimental results also emphasize the need for the individual threads to proceed unhindered for some time (e.g., until the first diversification move) before initiating exchanges of solutions. This ensures that local search histories can be established and good solutions can be found to establish the central memory as an *elite candidate* set. By contrast, early and frequent communications yielded a totally random search that was ineffective. The authors finally report that the co-operative multi-thread procedure also outperformed an independent search strategy that used the same search parameters and started from the same initial points. Other implementations of asynchronous co-operative multi-thread parallel tabu search methods are presented by Andreatta and Ribeiro (1994; see also Aiex et al., 1996; Martins et al., 1996) for the problem of partitioning integrated circuits for logical testing as well as by Cavalcante et al. (2002) for labor scheduling problems.

Crainic and Gendreau (1999) report the development of a hybrid search strategy combining their co-operative multi-thread parallel tabu search method with a genetic engine. The genetic algorithm initiates its population with the first elements from the central memory of the parallel tabu search. Asynchronous migration (migration rate = 1) subsequently transfers the best solution of the genetic pool to the parallel tabu search central memory, as well as solutions of the central memory towards the genetic population. The hybrid appears to perform well, especially on larger problems where the best known solutions are improved. It is noteworthy that the genetic algorithm alone was not performing well and that it was the parallel tabu search procedure that identified the best results once the genetic method contributed to the quality of the central memory.

Recently, Le Bouthiller and Crainic (2001) took this approach one step further and proposed a central memory parallel meta-heuristic for the VRP with time windows where several tabu search and genetic algorithm threads co-operate. In this model, the central memory constitutes the population common to all genetic threads. Each genetic algorithm has its own parent selection and crossover operators. The offspring are returned to the pool to be enhanced by a tabu search procedure. The tabu search threads follow the same rules as in the work of Crainic and Gendreau (2001). Only preliminary results are currently available, but they are extremely encouraging. Without any particular calibration, the parallel meta-heuristic obtains solutions whose quality is comparable to the best meta-heuristics available, and demonstrates almost linear speedups.

To conclude, Type 1 (and in some cases Type 2) parallelization strategies may still prove of value, especially for the evaluation of large neighbourhoods, or when used in hierarchical implementations to speedup computations of meta-heuristics involved in co-operative explorations. As illustrated above, Type 3, co-operative multi-thread strategies offer the most interesting perspectives. They do require, however, some care when they are designed and set up as will be discussed in the next section.

7 PERSPECTIVES AND RESEARCH DIRECTIONS

We have presented the main meta-heuristic parallelization strategies and their instantiation in the context of three major classes of methods: genetic algorithms, simulated annealing, and tabu search. Beyond the peculiarities specific to each methodology class and application domain, a number of general principles emerge:

- Meta-heuristics often have strong data dependencies. Therefore, straightforward data or functional parallelization techniques can identify only limited parallelism.
- Nevertheless, parallelization is very often beneficial. The evaluation of neighbouring solutions is a prime example of meta-heuristic algorithmic components that permit significant computational gains. Moreover, the concurrent exploration of the solution space by co-operating meta-heuristics often yields gains in solution quality, computational efficiency, and robustness of the search.
- Type 1 parallelization techniques are particularly useful for computation-intensive tasks, such as the evaluation of potential moves in the neighbourhood of a given solution. Moreover, such strategies may be advantageously incorporated into hierarchical parallel schemes where the higher level either explores partitions of the solution domain (Type 2 parallelism) or implements a co-operating multi-thread search (Type 3 parallelism).
- Hybridization, the incorporation of principles and strategies proper to one class of meta-heuristics into the algorithmic design of another, may improve performance of sequential and parallel meta-heuristics alike.
- When implemented properly, co-operating multi-thread parallel meta-heuristics appear to be the most promising strategy.

We have focussed on genetic methods, simulated annealing, and tabu search to reflect their wide-spread utilization in both sequential and parallel settings. Several other meta-heuristics have also been proposed in the literature, and some have proven quite successful for certain problem types. They include *scatter search* (Glover, 1994; Glover and Laguna, 1997), GRASP (Feo and Resende, 1995; Festa and Resende, 2002), *variable neighbourhood search* (Hansen and Mladenovic, 1997, 1999, 2002), *ant colony systems* (Colorni et al., 1991; Dorigo et al., 1996; Maniezzo and Carbonaro, 2002), as well as a host of ad-hoc methods based on neighbourhood exploration, which, in some surveys, are lumped together under the "local search" heading. Several of these methods also implement some sort of hybridization scheme where, typically, the current incumbent solution is enhanced by a local improvement procedure. In most cases, the basic working principle of the method may be cast in the generic meta-heuristic framework illustrated in Figure 17.2. Thus, the main principles and strategies described in this paper apply to the parallelization of these meta-heuristics as well, as illustrated by the parallelization efforts that have been reported for these methods (e.g., Kindervater et al., 1993; Pardalos et al., 1995; Sondergeld and Voß, 1999; Verhoeven and Severens, 1999). Of course, the most efficient applications of these principles and strategies to each of these meta-heuristics have yet to be established, which constitutes a most interesting research subject.

Co-operation and multi-thread parallelization appear to offer the most interesting perspectives for meta-heuristics. However, several issues and challenges remain to be addressed.

Synchronous implementations, where information is exchanged at regular intervals, have been reported for the three classes of meta-heuristics examined in this paper. In general, these implementations outperform the serial methods in solution quality. For tabu search (Crainic et al., 1995) and simulated annealing (Graffigne, 1992), synchronous co-operative methods appear to be outperformed, however, by independent search procedures. Yet, the study by Lee and Lee (1992) contradicts this trend. Their results show the independent thread approach to be outperformed by two strategies of synchronous co-operating parallel threads. Similar finds have been reported for genetic algorithms: Cohoon et al. (1987) and Cohoon et al. (1991a,b) report that parallel search with migration operators applied at regular intervals outperforms the same method without migration. These results point to interesting research issues that should be further investigated, especially since Lee and Lee used a dynamically adjusted synchronization interval that modified the traditional synchronous parallelism paradigm.

Asynchronous co-operative multi-thread search strategies appear to have been less studied but are being increasingly proposed. In fact, this strategy is probably the strongest current trend in parallel meta-heuristics, as illustrated by the development of memetic algorithms, of methods that evolve populations of simulated annealing threads, of the adaptive and central memory concepts initiated within the tabu search community but displaying general applicability characteristics. The results reported in the literature seem to indicate that asynchronous co-operative multi-thread parallelization strategies offer better results than synchronous and independent searches. More theoretical and empirical work is still required in this field, however.

An important issue for parallel meta-heuristic development and success concerns the definition of co-operation schemes and their impact on search behaviour and performance. A number of basic communication issues in designing multi-thread parallel meta-heuristics are discussed by Toulouse et al. (1996). More thorough analysis (Toulouse et al., 1997, 1999, 2000; Toulouse et al., 1998) shows that co-operative parallel meta-heuristics form dynamic systems and that the evolution of these systems may be more strongly determined by the co-operation scheme than by the optimization process. The selection of the information exchanged, and the part of it that is propagated past the initial exchange, significantly impacts the performance of co-operating procedures. The determination for each search thread of both when to import external information and how to use it (e.g., restart from the imported solution and clean up all history contrasted to the use of imported global information to bias local selection mechanisms) are equally important ingredients in the design of co-operating multi-thread parallel schemes. The application of these principles forms the basis of a new co-operation scheme, called *multi-level co-operative search* (Toulouse et al., 1999; see also Toulouse et al., 1998; Ouyang et al., 2000, 2000a), which has proven very successful for graph partitioning problems.

The solutions exchanged by co-operating search threads form populations that do not evolve according to genetic principles, but rather follow the information exchange mechanisms that define the co-operation. Of course, genetic operators may be used to control this evolution, as seen in some co-operative simulated annealing schemes. Scatter search and ant colony systems offer alternate co-operation mechanisms. In this respect, it is noteworthy that ant systems and, more generally, swarm-based methods (Bonabeau et al., 1999) appear as one of the first nature-inspired co-operation mechanisms. Yet, for now, despite the interest of its fundamental idea of trend enforcement/dilution, the co-operation principle underlying these methods appears much too rigid to offer a general purpose method. Scatter search, on the other hand, offers context-related combination mechanisms and memories for medium and long term steering of the evolution of the population. New mechanisms need to be developed, however, to relate this information to the search threads that make up the co-operating parallel algorithm. In fact, we believe that no single mechanism may adequately cover all possibilities and hybrid mechanisms will have to be defined.

Parallel co-operating methods do not have to include strategies belonging to only one meta-heuristic class. In fact, a number of recent studies (e.g., Le Bouthillier and Crainic, 2001) tend to demonstrate that combining different meta-heuristics yields superior results. At the parallel computing level, this approach generalizes the trend towards hybrid development observed in meta-heuristic communities. It also opens up an exciting field of enquiry. What meta-heuristics to combine? What role can each type of meta-heuristic play? What information is exchanged and how it is used in this context? These are only a few of the questions that need to be answered.

Last but not least, recall an earlier remark that co-operating parallel mechanisms bear little, if any, resemblance to the initial meta-heuristic one attempts to parallelize. This remark is even more true when different meta-heuristics combine their efforts. In fact, more and more authors argue that co-operating multi-thread parallel methods should form a "new", very broadly defined, class of meta-heuristics. If true, which we believe it is, we are left with the challenge of properly defining this meta-heuristic class. For example, memory mechanisms appear appropriate to record statistics on both the attributes of the solutions exchanged (or present in the solution population) and the performance of individual searches. How can this information be used to globally direct the search? What interactions are most appropriate between global and local (for each thread) information (memories)? These are among the most interesting challenges we face in this respect.

To conclude, parallel meta-heuristics offer the possibility to address problems more efficiently, both in terms of computing efficiency and solution quality. A rather limited number of strategies exist and this paper aims to both put these strategies into perspective and to briefly describe them. The study of parallel meta-heuristics design and performance still constitutes an exciting and challenging research domain with much opportunity for experimentation and development of important applications.

ACKNOWLEDGMENTS

Funding for this project has been provided by the Natural Sciences and Engineering Council of Canada, and by the Fonds F.C.A.R. of the Province of Québec.

REFERENCES

Aarts, E. and Korst, J. (2002) Selected topics in simulated annealing. In: C. Ribeiro and P. Hansen (eds.), *Essays and Surveys in Metaheuristics*. Kluwer Academic Publishers, Norwell, MA, pp. 1–57.

Aarts, E.H.L, de Bont, F.M.J., Habers, J.H.A. and van Laarhoven, P.J.M. (1986) Parallel implementations of statistical cooling algorithms. *Integration, The VLSI Journal*, **3**, 209–238.

Aarts, E.H.L. and Korst, J.H.M. (1989) *Simulated Annealing and Boltzmann Machines*. John Wiley & Sons, New York, NY.

Abramson, D. and Abela, J. (1992) A parallel genetic algorithm for solving the school timetabling problem. In: G. Gupta and C. Keen (eds.), *15th Australian Computer Science Conference*. Department of Computer Science, University of Tasmania, pp. 1–11.

Abramson, D., Mills, G. and Perkins, S. (1993) Parallelization of a genetic algorithm for the computation of efficient train schedules. In: D. Arnold, R. Christie, J. Day and P. Roe (eds.), *Proceedings of the 1993 Parallel Computing and Transputers Conference*. IOS Press, pp. 139–149.

Aiex, R.M., Martins, S.L., Ribeiro, C.C. and Rodriguez, N.R. (1996) Asynchronous parallel strategies for tabu search applied to the partitioning of VLSI circuits. Monografias em ciência da computação, Pontifícia Universidade Católica de Rio de Janeiro.

Andreatta, A.A. and Ribeiro C.C. (1994) A graph partitioning heuristic for the parallel pseudo-exhaustive logical test of VLSI combinational circuits. *Annals of Operations Research*, **50**, 1–36.

Azencott, R. (1992) *Simulated Annealing Parallelization Techniques*. John Wiley & Sons, New York, NY.

Badeau, P., Guertin, F., Gendreau, M., Potvin, J.-Y. and Taillard, É.D. (1997) A parallel tabu search heuristic for the vehicle routing problem with time windows. *Transportation Research C: Emerging Technologies*, **5**(2), 109–122.

Baluja, S. (1993) Structure and performance of fine-grain parallelism in genetic algorithms. In: S. Forrest (ed.), *Proceedings of the Fifth International Conference on Genetic Algorithms*. Morgan Kaufmann, San Mateo, CA, pp. 155–162.

Barr, R.S. and Hickman, B.L. (1993) Reporting computational experiments with parallel algorithms: issues, measures, and experts opinions. *ORSA Journal on Computing*, **5**(1), 2–18.

Battiti, R. and Tecchiolli, G. (1992) Parallel based search for combinatorial optimization: genetic algorithms and TABU. *Microprocessors and Microsystems*, **16**(7), 351–367.

Bhandarkar, S.M. and Chirravuri, S. (1996) A study of massively parallel simulated annealing algorithms for chromosome reconstruction via clone ordering. *Parallel Algorithms and Applications*, **9**, 67–89.

Bonabeau, E., Dorigo, M. and Theraulaz, G. (eds.) (1999) *Swarm Intelligence—From Natural to Artificial Systems*. Oxford University Press, New York, NY.

Cantú-Paz, E. (1995) A summary of research on parallel genetic algorithms. Report 95007, University of Illinois at Urbana-Champain.

Cantú-Paz, E. (1998) A survey of parallel genetic algorithms. *Calculateurs Parallèles, Réseaux et Systèmes répartis*, **10**(2), 141–170.

Cavalcante, C.B.C., Cavalcante, V.F., Ribeiro, C.C. and de Souza, C.C. (2002) Parallel cooperative approaches for the labor constrained scheduling problem. In: C. Ribeiro and P. Hansen (eds.), *Essays and Surveys in Metaheuristics*. Kluwer Academic Publishers, Norwell, MA, pp. 201–225.

Chakrapani, J. and Skorin-Kapov, J. (1992) A connectionist approach to the quadratic assignment problem. *Computers & Operations Research*, **19**(3/4), 287–295.

Chakrapani, J. and Skorin-Kapov, J. (1993) Massively parallel tabu search for the quadratic assignment problem. *Annals of Operations Research*, **41**, 327–341.

Chakrapani, J. and Skorin-Kapov, J. (1993a) Connection machine implementation of a tabu search algorithm for the traveling salesman problem. *Journal of Computing and Information Technology*, **1**(1), 29–36.

Chakrapani, J. and Skorin-Kapov, J. (1995) Mapping tasks to processors to minimize communication time in a multiprocessor system. In: *The Impact of Emerging Technologies of Computer Science and Operations Research*. Kluwer Academic Publishers, Norwell, MA, pp. 45–64.

Chen, Y.-W., Nakao, Z. and Fang, X. (1996) Parallelization of a genetic algorithm for image restoration and its performance analysis. In: *IEEE International Conference on Evolutionary Computation*, pp. 463–468.

Christofides, N., Mingozzi A. and Toth, P. (1979) The vehicle routing problem. In: N. Christofides, A. Mingozzi, P. Toth and C. Sandi (eds.), *Combinatorial Optimization*. John Wiley, New York, pp. 315–338.

Chu, K., Deng, Y. and Reinitz, J. (1999) Parallel simulated annealing algorithms by mixing states. *Journal of Computational Physics*, **148**, 646–662.

Cohoon, J., Hedge, S., Martin, W. and Richards, D. (1987) Punctuated equilibria: a parallel genetic algorithm. In: J. Grefenstette (ed.), *Proceedings of the Second International Conference on Genetic Algorithms and their Applications*. Lawrence Erlbaum Associates, Hillsdale, NJ, pp. 148–154.

Cohoon, J., Martin, W. and Richards, D. (1991a) Genetic algorithm and punctuated equilibria in VLSI. In: H.-P. Schwefel and R. Männer (eds.), *Parallel Problem Solving from Nature, Lecture Notes in Computer Science 496*. Springer-Verlag, Berlin, pp. 134–144.

Cohoon, J., Martin, W. and Richards, D. (1991b) A multi-population genetic algorithm for solving the k-partition problem on hyper-cubes. In: R. Belew and L. Booker (eds.), *Proceedings of the Fourth International Conference on Genetic Algorithms*. Morgan Kaufmann, San Mateo, CA, pp. 134–144.

Colorni, A., Dorigo, M. and Maniezzo, V. (1991) Distributed optimization by ant colonies. In: *Proceedings of the 1991 European Conference on Artificial Life*. North-Holland, Amsterdam, pp. 134–142.

Crainic, T.G. (2002) Parallel computation, co-operation, tabu search. In: C. Rego and B. Alidaee (eds.), *Adaptive Memory and Evolution: Tabu Search and Scatter Search*. Kluwer Academic Publishers, Norwell, MA (forthcoming).

Crainic, T.G. and Gendreau, M. (1999) Towards an evolutionary method—cooperating multi-thread parallel tabu search hybrid. In: S. Voß, S. Martello, C. Roucairol and I.H. Osman (eds.), *Meta-Heuristics 98: Theory & Applications*. Kluwer Academic Publishers, Norwell, MA, pp. 331–344.

Crainic, T.G. and Gendreau, M. (2001) Cooperative parallel tabu search for capacitated network design. *Journal of Heuristics* (forthcoming).

Crainic, T.G. and Toulouse, M. (1998) Parallel metaheuristics. In: T.G. Crainic and G. Laporte (eds.), *Fleet Management and Logistics*. Kluwer Academic Publishers, Norwell, MA, pp. 205–251.

Crainic, T.G., Toulouse, M. and Gendreau, M. (1995a) Parallel asynchronous tabu search for multicommodity location–allocation with balancing requirements. *Annals of Operations Research*, **63**, 277–299.

Crainic, T.G., Toulouse, M. and Gendreau, M. (1995b) Synchronous tabu search parallelization strategies for multicommodity location–allocation with balancing requirements. *OR Spektrum*, **17**(2/3), 113–123.

Crainic, T.G., Toulouse, M. and Gendreau, M. (1997) Towards a taxonomy of parallel tabu search algorithms. *INFORMS Journal on Computing*, **9**(1), 61–72.

Cung, V.-D., Martins, S.L., Ribeiro, C.C. and Roucairol, C. (2002) Strategies for the parallel implementations of metaheuristics. In: C. Ribeiro and P. Hansen (eds.), *Essays and Surveys in Metaheuristics*. Kluwer Academic Publishers, Norwell, MA, pp. 263–308.

Darema, F., Kirkpatrick, S. and Norton, V.A. (1987) Parallel algorithms for chip placement by simulated annealing. *IBM Journal of Research and Development*, **31**, 391–402.

De Falco, I., Del Balio, R. and Tarantino, E. (1995) Solving the mapping problem by parallel tabu search. Report, Istituto per la Ricerca sui Sistemi Informatici Paralleli-CNR.

De Falco, I., Del Balio, R., Tarantino, E. and Vaccaro, R. (1994) Improving search by incorporating evolution principles in parallel tabu search. In: *Proceedings International Conference on Machine Learning*, pp. 823–828.

Dorigo, M., Maniezzo, V. and Colorni, A. (1996) The ant system: optimization by a colony of cooperating agents. *IEEE Transactions on Systems, Man, and Cybernetics—Part B*, **26**(1), 29–41.

Du, Z., Li, S., Li, S., Wu, M. and Zhu, J. (1999) Massively parallel simulated annealing embedded with downhill—a SPMD algorithm for cluster computing.

In: *Proceedings of the 1st IEEE Computer Society International Workshop on Cluster Computing*. IEEE Computer Society Press, Washington, DC.

Durand, M.D. (1989) Parallel simulated annealing: accuracy vs. speed in placement. *IEEE Design & Test of Computers*, **6**(3), 8–34.

Durand, M.D. (1989a) Cost function error in asynchronous parallel simulated annealing algorithms. Technical Report CUCS-423-89, University of Columbia.

Felten, E., Karlin, S. and Otto, S.W. (1985) The traveling salesman problem on a hypercube, MIMD computer. In *Proceedings 1985 of the International Conference on Parallel Processing*, pp. 6–10.

Feo, T.A. and Resende, M.G.C. (1995) Greedy randomized adaptive search procedures. *Journal of Global Optimization*, **6**(2), 109–133.

Festa, P. and Resende, M.G.C. (2002) GRASP: an annotated bibliography. In: C. Ribeiro and P. Hansen (eds.), *Essays and Surveys in Metaheuristics*. Kluwer Academic Publishers, Norwell, MA, pp. 325–367.

Fiechter, C.-N. (1994) A parallel tabu search algorithm for large travelling salesman problems. *Discrete Applied Mathematics*, **51**(3), 243–267.

Fogarty, T.C. and Huang, R. (1990) Implementing the genetic algorithm on transputer based parallel systems. In: H.-P. Schwefel and R. Männer (eds.), *Proceedings of the 1st Workshop on Parallel Problem Solving from Nature*. Springer-Verlag, Berlin, pp. 145–149.

Fogel, D.B. (1994) Evolutionary programming: an introduction and some current directions. *Statistics and Computing*, **4**, 113–130.

Garcia, B.L., Potvin, J.-Y. and Rousseau, J.M. (1994) A parallel implementation of the tabu search heuristic for vehicle routing problems with time window constraints. *Computers & Operations Research*, **21**(9), 1025–1033.

Gendreau, M. (2002) Recent advances in tabu search. In: C. Ribeiro and P. Hansen (eds.), *Essays and Surveys in Metaheuristics*. Kluwer Academic Publishers, Norwell, MA, pp. 369–377.

Gendreau, M., Guertin, F., Potvin, J.-Y. and Taillard, É.D. (1999) Tabu search for real-time vehicle routing and dispatching. *Transportation Science*, **33**(4), 381–390.

Glover, F. (1986) Future paths for integer programming and links to artificial intelligence. *Computers & Operations Research*, **1**(3), 533–549.

Glover, F. (1989) Tabu search—part I. *ORSA Journal on Computing*, **1**(3), 190–206.

Glover, F. (1990) Tabu search—part II. *ORSA Journal on Computing*, **2**(1), 4–32.

Glover, F. (1994) Genetic algorithms and scatter search: unsuspected potentials. *Statistics and Computing*, **4**, 131–140.

Glover, F. (1996) Tabu search and adaptive memory programming—advances, applications and challenges. In: R. Barr, R. Helgason and J. Kennington (eds.), *Interfaces in Computer Science and Operations Research*. Kluwer Academic Publishers, Norwell, MA, pp. 1–75.

Glover, F. and Laguna, M. (1993) Tabu search. In: C. Reeves (ed.), *Modern Heuristic Techniques for Combinatorial Problems*. Blackwell Scientific Publications, Oxford, pp. 70–150.

Glover, F. and Laguna, M. (1997) *Tabu Search*. Kluwer Academic Publishers, Norwell, MA.

Goldberg, D.E. (1989) *Genetic Algorithms in Search, Optimization and Machine Learning*. Addison-Wesley, Reading, MA.

Graffigne, C. (1992) Parallel annealing by periodically interacting multiple searches: an experimental study. In: R. Azencott (ed.), *Simulated Annealing Parallelization Techniques*. John Wiley & Sons, New York, NY, pp. 47–79.

Greening, D.R. (1990) Parallel simulated annealing techniques. *Physica D*, **42**, 293–306.

Grefenstette, J. (1981) Parallel adaptive algorithms for function optimization. Technical Report CS-81-19, Vanderbilt University, Nashville.

Hansen, P. and Mladenovic, N. (1997) Variable neighborhood search. *Computers & Operations Research*, **24**, 1097–1100.

Hansen, P. and Mladenovic, N. (1999) An introduction to variable neighborhood search. In: S. Voß, S. Martello, C. Roucairol and I.H. Osman (eds.), *Meta-Heuristics 98: Theory & Applications*. Kluwer, Norwell, MA, pp. 433–458.

Hansen, P. and Mladenovic, N. (2002) Developments of variable neighborhood search. In: C. Ribeiro and P. Hansen (eds.), *Essays and Surveys in Metaheuristics*. Kluwer Academic Publishers, Norwell, MA, pp. 415–439.

Hauser, R. and Männer, R. (1994) Implementation of standard genetic algorithm on MIMD machines. In: Y. Davidor, H.-P. Schwefel and R. Männer (eds.), *Parallel Problem Solving from Nature III, Lecture Notes in Computer Science 866*. Springer-Verlag, Berlin, pp. 504–514.

Herdy, M. (1992) Reproductive isolation as strategy parameter in hierarchical organized evolution strategies. In: R. Männer and B. Manderick (eds.), *Parallel Problem Solving from Nature, 2*. North-Holland, Amsterdam, pp. 207–217.

Hillis, D.W. (1992) Co-evolving parasites improve simulated evolution as an optimization procedure. In: C.E.A. Langton (ed.), *Artificial Life II*. Addison-Wesley, pp. 313–324.

Holland, J.H. (1975) *Adaptation in Natural and Artificial Systems*. University of Michigan Press, Ann Arbor, MI.

Holmqvist, K. and Migdalas, A. and Pardalos, P.M. (1997) Parallelized heuristics for combinatorial search. In: A. Migdalas, P. Pardalos and S. Storoy (eds.), *Parallel Computing in Optimization*. Kluwer Academic Publishers, Norwell, MA, pp. 269–294.

Jayaraman, R. and Darema, F. (1988) Error tolerance in parallel simulated techniques. In: *Proceedings of the IEEE International Conference on Computer-Aided Design: ICCAD-88*. IEEE Computer Society Press, Washington, DC, pp. 545–548.

Kindervater, G.A.P, Lenstra, J.K. and Savelsberg, M.W.P. (1993) Sequential and parallel local search for the time constrained traveling salesman problem. *Discrete Applied Mathematics*, **42**, 211–225.

Kirkpatrick, S., Gelatt, C.D. and Vecchi, M.P. (1983) Optimization by simulated annealing. *Science*, **220**, 671–680.

Kliewer, G. and Tschöke, S. (2000) A general parallel simulated annealing library and its application in airline industry. In: *Proceedings of the 14th International Parallel and Distributed Processing Symposium (IPDPS 2000)*. Cancun, Mexico, pp. 55–61.

Kohlmorgen, U., Schmeck, H. and Haase, K. (1999) Experiences with fine-grained parallel genetic algorithms. *Annals of Operations Research*, **90**, 203–219.

Kurbel, K., Schneider, B. and Singh, K. (1995) VLSI standard cell placement by parallel hybrid simulated annealing and genetic algorithm. In: D.W. Pearson, N.C. Steele and R. F. Albrecht (eds.), *Proceedings of the Second International Conference on Artificial Neural Networks and Genetic Algorithms*. Springer-Verlag, Berlin, pp. 491–494.

Laarhoven, P. and Aarts, E.H.L. (1987) *Simulated Annealing: Theory and Applications*. Reidel, Dordrecht.

Laursen, P.S. (1994) Problem-independent parallel simulated annealing using selection and migration. In: Y. Davidor, H.-P. Schwefel and R. Männer (eds.), *Parallel Problem Solving from Nature III, Lecture Notes in Computer Science 866*. Springer-Verlag, Berlin, pp. 408–417.

Laursen, P.S. (1996) Parallel heuristic search—introductions and a new approach. In: A. Ferreira and P. Pardalos (eds.), *Solving Combinatorial Optimization Problems in Parallel, Lecture Notes in Computer Science 1054*. Springer-Verlag, Berlin, pp. 248–274.

Le Bouthillier, A. and Crainic, T.G. (2001) Parallel co-operative multi-thread meta-heuristic for the vehicle routing problem with time window constraints. Publication, Centre de recherche sur les transports, Université de Montréal, Montréal, QC, Canada.

Lee, F.-H.A. (1995) *Parallel Simulated Annealing on a Message-Passing Multi-Computer*. Ph.D. thesis, Utah State University.

Lee, K.-G. and Lee, S.-Y. (1992a) Efficient parallelization of simulated annealing using multiple markov chains: an application to graph partitioning. In: T. Mudge (ed.), *Proceedings of the International Conference on Parallel Processing*, volume III: Algorithms and Applications. CRC Press, pp. 177–180.

Lee, K.-G. and Lee, S.-Y. (1995) Synchronous and asynchronous parallel simulated annealing with multiple markov chains. *Lecture Notes in Computer Science 1027*, pp. 396–408.

Lin, S.-C., Punch, W. and Goodman, E. (1994) Coarse-grain parallel genetic algorithms: categorization and new approach. In: *Sixth IEEE Symposium on Parallel and Distributed Processing*. IEEE Computer Society Press, pp. 28–37.

Lis, J. (1996) Parallel genetic algorithm with the dynamic control parameter. In: *IEEE 1996 International Conference on Evolutionary Computation*, pp. 324–328.

Mahfoud, S.W. and Goldberg, D.E. (1995) Parallel recombinative simulated annealing: a genetic algorithm. *Parallel Computing*, **21**, 1–28.

Malek, M., Guruswamy, M., Pandya, M. and Owens, H. (1989) Serial and parallel simulated annealing and tabu search algorithms for the traveling salesman problem. *Annals of Operations Research*, **21**, 59–84.

Maniezzo, V. and Carbonaro, A. (2002) Ant colony optimization: an overview. In: C. Ribeiro and P. Hansen (eds.), *Essays and Surveys in Metaheuristics*. Kluwer Academic Publishers, Norwell, MA, pp. 469–492.

Martins, S.L., Ribeiro, C.C. and Rodriguez, N.R. (1996) Parallel programming tools for distributed memory environments. Monografias em Ciência da Computação 01/96, Pontifícia Universidade Católica de Rio de Janeiro.

Metropolis, N., Rosenbluth, A., Rosenbluth, M., Teller, A. and Teller, E. (1953) Equation of state calculation by fast computing machines. *Journal of Chemical Physics*, **21**, 1087–1092.

Michalewicz, Z. (1992) *Genetic Algorithms + Data Structures = Evolution Programs*. Springer-Verlag, Berlin.

Michalewicz, Z. and Fogel, D.B. (2000) *How to Solve It: Modern Heuristics*. Springer-Verlag, Berlin.

Moscato, P. (1989) On evolution, search, optimization, genetic algorithms and martial arts: towards memetic algorithms. Publication Report 790, Caltech Concurrent Computation Program.

Moscato, P. and Norman, M.G. (1992) A "memetic" approach for the traveling salesman problem. Implementation of a computational ecology for combinatorial optimization on message-passing systems. In: M. Valero, E. Onate, M. Jane, J. Larriba and B. Suarez (eds.), *Parallel Computing and Transputer Applications*. IOS Press, Amsterdam, pp. 187–194.

Mühlenbein, H. (1991) Evolution in time and space—the parallel genetic algorithm. In: G. Rawlins (ed.), *Foundations of Genetic Algorithm & Classifier Systems*. Morgan Kaufman, San Mateo, CA, pp. 316–338.

Mühlenbein, H. (1992) Parallel genetic algorithms in combinatorial optimization. In: O. Balci, R. Sharda and S. Zenios (eds.), *Computer Science and Operations Research*. Pergamon Press, New York, NY, pp. 441–456.

Mühlenbein, H. (1992a) How genetic algorithms really work: mutation and hill-climbing. In: R. Männer and B. Manderick (eds.), *Parallel Problem Solving from Nature, 2*. North-Holland, Amsterdam, pp. 15–26.

Mühlenbein, H., Gorges-Schleuter, M. and Krämer, O. (1987) New solutions to the mapping problem of parallel systems—the evolution approach. *Parallel Computing*, **6**, 269–279.

Mühlenbein, H., Gorges-Schleuter, M. and Krämer, O. (1988) Evolution algorithms in combinatorial optimization. *Parallel Computing*, **7**(1), 65–85.

Mühlenbein, H. and Schlierkamp-Voosen, D. (1994) The science of breeding and its application to the breeder genetic algorithm BGA. *Evolutionary Computation*, **1**(4), 335–360.

Ouyang, M., Toulouse, M., Thulasiraman, K., Glover, F. and Deogun, J.S. (2000a) Multi-level cooperative search: application to the netlist/hypergraph partitioning problem. In: *Proceedings of International Symposium on Physical Design*. ACM Press, pp. 192–198.

Ouyang, M., Toulouse, M., Thulasiraman, K., Glover, F. and Deogun, J.S. (2000b) Multilevel cooperative search for the circuit/hypergraph partitioning problem. *IEEE Transactions on Computer-Aided Design*. (to appear).

Pardalos, P.M., Pitsoulis, L., Mavridou, T., and Resende, M.G.C. (1995) Parallel search for combinatorial optimization: genetic algorithms, simulated annealing, tabu search and GRASP. In: A. Ferreira and J. Rolim (eds.), *Proceedings of Workshop on Parallel Algorithms for Irregularly Structured Problems, Lecture Notes in Computer Science 980*. Springer-Verlag, Berlin, pp. 317–331.

Pardalos, P.M., Pitsoulis, L. and Resende, M.G.C. (1995) A parallel GRASP implementation for the quadratic assignment problem. In: A. Ferreira and J. Rolim (eds.), *Solving Irregular Problems in Parallel: State of the Art*. Kluwer Academic Publishers, Norwell, MA.

Porto, S.C.S. and Ribeiro, C.C. (1995) A tabu search approach to task scheduling on heteregenous processors under precedence constraints. *International Journal of High-Speed Computing*, **7**, 45–71.

Porto, S.C.S. and Ribeiro, C.C. (1996) Parallel tabu search message-passing synchronous strategies for task scheduling under precedence constraints. *Journal of Heuristics*, **1**(2), 207–223.

Potter, M. and De Jong, K. (1994) A cooperative coevolutionary approach to function optimization. In: Y. Davidor, H.-P. Schwefel and R. Männer (eds.), *Parallel Problem Solving from Nature III, Lecture Notes in Computer Science 866*. Springer-Verlag, Berlin, pp. 249–257.

Ram, D.J., Sreenivas, T.H. and Subramaniam, K.G. (1996) Parallel simulated annealing algorithms. *Journal of Parallel and Distributed Computing*, **37**, 207–212.

Rego, C. and Roucairol, C. (1996) A parallel tabu search algorithm using ejection chains for the VRP. In: I. Osman and J. Kelly (eds.), *Meta-Heuristics: Theory & Applications*. Kluwer Academic Publishers, Norwell, MA, pp. 253–295.

Rochat, Y. and Taillard, É.D. (1995) Probabilistic diversification and intensification in local search for vehicle routing. *Journal of Heuristics*, **1**(1), 147–167.

Schlierkamp-Voosen, D. and Mühlenbein, H. (1994) Strategy adaptation by competing subpopulations. In: Y. Davidor, H.-P. Schwefel and R. Männer (eds.), *Parallel Problem Solving from Nature III, Lecture Notes in Computer Science 866*. Springer-Verlag, Berlin, pp. 199–208.

Schnecke, V. and Vornberger, O. (1996) An adaptive parallel genetic algorithm for VLSI-layout optimization. In: Y. Davidor, H.-P. Schwefel and R. Männer (eds.), *Parallel Problem Solving from Nature III, Lecture Notes in Computer Science 866*. Springer-Verlag, Berlin, pp. 859–868.

Schulze, J. and Fahle, T. (1999) A parallel algorithm for the vehicle routing problem with time window constraints. *Annals of Operations Reseach*, **86**, 585–607.

Schwehm, M. (1992) Implementation of genetic algorithms on various interconnection networks. In: M. Valero, E. Onate, M. Jane, J. Larriba and B. Suarez (eds.), *Parallel Computing and Transputers Applications*. IOS Press, Amsterdam, pp. 195–203.

Shonkwiler, R. (1993) Parallel genetic algorithms. In: S. Forrest (ed.), *Proceedings of the Fifth International Conference on Genetic Algorithms*. Morgan Kaufmann, San Mateo, CA, pp. 199–205.

Sondergeld, L. and Voß, S. (1999) Cooperative intelligent search using adaptive memory techniques. In: S. Voß, S. Martello, C. Roucairol and I.H. Osman (eds.), *Meta-Heuristics 98: Theory & Applications*. Kluwer, Norwell, MA, pp. 297–312.

Starkweather, T., Whitley, D. and Mathias, K. (1991) Optimization using distributed genetic algorithms. In: H.-P. Schwefel and R. Männer (eds.), *Parallel Problem Solving from Nature, Lecture Notes in Computer Science 496*. Springer-Verlag, Berlin, pp. 176–185.

Taillard, É.D. (1991) Robust taboo search for the quadratic assignment problem. *Parallel Computing*, **17**, 443–455.

Taillard, É.D. (1993a) Parallel iterative search methods for vehicle routing problems. *Networks*, **23**, 661–673.

Taillard, É.D. (1993b) *Recherches itératives dirigées parallèles*. Ph.D. thesis, École Polytechnique Fédérale de Lausanne.

Taillard, É.D. (1994) Parallel taboo search techniques for the job shop scheduling problem. *ORSA Journal on Computing*, **6**(2), 108–117.

Taillard, É.D., Badeau, P., Gendreau, M., Guertin, F. and Potvin, J.-Y. (1997) A tabu Search heuristic for the vehicle routing problem with soft time windows. *Transportation Science*, **31**, 170–186.

ten Eikelder, H.M.M., Aarts, B.J.M., Verhoeven, M.G.A. and Aarts, E.H.L. (1999) Sequential and parallel local search for job shop scheduling. In: S. Voß, S. Martello, C. Roucairol and I.H. Osman (eds.), *Meta-Heuristics 98: Theory & Applications*. Kluwer, Norwell, MA, Montréal, QC, Canada, pp. 359–371.

Toulouse, M., Crainic, T.G. and Gendreau, M. (1996) Communication issues in designing cooperative multi thread parallel searches. In: I.H. Osman and J.P. Kelly (eds.), *Meta-Heuristics: Theory & Applications*. Kluwer Academic Publishers, Norwell, MA, pp. 501–522.

Toulouse, M., Crainic, T.G. and Sansó, B. (1997) Systemic behavior of cooperative search algorithms. Publication CRT-97-55, Centre de recherche sur les transports, Université de Montréal, Montréal, QC, Canada.

Toulouse, M., Crainic, T.G. and Sansó, B. (1999a) An experimental study of systemic behavior of cooperative search algorithms. In: S. Voß, S. Martello, C. Roucairol and I.H. Osman (eds.), *Meta-Heuristics 98: Theory & Applications*. Kluwer Academic Publishers, Norwell, MA, pp. 373–392.

Toulouse, M., Crainic, T.G., Sansó, B. and Thulasiraman, K. (1998a) Self-organization in cooperative search algorithms. In: *Proceedings of the 1998 IEEE International Conference on Systems, Man, and Cybernetics*. Omnipress, pp. 2379–2385.

Toulouse, M., Crainic, T.G. and Thulasiraman, K. (2000) Global optimization properties of parallel cooperative search algorithms: a simulation study. *Parallel Computing*, **26**(1), 91–112.

Toulouse, M., Glover, F. and Thulasiraman, K. (1998b) A multi-scale cooperative search with an application to graph partitioning. Report, School of Computer Science, University of Oklahoma, Norman, OK.

Toulouse, M., Thulasiraman, K. and Glover, F. (1999b) Multi-level cooperative search. In: P. Amestoy, P. Berger, M. Daydé, I. Duff, V. Frayssé, L. Giraud and D. Ruiz (eds.), *5th International Euro-Par Parallel Processing Conference*, volume 1685 of *Lecture Notes in Computer Science*. Springer-Verlag, Berlin, pp. 533–542.

Verhoeven, M.G.A. and Severens, M.M.M. (1999) Parallel local search for steiner trees in graphs. *Annals of Operations Research*, **90**, 185–202.

Verhoeven, M.G.A. and Aarts, E.H.L (1995) Parallel local search. *Journal of Heuristics*, **1**(1), 43–65.

Voß, S. (1993) Tabu search: applications and prospects. In: D.-Z. Du and P. Pardalos (eds.), *Network Optimization Problems*. World Scientific Publishing Co., Singapore, pp. 333–353.

Whitley, D. (1993) Cellular genetic algorithms. In: S. Forrest (eds.), *Proceedings of the Fifth International Conference on Genetic Algorithms*. Morgan Kaufmann, San Mateo, CA, pp. 658–658.

Whitley, D. and Starkweather, T. (1990a) Optimizing small neural networks using a distributed genetic algorithm. In: *Proceedings of the International Conference on Neural Networks*. IEEE Press, pp. 206–209.

Whitley, D. and Starkweather, T. (1990b) GENITOR II: a distributed genetic algorithm. *Journal of Experimental and Theoretical Artificial Intelligence*, **2**(3), 189–214.

Whitley, L.D. (1994) A genetic algorithm tutorial. *Statistics and Computing*, **4**, 65–85.

Chapter 18

METAHEURISTIC CLASS LIBRARIES

Andreas Fink and Stefan Voß
Technische Universität Braunschweig
Institut für Wirtschaftswissenschaften
Abt-Jerusalem-Straße 7
D-38106 Braunschweig, Germany
E-mail: stefan.voss@tu-bs.de

David L. Woodruff
Graduate School of Management
University of California at Davis
Davis, California 95616, USA
E-mail: dlwoodruff@ucdavis.edu

Abstract Just as metaheuristics provide flexible and extensible methods of solving optimization problems, class libraries provide flexible and extensible building blocks for creating software to implement metaheuristics. In this chapter we provide a brief introduction to the class libraries and object oriented programming techniques. We also provide a short summary of some of the more general metaheuristics class libraries and the literature that describes them. We conclude with a detailed example of library design.

1 INTRODUCTION

The period of development of modern metaheuristics approximately coincided with the extensive deployment of object oriented programming techniques (OOP) [23,25,27]. A key feature of programming languages that support OOP is the ability to specify classes of objects using a mixture of abstract interface specification and concrete operational and data storage details. The object class specifications for a particular family of tasks are often organized as a library for subsequent use and reuse.

In a library, each class or template corresponds to a component of the methods for the task at hand. For example, a library for local search would provide one or more class definitions that capture the data and functionality associated with a neighborhood. The classes are instantiated as *objects* that encapsulate data and executable code.

Classes in a library are typically organized in a hierarchy with more abstract definitions at the top of the hierarchy and at the bottom more specific classes. The abstract classes may not define all necessary executable code or data structures. Instead, they provide interfaces and may provide some default code and data structures. More specific classes can provide details left out of the abstract class definitions and may also override definitions. A local search library, to continue the example, would contain abstract class definitions that provide information about the interface to a neighborhood object while a more specific class would provide implementation details for this interface for a neighborhood induced by 2-opt moves.

Libraries facilitate combinations of classes instantiated as objects to form an executable problem. Most class libraries also provide for *extensibility*, which means that some or all of the executable code and data structure implementation can be overridden by programmers who make use of all or part of the original object class specifications. Thus, OOP provides extreme levels of flexibility in the combination of components provided by a library.

At the highest levels of abstraction, a class library provides a unifying framework for the problem at hand. At the lowest level, a mixable set of extensible components is provided. Metaheuristic algorithms are a particularly ripe area for each aspect of OOP. After even a brief study of metaheuristics, one cannot help but notice that a unifying framework is lurking. OOP techniques provide a way of expressing the concepts and their interconnections, which captures the framework in an extremely useful way. An important aspect of modern metaheuristics is the flexibility that they offer that facilitates tailoring them for the task at hand. Metaheuristic class libraries enable exploitation of problem-specific knowledge or new ideas for algorithmic components by allowing the library classes to be extended or replaced.

To illustrate the class library concepts that we have introduced, Figure 18.1 shows a simplified version of the class hierarchy described in [36]. The figure illustrates the relationships between some of the classes provided in the library. Of course, most metaheuristic libraries are much more complex, but Figure 18.1 illustrates some of the important concepts. If class B is connected to class A by a line with an open triangle pointing toward A, then B is derived from A. We say that B "is" an A. We might also say that B is a child of A. Children inherit the interface and properties of their base classes. For example, in Figure 18.1, TABU_SEARCH is a LOCAL_SEARCH, which is a child of METAHEURISTIC.

If class A has a filled arrow pointing to class B, then an instance of class A "has" one or more instances of an object of class B. For example, in Figure 18.1, LOCAL_SEARCH has a NEIGHBORHOOD. The properties of a neighborhood are defined in the class shown as NEIGHBORHOOD and these properties are exploited by local search algorithms. However, the functionality must be provided by classes derived from NEIGHBORHOOD for a particular type of problem. Such a derived class would inherit the interface properties of a neighborhood and would supply implementation details as well as, perhaps, some additional interface definition. We have shown abstract classes in rectangles, and less abstract (i.e., instantiable) classes in squares with rounded corners which conforms the well-known UML notation.

The interplay between abstract and specific permeates both OOP and metaheuristics, which is why the connection between the two is so natural. As illustrated in Figure 18.1, a class library can capture at various levels the concepts shared by metaheuristics. It also enables creation of new code and data structures to implement novel extensions of the methods as well as application-specific details.

To be a little more specific, consider the sub-hierarchy in Figure 18.1 that begins with the abstract class PROBLEM. The class PROBLEM would provide interface specifications that would apply to any problem, such as a function to compute an objective function value. These specifications would be assumed to be available to the class

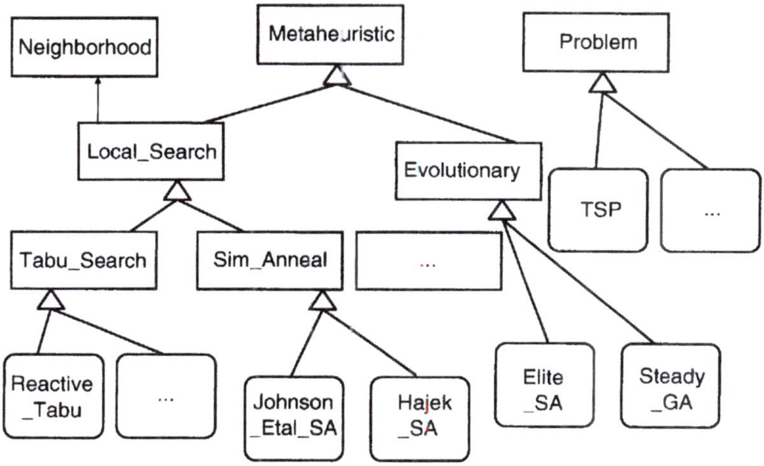

Figure 18.1. A simple class diagram for heuristic search.

METAHEURISTIC and, by inheritance, all of its child classes. Of course, it is not possible to specify how the objective function value would be computed without specifying the problem of interest. However, the existence of an interface can be exploited by writers of classes that are children of the base class METAHEURISTIC and that can access specific instantiations of children of the base class PROBLEM.

A class for a particular problem such as the Traveling Salesman Problem (TSP) would specify how the objective function would be computed along with methods for storing and accessing the data that describes the problem. The TSP is the problem of finding the shortest tour that visits a list of cities exactly once. Hence, a TSP problem class would describe methods for storing and accessing distances between cities as well for computing the length of a particular tour.

To keep Figure 18.1 simple, we show TSP as a direct descendent of PROBLEM. In a sophisticated library, some of the direct descendents of PROBLEM might be for classes that share important characteristics such as problems based on graphs. The TSP would then be a direct descendent of the graph problem class. The act of finding the commonalities in a group of objects and then creating a parent class that defines the common components is referred to as *factoring*. In Figure 18.1, e.g., the elements common to local search metaheuristics such as tabu search and simulated annealing have been factored into the class LOCAL_SEARCH.

Decisions regarding the proper construction of a hierarchy are intellectually stimulating, as are many other aspects of the design of class libraries. The discovery and articulation of abstractions and hierarchies is particularly rewarding when it results in software constructs that can be assembled into a functioning program. The allure of the library design process explains some of the appeal of creating a new class library as opposed to using one that has previously been created.

A design methodology primarily devoted to the goal of achieving a higher degree of reuse is the *framework* approach; see, e.g., [4,9,22]. Taking into account that for the development of application systems for given domains quite similar software is needed, it is reasonable to implement such common aspects by a generic design and embedded reusable software components. Here, one assumes that reuse on a large scale cannot only be based on individual components, but there also must be some reuse of design. Thus, the components have to be embedded in an architecture, which defines the collaboration between the components. Such a *framework*, or *frame* may be defined as a set of classes that embody an abstract design for solutions to a family of related problems (e.g., heuristics for discrete optimization problems), and thus provides us with abstract applications in a particular domain, which may be tailored for individual applications. A framework establishes a reference application architecture ("skeleton"), providing not only reusable software elements but also some type of reuse of architecture and design patterns [5,14], which may simplify software development considerably. Thus, frameworks represent implementation-oriented generic models for specific domains.

There is no clear dividing line between class libraries and frameworks. Whereas class libraries may be more flexible, frameworks often impose a broader structure on the whole system. Frameworks, which are sometimes called component libraries, may be subtly differentiated from class libraries by the "activeness" of components, i.e., components of the framework define application logic and call application-specific code. This generally results in a bi-directional flow of control.

We should mention that the phrase frame is also frequently used in the area of metaheuristics. For instance, the phrase *adaptive memory programming* (AMP) was coined to encompass a general approach (or philosophy) within heuristic search focusing on exploiting a collection of memory components [17,29]. That is, iteratively constructing (new) solutions based on the exploitation of some sort of memory may be viewed as an AMP process, especially when combined with learning mechanisms, that helps to adapt the collection and use of the memory. Based on the simple idea of initializing the memory and then iteratively generating new solutions (utilizing the given memory) while updating the memory based on the search, most of the metaheuristics described can be subsumed by the AMP approaches. This also includes the idea of exploiting provisional solutions in population based methods that are improved by a local search approach.

In the next section we give a brief overview of class libraries that have been made available for metaheuristics. Generally, the users of a library follow a path of adoption that leads to full exploitation of the capabilities and flexibility afforded. The adoption path is sketched in section 3. In order to provide readers with a more specific sense of what is involved in the design and use of a class library, some of the metaheuristic components of one particular library are described in Section 4. The chapter concludes with remarks on the outlook for future developments.

2 REVIEWS OF SELECTED LIBRARIES

In this section, we provide brief reviews of some of the prominent general class libraries for metaheuristics to give the reader a sense of the current state of the art. References are given to enable further study or use of the libraries. Readers interested in a book length overview of class libraries for optimization should consult [33].

2.1 HotFrame

HOTFRAME, a Heuristic OpTimization FRAMEwork implemented in C++, provides both adaptable components that incorporate different metaheuristics and an architectural description of the collaboration among these components and problem-specific complements. All typical application-specific concepts are treated as objects or classes: problems, solutions, neighbors, solution and move attributes. On the other side, metaheuristics concepts such as different methods and their building-blocks such as tabu criteria and diversification strategies are also treated as objects. HOTFRAME uses genericity as the primary mechanism to make these objects adaptable. That is, common behavior of metaheuristics is factored out and grouped in generic classes, applying static type variation. Metaheuristic template classes are parameterized by aspects such as solution spaces and neighborhood structures.

HOTFRAME defines an architecture for the interplay between heuristic classes and application-specific classes, and provides several such classes, which implement classic methods that are applicable to arbitrary problem types, solution spaces and neighborhood structures. All heuristics are implemented in a consistent way, which facilitates easy embedding of arbitrary methodss into application systems. New metaheuristics as well as new applications can be added to the framework. HOTFRAME includes built-in support for solution spaces representable by binary vectors or permutations, in connection with corresponding standard neighborhood structures, solution and move attributes, and recombination operators. Otherwise, the user may derive specialized classes from suitable built-in classes or implement corresponding classes from scratch according to a defined interface.

Metaheuristics included in HotFrame include (iterated) local search, simulated annealing (and variations), tabu search, some candidate list strategies, evolutionary methods and the pilot method. For further information about HOTFRAME see [11,19].

2.2 Templar

The Templar framework, implemented in C++, provides a method, and software components, for constructing systems to solve optimization problems. An important feature of the system is that it provides support for distribution using a variant of the Message Passing Interface (MPI) standard. This allows users of the library to distribute the search across heterogenous computers. The asynchronous message passage supported by Templar is ideal for creating hybrid strategies based on cooperation between objects implementing a variety of search methods.

The Templar framework is based on problem classes and engine classes. The engine's source code is representation and problem-independent. The problem class is an abstract base class, which is used to derive specific problem classes. These specific problem classes embody the characteristics of the problem type and are capable of loading problem instances. Problems make operators, such as neighborhood operators or genetic operators, available for use by an engine. The representation used by the problem (e.g., permutation) also provides its own operators. Engines can then use these operators to produce good solutions to a given problem instance.

The abstract base engine class is also used for deriving specific engine classes. Examples include simulated annealing engines, tabu search engines, and genetic algorithm engines. Although a single instance of an engine is connected to a particular problem, each specific engine class is problem, and representation, independent. As well as being easy to control, engines are designed so that a driving process can make a group of engines work cooperatively. For further information, see [30].

2.3 EasyLocal

EasyLocal [8] is a C++ class library that supports metaheuristics based on local search such as tabu search and simulated annealing. The library provides explicit support for so-called *kick moves*. A local search is typically based on a particular neighborhood structure induced by a corresponding type of move. A kick move may be made at strategically defined epochs by temporarily adopting a different neighborhood.

Another important feature of the library is inclusion of classes to support algorithmic testing. A batch control language EXPSPEC is supported by classes that collect run-time statistics. The output is generated both in machine- and human-readable format.

2.4 NeighborSearcher

NeighborSearcher [2] is an object-oriented framework for local search heuristics. The main goal is to provide an architectural basis for the implementation and comparison of different local search heuristics. Accordingly, a coarse-grained modularization is defined. The library is based on abstract classes that encapsulate concepts such as the construction of the initial solution, the local search algorithm, the solution, or the movement model. With this, one may implement a specific search strategy by selecting derived classes that provide the respective functionality. This provides an adaptation mechanism to the higher level client code.

2.5 iOpt

The Java class library iOpt [34] has support for population based evolutionary algorithms (EA) as well as local search based methods. Such libraries facilitate hybrid algorithms, e.g., the use of tabu search as a mutation operator (this is sometimes referred to as "learning" by EA researchers who yearn for metaphor). Classes for some problem domains such as scheduling are also provided.

A key distinguishing feature of iOpt is built-in support for propagation of one-way constraints [37]. Thus the library provides connections between metaheuristics and constraint logic programming. A one-way constraint is based on a general function C, which is a function of one or more variables and whose value constrains the value of some other variable. For example,

$$x_i = C(x_j, x_k, \ldots, x_m)$$

fixes the value of x_i as a function of some other variables whose values are not directly affected by the constraint. The iOpt library supports sophisticated methods of propagating changes in variables taking advantage of the fact that if there are many one-way constraints, a change in one variable can result in cascading changes in many variables. The use of one-way constraints requires some modeling sophistication, but the result is that local search neighborhoods involving changes in a single variable become much more powerful.

2.6 ILOG Local Search Classes

The ILOG Solver [20] is a sophisticated, commercially available class library with modeling and solution support [21]. The main focus of the library is constraint logic programming and the bulk of the classes in the library support modeling and classic tree based solution methodologies. Recently, classes to support local search and metaheuristics have been added. An important feature is the ability to integrate the metaheuristics with a tree based search so that provably optimal solutions can be obtained. As was the case with iOpt, the ability to propagate constraints adds significantly to the power of local search methods. For a detailed description of the combination of metaheuristics and constraint programming, see the chapter "Local Search and Constraint Programming" in this Handbook by Focacci, Laburthe, and Lodi.

2.7 Evolutionary Algorithms

There exist several libraries for genetic algorithms; surveys, as well as software, can be found in [18]. In principle, an advantage of using classic genetic algorithm libraries such as [15] or [13] is that no neighborhood must be specified. If the built-in genomes of a genetic algorithm library adequately represent one's problem, a user-specified objective function may be the only problem-specific code that must be written. Unfortunately, genetic algorithms without a local search component have not generally proven to be very effective. For a comprehensive overview of genetic algorithm libraries the reader is referred to [24].

As an example we briefly discuss the functionality of the GAlib library, a C++ library, which provides the application programmer with a set of genetic algorithms objects; cf. [13]. GAlib is flexible with respect to arbitrary data representations and standard or custom selection, crossover, mutation, scaling, replacement, and termination methods. Overlapping (steady-state GA) and non-overlapping (simple GA) populations are supported. Built-in selection methods include rank, roulette wheel, tournament, stochastic remainder sampling, stochastic uniform sampling, and deterministic sampling. One can use chromosome types built-in to the library (bit-string, array, list, tree) or derive a chromosome based on user-defined objects. All chromosome initialization, mutation, crossover, and comparison methods can be customized. Built-in mutation operators include random flip, random swap, Gaussian, destructive, swap subtree, swap node. Built-in crossover operators include arithmetic, blend, partial match, ordered, cycle, single point, two point, even, odd, uniform, node- and subtree-single point.

3 AN INCREMENTAL ADOPTION PATH

To fully grasp the rules and mechanisms to apply a framework one may have to navigate a long learning curve. Therefore, software libraries and frameworks should enable an incremental application process, or *adoption path*; cf. [10]. That is, the user may start with a simple scenario, which can be successively extended, if needed, after having learned about more complex application mechanisms. Such an evolutionary problem solving approach corresponds to the general tendency of a successive diffusion of knowledge about a new technology and its application; see [1,26].

In the following we sketch a typical adoption path for the case that some of the problem-specific standard components are appropriate for the considered application. In this process, we quickly—after completing the first step—arrive at being able to apply several kinds of metaheuristics for the considered problem, while efficiently obtaining high-quality results may require following the path to a higher level.

1. *Objective Function*: After selecting an appropriate solution component, one has to derive a new class and to code the computation of the objective function. Of course, one also needs some problem component, which provides problem instance data. All other problem-specific components may be reused without change.
2. *Efficient Neighborhood Evaluation*: In most cases, the system that results from step 1 has significant potential for improvement in its run-time efficiency. In particular, one should implement an adaptive move evaluation (which replaces the default evaluation by computing objective function values for neighbor solutions from scratch). In this context, one may also implement some move evaluation that differs from the default evaluation (which is the implied change of the objective function value).
3. *Problem-Specific Adaptation*: Obtaining high-quality solutions may require the exploitation of problem-specific knowledge. For example, this may entail defining and implementing a new neighborhood structure or an adapted tabu criterion in relation to specific solution information or attribute components.
4. *Extension of Metaheuristics*: While the preceding steps only involve problem-specific adaptations, one may eventually want to extend or combine some metaheuristics.

4 HOTFRAME METAHEURISTIC IMPLEMENTATIONS

In Section 2, we provided a broad overview of extant class libraries for metaheuristics. To provide a deeper view of some aspects of library design and use, we borrow from the description of HOTFRAME provided by [12].

4.1 Metaheuristic Concepts

The HOTFRAME class library provides a rich framework for applying a very wide range of metaheuristics. In order to provide some depth to our exposition we must limit our scope significantly. We provide descriptions only of iterated local search and tabu search. We ignore many potentially important features such as candidate lists. Even with this narrow scope, we must limit ourselves to pseudo-code sketches of the algorithms in the library. However, these sketches enable us to convey many important aspects of the design of the library and its use. Of particular importance is the articulation of common and variable constructs.

We rely on some notation concerning optimization problems in metaheuristics, which is developed before we proceed with the algorithm sketches. For every problem P, there is (at least) one

$$\text{solution space} \quad S$$

with

$$\text{solutions} \quad s \in S.$$

Solutions are evaluated by an

$$\text{objective function} \quad f : S \to \mathsf{R}.$$

We generally formulate problems as minimization problems; i.e., we seek to minimize f.

To efficiently manage information about solutions (in connection with some tabu search method), we may need a function

$$h : S \to S_h$$

that maps solutions to elements of a suitable set S_h. With respect to efficiency, one mostly uses non-injective ("approximate") functions (e.g., hash-codes).

For every solution space S, there are one or more

$$\text{neighborhoods} \quad N_S,$$

which define for each solution $s \in S$ an ordered set of neighboring solutions

$$N_S(s) = \{n_1(s), \ldots, n_{|N_S(s)|}(s)\}.$$

From a transformation point of view every neighbor $n(s) \in N_S(s)$ of a solution $s \in S$ corresponds to a

$$\text{move} \quad \mu(s, n(s)).$$

So we can also define a neighborhood as

$$N_S^{\mu}(s) = \{\mu_1 = \mu(s, n_1(s)), \ldots, \mu_{|N_S(s)|} = \mu(s, n_{|N_S(s)|}(s))\}.$$

Moves $\mu(s, n(s)) \in N_S^{\mu}(s)$ are evaluated by a

$$\text{move evaluation} \quad \hat{f}(\mu(s, n(s))),$$

which is often defined as $f(s) - f(n(s))$. Other kinds of move evaluations, e.g., based on measures indicating structural differences between solutions, are possible. In general, positive values should indicate "improving" moves. Move evaluations provide the basis for guiding of the search process.

Both solutions and moves can be decomposed into

$$\text{attributes} \quad \psi \in \Psi,$$

with Ψ representing some attribute set, which may depend on the neighborhood. A solution s corresponds to a set

$$\psi(s) = \{\psi_1(s), \ldots, \psi_{|\psi(s)|}(s)\} \quad \text{with } \psi_j(s) \in \Psi \; \forall j = 1, \ldots, |\psi(s)|.$$

There are two kinds of variabilities to be considered for metaheuristics. On the one hand, metaheuristics are generic regarding the type of problem. On the other hand, metaheuristics usually encompass some variation in their subordinate algorithms (e.g, move selection rules or cooling schedules) and simple parameters (e.g., to control the termination criterion). Accordingly, a configuration C of a metaheuristic H is composed of a definition of a subset C_P of the problem-specific abstractions (S, N, h, Ψ) discussed in the previous subsection, and of a configuration C_H that is specific to the metaheuristic. Given such a configuration C, a metaheuristic defines a transformation that maps an initial solution s to a solution s': $H_C : s \to s'$.

To provide a clear exposition, we use straightforward pseudo-code with simple data structures in our algorithm statements. In the interest of efficiency and flexibility, the actual implementations are somewhat more complicated. A simple example is that many lists that are implemented as general containers are shown here as fixed length structures.

In the pseudo-code description of algorithms, we generally denote a parameterization by "< ··· >" to define algorithmic variation points, and we use "(···)" to define simple value parameters (e.g., the initial solution or numeric parameters). When appropriate, we specify standard definitions, which allows using algorithms without explicitly defining all variation points. By the symbol ϕ we denote non-relevance or invalidity. To simplify the presentation, we use Ω to denote an additional termination criterion, which is implicitly assumed to be checked after each iteration of the local search process. By ω we include a means to specify external termination, which is useful, e.g., in online settings. Using our notation, the interface of a metaheuristic H may be defined by $H < C > (s, T_{\max}, I_{\max}, \omega)$. Such a metaheuristic with configuration C transforms an initial solution s into a series of successor solutions, given a maximum computation time $T_{\max}$, a maximum iteration number $I_{\max}$, and an external termination criterion ω.

In our discussion of the algorithms we make use of feature diagrams that are based on a concise methodology for describing the variation points of concepts in a manner independent from any implementation concerns. Conventions are introduced as needed. For a full description of feature diagrams see [6] or [28].

4.1.1 *Iterated Local Search*

Figure 18.2 shows a feature diagram for simple local search methods (IteratedLocalSearch). In principle, S and N are mandatory features (denoted by the filled circles). The crowsfeet indicate that S and N are considered to be abstract features, which have to be instantiated by specific definitions/implementations. A critical feature of any iteration of a local search procedure is the mechanism for selecting a neighbor. The diagram shows that there are four alternatives for instantiating this feature (denoted by the arc): select the best neighbor out of $N(s)$ with a positive move evaluation, select

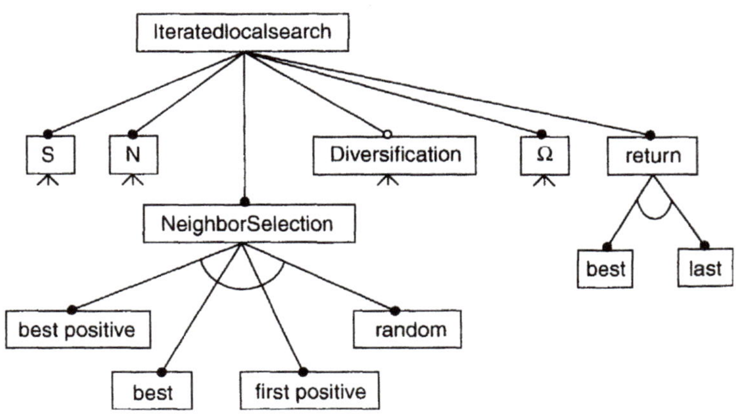

Figure 18.2. Feature diagram for simple local search methods.

the best neighbor even if its evaluation is non-positive, select the first neighbor with a positive move evaluation, or select a random neighbor. The diversification feature is optional (denoted by the open circle), since we do not need a diversification mechanism, e.g., when we specialize IteratedLocalSearch as a one-time application of a greedy local search procedure. The termination criterion Ω is a mandatory feature, which has to be defined in a general way. Finally, there is a feature that allows specification of whether the search procedure should return the best solution found or the last solution traversed. The latter option is useful, e.g., when we use a local search procedure with a random neighbor selection rule as a subordinate diversification component of another metaheuristic.

A sketch of the library's implementation of IteratedLocalSearch is shown by Algorithm 1. We do not provide formal statements of all the features, but Algorithm 2 provides a sketch of one sub-algorithm By using BestPositiveNeighbor, instantiate IteratedLocalSearch can be instantiated to generate SteepestDescent (as shown in Algorithm 3). In a similar manner, a RandomWalk procedure can be used to generate a more complex procedure as shown in Algorithm 4.

4.1.2 Tabu Search

A feature diagram of tabu search as implemented in HOTFRAME is shown in Figure 18.3. Besides S and N, the primary features of tabu search are the tabu criterion and the rule to select neighbors. Moreover, there may be an explicit diversification scheme. We explicitly model the strict tabu criterion (i.e., defining a move as tabu if and only if it would lead to an already traversed neighbor solution), the static tabu criterion (i.e.,

Algorithm 1 IteratedLocalSearch

IteratedLocalSearch
    $< S, N, NeighborSelection, Diversification >$
    $(s, T_{\max} = \infty, I_{\max} = \infty, R_{\max} = 1, \omega = \text{false}, returnBest = \text{true})$:

$\Omega : (t \geq T_{\max})$ **or** (ω)

$s_{\text{best}} = s$;
for $r = 1$ **to** $R_{\max}$
    if $r > 1$
        $Diversification(s)$;
    $i = 0$;
    do
        $i = i + 1$;
        $s' = NeighborSelection < S, N > (s)$;
        if s' is valid
            $s = s'$;
            if $f(s) < f(s_{\text{best}})$
                $s_{\text{best}} = s$;
    while (s' is valid) **and** ($i < I_{\max}$);
if $returnBest$
    $s = s_{\text{best}}$;

Algorithm 2 BestPositiveNeighbor

BestPositiveNeighbor $< S, N > (s)$:

$j = \text{argmax}\{\hat{f}(\mu(s, n_j(s))) | j = 1, \ldots, |N(s)|\}$;
if $\hat{f}(\mu(s, n_j(s))) > 0$
  return $n_j(s)$;
else
  return ϕ;

Algorithm 3 SteepestDescent

SteepestDescent $< S, N > (s, T_{\text{max}}, I_{\text{max}}, \omega)$:

IteratedLocalSearch $< S, N,$ BestPositiveNeighbor, $\phi >$
  $(s, T_{\text{max}}, I_{\text{max}}, 1, \omega,$ true);

Algorithm 4 IteratedSteepestDescentWithPerturbationRestarts

IteratedSteepestDescentWithPerturbationRestarts $< S, N >$
  $(s, T_{\text{max}}, I_{\text{max}}, R_{\text{max}}, \textit{perturbations}, \omega)$:

IteratedLocalSearch
  $< S, N,$ BestPositiveNeighbor,
   RandomWalk $< S, N > (\phi, \infty, \textit{perturbations}, \omega,$ false) $>$
  $(s, T_{\text{max}}, I_{\text{max}}, R_{\text{max}}, \omega,$ true);

storing attributes of performed moves in a tabu list of a fixed size and prohibiting these attributes from being inverted), and the reactive tabu criterion according to [3].

The factors common to tabu search metaheuristic are shown in Algorithm 5. In classic tabu search approaches the search is controlled by dynamically classifying neighbors and corresponding moves as tabu. To implement tabu criteria, one uses information about the search history: solutions visited and/or attributes of moves performed. Using such information, a tabu criterion defines whether neighbors and corresponding moves are classified as tabu. A move is admissible if it is not tabu or an aspiration criterion is fulfilled. That is, aspiration criteria may invalidate a tabu classification (e.g., if the move under consideration leads to a neighboring solution with a new best objective function value). The tabu criterion may also signal that an explicit diversification seems to be reasonable. In such a case, a diversification procedure is applied (e.g., a random walk).

A simple approach for applying the tabu criterion as part of the neighbor selection procedure is to choose the best admissible neighbor (Algorithm 6). Alternatively, some measure of the tabu degree of a neighbor can be used to compute a penalty value that is added to the move evaluation (Algorithm 7). With regard to the latter option, the tabu criterion provides for each move a tabu degree value (between 0 and 1). Multiplying the tabu degree with a parameter σ results in the penalty value.

Some tabu criteria are defined in Algorithms 8–11. In each case, the tabu memory is modeled by state variables using simple container data structures such as lists or

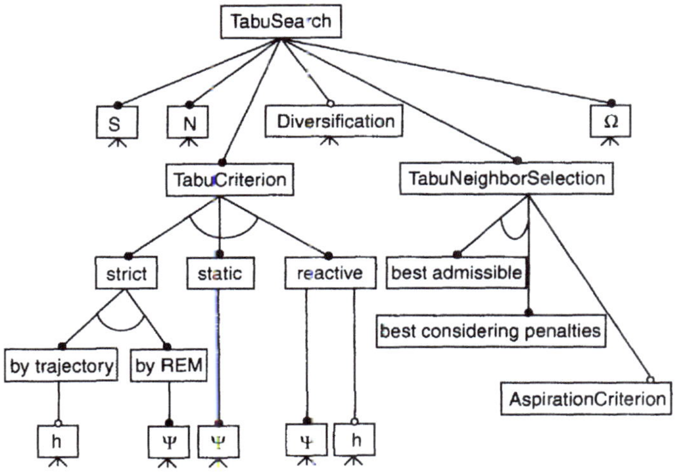

Figure 18.3. Feature diagram for tabu search.

Algorithm 5 TabuSearch

TabuSearch
$< S, N, TabuCriterion, TabuNeighborSelection, Diversification >$
$(s, T_{\max} = \infty, I_{\max} = \infty, \omega =$ false):

$\Omega : (t \geq T_{\max})$ **or** (ω)

$s_{\text{best}} = s;$
$i = 0;$
while $i < I_{\max}$
$i = i + 1;$
$s' = TabuNeighborSelection< S, N, TabuCriterion >(s);$
$TabuCriterion.add(\mu(s, s'), i);$
$s = s';$
$TabuCriterion.add(s, i);$
if $f(s) < f(s_{\text{best}})$
$s_{\text{best}} = s;$
if $TabuCriterion.escape()$
$Diversification(s);$
$s = s_{\text{best}};$

sets, which are parameterized by the type of the respective objects. If lists are defined as having a fixed length, objects are inserted in a first-in first-out manner. Not all tabu criteria implement all functions. For instance, most of the tabu criteria do not possess means to detect and signal situations when an explicit diversification seems to be reasonable.

The strict tabu criterion can be implemented by storing information about all traversed solutions (using the function h). In Algorithm 8, we do not apply frequency or recency information to compute a relative tabu degree but simply use an absolute

Algorithm 6 BestAdmissibleNeighbor

BestAdmissibleNeighbor $< Aspiration >< S, N, TabuCriterion > (s)$:

if $\exists j \in \{1, \ldots, |N(s)|\}$: (**not** $TabuCriterion.tabu(\mu(s, n_j(s)), n_j(s))$)
or ($Aspiration(\mu(s, n_j(s)), TabuCriterion)$)
$j = \text{argmax}\{\hat{f}(\mu(s, n_j(s))) | j = 1, \ldots, |N(s)|,$
(**not** $TabuCriterion.tabu(\mu(s, n_j(s)), n_j(s))$)
or ($Aspiration(\mu(s, n_j(s)), TabuCriterion)$)$\}$;
return $n_j(s)$;
else
return RandomNeighbor $< S, N >(s)$;

Algorithm 7 BestNeighborConsideringPenalties

BestNeighborConsideringPenalties$< Aspiration, \sigma >$
$< S, N, TabuCriterion > (s)$:

if $Aspiration(\mu(s, n_j(s)), TabuCriterion)$
$j = \text{argmax}\{\hat{f}(\mu(s, n_j(s))) | j = 1, \ldots, |N(s)|\}$;
else
$j = \text{argmax}\{\hat{f}(\mu(s, n_j(s)))$
$+ \sigma \cdot TabuCriterion.tabuDegree(\mu(s, n_j(s)), n_j(s))$
$| j = 1, \ldots, |N(s)|\}$;
return $n_j(s)$;

Algorithm 8 StrictTabuCriterionByTrajectory

StrictTabuCriterionByTrajectory $< S, h >$:

State: Set$< S_h >$ *trajectory*;

add(s, ϕ):
trajectory.insert($h(s)$);

tabu(ϕ, s'):
if $s' \in$ *trajectory*
return true;
else
return false;

tabuDegree(ϕ, s'):
if $s' \in$ *trajectory*
return 1;
else
return 0;

Algorithm 9 REMTabuCriterion

REMTabuCriterion$< S, N, \Psi >$
    $(tabuDuration, rcsLengthParameter)$:

State: List$< \Psi >$ $runningList$;
    Set$<$(Set$< \Psi >$, Integer)$>$ $tabuList$;

add(μ, i):
    for $j = 1$ **to** $|\psi(\mu)|$
        $runningList.append(\psi_j(\mu))$;
    $RCS = \emptyset$;
    for $j = |runningList|$ **downto** 1
        if $\overline{runningList[j]} \in RCS$
            $RCS = RCS \setminus \overline{runningList[j]}$;
        else
            $RCS.insert(runningList[j])$;
        if $|RCS| \leq rcsLengthParameter$
            $tabuList.append(\overline{RCS}, i + tabuDuration)$;

tabu(μ, ϕ):
    if $\psi(\mu) \in tabuList$ **and** μ was set tabu not longer than
        $tabuDuration$ iterations before
        return true;
    else
        return false;

tabuDegree(μ, ϕ):
    if $\psi(\mu) \in tabuList$ **and** μ was set tabu not longer than
        $tabuDuration$ iterations before
        return (remaining tabu duration of μ)$/tabuDuration$;
    else
        return 0;

tabu classification. In principle, the strict tabu criterion is a necessary and sufficient condition to prevent cycling in the sense that it classifies exactly those moves as tabu that would lead to an already traversed neighbor. However, as one usually applies a non-injective (approximate) function h, moves might unnecessarily be set tabu (when collisions occur); see [35].

As an alternative implementation of the strict tabu criterion, the reverse elimination method (REM, Algorithm 9) exploits logical interdependencies among moves, their attributes and respective solutions (see [7,16,31,32]). A *running list* stores the sequence of the attributes of performed moves (i.e., the created and destroyed solution attributes). In every iteration, a *residual cancelation sequence* (RCS) is computed, which includes those attributes that separate the current solution from a formerly traversed solution. If the RCS exactly constitutes a move, the corresponding inverse move must be classified as tabu (for one iteration). It should be noted that it is not quite clear how to generally implement the REM tabu criterion for multi-attribute moves in an efficient way. For

Algorithm 10 StaticTabuCriterion

StaticTabuCriterion $< S, N, \Psi > (\nu, \kappa)$:

State: List$< \Psi, \kappa >$ *tabuList*;

add(μ, ϕ):
    for $j = 1$ **to** $|\psi^{+}(\mu)|$
        tabuList.append$(\psi_j^{+}(\mu))$;

tabu(μ, ϕ):
    $k = 0$;
    for $j = 1$ **to** $|\psi^{-}(\mu)|$
        if $\psi_j^{-}(\mu) \in$ *tabuList*
            $k = k + 1$;
    if $k \geq \nu$
        return true;
    else
        return false;

tabuDegree(μ, ϕ):
    $k = 0$;
    for $j = 1$ **to** $|\psi^{-}(\mu)|$
        if $\psi_j^{-}(\mu) \in$ *tabuList*
            $k = k + 1$;
    return $\min\{k/\nu, 1\}$;

this reason, the REM component of HOTFRAME is restricted to single attribute moves that can be coded by natural numbers.

The strict tabu criterion is often too specific to provide a sufficient search diversification. We consider two options to broaden the tabu criterion of the REM. The first alternative uses the parameter *tabuDuration* to define a tabu duration longer than one iteration. The second uses the parameter *rcsLengthParameter* to define a threshold for the length of the RCS, so that all possibilities to combine (subsets of) the attributes of the RCS of a corresponding maximum length as a move are classified as tabu.

The static tabu criterion is defined in Algorithm 10. The parameter Ψ represents the decomposition of moves into attributes. The parameter κ defines the capacity of the tabu list (as the number of attributes). The parameter ν defines the number of attributes of a move, for which there must be inverse correspondents in the tabu list to classify this move as tabu. Furthermore, ν is also the reference value to define a proportional tabu degree.

Algorithm 11 shows the mechanism of the tabu criterion for reactive tabu search. With respect to the dynamic variation of the length of the tabu list, a history stores information about the moves that have been made. This includes the iteration number of the last visit and the frequency. The actual tabu status/degree is defined in the same way as for the static tabu criterion using the parameter ν. The tabu list length is

Algorithm 11 ReactiveTabuCriterion

ReactiveTabuCriterion < S, N, Ψ, h > $(\nu, \delta_1, \delta_2, \kappa, \zeta_1, \zeta_2, \theta)$:

State: List < Ψ > *tabuList*;
  Set< (S_h, Integer, Integer) > *trajectory*;
  movingAverage = *lastReaction* = 0;

add(μ, ϕ):
  for $j = 1$ **to** $|\psi^+(\mu)|$
    tabuList.append($\psi_j^+(\mu)$);

add(s, i):
  if $h(s) \in$ *trajectory* (with corresponding iteration k)
    extend length l of *tabuList* to $\min\{\max\{\lceil l \cdot \delta_1 \rceil, l + \delta_2\}, \kappa\}$;
    lastReaction $= i$;
    movingAverage $= \theta \cdot (i - k) + (1 - \theta) \cdot$ *movingAverage*;
    update iteration to i and increment frequency of $h(s)$;
    if there are ζ_1 solutions in *trajectory*
      that have been traversed ζ_2 times or more
        trigger escape;
  else
    trajectory.insert($h(s), i, 1$);
    if i − *lastReaction* > *movingAverage*
      reduce length l of *tabuList* to $\max\{\min\{\lfloor l/\delta_1 \rfloor, l - \delta_2\}, \delta_2\}$;
      lastReaction $= i$;

tabu(μ, ϕ):
  $k = 0$;
  for $j = 1$ **to** $|\psi^-(\mu)|$
    if $\psi_j^-(\mu) \in$ *tabuList*
      $k = k + 1$;
  if $k \geq \nu$ **return** true;
  else return false;

tabuDegree(μ, ϕ):
  $k = 0$;
  for $j = 1$ **to** $|\psi^-(\mu)|$
    if $\psi_j^-(\mu) \in$ *tabuList*
      $k = k + 1$;
  return $\min\{k/\nu, 1\}$;

computed in dependence of the parameters δ_1 and δ_2. When a solution is revisited, the list is enlarged provided it has not reached its a maximum length κ.

The length of the tabu list is reduced if there has not been any re-visit for some number of iterations which is a function of the parameter θ and an exponentially smoothed average number of iterations between re-visits. If there are ζ_1 solutions that each have been visited at least ζ_2 times, the need for an explicit diversification is signalled.

Algorithm 12 StrictTabuSearch

StrictTabuSearch $< S, N > (s, T_{\max}, I_{\max}, \omega)$:

TabuSearch $<S, N,$ StrictTabuCriterionByTrajectory $< S, id >$,
BestAdmissibleNeighbor $< \phi >, \phi >$
$(s, T_{\max}, I_{\max}, \omega)$;

Algorithm 13 REMpen

REMpen $< S, N, \Psi >$
$(s, T_{\max}, I_{\max}, \omega, tabuDuration, rcsLengthParameter, \sigma)$:

TabuSearch $<S, N,$
REMTabuCriterion
$< S, N, \Psi > (tabuDuration, rcsLengthParameter)$,
BestNeighborConsideringPenalties $< \phi, \sigma >, \phi >$
$(s, T_{\max}, I_{\max}, \omega)$;

Algorithm 14 StaticTabuSearch

StaticTabuSearch $< S, N, \Psi > (s, T_{\max}, I_{\max}, \omega, \nu, \kappa)$:

TabuSearch $<S, N,$ StaticTabuCriterion $< S, N, \Psi > (\nu, \kappa)$,
BestAdmissibleNeighbor $< \phi >, \phi >$
$(s, T_{\max}, I_{\max}, \omega)$;

Algorithm 15 ReactiveTabuSearch

ReactiveTabuSearch $< S, N, \Psi, h >$
$(s, T_{\max}, I_{\max}, \omega, \nu, \kappa, perturbations)$:

TabuSearch
$< S, N,$ ReactiveTabuCriterion $< S, N, \Psi, h > (\nu, 1.2, 2, \kappa, 3, 3, 0.5)$,
BestAdmissibleNeighbor $< \phi >$,
RandomWalk $< S, N > (\phi, \infty, perturbations, \omega,$ false$) >$
$(s, T_{\max}, I_{\max}, \omega)$;

TabuSearch and the modules it uses are parameterized so as to enable various possibilities for constructing specific tabu search heuristics. For example, Algorithm 12 (StrictTabuSearch) encodes the simplest form of strict tabu search: All solutions visited are stored explicitly (*id* represents the identity function), which means that they are classified as tabu in the subsequent search process. Algorithm 13 (REMpen) shows the enhanced reversed elimination method in combination with the use of penalty costs. Static tabu search is shown in Algorithm 14. Algorithm 15 defines reactive tabu search in combination with the use of RandomWalk as diversification mechanism, with most of the parameter values at reasonable default settings.

5 CONCLUSIONS AND OUTLOOK

Class libraries capture the connections between various metaheuristics in a way that is both intellectually interesting and practically useful. From a research perspective libraries can be thought of as providing a concrete taxonomy for heuristic search. So concrete, in fact, that they be compiled into machine code. These taxonomies shed light on the relationships between metaheuristic methods for optimization and on ways in which they can be combined.

From a practical and empirical research perspective, the libraries provide vehicles for using and testing metaheuristics. A user of a library need only provide a problem definition and perhaps a neighborhood structure in order to have available a number of techniques. The classes in the library can be extended and/or combined to produce new search strategies. For application domains such as optimization metaheuristics, OOP techniques provide an excellent means of organizing the software components in a way that encourages reuse without sacrificing flexibility. A class library environment enhances the testing of metaheuristic components by providing a framework for varying some aspects of an algorithm while holding all other aspects constant. For example, multiple neighborhood structures can be tested by simply using a different class for each neighborhood while all other classes, such as the problem class and metaheuristic class, remain fixed. Conversely, a number of metaheuristic strategies can be tested with the neighborhood and all other components fixed.

Just as metaheuristics provide flexible and extensible methods of solving optimization problems, class libraries provide flexible and extensible building blocks for creating software to implement metaheuristics. We have undertaken to give a sense of the intellectual challenges and the rich potential of this burgeoning field, and of the state-of-the-art that underlies some of the more general metaheuristics class libraries.

REFERENCES

[1] M. Allard (1998) Object technology: overcoming cultural barriers to the adoption of object technology. *Information Systems Management*, **15**(3), 82–85.

[2] A.A. Andreatta, S.E.R. Carvalho and C.C. Ribeiro (1998) An object-oriented framework for local search heuristics. In: *Proceedings of the 26th Conference on Technology of Object-Oriented Languages and Systems (TOOLS USA'98)*. IEEE, Piscataway, pp. 33–45.

[3] Roberto Battiti and Giampietro Tecchiolli (1994) The reactive tabu search. *ORSA Journal on Computing*, **6** 126–140.

[4] J. Bosch, P. Molin, M. Mattsson, P. Bengtsson, and M.E. Fayad (1999) Framework problems and experiences. In: M.E. Fayad, D.C. Schmidt and R.E. Johnson (eds.), *Building Application Frameworks: Object-Oriented Foundations of Framework Design*. Wiley, Chichester, pp. 55–82.

[5] F. Buschmann, R. Meunier, H. Rohnert, P. Sommerlad and M. Stal (1996) *Pattern-Oriented Software Architecture*. Wiley, Chichester.

[6] K. Czarnecki and U.W. Eisenecker (2000) *Generative Programming: Methods, Tools, and Applications*. Addison-Wesley, Reading.

[7] F. Dammeyer and S. Voß (1993) Dynamic tabu list management using the reverse elimination method. *Annals of Operations Research*, **41**, 31–46.

[8] Luca Di Gaspero and Andrea Schaerf (2002) Writing local search algorithms using EasyLocal++. In: Stefan Voß and David L. Woodruff (eds.), *Optimization Software Class Libraries*, OR/CS Interfaces Series. Kluwer Academic Publishers, Boston, pp. 155–175.

[9] M.E. Fayad and D.C. Schmidt (eds.) (1997), *Special Issue: Object-Oriented Application Frameworks*. Communications of the Association of Computing Machinery **40**(10), 32–87. ACM.

[10] A. Fink, S. Voß and D. L. Woodruff (1999) An adoption path for intelligent heuristic search componentware. In: E. Rolland and N.S. Umanath (eds.), *Proceedings of the 4th INFORMS Conference on Information Systems and Technology*. INFORMS, Linthicum, pp. 153–168.

[11] Andreas Fink and Stefan Voß (1999) Generic metaheuristics application to industrial engineering problems. *Computers & Industrial Engineering*, **37**, 281–284.

[12] Andreas Fink and Stefan Voß (2002) HotFrame: a heuristic optimization framework. In: Stefan Voß and David L. Woodruff (eds.), *Optimization Software Class Libraries*, OR/CS Interfaces Series. Kluwer Academic Publishers, Boston, pp. 81–154.

[13] GAlib (2001) A C++ Library of Genetic Algorithm Components, http://lancet.mit.edu/ga/.

[14] E. Gamma, R. Helm, R. Johnson and J. Vlissides (1995) *Design Patterns—Elements of Reusable Object-Oriented Software*. Addison-Wesley, Reading.

[15] Genitor (2001) A genetic algorithm library, http://www.cs.colostate.edu/~genitor/.

[16] F. Glover (1990) Tabu search—Part II. *ORSA Journal on Computing*, **2**, 4–32.

[17] F. Glover (1997) Tabu search and adaptive memory programming—Advances, applications and challenges. In: R.S. Barr, R.V. Helgason and J.L. Kennington (eds.), *Interfaces in Computer Science and Operations Research: Advances in Metaheuristics, Optimization, and Stochastic Modeling Technologies*. Kluwer, Boston, pp. 1–75.

[18] Jörg Heitkötter and David Beasley (2001) The hitch-hiker's guide to evolutionary computation (FAQ for comp.ai.genetic), Issue 9.1, 12 April 2001, http://surf.de.uu.net/encore/www/.

[19] HotFrame (2001) Heuristic OpTimization FRAMEwork. http://www.winforms.phil.tu-bs.de/winforms/research/hotframe.html.

[20] ILOG (2000) *ILOG Solver User's Manual, Version 5.1*. ILOG S.A., 9, Rue de Verdun, Gentilly, France.

[21] ILOG (2001) Constraint logic programming libraries, http://www.ilog.com.

[22] R.E. Johnson and B. Foote (1988) Designing reusable classes. *Journal of Object-Oriented Programming*, **1**(2), 22–35.

[23] B. Meyer (1997) *Object-Oriented Software Construction*. Prentice Hall.

[24] Andrew R. Pain and Colin R. Reeves (2002) Genetic algorithm optimization software class libraries. In Stefan Voß and David L. Woodruff (eds.), *Optimization Software Class Libraries*, OR/CS Interfaces Series. Kluwer Academic Publishers, Boston, pp. 295–329.

[25] Arthur J. Riel (1996) *Object-oriented Design Heuristics.* Addison-Wesley, Reading, MA.

[26] E.M. Rogers (1995) *Diffusion of Innovations*, 4th edition. The Free Press, New York.

[27] J. Rumbaugh (1995) OMT: The object model, *Journal of Object Oriented Programming*, **7** 21–27.

[28] M. Simos and J. Anthony (1998) Weaving the model web: a multi-modeling approach to concepts and features in domain engineering. In: P. Devanbu and J. Poulin (eds.), *Proceedings of the Fifth International Conference on Software Reuse*. IEEE Computer Society Press, Los Alamitos, pp. 94–102.

[29] E.D. Taillard, L.M. Gambardella, M. Gendreau and J.-Y. Potvin (2001) Adaptive memory programming: a unified view of meta-heuristics. *European Journal of Operational Research*, **135**, 1–16.

[30] Templar (2001) The Templar Framework. http://www.sys.uea.ac.uk/~msj/templar/.

[31] S. Voß (1993) Intelligent search. Manuscript, Technische Hochschule Darmstadt.

[32] S. Voß (1995) Solving quadratic assignment problems using the reverse elimination method. In: S.G. Nash and A. Sofer (eds.), *The Impact of Emerging Technologies on Computer Science and Operations Research*. Kluwer, Boston, pp. 281–296.

[33] Stefan Voß and David L. Woodruff (2002) *Optimization Software Class Libraries*, OR/CS Interfaces Series. Kluwer Academic Publishers, Boston.

[34] Christos Voudouris and Raphaël Dorne (2002) Integrating heuristic search and one-way constraints in the iOpt toolkit. In: Stefan Voß and David L. Woodruff (eds.), *Optimization Software Class Libraries*, OR/CS Interfaces Series. Kluwer Academic Publishers, Boston, pp. 177–191.

[35] D.L. Woodruff and E. Zemel (1993) Hashing vectors for tabu search. In: F. Glover, E.D. Taillard, M. Laguna and D. de Werra (eds.), *Tabu Search*, Annals of Operations Research 41. Baltzer, Amsterdam, pp. 123–137.

[36] David L. Woodruff (1997) A class library for heuristic search optimization. *INFORMS Computer Science Technical Section Newsletter*, **18**(2), 1–5.

[37] B.V. Zanden, B.A. Myers and D. Giuse P. Szekely (1994) Integrating pointer variables into one-way constraint models. *ACM Transactions on Computer–Human Interaction*, **1**, 161–213.

Chapter 19

ASYNCHRONOUS TEAMS

Sarosh Talukdar
Carnegie Mellon University

Sesh Murthy and Rama Akkiraju
T. J. Watson Labs, IBM

1 INTRODUCTION

Obtaining a good solution to a complex problem, such as the operation of an electric grid or the scheduling of a job-shop, can require a large and diverse set of skills. (An electric grid has millions of devices that must be controlled and coordinated. In job-shop-scheduling, much of the problem can often be solved by a single optimization procedure, such as linear programming. But the remainder invariably requires a variety of situation-specific heuristics and insights.) The questions for the designer of a problem-solving system are which skills to include and how to package them.

In an Asynchronous Team (A-Team), skills are packaged as agents. More specifically, an A-Team is a multi-population, multi-agent system for solving optimization problems. The calculation process is as follows: The problem-to-be-solved is decomposed into sub-problems. A population of candidate-solutions is maintained for each sub-problem. These populations are iteratively changed by a set of agents. Also, solutions are made to circulate among the populations. Under conditions to be discussed, some of the solutions in each population will improve. The calculation process is terminated when the improvements cease.

The agents in an A-Team are identical in two respects. First, they are all completely autonomous (they work without any supervision). Second, they all have the same work-cycle (in every iteration, every agent performs the same three steps: select a solution from a population; modify the selected solution; then insert the modified solution in a population). In all but these two respects, the agents can, and usually should, be diverse. For instance, some may be computer programs while others are human. Some may iterate quickly while others are slow. Some may make small improvements while others make radical or even innovative changes to the candidate-solutions.

	Issues-of-concern			
Structural Features	Design & Assembly	Solution-Quality	Solution-Speed	Robustness
Problem Decomposition	−	±	±	±
Solution Populations		+	±	+
Complete Autonomy	+	−	±	+
Common Work Cycle	+			+
Solution Circulation	−	±	±	±

Figure 19.1. The qualitative relationships between the structural features of A-Teams and four issues-of-concern. +: the structural feature has a beneficial effect on the issue-of-concern; − : the structural feature has a detrimental effect on the issue-of-concern; ± : the effect of the structural feature on the issue-of-concern could be beneficial or otherwise, depending on the design of the A-Team.

In designing a problem-solving system there are four conflicting issues of concern:

- Design and assembly: How much effort will be needed to produce and upgrade the system?
- Solution-quality: How close will the solutions obtained by the system be to the best possible solutions?
- Solution-speed: How quickly will the system complete its calculations?
- Robustness: How will solution-quality and speed be affected by failures and disturbances?

A qualitative assessment of the principal relationships between the structural features of A-Teams and these issues-of-concern is given in Figure 19.1.

The remainder of the chapter is organized as follows. Section 2 lists the terminology that will be used in describing problems and their solutions. Sections 3–6 elaborate on the cause–effect-relationships indicated in Figure 19.1. Section 7 describes the structure of an A-Team. Section 8 describes how the quality of the solutions produced by an A-Team can, when necessary, be improved. Section 9 contains some guidelines for designing A-Teams. And, Section 10 describes an A-Team that is used for solving job-shop-scheduling problems from paper mills.

2 OPTIMIZATION TERMINOLOGY

Optimization is not the best language for expressing every problem, but it is relatively simple and quite general (all problems can, at least in principle, be expressed as optimization problems). It is used here for these reasons.

An optimisation problem is a search specified in terms of three components—objectives, constraints and decision variables. The objectives specify the goals of the search, the constraints specify the limits of the search, and the decision variables specify the space over which the search is to be conducted. In other words, an optimisation problem has the form: $\{\Phi, C, X\}$, where Φ is a set of objectives, C is a set of constraints, and X is a set of decision variables. The values that the decision variables can take constitute a space, S, that is called decision- or solution-space. As an example, consider the problem: simultaneously solve the equations, $x^2 + y^2 = 4$ and $x = y$. One of many ways of expressing this problem in optimisation terms, is:

$$\underset{x,y}{\text{Minimize}} |x^2 + y^2 - 4|$$

$$\text{subject to } x = y$$

Here, "x" and "y" are decision variables, "$|x^2 + y^2 - 4|$" is an objective, "$x = y$" is a constraint, and the solution-space is the plane whose axes are the decision-variables, x and y.

The solution space can be divided into three subsets: infeasible solutions (those violating one or more constraints), feasible solutions (those meeting all the constraints), and optimal solutions (the best of the feasible solutions).

Real problems often contain either no objectives or multiple conflicting objectives. In the former case, all the feasible solutions are optimal, in the latter case, the Pareto solutions are optimal. (Two objectives conflict—as opposed to being commensurate—when one objective cannot be improved without degrading the other. A Pareto solution is a feasible solution that is not dominated by any other feasible solution. Solution-X dominates solution-Y, if X is at least as good as Y with respect to all the objectives, and X is better than Y with respect to at least one objective.)

Two optimization problems are coupled if they share decision variables, constraints or objectives.

Consider a set of coupled optimization problems, $B = \{\beta_1, \beta_2, \ldots, \beta_M\}$. The point X is a Pareto solution of this set, if X is feasible (with respect to the constraints of all the problems) and X is not dominated (with respect to the objectives of all the problems) by any other feasible point.

Suppose that Ω, the problem-to-be-solved, is decomposed into B, a set of coupled sub-problems. Suppose that a set of agents, V, is assigned the task of iteratively solving the sub-problems. The result is a dynamic system: $(B, V, \Gamma, \Lambda, \Psi)$, where Γ is the assignment of agents to sub-problems, Λ is the schedule of communications among the agents, and Ψ is the schedule of the iterations by the agents. If the results of this dynamic system converge, they will do so, not to optimal solutions of Ω, but rather, to equilibria and other attractors of the dynamic system. (These attractors are features, such as points and limit cycles, in the solution space of the Ω.) Note: changing any of the five components of the dynamic system will change its attractors.

If: Γ is one-to-one (sub-problem-m is assigned to agent-m),

Ψ prescribes sequential iterations (the agents take turns working, each completing one iteration per turn),

V contains only optimizing agents (each agent solves its sub-problem to optimality in each iteration),

Λ prescribes information broadcasts (each agent sends the results of each of its iterations to all the other agents), and

the iteration-results converge to a single point in the solution space of Ω, then: this point, is called a Nash equilibrium. Note: a Nash equilibrium represents a stalemate (once all the agents get there, no agent can leave without suffering a degradation of its objective or a violation of its constraints). Also, the Pareto solutions of Ω, the Pareto solutions of B, and the Nash equilibria may all be distinct and quite different.

3 POPULATION METHODS

Consider the problem Ω with solution space S. A population method, when applied to this problem, generates a sequence, $\{P_n\}$, where $P_n \subset S$ is a population (set) of candidate solutions, and P_n is calculated from P_k, with $k < n$. The calculations cease when at least one member of the latest population is found to have come acceptably close to an optimal solution.

The original population method, called the Simplex method, was proposed by Spendley et al. [1]. The name comes from the method's use of a simplex to search the solution-space. A simplex in a space of dimension m, is a set of $m + 1$ points that can be thought of as the vertices of a body with planar faces. The method obtains P_n from P_{n-1} by replacing the worst vertex in P_{n-1} by its reflection in the centroid of the other vertices. Many variations of this method have been proposed, including methods called "swarms" that claim insect societies as their inspiration, but have more in common with simplex methods than insects. On the plus side, simplex-like methods have wide applicability and easy implementation; on the minus side, simplex-like methods can be very slow and tend not to handle constraints well.

Genetic algorithms are better known examples of population methods than the Simplex method. In the basic genetic algorithm, solutions are represented by binary strings. P_n is calculated from P_{n-1}. First, a "fitness function" is used to select the fittest members of P_{n-1}. Next, "crossover" and "mutation," two operators that mimic processes in natural reproduction, are repeatedly applied to the fittest members to produce the new population. On the plus side, genetic algorithms tend to produce good solutions and are fairly robust. On the minus side, they are painfully slow and are often difficult to apply—a consequence of using only string-representations and synchronous calculations (P_n must be calculated from P_{n-1}, and the faster operators to have to wait for the slowest to finish.) A-Teams attempt to eliminate the disadvantages of Simplex-like methods and Genetic algorithms by packaging skills in autonomous agents rather than in mathematical operators, such as "crossover" and "mutation."

4 AUTONOMOUS AGENTS

Agents are the modules from which problem-solving systems can be built. Structurally, an agent can be thought of as a bundle of sensors, decision-makers and actuators. Behaviorally, an agent can be thought of as a mapping from an in-space (all the things the agent can sense) to an out-space (all the things the agent can affect). These spaces may overlap. (Note that by these definitions, a great many, very different things qualify as agents, including thermostats, computer programs, mobile robots, insects, people, societies and corporations.)

An agent is autonomous to the extent that it is unsupervised. A non-autonomous agent does only what its supervisors tell it to do; it makes no decisions for itself. In contrast, a completely autonomous agent has no supervisors, rather, it decides entirely for itself what to do and when.

The exclusive use of completely autonomous agents has some advantages:

- No supervisory structure need be constructed.
- The mistakes and failures of an agent do not affect its supervisees (there are none).
- The speed of an agent's reactions are limited only by its capabilities, not by the delays in a chain-of-command.
- The agents can work asynchronously (in parallel, each at its own speed).
- The agents can improve themselves through learning (converting experiences into competence).

However, the work of completely autonomous agents cannot be externally coordinated. Unless such agents are self-coordinating, they will tend to work at cross-purposes, creating pandemonium rather than progress.

In an A-Team, self-coordination results from the agents' selection mechanisms. (These mechanisms are part of the first step of the work-cycle of each agent. This work-cycle is outlined below. The selection mechanisms and other details are described in later sections.)

The agents of an A-Team work on populations of candidate-solutions. These populations are stored in computer memories. Consider the j-th agent. In each of its iterations, this agent performs three steps: (a) it reads P_n, the incumbent population in its input-memory, selects some candidate-solutions, and removes the selected solutions from P_n; (b) it modifies the selected solutions; and (c) it inserts the modified solutions into Q_k, the incumbent population in its output-memory. Symbolically:

$$\Delta P_{nj} = f_j(P_n) \quad \text{and} \quad P_{n+1} = P_n \backslash \Delta P_{nj} \tag{1}$$

$$\Delta Q_{nj} = g_j(\Delta P_{nj}) \tag{2}$$

$$Q_{k+1} = Q_k \cup \Delta Q_{nj} \tag{3}$$

where f_j and g_j are mappings representing the selection and modification processes of the j-th agent, and $\{P_n\}$ and $\{Q_k\}$ are sequences of the populations in this agent's input and output memories.

The agents can be divided into two categories, creators and destroyers, by differences in the way they perform the second step. Creators add solutions to the populations in their output memories (ΔQ_{nj} is non-empty); destroyers remove solutions from the populations in their input memories (ΔQ_{nj} is empty). The creators can be further divided into constructors and improvers. The constructors produce new solutions from scratch (ΔP_{nj} is empty); the improvers modify existing solutions (ΔP_{nj} is non-empty).

Note that the output memory of an agent may be the same as its input memory, and several agents may share memories, resulting in configurations of the sort shown in Figure 19.2.

5 ASYNCHRONOUS WORK

Synchronous iteration schedules dictate when agents must work and when they must remain idle. They do this by specifying the order in which iterations are to occur, for

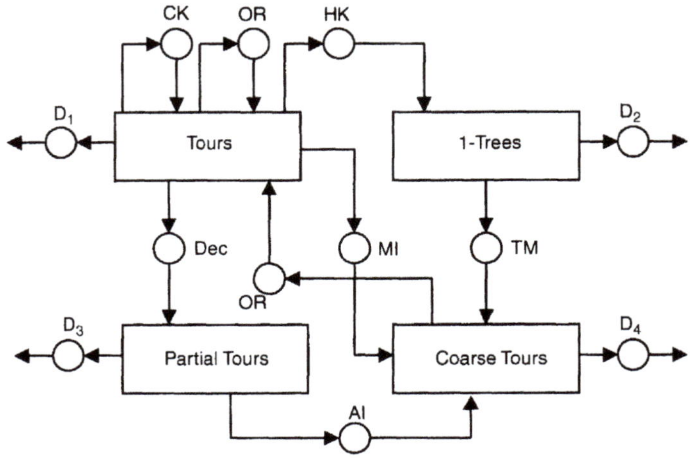

Figure 19.2. One of many possible A-Teams for solving Traveling Salesman Problems (TSPs). The TSP can be stated as follows: given the locations of N cities, find the shortest tour (closed path) through all these cities. The A-Team that is shown uses four memories. The first contains a population of tours, the second, a population of 1-Trees (members of a superset of the tours), the third, a population of partial tours (paths that go through only some of the N cities), and the fourth, a population of coarse tours (complete but relatively long tours). Each population is worked on by some of the agents in the team. Of these agents, D1–D4 are destroyers; CK, OR, HK, Dec, MI, TM and AI are improvement agents. There are no construction agents. The calculations are started by seeding the memories with randomly selected solutions for their sub-problems. For further details see [18].

instance, "no agent will begin iteration $m+1$ till all the agents have completed iteration m." Asynchronous schedules do not constrain the order of iterations. Rather, they allow all the agents to work in parallel all the time, each at its own speed. In other words, asynchronous schedules allow completely autonomous agents, if they so choose, to reduce their idle times to zero.

The relationship between idle times and overall performance is neither obvious nor monotone. But in certain circumstances, reducing idle times does increase performance. A qualitative explanation follows.

Consider a synchronous schedule in which the iterations of all the agents occur in lock-step, and there is just one memory that serves as both the input and the output memory for all the agents. Suppose that in its n-th iteration, each agent selects and modifies some members from the latest population, P_n, in this memory, and the aggregate of these modifications produces the next population, P_{n+1}. In other words:

$$\Delta P_{nj} = f_j(P_n), \quad j \in J \tag{4}$$

$$P = P_n \backslash \Delta P_n \tag{5}$$

$$\Delta Q_{nj} = g_j(\Delta P_{nj}), \quad j \in J \tag{6}$$

$$P_{n+1} = P \cup \Delta Q_n \tag{7}$$

where

$$\Delta P_n = \bigcup_j \Delta P_{nj}, \ j \in J \quad \Delta Q_n = \bigcup_j \Delta Q_{nj}, j \in J$$

and J is the set of agents. In this synchronous schedule, all the agents must wait till the slowest one has finished.

This synchronous schedule can be converted to an asynchronous schedule by relaxing the constraint that the agents work in lock-step and allowing each to iterate as fast as it can. As a result, one agent may be performing its 100-th iteration while another is just on its 10-th. Symbolically, the m-th iteration for the j-th agent is:

$$\Delta P_{nj} = f_j(P_n) \quad \text{and} \quad P_{n+1} = P_n \backslash \Delta P_{nj} \tag{8}$$

$$\Delta Q_{nj} = g_j(\Delta P_{nj}) \tag{9}$$

$$P_{k+1} = P_k \cup \Delta Q_{nj}, \quad k > n \tag{10}$$

where P_n is the latest population when agent-j starts its m-th iteration, and P_k is the latest population when agent-j completes its m-th iteration.

It happens that sufficient conditions for (4)–(7) to converge to a unique solution are only slightly less restrictive than sufficient conditions for (8)–(10) to converge to the same solution [2]. Therefore, one may expect that the reductions in idle time obtained through the use of asynchronous schedules will often leave solution-quality unaffected. What about solution-speed? It seems that populations are able to transmute reductions-in-idle-time into increases in solution-speed when many agents are involved. We do not understand the mechanism, but populations seem to be especially helpful when the agents differ considerably in speed.

Fast agents, such as those using hill-climbing techniques, iterate quickly and tend to be good at refining solutions through a succession of many incremental improvements. Slower agents, such as humans, are often better at making radical changes. These two types of changes can be combined by constraining the fast and slow agents to work in series, so a few iterations of a slow agent are always followed by many iterations of a fast agent. Populations provide a better way to combine fast and slow agents. Specifically, they free agents to work in parallel, each as fast as it can. In other words, populations seem to provide for the automatic blending of incremental and radical changes. (For experiments with mixes of very different software agents, see [3,4], for mixes of software and human agents, see [4,16].)

6 COMPETITION AND COOPERATION

In 1776, Adam Smith made a case for markets and competition [5]. His insight was that good solutions to certain large and difficult problems could be obtained by dividing the problem among a number of autonomous agents, and making these agents compete. This insight has become an article of faith for some designers of autonomous-agent-systems, particularly, designers of markets. But autonomy and competition do not, by themselves, guarantee optimality. Rather, a number of other, quite impractical conditions must also be met, such as the availability of perfect information and the absence of externalities [6]. These other conditions are invariably violated by real systems. Consequently, competitive arrangements often produce very poor solutions, such as the shortages and high prices of the California energy market of 2001.

Of course, agents do not have to compete. They can, instead, cooperate.

Competition requires conflicting goals. In optimization terms, two agents can compete if they are assigned distinct but coupled problems with conflicting objectives. Cooperation requires commensurate goals. In optimization terms, two agents can cooperate if they are assigned coupled problems with commensurate or almost commensurate objectives.

In other words, whether agents cooperate or compete depends, at least in part, on the problems they are assigned.

Let Ω be the problem-to-be-solved; and let $B = \{\beta_1, \beta_2, \ldots, \beta_M\}$ be a decomposition of Ω into sub-problems, such that Ω and B have the same Pareto solutions. If B is a competitive decomposition then, on rare occasions, its Nash equilibria will be the same as the Pareto solutions of Ω. More often, the Nash equilibria of B will be quite inferior to the Pareto solutions of Ω. In contrast, if B is a cooperative decomposition, its Nash equilibria will invariably be the same as the Pareto solutions of Ω.

In other words, cooperative decompositions produce better Nash equilibria than competitive decompositions. The generalization of this observation is: cooperation makes possible solutions of higher quality than can be obtained from competition. We suspect that this generalization is valid.

The structure of an A-Team allows for cooperation at two levels. First, the problem to be solved is decomposed into sub-problems. The team's designer can choose to make this a cooperative decomposition. Second, multiple agents are assigned to each sub-problem. These agents select and modify solutions from a population of candidate-solutions that is maintained for the sub-problem. The agents work asynchronously, each using its own methods for deciding when to work, which solutions to select, and how to modify them. The agents do not explicitly coordinate their activities with one another, nor is there a central controller. Rather, the agents cooperate through the products of their work—by modifying solutions that have been worked on by other agents. The final solution is invariably the result of contributions from many agents. (This form of cooperation was inspired by the work-styles of social insects—bees, ants and certain wasps. Although these insects live in close-knit colonies, they have no supervisors. Certainly there is a queen, but her function is strictly reproductive; she does not lead the other colony members, nor does she issue orders to the workers. Although different castes exist within the colony—drones, soldiers and workers, for instance—there is no hierarchical relationship among them. Rather, they act as autonomous agents and cooperate through the products of their work. For instance, the construction of the nest proceeds without centralized control or the benefit of a blueprint to show what the finished result should be; instead, "it is the product of work previously accomplished, rather than direct communication among nestmates, that induces the insects to perform further labor. Even if the work force is constantly renewed, the nest structure already completed determines, by its location, its height, its shape and probably also its odor, what further work will be done" [19].)

7 ORGANIZATIONS, SUPER-AGENTS AND ORGANIZATION SPACE

The previous sections have covered some aspects of the structure of A-Teams. This section specifies their structure more precisely, and lists the decisions that the designers of A-Teams must make.

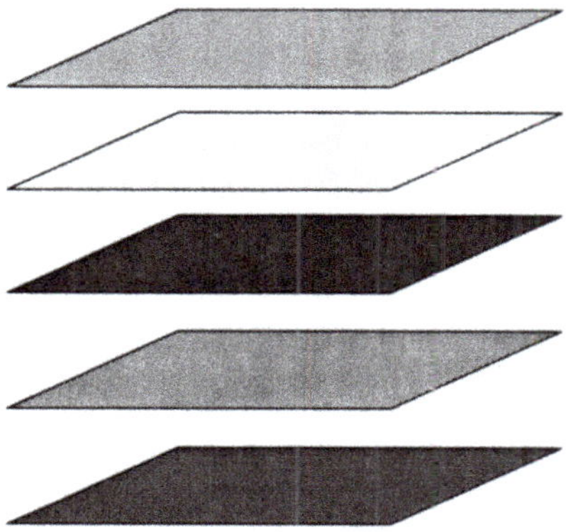

Figure 19.3. A stack of five networks models the structure of an organization.

Lesser agents can be organized into greater (super) agents, which can be organized into still greater agents, and so on, just as cells are organized into organs, which are organized into humans, which are organized into societies and nations. The capabilities of a super-agent depend on its organization, and can range from much less than the sum of the capabilities of its constituent-agents, to very much more. In other words, the design of the organization is at least as important as the choice of agents.

An organization is the "glue" that binds the constituent–agents together. Its purpose is two-fold: to divide the labor among the agents, and to coordinate their labor. Structurally, an organization can be thought of as a stack of five networks (Figure 19.3):

1. Control Flow: a tree-like network that divides the agents into layers, showing a) supervisory relationships (who reports to whom), and b) how much autonomy each agent has. Nodes in this network denote agents. Directed arcs denote supervisory relationships. A number from zero to one is appended to each arc to denote the degree of control the supervisor exercises over the "supervisee" (the larger this number, the greater the degree of control).
2. Problem Decomposition: a network that shows the results of decomposing Ω, the problem-to-be-solved, into B, a set of sub-problems. Nodes represent the sub-problems. Arcs represent the couplings among the sub-problems.
3. Sub-Problem Assignment: a bi-partite network that shows which agent is assigned to which sub-problem. Nodes are of two types; one type represents agents, the other, sub-problems. Arcs connect agents to the sub-problems they have been assigned.
4. Data Flow: a directed, bipartite network that shows who can "talk" to whom and how. There are two types of nodes; one type represents agents, the other, data stores. Arcs represent directed communication channels over which the agents can send messages to one another, post messages in the data stores, or obtain messages from the data stores. (These stores serve as bulletin boards for the agents connected to them.)
5. Work Schedule: a network that shows the order in which the agents' tasks—iterations and communications—are to be performed. Nodes in this network represent the tasks. Arcs represent the precedence constraints among the tasks.

The problem of designing an organization has a very large solution space, namely, the set of all the possible five-network-stacks of the sort shown in Figure 19.3. Of course, other problems, such as aircraft and computer design, also have large solution spaces. But many of these other problems benefit from extensive simulation and verification facilities. They can use "generate and test strategies," i.e., strategies that rely on the quick and accurate evaluation of many candidate-solutions. However, such simulation and verification facilities are not available for organizations. Therefore, it is necessary to prune their space, leaving a smaller and more easily searched sub-space. The set of A-Teams is one such sub-space. It is obtained from organization space by:

- Setting the control flow network to "null" (the agents in an A-Team are completely autonomous).
- Assigning multiple agents to each sub-problem.
- Eliminating all the arcs from the data flow that connect pairs of agents (the agents in an A-Team communicate only with computer memories, not other agents, and cooperate only by modifying one another's work).
- Making the data flow strongly cyclic, i.e., establishing paths, through agents, by which solutions can circulate among the memories.
- Setting the work schedule network to "null" (the agents in an A-Team work asynchronously).

Note that each memory in an A-Team contains a population of solutions. All the solutions in a population are expressed in the same representational form. This representation can vary from one memory to another. A memory used by humans may express solutions in diagrams, while a memory used by optimization software may use vectors. One way to allow for agents that require different representations but must work on the same population, is to maintain copies of the population, each in a different representation and memory.

8 SOLUTION-QUALITY AND CONVERGENCE

The following thought experiment helps explain how A-Teams work and how solution-quality can be improved.

Consider an A-Team that contains only one memory which is dedicated to storing a population of solutions to the problem, Ω. This memory is shared by C, a set of construction agents, I, a set of improvement agents, and D, a set of destruction agents. Suppose that the improvement agents may select solutions randomly, but make only deterministic modifications to the selected solutions. Suppose that the construction agents create an initial population of solutions, P_0, from which the agents in I and D, working asynchronously, produce a sequence of populations, $P_1, P_2, \ldots, P_N$. Suppose that the work is stopped when a population P_N is obtained, such that no agent in I can improve the quality of the best solution in P_N. What is the quality of P_N? And, is N finite?

To address these questions, we define a distance metric in terms of iterations by improvement agents. Consider: $s_0, s_1, s_2, \ldots, s_M$, a trajectory in S, the solution space of Ω. Each step, $s_m \rightarrow s_{m+1}$, in this trajectory is produced by one iteration of an agent drawn from I. Thus, we can think of the trajectory as being M iterations long.

Of course, there may be other trajectories from s_0 to s_M. We define the distance from s_0 to s_M as the length, in iterations, of the shortest trajectory from s_0 to s_M. Note that the distance from s_0 to s_M is: (a) infinite, if there is no trajectory from s_0 to s_M, and (b) dependent on I.

Let:

$q(s) : S \rightarrow [0, 1]$, be a calculable, scalar measure of the quality of every possible solution, s, with $q(s) = 1$ if s is an optimal solution, and $q(s) = 0$ if s is very different from an optimal solution.

$q^*(P_n) = \text{Max}[q(s)|s \in P_n]$ be the quality of population P_n.

$G(Q)$ be the set of all the solutions of quality Q or better, that is, $G(Q) = \{s|q(s) \geq Q\}$.

$f(s, g, I)$ be the distance, in iterations, from $s \in S$ to $g \in G(Q)$. Note: $f(s, g, I)$ is infinite if there is no trajectory from s to g; and $f(s, g, I_1) \geq f(s, g, I_2)$ if $I_1 \subset I_2$.

$h(P_n, Q, I) = \text{Min}[f(s, g, I)|s \in P_n, g \in G(Q)]$ be the distance, in iterations, from P_n to $G(Q)$. Note: $h(P_n, Q, I)$ is infinite when I does not contain the skills necessary to transform any point in P_n into a solution of quality Q or better.

If:

- $h(P_0, Q, I)$ is finite;
- the improvers select solutions randomly, with a bias for solutions of higher quality (the biasing details are given in [13]);
- the destroyers select solutions randomly, with a bias towards solutions of lower quality (the biasing details are given in [13]);
- $q^*(P_n)$ is made non-decreasing with n by saving a copy of the best solution in P_n;

then, $G(Q)$ is reachable (the expected value of N is finite, and the expected value of $q^*(P_N)$ is Q or better.). The proof can be found in [13,18].

The problem of calculating the distance, $h(P_0, Q, I)$, is intractable. Therefore, the above result cannot be used to predict the quality of the solutions that will be produced by an A-Team. But it does tell us that $G(Q)$ will be reachable, if the agents use relatively simple and random strategies for selection, and if there is at least one trajectory from a point in P_0 to a point in $G(Q)$. Furthermore, the lack of such a trajectory can be remedied by adding construction agents (thereby, changing P_0) or adding improvement agents (thereby, increasing the number of trajectories emanating from points in P_0 and the chances that one of them will pass through a point in $G(Q)$. For instance, if I allows for trajectories: $s \rightarrow a$ and $b \rightarrow g$, and if we happen to augment I with an agent that can bridge the gap from a to b, then a trajectory $s \rightarrow g$ will be obtained). In other words, solution-quality tends to increase as the number and diversity of the construction and improvement agents increases.

Solution-quality can also be increased by using better destroyers. (We suspect that creation—construction together with improvement—and destruction are duals, and that adept destruction can compensate for inept creation, and vice-versa.)

Of course, increasing the number of agents could increase the total computing time. But the agents work asynchronously. Therefore, providing more computers, so the additional agents can work in parallel with all the other agents, is likely to make any increases in total computing time slight, if not negligible.

In other words, one might expect, from the above analysis, that A-Teams are scale-effective: adding agents to an A-Team tends to improve its solution-quality, and adding computers tends to improve its solution-speed. There is some empirical evidence in support of this conclusion [3], but not enough has to be completely convincing.

9 DESIGN GUIDELINES

The design of A-Teams, like the design of many artifacts, is a craft whose procedural knowledge does not extend beyond very general guidelines. In such situations, the designer has little choice but to build, test and modify prototypes, till one with acceptable behavior is found. Since A-Teams tend to be scale effective, it makes sense to start with a small team, adding agents when solution-quality is to be improved, and adding computers when solution-speed is to be improved.

Some guidelines and observations for designers to keep in mind are:

- Problem decomposition: The decomposition of the problem-to-be-solved is critical; only the number and diversity of the agents assigned to each sub-problem affects overall performance to a greater extent. Let Ω be the problem to be solved, and β_{m} be a sub-problem. Then, β_{m} can be of three types:

  1. $\beta_{\mathrm{m}} = \Omega$.
  2. β_{m} is related to Ω, just as the 1-Tree problem is related to the traveling salesman problem (Figure 19.2).
  3. β_{m} is a component of Ω, that is, $\beta_{\mathrm{m}} = \{\Phi', \mathrm{C}', \mathrm{X}'\}$, where $\Omega = \{\Phi, \mathrm{C}, \mathrm{X}\}; \Phi' \subset \Phi, \mathrm{C}' \subset \mathrm{C}, \mathrm{X}' \subset \mathrm{X}; \Phi$ is a set of objectives, C is a set of constraints and X is a set of decision variables.

  (Notice that the A-Team of Figure 19.2 contains sub-problems of all three types, while the team of Figure 19.4 contains only types 1 and 3.) All A-Teams should contain at least one sub-problem of type-1 so that solutions to Ω are automatically available. The quality of the final solutions to this sub-problem can usually be improved by adding sub-problems of types 2 and 3.

- Agent assignment: The greater the variety of skills that are brought to bear on a sub-problem, the greater the quality of the solutions that will be obtained for it. In other words, the solution-quality of an A-Team can be improved by adding agents with new and relevant skills. In making these additions, one should keep in mind that creation and destruction appear to be duals: adept destruction can probably compensate for inept creation, and vice-versa. In other words, adding adept destroyers is as good as adding adept creators.
- Data flow: Empirical evidence suggests that the circulation of solutions among the populations has a beneficial effect on both solution-quality and speed.
- Population size: In our experience, solution-quality increases with population size, but at a rapidly diminishing rate.

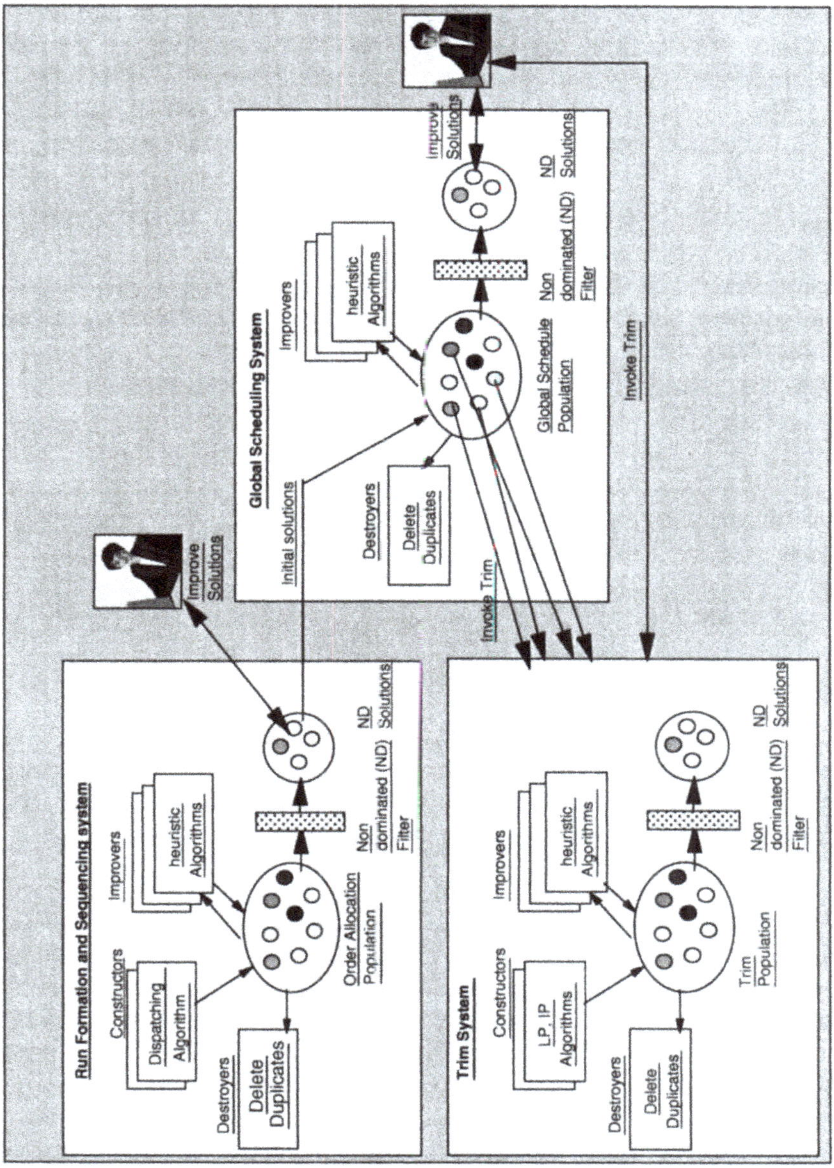

Figure 19.4. An A-Team for scheduling paper production. The team includes human and software agents.

- Selection strategies: We have experimented with only two selection strategies for improvement agents: (a) randomly select solutions with a bias that makes the better solutions more likely to be selected, and (b) randomly select solutions with a bias towards solutions the agent is more likely to be able to improve. Both seem to work well. For destroyers, we have also tried two strategies: (a) randomly select solutions with a bias towards the poorer solutions, and (b) select duplicates. Both seem to work.

10 A CASE STUDY

This section describes an A-Team, developed at IBM, for scheduling the production and distribution of paper products.

Paper production is a complex task involving multiple objectives: maximize profit, maximize customer satisfaction, and maximize production efficiency. These objectives are often in conflict and their relative importance varies with the state of the production environment and with market conditions. Process interactions further increase the difficulty of the problem [9]. For instance, a scheduling improvement in one stage of the production process may negatively impact downstream processes.

In paper production and distribution, the key decisions are:

(a) Order allocation: Allocating orders to paper machines across multiple paper mills in different geographical locations

(b) Run formation and sequencing: Forming and sequencing batches of similar types of paper on each paper machine

(c) Trimming: Cutting large reels of paper produced by paper machines into smaller rolls of customer-specified widths and diameters

(d) Load planning: Loading paper rolls onto vehicles for shipment

The traditional approach to paper mill scheduling is to schedule each stage in the process independently. Typically, paper manufacturers allocate orders to paper machines and sequence them manually. Then they use one software-package for trim scheduling and another package for outbound logistics scheduling, and so on. Each of these packages focuses on a single process-step and attempts to create an optimized schedule based on local objectives. Since there is no interaction between applications, the complete schedule obtained by combining the sub-schedules is usually of very low quality. For example, a trim schedule that minimizes trim-loss may cause vehicles to be loaded inefficiently, unacceptably increasing shipping costs. This piecemeal approach presents schedulers[1] with a single take-it-or-leave-it choice and does not illustrate the tradeoffs between competing objectives that are needed to make well informed decisions.

Realizing the shortcomings of the existing approaches, we at IBM Research, have built a new scheduling system that considers all stages of paper production and distribution simultaneously, and generates multiple enterprise-wide schedules. These schedules are created by algorithms that take into account the interactions between the process stages and focus on enterprise-wide objectives. The algorithms that we have developed use approaches such as linear programming, integer programming with and without

[1] Schedulers are the people who perform the task of scheduling in an organization. The software in our model assists these schedulers in performing their tasks.

randomized rounding, network flow and various heuristic methods. We combine these multiple problem-solving approaches in an A-Team to achieve iterative improvements in the solutions.

For scheduling as well as other manufacturing applications, we have found four categories of attributes to be important: *Timeliness, Product-Quality, Profitability,* and *Disruptions*. Not coincidentally, these categories reflect the concerns of the people affected by the schedules: customer service representatives, quality engineers, accountants and manufacturing supervisors. As the iterations by the software agents proceed, the schedules with the best tradeoffs among the categories are displayed to the scheduler. By examining these schedules, the human scheduler gains an understanding of the tradeoffs. She can select schedules, drastically modify them, and return them to the populations being worked on by software agents. Thereby, she can dramatically alter the course of the software agents' calculations. She can also negotiate compromises with other interested parties. When she sees a solution she feels is acceptable, she terminates the calculations.

We cannot overemphasize the importance of intimately involving human schedules in the calculations. The asynchronous mode of work and a number of filters make such involvements possible and practical. Specifically, asynchronous work allows fast software agents to work in parallel with much slower humans. The filters allow only the solutions with the best tradeoffs to be viewed by humans, thereby, keeping the humans from being overloaded with information. The filter most often used allows only non-dominated solutions to pass through it.

10.1 An Application

This section describes the construction of an A-Team for solving an instance of a paper-manufacturing problem. This problem consists of more than 5000 orders (which constitute about 8 weeks of production for 15 product types) to be scheduled on 9 machines that are located in 4 different mills.[2] Once orders are allocated to machines and grouped according to their grades, these groupings (known as *runs*) have to be trimmed to fit the roll size requirements of the orders. The efficiency of trimming is dependent on the order composition in the groups. However, orders may not be grouped solely to increase trim efficiency, since such groupings may incur unacceptable delays of orders with tight deadlines.

A partial set of evaluation metrics used in our implementation, and some of their values, are presented in columns 2–10 of Table 19.1. These evaluation metrics are customer dependent and are configurable. We obtained these metrics during the initial design study and incorporated them into the system during the benchmarking process.

The manufacturing process contains two distinct stages: (1) run formation and sequencing, and (2) trimming.[3] These stages are coupled: the quality of trim optimization depends on how the runs are formed and sequenced. Therefore, it is important to consider the overall global optimization problem while generating schedules.

The A-Team we use is depicted in Figure 19.4. First, we create a *run formation team* with its own set of constructors, improvers and destroyers for creating runs and

[2]The data were provided by one of the largest paper manufacturers in the U.S.

[3]A third optimization problem, namely transportation scheduling or load planning, is eliminated from this analysis for simplicity. A more detailed description of the problem and our solution approach can be found in [16].

Table 19.1. Summary of results for an instance with 5,000 orders, 4 mills and 9 machines

	Performance on schedule criteria								
Schedule	Tardiness (1000 ton-days)	Earliness (1000 ton-days)	# of orders late	# of orders early	Transp. cost ($1,000)	# of run size violations	Trim efficiency[2]%	Total production tons	Trim loss tons
01	465	217	621	537	$6,504	11	97.68	239,892	5,693
02	464	160	644	436	$6,504	9	97.75	240,002	5,522
03	659	1,231	882	1,863	$6,500	2	98.52	240,094	3,600
04	460	223	616	571	$6,500	13	97.68	239,911	5,693
05	466	904	793	1,487	$6,507	5	98.26	243,433	4,310
06	674	1,079	909	1,737	$6,505	2	98.51	239,975	3,628
07	466	200	587	548	$6,506	10	97.74	243,260	5,623
08	650	1,162	851	1,859	$6,506	2	98.54	243,220	3,617
09	656	1,095	856	1,770	$6,510	2	98.56	243,277	3,565
10	560	252	717	650	$6,512	9	98.86	243,204	2,801

[a]Without using help rolls.

sequencing the orders within those runs. This team generates a set of non-dominated schedules[4] that serve as starting points for trim optimization. While these solutions can be evaluated at a high level based on transportation cost, due dates, and order-machine restrictions, the goodness of these solutions cannot be determined until each run in each schedule is trimmed and the amount of waste is compared. However, trimming in itself is a multi-objective optimization. Therefore, we next construct a *trim team* for trim optimization with suitable constructors, improvers and destroyers. This trim team creates near optimal trim solutions, given a single run and a sequence of orders within that run. This trim team can be invoked multiple times to trim each run in a given schedule. However, there are many such schedules, generated by the run formation team, that need to be explored for overall trim efficiency. Therefore, in order to explore the best possible solutions, we employ a third team, the *global optimization team*, that changes the run formation and sequencing in an effort to achieve better overall solutions (including better trim) as defined by the evaluation metrics. In essence, the run formation and sequencing team and trim team are super-agents within the global optimization team.[5] Trimming is computationally intensive. Therefore, it is important to be judicious in selecting the schedules for trimming. Below, we briefly describe the algorithms that we used in the run formation and sequencing team and the trim team. The global optimization team uses a combination of constructors, improvers and destroyers from the run formation team and the trim team.

10.2 Run Formation and Initial Sequencing Stage

In our system, orders are allocated to machines based on considerations such as transportation cost, due dates, trim preferences and order-machine restrictions. The constructors, improvers and destroyers used in run formation team are:

- *Constructors*: Many approaches such network flow, dispatch algorithms, linear programming models, and greedy approaches can be used to create initial population of solutions for order sequencing in an A-Team. In solving NP hard problems such as these, the A-Team framework encourages the use of multiple approaches for generating initial solutions in the population. Multiple approaches could potentially cover more search space than one approach. In our implementation, we use *dispatch algorithms* for order allocation. The idea is to select one order at a time from the sorted list of remaining orders (several sorting heuristics could be used) and schedule it as the next order on a given machine. These methods create partial solutions that have order allocation information, and sometimes an initial sequencing of the orders (and hence a sequence of runs) on each machine.
- *Improvers*: Improvement algorithms take an existing schedule and try to improve it in several different dimensions. For example, an improver may move orders between runs to reduce tardiness and improve trim efficiency, may merge runs in order to decrease the number of small runs, may resequence runs by moving subsets of orders in a run to improve the solution of a downstream problem (e.g.,

[4]A schedule at this point is a sequence of runs in which each run is a grouping of similar orders in a specific sequence.

[5]The same approach can be extended to include additional down-stream processes, such as sheeting and transportation planning.

moving a set of orders to a different run to improve trim efficiency) etc. Since improvers have inherent knowledge about what aspects of a solution they intend to improve, they are programmed to pick those solutions that exhibit weakness in those specific aspects. For example, an improver that intends to improve the tardiness of a solution would pick solutions that have many late orders.

- *Destroyers*: In our A-Team we used a simple "delete duplicates" destruction approach. More intelligent destruction agents could be created as well.

Further details on the algorithms can be found in [8].

10.3 Trim Stage

A paper machine produces large reels of paper. The process of cutting the reels into rolls of paper (based on customer specifications) is called *trimming*. The main objective in trimming is to minimize the *trim loss* (the unused portion of the reel that cannot be used to fill an order) while considering other manufacturing objectives such as on-time delivery and customer satisfaction. This again, is a multi-objective optimization problem. The constructors, improvers and destroyers used in the trim team are:

- *Constructors*: Trimming paper rolls can be cast as a one-dimensional *cutting stock* problem (CSP). This problem has been studied by Gilmore and Gomory in 1961 in their seminal work [11]. Past work in this area [10,12,13] indicates that linear and integer-programming models work fairly well for generating initial trim patterns. Therefore, we use linear programming and integer-programming approaches to generate initial trim solutions. However, these solutions can be improved further by using iterative heuristic techniques.
- *Improvers*: Trim efficiency can be improved by modifying the sequence of orders within a run or by exchanging orders with other runs in the schedule. This can be done either randomly or based on some heuristics that have specific information about the required widths that could improve trim. For example, if a paper mill knows that there is a constant demand for certain standard widths such as 25″ and 30″, it may not mind making rolls of that size and stocking them, if it improves the trim efficiency, and even if there are no immediate orders for those rolls (these are sometimes called "help rolls"). Trim efficiency improvement heuristics can embody these types of domain details to improve overall trim efficiency. In deciding which solutions to improve, we use simple selection mechanisms such as random selection with a bias. For example, improvers that specialize in improving trim efficiency are programmed to randomly pick those solutions from the populations that do not have perfect trim.
- *Destroyers*: In our trim team we used a simple "delete duplicates" destruction approach. More intelligent destruction agents could be created as well.

More details on the algorithms can be found in [14].

The global A-Team in this application consisted of 15 constructors and 5 improvers. Each agent (constructor or an improver) is an embodiment of the algorithms described above (run with various parameter settings). For our sample problem, an A-Team invocation of $1\frac{1}{2}$ h of CPU time on a single processor IBM RS/6000 Model 591 (256 MB of memory generated approximately 100 solutions; of these solutions, 10 solutions were non-dominated. Table 19.1 shows the evaluations for these 10 solutions illustrating the tradeoffs among the objectives. For example, solutions 3 and 4 have the same transportation cost, which suggests that they have almost the same allocation of orders to mills. However, they differ significantly in their tardiness and trim efficiencies. Solution 3 sacrifices order tardiness for better trim while solution 4 sacrifices trim for better on-time order delivery.

Comparison of solution 10 with the schedule generated by our customer by their traditional methods showed that our system could provide significant reductions in costs (6% savings in transportation costs and improvements in customer satisfaction through reduced tardiness were reported). In this company, as in many other paper companies, an experienced team of schedulers worked on generating the traditional schedules. They allocated and sequenced orders manually or by using a stand-alone scheduling program, and used another computer program for trimming. For fine-tuning the schedules, they used numerous heuristics they had developed over the years.

10.4 Business Impact

Our paper mill scheduling system has been fielded at several paper mills in the United States and is being used in their day-to-day operations. The system significantly improves the scheduling and decision making process for our customers, giving them substantial monetary returns. Improvements come both from the higher quality of the solutions that our system generates and from positive changes in the business processes that our approach to decision support fosters. In terms of solution quality, one of our customers, Madison Paper Industries, reports a reduction in trim loss by 6 tons per day and a 10% reduction in freight costs [17]. Each of these savings amounts to millions of dollars per year.

Adam Stearns of Madison Paper Industries, our pilot customer, reports "We would use our old trim package, throw orders into it and let it trim them the best it could. Then we would let the IBM module take the same orders and manipulate them to come up with a trim. We saw that the IBM package was consistently saving over two inches [of trim loss]—just an incredible amount." [17, page 74] "Testing shows that we are getting about 10% savings annually on distribution costs from the load planning piece alone, which amounts to millions of dollars. . . .We expected the system's GUI [graphical user interface] to make load planning easier, but we didn't expect to gain these efficiencies." [17].

By 1999, 31 mills were either using or were in the process of implementing the IBM mill scheduling system. The users are primarily roll manufacturers in the corrugated and publishing paper markets.

In summary, we have developed a system for enterprise-wide scheduling of paper manufacturing and distribution using the iterative improvement framework facilitated by Asynchronous Teams. The A-Team architecture facilitated cooperation between the various computer algorithms and the human scheduler, resulting in better solutions than any one of the implemented algorithms could have achieved by working alone.

REFERENCES

[1] P.E. Gill and W. Murray (eds.) (1974) *Numerical Methods for Constrained Optimization*. Academic Press.

[2] S.S. Pyo (1985) Asynchronous Procedures for Distributed Processing. Ph.D. dissertation, CMU.

[3] P.S. deSouza (1993) Asynchronous Organizations for Multi-Algorithm Problems. Ph.D. dissertation, CMU.

[4] S. Sachdev (1998) An exploration of A-teams. Ph.D. dissertation, CMU.

[5] A. Smith (1776) *The Wealth of Nations.*

[6] G. Debreu (1959) *The Theory of Value.* Wiley, New York.

[7] S.N. Talukdar (1999) Collaboration rules for autonomous software agents. *The International Journal of Decision Support Systems*, **24**, 269–278.

[8] R. Akkiraju, P. Keskinocak, S. Murthy and F. Wu (1998) An agent-based approach for multi-machine scheduling. *Proceedings of the Tenth Annual Conference on Innovative Applications of Artificial Intelligence*, Menlo Park, CA.

[9] C. Biermann (1993) *Essentials of Pulping and Papermaking*. Academic Press, San Diego.

[10] H. Dyckhoff (1990) A typology of cutting and packing problems. *European Journal of Operational Research*, **44**, 145–159.

[11] P.C. Gilmore and R.E. Gomory (1961) A linear programming approach to the cutting stock problem. *Operations Research*, **9**, 849–859.

[12] R.W. Haessler (1980) A note on the computational modifications to the Gilmore-Gomory cutting stock algorithm. *Operations Research*, **28**, 1001–1005.

[13] R.W. Haessler and P.E. Sweeney (1991) Cutting stock problems and solution procedures. *European Journal of Operational Research*, **54**, 141–150.

[14] P. Keskinocak, F. Wu, R. Goodwin, S. Murthy, R. Akkiraju, S. Kumaran and A. Derebail (2002) Scheduling solutions for the paper industry. *Operations Research*, **50**(2), 249–259.

[15] S. Murthy (1992) Synergy in cooperating agents: designing manipulators from task specifications. Ph.D. thesis. Carnegie Mellon University, Pittsburgh, Pennsylvania.

[16] S. Murthy, R. Akkiraju, R. Goodwin, P. Keskinocak, J. Rachlin, F. Wu, S. Kumaran, J. Yeh, R. Fuhrer, A. Agarwal, M. Sturzenbecker, R. Jayaraman and R. Daigle (1999) Cooperative multi-objective decision-support for the paper industry. *Interfaces*, **29**, 5–30.

[17] M. Shaw (1998) Madison streamlines business processes with integrated information system. *Pulp & Paper*, **72**(5), 73–81.

[18] S. Talukdar, L. Baerentzen, A. Gove and P. deSouza (1998) Asynchronous teams: cooperation schemes for autonomous agents. *Journal of Heuristics*, **4**, 295–321.

[19] P.P. Grassé (1967) Nouvelle expériences sur le termite de Muller (macrotermes mülleri) et considérations sur la théorie de la stigmergie. *Insectes Sociaux*, **14**(1), 73–102.

SUBJECT INDEX

GPSR Compliance
The European Union's (EU) General Product Safety Regulation (GPSR) is a set of rules that requires consumer products to be safe and our obligations to ensure this.

If you have any concerns about our products, you can contact us on

ProductSafety@springernature.com

In case Publisher is established outside the EU, the EU authorized representative is:

Springer Nature Customer Service Center GmbH
Europaplatz 3
69115 Heidelberg, Germany

www.ingramcontent.com/pod-product-compliance
Ingram Content Group UK Ltd.
Pitfield, Milton Keynes, MK11 3LW, UK
UKHW020129290926
15471UKWH00003B/382

* 9 7 8 1 4 7 5 7 7 7 7 5 8 *